Second Edition

ORGANIC POLLUTANTS

An Ecotoxicological Perspective

Second Edition

ORGANIC POLLUTANTS

An Ecotoxicological Perspective

C. H. Walker

With contribution from Charles Tyler

CRC Press
Taylor & Francis Group
Boca Raton London New York

CRC Press is an imprint of the
Taylor & Francis Group, an **informa** business

CRC Press
Taylor & Francis Group
6000 Broken Sound Parkway NW, Suite 300
Boca Raton, FL 33487-2742

First issued in paperback 2019

© 2009 by Taylor & Francis Group, LLC
CRC Press is an imprint of Taylor & Francis Group, an Informa business

No claim to original U.S. Government works

ISBN-13: 978-1-4200-6258-8 (hbk)
ISBN-13: 978-0-367-38640-5 (pbk)

This book contains information obtained from authentic and highly regarded sources. Reasonable efforts have been made to publish reliable data and information, but the author and publisher cannot assume responsibility for the validity of all materials or the consequences of their use. The authors and publishers have attempted to trace the copyright holders of all material reproduced in this publication and apologize to copyright holders if permission to publish in this form has not been obtained. If any copyright material has not been acknowledged please write and let us know so we may rectify in any future reprint.

Except as permitted under U.S. Copyright Law, no part of this book may be reprinted, reproduced, transmitted, or utilized in any form by any electronic, mechanical, or other means, now known or hereafter invented, including photocopying, microfilming, and recording, or in any information storage or retrieval system, without written permission from the publishers.

For permission to photocopy or use material electronically from this work, please access www.copyright.com (http://www.copyright.com/) or contact the Copyright Clearance Center, Inc. (CCC), 222 Rosewood Drive, Danvers, MA 01923, 978-750-8400. CCC is a not-for-profit organization that provides licenses and registration for a variety of users. For organizations that have been granted a photocopy license by the CCC, a separate system of payment has been arranged.

Trademark Notice: Product or corporate names may be trademarks or registered trademarks, and are used only for identification and explanation without intent to infringe.

Library of Congress Cataloging-in-Publication Data

Walker, C. H. (Colin Harold), 1936-
    Organic pollutants : an ecotoxicological perspective / Colin H. Walker. -- 2nd ed.
        p. cm.
    Includes bibliographical references and index.
    ISBN 978-1-4200-6258-8 (alk. paper)
    1. Organic compounds--Toxicology. 2. Organic compounds--Environmental aspects. 3. Environmental toxicology. I. Title.

RA1235.W35 2009
615.9'5--dc22
                                                                    2008030227

Visit the Taylor & Francis Web site at
http://www.taylorandfrancis.com

and the CRC Press Web site at
http://www.crcpress.com

Contents

Preface to First Edition xiii

Preface to Second Edition... xv

Acknowledgments in First Edition xvii

PART 1   Basic Principles

Chapter 1   Chemical Warfare 3

    1.1   Introduction 3
    1.2   Plant–Animal Warfare 4
        1.2.1   Toxic Compounds Produced by Plants....................... 4
        1.2.2   Animal Defense Mechanisms against Toxins
                Produced by Plants... 8
    1.3   Toxins Produced by Animals and Microorganisms 10
        1.3.1   Toxins Produced by Animals 10
        1.3.2   Microbial Toxins 11
    1.4   Human-Made Chemical Weapons................................... 13
    1.5   Summary 15
    Further Reading 15

Chapter 2   Factors Determining the Toxicity of Organic Pollutants to
Animals and Plants 17

    2.1   Introduction 17
    2.2   Factors That Determine Toxicity and Persistence................... 19
    2.3   Toxicokinetics... 21
        2.3.1   Uptake and Distribution 21
        2.3.2   Metabolism... 24
                2.3.2.1   General Considerations...............................24
                2.3.2.2   Monooxygenases..26
                2.3.2.3   Esterases and Other Hydrolases 36
                2.3.2.4   Epoxide Hydrolase (EC 4.2.1.63)...............40
                2.3.2.5   Reductases 41
                2.3.2.6   Conjugases... 42
                2.3.2.7   Enzyme Induction.......................................48
        2.3.3   Storage... 50
        2.3.4   Excretion 51
                 2.3.4.1   Excretion by Aquatic Animals 52
                2.3.4.2   Excretion by Terrestrial Animals 52

2.4   Toxicodynamics...54
2.5   Selective Toxicity...60
2.6   Potentiation and Synergism..62
2.7   Summary 64
Further Reading 65

Chapter 3   Influence of the Properties of Chemicals on Their
Environmental Fate 67

3.1   Properties of Chemicals That Influence Their Fate in the
Gross Environment..68
3.2   Models of Environmental Fate 70
3.3   Influence of the Properties of Chemicals on Their
Metabolism and Disposition...71
3.4   Summary 72
Further Reading 73

Chapter 4   Distribution and Effects of Chemicals in Communities and
Ecosystems 75

4.1   Introduction 75
4.2   Movement of Pollutants along Food Chains 75
4.3   Fate of Pollutants in Soils and Sediments 81
4.4   Effects of Chemicals upon Individuals—the Biomarker
Approach 84
4.5   Biomarkers in a Wider Ecological Context.......................89
4.6   Effects of Chemicals at the Population Level 90
       4.6.1   Population Dynamics 90
       4.6.2   Population Genetics..93
4.7   Effects of Pollutants upon Communities and
Ecosystems—the Natural World and Model Systems.............96
4.8   New Approaches to Predicting Ecological Risks
Presented by Chemicals 97
4.9   Summary 98
Further Reading 98

PART 2   *Major Organic Pollutants*

Chapter 5   The Organochlorine Insecticides 101

5.1   Background..101
5.2   DDT [1,1,1,-trichloro-2,2-bis (p-chlorophenyl) ethane].........102
       5.2.1   Chemical Properties 102
       5.2.2   Metabolism of DDT 104
       5.2.3   Environmental Fate of DDT..................................105

5.2.4   Toxicity of DDT 109

5.2.5   Ecological Effects of DDT....................................... 112

        5.2.5.1   Effects on Population Numbers 112

        5.2.5.2   Effects on Population Genetics (Gene

                  Frequencies).. 115

5.3   The Cyclodiene Insecticides... 116

     5.3.1   Chemical Properties 116

     5.3.2   The Metabolism of Cyclodienes.............................. 117

     5.3.3   Environmental Fate of Cyclodienes 119

     5.3.4   Toxicity of Cyclodienes.................................. 122

     5.3.5   Ecological Effects of Cyclodienes........................... 124

        5.3.5.1   Effects on Population Numbers 124

        5.3.5.2   Development of Resistance to

                  Cyclodienes... 130

5.4   Hexachlorocyclohexanes 131

5.5   Summary 132

Further Reading 132

Chapter 6   Polychlorinated Biphenyls and Polybrominated Biphenyls.............. 133

6.1   Background.. 133

6.2   Polychlorinated Biphenyls 134

     6.2.1   Chemical Properties 134

     6.2.2   Metabolism of PCBs 136

     6.2.3   Environmental Fate of PCBs................................ 140

     6.2.4   The Toxicity of PCBs 143

     6.2.5   Ecological Effects of PCBs 146

        6.2.5.1   Physiological and Biochemical Effects

                  in the Field 146

        6.2.5.2   Population Effects.................................... 146

        6.2.5.3   Population Genetics 149

6.3   Polybrominated Biphenyls.. 149

6.4   Summary 150

Further Reading 150

Chapter 7   Polychlorinated Dibenzodioxins and Polychlorinated

Dibenzofurans 151

7.1   Background.. 151

7.2   Origins and Chemical Properties 151

7.3   Metabolism 153

7.4   Environmental Fate 153

7.5   Toxicity... 154

7.6   Ecological Effects Related to TEQs for 2,3,7,8-TCDD......... 158

7.7   Summary 160

Further Reading 161

Chapter 8 Organometallic Compounds 163

    8.1 Background.. 163
    8.2 Organomercury Compounds 163
        8.2.1 Origins and Chemical Properties............................ 163
        8.2.2 Metabolism of Organomercury Compounds........... 165
        8.2.3 Environmental Fate of Organomercury 166
        8.2.4 Toxicity of Organomercury Compounds................. 168
        8.2.5 Ecological Effects of Organomercury
              Compounds.. 170
    8.3 Organotin Compounds 172
        8.3.1 Chemical Properties.. 172
        8.3.2 Metabolism of Tributyltin 173
        8.3.3 Environmental Fate of Tributyltin.......................... 173
        8.3.4 Toxicity of Tributyltin 174
        8.3.5 Ecological Effects of TBT...................................... 176
    8.4 Organolead Compounds.. 177
    8.5 Organoarsenic Compounds 178
    8.6 Summary 179
    Further Reading 180

Chapter 9 Polycyclic Aromatic Hydrocarbons... 181

    9.1 Background.. 181
    9.2 Origins and Chemical Properties 182
    9.3 Metabolism 183
    9.4 Environmental Fate 185
    9.5 Toxicity 187
    9.6 Ecological Effects... 189
    9.7 Summary 191
    Further Reading 191

Chapter 10 Organophosphorus and Carbamate Insecticides............................. 193

    10.1 Background.. 193
    10.2 Organophosphorus Insecticides... 194
        10.2.1 Chemical Properties.. 194
        10.2.2 Metabolism.. 197
        10.2.3 Environmental Fate 200
        10.2.4 Toxicity... 202
        10.2.5 Ecological Effects 208
             10.2.5.1 Toxic Effects in the Field......................... 208
             10.2.5.2 Population Dynamics................................ 209
             10.2.5.3 Population Genetics 211
    10.3 Carbamate Insecticides... 212
        10.3.1 Chemical Properties.. 212
        10.3.2 Metabolism.. 213

10.3.3   Environmental Fate 213
10.3.4   Toxicity 215
10.3.5   Ecological Effects 217
10.4   Summary 218
Further Reading 218

Chapter 11   Anticoagulant Rodenticides 219

11.1   Background 219
11.2   Chemical Properties 219
11.3   Metabolism of Anticoagulant Rodenticides 221
11.4   Environmental Fate 222
11.5   Toxicity 224
11.6   Ecological Effects 226
11.6.1   Poisoning Incidents in the Field 226
11.6.2   Population Genetics 228
11.7   Summary 228
Further Reading 229

Chapter 12   Pyrethroid Insecticides 231

12.1   Background 231
12.2   Chemical Properties 231
12.3   Metabolism of Pyrethroids 232
12.4   Environmental Fate of Pyrethroids 234
12.5   Toxicity of Pyrethroids 236
12.6   Ecological Effects of Pyrethroids 237
12.6.1   Population Dynamics 237
12.6.2   Population Genetics 238
12.7   Summary 238
Further Reading 239

PART 3   Further Issues and Future Prospects

Chapter 13   Dealing with Complex Pollution Problems 243

13.1   Introduction 243
13.2   Measuring the Toxicity of Mixtures 244
13.3   Shared Mechanism of Action—an Integrated Biomarker
          Approach to Measuring the Toxicity of Mixtures 245
13.4   Toxic Responses That Share Common Pathways of
          Expression 250
13.5   Bioassays for Toxicity of Mixtures 251
13.6   Potentiation of Toxicity in Mixtures 253
13.7   Summary 254
Further Reading 254

Chapter 14 The Ecotoxicological Effects of Herbicides 257

14.1   Introduction 257
14.2   Some Major Groups of Herbicides and Their Properties 258
14.3   Impact of Herbicides on Agricultural Ecosystems 258
14.4   Movement of Herbicides into Surface Waters and
         Drinking Water 261
14.5   Summary 263
Further Reading 264

Chapter 15 Endocrine-Disrupting Chemicals and Their Environmental
Impacts 265

R. M. Goodhead and C. R. Tyler

15.1   Introduction 265
15.2   The Emergence of Endocrine Disruption as a Research
         Theme 266
15.3   Modes of Action of Endocrine-Disrupting Chemicals 266
15.4   Case Studies of Endocrine Disruption in Wildlife 270
         15.4.1   DDT (and Its Metabolites) and Developmental
                    Abnormalities in Birds and Alligators 270
         15.4.2   TBT and Imposex in Mollusks 272
         15.4.3   Estrogens and Feminization of Fish 273
         15.4.4   Atrazine and Abnormalities in Frogs 275
         15.4.5   EDCs and Health Effects in Humans 276
15.5   Screening and Testing for EDCs 276
15.6   A Lengthening List of EDCs 278
         15.6.1   Natural and Pharmaceutical Estrogens 279
         15.6.2   Pesticides 279
         15.6.3   PCBs 279
         15.6.4   Dioxins 280
         15.6.5   Polybrominated Diphenyl Ethers 280
         15.6.6   Bisphenols 281
         15.6.7   Alkylphenols 281
         15.6.8   Phthalates 282
         15.6.9   Natural EDCs 283
15.7   Effects of Mixtures 283
15.8   Windows of Life with Enhanced Sensitivity 284
15.9   Species Susceptibility 286
15.10 Effects of EDCs on Behavior 288
15.11 Lessons Learned from Endocrine Disruption and Their
         Wider Significance in Ecotoxicology 290
15.12 Summary 292
Further Reading 292

Chapter 16 Neurotoxicity and Behavioral Effects of Environmental
Chemicals... 293

16.1 Introduction 293
16.2 Neurotoxicity and Behavioral Effects 295
16.3 The Mechanisms of Action of Neurotoxic Compounds......... 296
16.4 Effects on the Functioning of the Nervous System 302
16.4.1 Effects on the Peripheral Nervous System.............. 302
16.4.2 Effects on the Central Nervous System................... 305
16.5 Effects at the Level of the Whole Organism 306
16.6 The Causal Chain: Relating Neurotoxic Effects at
Different Organizational Levels.. 308
16.6.1 Chemicals Sharing the Same Principal Mode
of Action... 308
16.6.2 Effects of Combinations of Chemicals with
Differing Modes of Action 310
16.7 Relating Neurotoxicity and Behavioral Effects to
Adverse Effects upon Populations.. 311
16.8 Concluding Remarks 313
16.9 Summary 316
Further Reading... 317

Chapter 17 Organic Pollutants: Future Prospects....................................... 319

17.1 Introduction 319
17.2 The Adoption of More Ecologically Relevant Practices
in Ecotoxicity Testing... 321
17.3 The Development of More Sophisticated Methods of
Toxicity Testing: Mechanistic Biomarkers 323
17.4 The Design of New Pesticides.. 324
17.5 Field Studies 326
17.6 Ethical Questions.. 328
17.7 Summary 328
Further Reading... 329

Glossary 331

References... 337

Index.. 377

Preface to First Edition

This book is intended to be a companion volume to *Principles of Ecotoxicology*, first published in 1996 and now in its second edition. Both texts have grown out of teaching material used for the M.Sc. course, Ecotoxicology of Natural Populations, taught at Reading University between 1991 and 1997. At the time that both of these books were written, a strong driving force was the lack of suitable teaching texts in the respective areas. Although this shortcoming is beginning to be redressed in the wider field of ecotoxicology, with the recent appearance of some valuable new teaching texts, this is not evident in the more focused field of the ecotoxicology of organic pollutants viewed from a mechanistic biochemical point of view. Matters are further advanced in the field of medical toxicology, where there are now some very good teaching texts in biochemical toxicology.

Principles of Ecotoxicology deals in broad brush strokes with the whole field, giving due attention to the top-down approach—considering adverse changes at the levels of population, community, and ecosystem, and relating them to the effects of both organic and inorganic pollutants. The present text gives a much more detailed and focused account of major groups of organic pollutants, and adopts a bottom-up approach. The fate and effects of organic pollutants are seen from the point of view of the properties of the chemicals and their biochemical interactions. Particular attention is given to comparative metabolism and mechanism of toxic action, and these are related, where possible, to consequent ecological effects. Biomarker assays that provide measures of toxic action are given some prominence, because they have the potential to link the adverse effects of particular types of pollutant at the cellular level to consequent effects at the levels of population and above. In this way the top-down approach is complementary to the bottom-up approach; biomarker assays can provide evidence of causality when adverse ecological effects in the field are associated with measured levels of pollutants. Under field conditions, the discovery of a relationship between the level of a pollutant and an adverse effect upon a population is no proof of causality. Many other factors (including other pollutants not determined in the analysis) can have ecological effects, and these factors may happen to correlate with the concentrations of pollutants determined in ecotoxicological studies. The text will also address the question "To what extent can ecological effects be predicted from the chemical properties and the biochemistry of pollutants?," which is relevant to the utility, or otherwise, of the use of Quantitative Structure Activity Relationships (QSARs) of chemicals in ecotoxicology.

The investigation of the effects of chemicals upon the numbers and genetic composition of populations has inevitably been a long-term matter, the fruits of which are now becoming more evident with the passage of time. The emergence of resistant strains in response to the selective pressure of pesticides and other pollutants has given insights into the evolutionary process. The evolution of detoxifying enzymes such as the monooxygenases, which have cytochrome P450 at their active center, is believed to have occurred in herbivores and omnivores with their movement from water to land. The development of detoxifying mechanisms to protect animals against plant toxins is a feature of plant-animal "warfare," and is mirrored in the resistance mechanisms developed by vertebrates against pesticides. In the present text, the ecological effects of organic pollutants are seen against the background of the evolutionary history of chemical warfare.

The text is divided into three parts. The first deals with the basic principles underlying the environmental behavior and effects of organic pollutants; the second describes the properties and ecotoxicology of major pollutants in reasonable detail; the last discusses some issues that arise after consideration of the material in the second part of the text, and looks at future prospects. The groups of compounds represented in the second part of the book are all regarded as pollutants rather than simply contaminants, because they have the potential to cause adverse biological effects at realistic environmental levels. In most cases these effects have been well documented under environmental conditions. The term *adverse effects* includes harmful effects upon individual organisms, as well as effects at the level of population and above.

The layout of Chapters 5 through 12, which constitute Part 2, follows the structure of the text as far as possible. Where there is sufficient evidence to do so, the presentations for individual groups of pollutants are arranged as follows:

Layout in present text	**Book divisions in *Principles of Ecotoxicology***
1. Chemical properties	1. Pollutants and Their Fate in Ecosystems
2. Metabolism	
3. Environmental Fate	
4. Toxicity	2. Effects of Pollutants on Individual Organisms
5. Ecological effects	3. Effects on Populations and Communities

C. H. Walker
Colyton

Preface to Second Edition

The first edition of this text was written as a companion volume to *Principles of Ecotoxicology*, first published in 1996 and now in its third edition. Both books grew out of an M.Sc. course, Ecotoxicology of Natural Populations, taught at Reading University between 1991 and 1997. The aim of the first edition was to deal in greater depth and detail with mechanistic aspects of ecotoxicology than had been appropriate for the broad introduction to the subject given in *Principles of Ecotoxicology*.

This second edition has retained the overall structure of the original text and is intended to be a companion volume for the third edition of *Principles of Ecotoxicology*. In producing it there have been two major aims. First, the entire text has been updated to take into account recent developments in the field. Secondly, the third part of the text has been considerably expanded: this section deals with the problems of complex pollution and the exploitation of recent scientific and technological advances to investigate them. In the first edition, the main focus was upon the environmental effects caused by major groups of pollutants, which were described in the second part of the text. More complex pollution patterns were dealt with only briefly, in the third part of the text. Here, two new chapters have been added to strengthen Part 3 of the text, "Endocrine-Disrupting Chemicals and Their Environmental Impacts" and "Neurotoxicity and Behavioral Effects," as well as expanding Chapters 13 and 17.

Professor Charles Tyler has made an important contribution to this text—first by writing, in collaboration with his colleague R. W. Goodhead, a new chapter on endocrine disruptors, which greatly strengthens the third part of this book, but also by giving much valuable discussion and advice on other aspects of the subject relevant to the present book. He is head of a research group at the University of Exeter that investigates endocrine disruption in fish and is particularly well qualified to make this contribution because, as well as being at the cutting edge of this area of research, he runs a course in ecotoxicology for final year undergraduates at the University of Exeter. I am also very grateful to my former colleague at Reading University, Dr. Richard Sibly, for much valuable discussion on population biology and the employment of new techniques, including those of genomics, in studies on the population effects of pesticides and other pollutants.

It is hoped that this text will prove useful to final-year undergraduates, higher degree students, and to researchers in the field of ecotoxicology.

C. H. Walker
Colyton

Acknowledgments in First Edition

Many people have contributed in various ways to this book, and it is not feasible in limited space to mention them all. Over a period of nearly 40 years, colleagues at Monks Wood have given valuable advice on a variety of subjects. At Reading, colleagues and students have given much good advice, critical discussion, and encouragement over many years. Working visits to the research group of Prof. Franz Oesch in the Pharmacology Institute at the University of Mainz were stimulating and productive. Advanced courses such as the ecotoxicology course run at Ecomare, Texel, the Netherlands, by the European Environmental Research Organisation (Prof. and Mrs. Koeman), and the Summer School on Multidisciplinary Approaches in Environmental Toxicology at the University of Siena, Italy (Prof. Renzoni), did much to advance knowledge of the subject—not least for those who were fortunate enough to be invited to contribute! To all of these, grateful thanks are due.

David Peakall has been a continuing source of good advice and critical comment throughout the writing of this book—not least for compensating for some of the inadequacies of my computer system! Richard Sibly and Steve Hopkin continued to give advice and encouragement after completion of *Principles of Ecotoxicology*. I have benefited from the expert knowledge of the following in the stated areas: Gerry Brooks (organochlorine insecticides), Martin Johnson (organophosphorous insecticides), Ian Newton (ecology of raptors), David Livingstone and Peter Donkin (marine pollution), Frank Moriarty (bioaccumulation and kinetic models), Ken Hassall (biochemistry of herbicides), Mike Depledge (biomarkers), and Demetris Savva (DNA technology). My gratitude to all of them.

Last but not least, I am grateful to all the research students and postdoctoral research workers at Reading who have contributed in so many ways to the production of this text.

Part 1

Basic Principles

The first part of the book will deal with the basic principles that determine the environmental distribution and toxic effects of organic pollutants in order to set the stage for Part 2, which describes the properties and behavior of major groups of compounds that fall into this category. The first chapter puts this issue into evolutionary perspective. From a toxicological point of view, organisms have been exposed to toxic xenobiotics for much of the evolutionary history of this planet—much longer than they have encountered human-made organic pollutants, which are the main subject of this book. Indeed, many such compounds have functioned as chemical weapons of both defense and attack, and animals have been found to possess detoxication systems for them, which also work against products of the chemical industry to which they could have had no previous exposure. Also, natural products with "biological activity" have often been used as models in the development of new pesticides and drugs.

Following this introduction, the next three chapters will describe the principles and processes that determine the fate and behavior of organic pollutants in the natural environment. Throughout, emphasis will be given to the importance of the physical, chemical, and biological properties of the compounds themselves in determining their fate and behavior. Chapter 2 is concerned with the factors that determine the distribution and toxicity of these compounds in individual organisms. Chapter 3 will describe the factors that determine the distribution of chemicals through the major compartments of the gross environment and attempts to develop descriptive and predictive models for this. Chapter 4 will focus on distribution and effects in communities and ecosystems.

1 Chemical Warfare

1.1 INTRODUCTION

Chemical warfare has been taking place since very early in the history of life on earth, and the design of chemical weapons by humans is an extremely recent event on the evolutionary scale. The synthesis by plants of secondary compounds ("toxins"), which are toxic to invertebrates and vertebrates that feed upon them, together with the development of detoxication mechanisms by the animals in response, has been termed a "coevolutionary arms race" (Ehrlich and Raven 1964, Harborne 1993). Animals, too, have developed chemical weapons, both for attack and defense. Spiders, scorpions, wasps, and snakes possess venoms that paralyze their prey; bombardier beetles and certain slow-moving herbivorous tropical fish produce chemicals that are toxic to other organisms that prey upon them (see Agosta 1996). Microorganisms produce compounds that are toxic to other microorganisms that compete with them (e.g., penicillin produced by the mold *Penicillium notatum*). Thus, chemical weapons of both attack and defense are widely distributed in nature, and are found in many different species of animals, plants, and microorganisms.

A very large number of natural chemical weapons have already been identified and characterized, and many more are being discovered with each passing year—in plants, marine organisms, etc. They, just as much as pesticides and other human-made chemicals, are in a biochemical sense "foreign compounds" (xenobiotics) to the organisms that they are directed against. They are "normal" to the organism that synthesizes them but "foreign" to the organism against which they express toxicity. During the course of evolution, defense mechanisms have been evolved to give protection against the toxins of plants and other naturally occurring xenobiotics.

The use of pesticides and "chemical warfare agents" by humans should be seen against this background. Many defense mechanisms already exist in nature, mechanisms that have evolved to give protection against natural xenobiotics. These systems may work, to a greater or lesser extent, against human-made pesticides from the moment they are first introduced into the environment, despite the fact that the living environment has had no previous exposure to the chemicals.

Many pesticides are not as novel as they may seem. Some, such as the pyrethroid and neonicotinoid insecticides, are modeled on natural insecticides. Synthetic pyrethroids are related to the natural pyrethrins (see Chapter 12), whereas the neonicotinoids share structural features with nicotine. In both cases, the synthetic compounds have the same mode of action as the natural products they resemble. Also, the synthetic pyrethroids are subject to similar mechanisms of metabolic detoxication as natural pyrethrins (Chapter 12). More widely, many detoxication mechanisms are relatively nonspecific, operating against a wide range of compounds that have common structural features (e.g., benzene rings, methyl groups, or ester bonds). Thus, they can metabolize both human-made and natural xenobiotics even where overall structures of the compounds are not very closely related.

We are dealing here with an area of science in which pure and applied approaches come together. The discovery of natural products with high biological activity (toxicity in the present case), and the elucidation of their modes of action and the defense systems that operate against them, can all provide knowledge that aids the development of new pesticides. They can also aid the development of mechanistic biomarker assays that can establish their side effects on nontarget organisms and provide an understanding of the mechanisms of resistance that operate against them. Whether compounds are natural or human-made, the molecular basis of toxicity remains a fundamental issue; whether biocides are natural or "unnatural," similar mechanisms of action and metabolism apply. Much of what is now known about the structure and function of enzymes that metabolize xenobiotics has been elucidated during the course of "applied" research with pesticides and drugs, and the knowledge gained from this is immediately relevant to the metabolism of naturally occurring compounds. The development of this branch of science illustrates how misleading the division between "pure" and "applied" science can be. Here, at a fundamental scientific level, they are one and the same; the difference has to do with motivation—whether research is undertaken with a view to some "practical" outcome (e.g., development of a new pesticide) or not. The phenomenon of plant–animal warfare will now be discussed, before moving on to a brief review of toxins produced by animals.

1.2 PLANT–ANIMAL WARFARE

1.2.1 Toxic Compounds Produced by Plants

A formidable array of compounds of diverse structure that are toxic to invertebrates or vertebrates or both have been isolated from plants. They are predominately of lipophilic character. Some examples are given in Figure 1.1. Many of the compounds produced by plants known to be toxic to animals are described in Harborne and Baxter (1993); Harborne, Baxter, and Moss (1996); Frohne and Pfander (2006); D'Mello, Duffus, and Duffus (1991); and Keeler and Tu (1983). The development of new pesticides using some of these compounds as models has been reviewed by Copping and Menn (2000), and Copping and Duke (2007). Information about the mode of action of some of them are given in Table 1.1, noting cases where human-made pesticides act in a similar way.

The compounds featured in Table 1.1 are considered briefly here. Pyrethrins are lipophilic esters that occur in *Chrysanthemum* spp. Extracts of flower heads of *Chrysanthemum* spp. contain six different pyrethrins and have been used for insect control (Chapter 12). Pyrethrins act upon sodium channels in a manner similar to *p,p'*-DDT. The highly successful synthetic pyrethroid insecticides were modeled on natural pyrethrins.

Veratridine is a complex lipophilic alkaloid that also binds to sodium channels, causing them to stay open and thereby disrupting the transmission of nerve action potential. It is found in the seeds of a member of the Liliaceae, *Schoenocaulon*

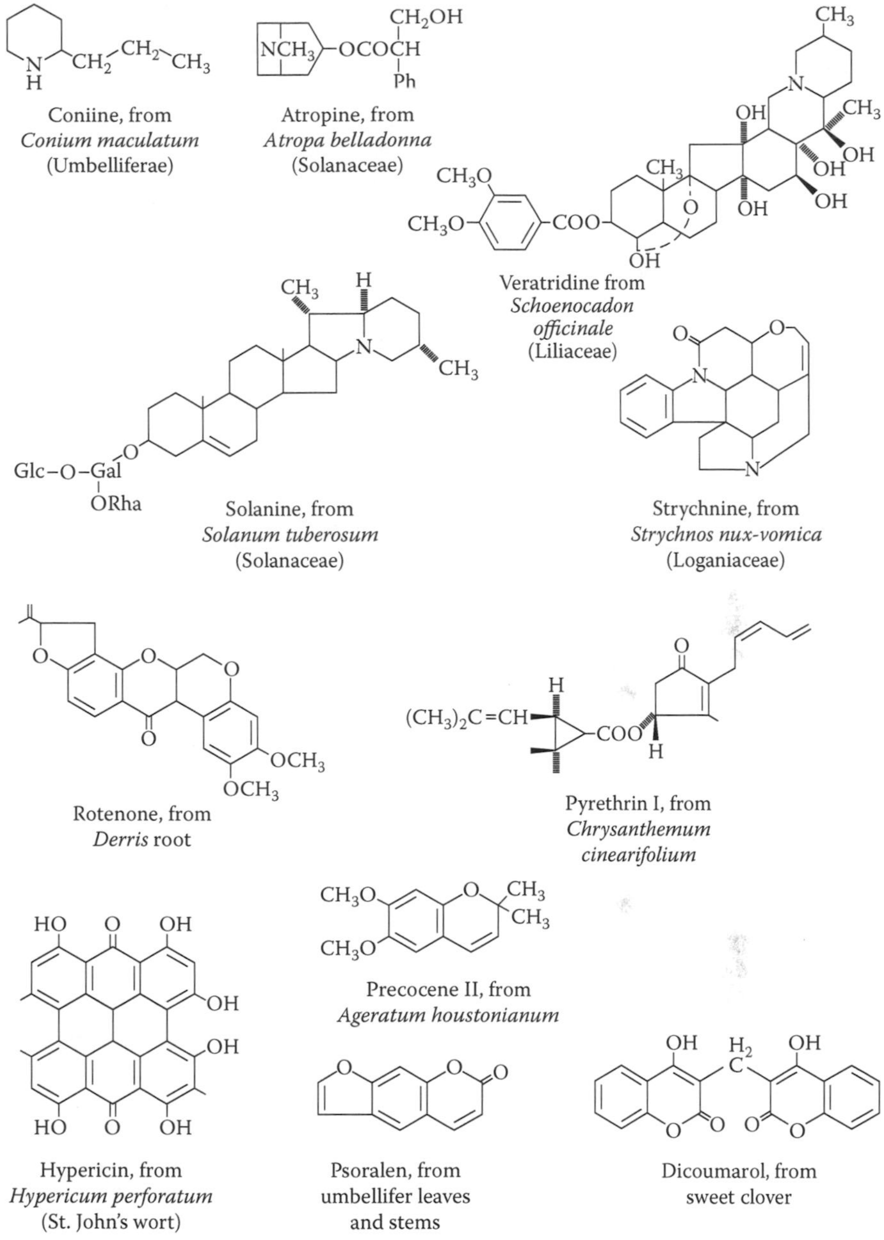

Coniine, from
Conium maculatum
(Umbelliferae)

Atropine, from
Atropa belladonna
(Solanaceae)

Veratridine from
*Schoenocadon
officinale*
(Liliaceae)

Solanine, from
Solanum tuberosum
(Solanaceae)

Strychnine, from
Strychnos nux-vomica
(Loganiaceae)

Rotenone, from
Derris root

Pyrethrin I, from
*Chrysanthemum
cinearifolium*

Hypericin, from
Hypericum perforatum
(St. John's wort)

Precocene II, from
Ageratum houstonianum

Psoralen, from
umbellifer leaves
and stems

Dicoumarol, from
sweet clover

FIGURE 1.1   Some toxins produced by plants.

officinale, formerly known as *Veratrum sabadilla*. Retr. Sabadilla is an insecticidal preparation derived from this source, which also contains another alkaloid, ceva-dine. Further, *pp'*-DDT and pyrethroid have similar effects to veratridine but evidently bind to different sites on the sodium channel (Eldefrawi and Eldefrawi 1990,

TABLE 1.1

Some Toxins Produced by Plants

Compounds	Mode of Action	Pesticides Acting against Same Target	Comments
Pyrethrins	Upon Na+ channel of axonal membrane	Pyrethroids p,p'-DDT	Pyrethroids modeled on pyrethrins
Veratridine	Na+ channel of axonal membrane	Pyrethroids p,p'-DDT	Binding site appears to differ from the one occupied by pesticides
Physostigmine	Anticholinesterase	Insecticidal carbamates	OP insecticides also act in this way
Coumarol	Vitamin K antagonist	Warfarin Superwarfarins	All act as anticoagulants (Chapter 11)
Strychnine	Acts on central nervous system (CNS)		Used to control some vertebrate pests
Rotenone	Inhibits mitochondrial electron transport		Used as an insecticide (Derris powder)
Nicotine	Acts on nicotinic receptor for acetyl choline	Neonicotinoids	An insecticide in its own right and a model for neonicotinoids
Precocenes	Inhibit synthesis of juvenile hormone in some insects		A model for the development of novel insecticides
Ryanodine	Acts upon channels that regulate Ca_2^+ release	Phthalic acid Diamides	Insecticides in the course of development

Sources: Data from Harborne (1993), Eldefrawi and Eldefrawi (1990), Ballantyne and Marrs (1992), Brooks, Pratt, and Jennings (1979), Salgado (1999), Copping and Duke (2007).

Copping and Duke 2007). Sabadilla was used by the native people of South and Central America as an insecticide for many years.

Physostigmine (eserine) is a carbamate found in the calabar bean (*Physostigma benenosum*), which acts as an anticholinesterase. It was used in West Africa in witchcraft trials by ordeal. It has also been used in human medicine. Insecticidal carbamates are structurally related to it and also act as anticholinesterases (Ballantyne and Marrs 1992).

Dicoumarol is found in sweet clover and can cause hemorrhaging in cattle because of its anticoagulant action. It acts as a vitamin K antagonist and has served as a model for the development of warfarin and related anticoagulant rodenticides.

Strychnine is a complex lipophilic alkaloid from the plant *Strychnos nux-vomica*, which acts as a neurotoxin. It has been used to control vertebrate pests, including moles. The acute oral LD_{50} to the rat is 2 mg/kg.

Rotenone is a complex flavonoid found in the plant *Derris ellyptica*. It acts by inhibiting electron transport in the mitochondrion. Derris powder is an insecticidal preparation made from the plant, which is highly toxic to fish.

Nicotine is a component of *Nicotiana tabacum*, the tobacco plant. It is toxic to many insects because of its action upon the nicotinic receptor of acetyl choline. It has served as a model for a new range of insecticides, the neonicotinoids, which also act upon the nicotinic receptor (Salgado 1999).

Precocenes are found in *Ageratum houstonianum*, and can cause premature molting in milkweed bugs (*Oncopeltus fasciatus*) and locusts (*Locusta migratoria*). Precocene 2 is activated by monooxygenase attack to form a reactive metabolite (evidently an epoxide), which inhibits synthesis of juvenile hormone by the *corpora allata* of milkweed bugs and locusts, leading to atrophy of the organ itself (Brooks, Pratt, and Jennings 1979).

Flubendiamide is an example of a new chemical class of insecticides that have been termed phthalic acid diamides (Nauen 2006, Copping and Duke 2007). They are related to the alkaloid ryanodine, which is extracted from *Ryania* species. Ryanodine affects muscles by binding to calcium channels of the sarcoplasmic reticulum. Ca^{2+} ions act as intracellular messengers, and their flux is modulated by calcium channels of this type. The toxic action of ryanodine and synthetic insecticides related to it is due to the disturbance of calcium flux.

Widening the range of examples, some further toxic compounds are shown in Figure 1.1. Coniine is a toxic compound in hemlock (*Conium maculaatum*), and solanine is the toxic component of green potatoes. Atropine is the principal toxin of deadly nightshade (*Atropa belladonna*). It acts as an antagonist of acetylcholine at muscarinic receptors and, in small quantities, is used as an antidote for poisoning by organophosphorous compounds (see Box 10.1, Chapter 10). Hyperiicin is a toxic compound found in St. John's wort (*Hypericum* spp.); psoralen is a toxin found in the stems and leaves of umbellifers.

These are just a few examples among many, and further examples are given in the references quoted at the end of this chapter. They are intended to illustrate the remarkable range of chemical structures among the toxic compounds produced by plants, which give evidence of the intensity of plant–animal warfare during the course of evolution. In some cases, they provide examples of how natural compounds have served—and continue to serve—as models for the development of new pesticides.

Many of the compounds mentioned previously are toxic—sometimes highly toxic—to humans, and this gives cause to reflect on the currently widely held popular belief in the quality and safety of what is termed "organic" produce, that is, food that is relatively low in synthetic products such as pesticides, pharmaceuticals, etc. (In fact, the total exclusion of many such compounds from foods is effectively impossible because of their wide distribution in the natural environment and the movement of many of them in air and water.) Apart from the examples given, many other natural products are known to be highly toxic to humans. Ricin from the castor oil plant (*Ricinus communis*) is highly toxic and has been used for political assassination. Ergot alkaloids from a fungus that attacks rye has caused poisoning incidents in human populations over a long period of history. The intoxication they cause used to be termed St. Anthony's fire. Aflatoxin is a hepatocarcinogen produced by a fungus found on poorly stored groundnuts and other products. There have been cases where foodstuffs contaminated by it have been withdrawn from the market. Botulinum toxin, produced by the bacterium *Clostridium botulinum* in contaminated meat, is one of the most toxic compounds known and has produced many human fatalities. In short, there are many examples of human poisoning caused by natural products. So-called "organic" produce may not be as safe as some people imagine. At least, synthetic pesticides and pharmaceuticals have been subject to rigorous testing procedures and consequent bans and restrictions, which take into account human health risks. Contamination of food by natural toxins, not least by mycotoxins (see Section 1.3.2), remains a human health problem and a subject of ongoing research.

1.2.2   ANIMAL DEFENSE MECHANISMS AGAINST TOXINS PRODUCED BY PLANTS

The toxicity of chemicals to living organisms is determined by the operation of both toxicokinetic and toxicodynamic processes (Chapter 2). The evolution of defense mechanisms depends upon changes in toxicokinetics or toxicodynamics or both, which will reduce toxicity. Thus, at the toxicokinetic level, increased storage or metabolic detoxication will lead to reduced toxicity; at the toxicodynamic level, changes in the site of action that reduce affinity with a toxin will lead to reduced toxicity.

Some insects can protect themselves against the toxins present in their food plants by storing them. One example is the monarch butterfly, the caterpillars of which store potentially toxic cardiac glycosides obtained from a food plant, the milkweed (see Harborne 1993). Subsequently, the stored glycosides have a deterrent effect upon blue jays that feed upon them.

Both direct and indirect evidence points to the importance of enzymic detoxication in protecting animals, vertebrates or invertebrates against toxic chemicals produced by plants. In the first place, it is known that detoxication mechanisms operate against natural as well as human-made xenobiotics. Nicotine and pyrethrins, for example, undergo metabolic detoxication in the housefly. Gray kangaroos (*Macropus* spp.) defluorinate fluoracetate, a natural plant product that occurs in some 34 species of *Gastrolobium* and *Oxylobium*, which grow in Western Australia. Fluoracetate inhibits respiration, being converted to fluorcitrate, a competitive inhibitor of aconitase hydrase. Inhibition of aconitase hydrase causes blockage of the Krebs tricarboxylic acid cycle at the citrate stage. Rat kangaroos (*Bettongia* spp.) also appear to have developed resistance to fluoracetate in Western Australia, where they are exposed to relatively high levels of the compound in the plants that they consume. However, this is not the case in Eastern Australia, where plants containing fluoracetate are not found and the rat kangaroos do not show tolerance to fluoracetate (see Harborne 1993). It is suggested that rat kangaroos originated in the east and radiated westward about 1000 years ago, developing resistance when they came into contact with fluoracetate in their food.

There is increasing evidence that microsomal monooxygenases with cytochrome P450 as their active center have a dominant role in the detoxication of the great majority of lipophilic xenobiotics be they naturally occurring or human-made (Lewis 1996; Chapter 2 of this book). CYP gene families 1, 2, 3, and 4 are all involved in xenobiotic metabolism. They have wide-ranging yet overlapping substrate specificities and, collectively, can detoxify nearly all lipophilic xenobiotics below a certain size. Some of them are inducible, so they can be "upregulated" when there is exposure to unduly high levels of xenobiotics.

The wide distribution of cytochrome P450 enzymes throughout all aerobic organisms clearly indicates a prokaryotic origin with increasing diversification of forms during the course of evolution of vertebrates. Attempts have been made to relate the appearance of different forms of members of P450 families 1–4 to evolutionary events, represented as an evolutionary tree originating from a primordial P450 gene (Nelson and Strobel 1987; see Lewis 1996). Particular interest centers on the radiation of the cytochrome P450 family 2 (CYP2; see Figure 1.2), which is believed to have commenced about 400 million years ago, thus coinciding with the movement of animals to land (Nebert and Gonzalez 1987). Whereas most aquatic organisms can lose lipophilic compounds obtained in their food by diffusion across permeable membranes (especially respiratory membranes) into ambient water, this simple detoxication mechanism is not available to terrestrial animals. They have evolved detoxication systems (predominantly monooxygenases), which can convert lipophilic compounds to water-soluble products that are readily excreted into urine and feces (Chapter 2). Therefore, it seems reasonable to suggest that the radiation of CYP2 represents an adaptation of herbivorous/omnivorous animals to life on land, where survival became dependent upon the ability to detoxify lipophilic toxins produced by plants.

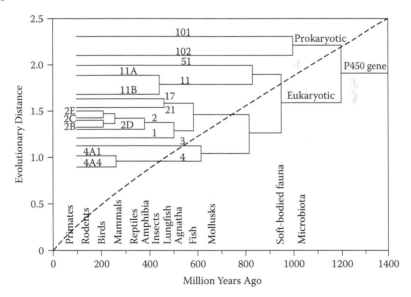

FIGURE 1.2 An abbreviated version of the P450 phylogenetic tree compared with an evolutionary timescale (Lewis 1996). The dashed line represents a plot of evolutionary distance (Nelson and Strobel 1987).

This argument gains strength from a comparison of monooxygenase activities in different groups of vertebrates (see Chapter 2, and Walker 1978, Walker 1980, and Ronis and Walker 1989). Considering the activities of microsomal monooxygenase toward a range of lipophilic xenobiotics, fish have much lower activities than herbivorous or omnivorous mammals. Among birds, fish-eating birds and other specialist predators tend to have much lower activities than omnivorous/herbivorous birds or mammals (see also Chapter 2, Section 2.3.2.2, and Walker 1998a). Most of the xenobiotics used in the assays are substrates for enzymes containing P450s belonging to CYP2. It therefore appears that certain P450s (principally CYP2 isoforms) have evolved in omnivorous/herbivorous vertebrates with adaptation to land, and do not occur in fish or specialized predatory birds. The fish-eating birds in the study cannot use diffusion into the water as mechanisms of detoxication to any important extent; they do not have permeable respiratory membranes, such as the gills of fish, across which lipophilic compounds can diffuse into ambient water. Also, most of them spend long periods out of water anyway. It appears that they have not developed certain P450-based detoxication systems because there have been very few lipophilic xenobiotics in their food in comparison to the plants eaten by herbivores/omnivores.

There is growing evidence that different P450 forms of families 1 and 2 do have some degree of specialization, notwithstanding the rather wide range of compounds that most of them can metabolize. Members of CYP1 specialize in the metabolism of planar compounds, a characteristic that is due to the structure of the binding sites at the active center (Chapter 2, and Lewis 1996). CYP1 family can metabolize flavones and safroles; coumarins are metabolized by CYP2A, pyrazines by CYP2E, and quinoline alkaloids by CYP2D. The evolution of P450 forms within the general scenario of plant–animal warfare is a rich field for investigation.

The extent to which the evolution of defense systems against natural xenobiotics has been based on alterations in toxicodynamics is an open question. Studies on the development of resistance by insects to insecticides (see Chapter 2, Section 2.4) have frequently established the existence of resistant strains possessing insensitive "aberrant" forms of the target, frequently differing from the normal sensitive forms by only a single amino acid substitution. Included here are forms of acetylcholinesterase, axonal sodium channel, and GABA receptor, which are insensitive to organophosphorous insecticides, pyrethroids, and cyclodienes, respectively. This indicates the existence of considerable genetic diversity in insect populations and the possibility of the emergence of resistant strains carrying genes coding for insensitive forms of target proteins under the selective pressure of toxic chemicals. Because at least two of these targets are common to both human-made insecticides and naturally occurring ones, it seems probable that resistance of this type evolved in nature long before the appearance of commercial insecticides.

1.3 TOXINS PRODUCED BY ANIMALS AND MICROORGANISMS

1.3.1 Toxins Produced by Animals

Animals use chemical weapons for both defense and attack. Considering defensive tactics first, bombardier beetles (*Brachinus* spp.) can fire a hot solution of irritant quinones at their attackers (see Agosta 1996). The quinones are generated in abdominal glands by mixing phenols and hydrogen peroxide with catalases and peroxidases. Heat is generated by the reaction, and the cocktail is fired at the assailant with an audible pop. Silent but more deadly is the action of tetrodotoxin (Figure 1.3), found in the puffer fish (*Fugu vermicularis*). Tetrodotoxin is an organic cation that can bind to and consequently block sodium channels (Eldefrawi and Eldefrawi 1990). Interestingly, tetrodotoxin is synthesized by microorganisms that exist on reefs, and is evidently taken up and stored by puffer fish. Humans as well as other predators of puffer fish have died from tetrodotoxin poisoning. Saxitoxin, a toxin found in red tide, acts in the same way as tetrodotoxin (Figure 1.3). The use of chemical defense is not uncommon in small, slow-moving herbivorous fish that live in enclosed spaces such as reefs. For them, avoidance of predation by rapid movement is not a viable strategy, and chemical weapons can be important for survival. On the other hand, chemical defense is not usually found in the fast-swimming fish, especially predators, of open oceans upon which many humans feed. Chemical defense is also important in immobile invertebrates such as sea anemones.

Turning now to chemical attack, many predators immobilize their prey by injecting toxins, often neurotoxins, into them. Examples include venomous snakes, spiders, and scorpions. Some spider toxins (Quick and Usherwood 1990; Figure 1.3) are neurotoxic through antagonistic action upon glutamate receptors. The venom of some scorpions contains polypeptide neurotoxins that bind to the sodium channel.

A striking feature of the toxic compounds considered so far is that many of them are neurotoxic to vertebrates or invertebrates or both. The nervous system of animals appears to be a particularly vulnerable target in chemical warfare. Not altogether surprisingly, all the major types of insecticides that have been commercially successful are also neurotoxins. Indeed, in 2003, neurotoxic insecticides accounted for over 70% of total insecticide sales globally (Nauen 2006).

1.3.2 MICROBIAL TOXINS

This is a large subject that can only be dealt with in the barest outline in the present text. Many antibacterial and antifungal compounds have been discovered in microorganisms, and some of them have been successfully developed as antibiotics for use in human and veterinary medicine. They lie outside the scope of this book. A considerable number of other microbial compounds act as insecticides, acaricides, or herbicides, although few of them have been developed commercially (Copping and Menn 2000, Copping and Duke 2007).

Avermectins (Figure 1.3) are complex molecules synthesized by the bacterium *Streptomyces avermitilis*, which have strong insecticidal, acaricidal, and antihelminthic properties. Eight forms have been found to occur naturally, and the commercial product abamectin consists of two of these forms. Emamectin benzoate, a synthetic product derived from abamectin (Copping and Duke 2007), has been synthesized and marketed commercially. They are toxic because they stimulate the release of gamma amino butyric acid (GABA) from nerve endings and so cause overstimulation of GABA receptors (Copping and Menn 2000). Avermectins have been used as insecticides, for mite control, and for deworming cattle. In the latter case, they

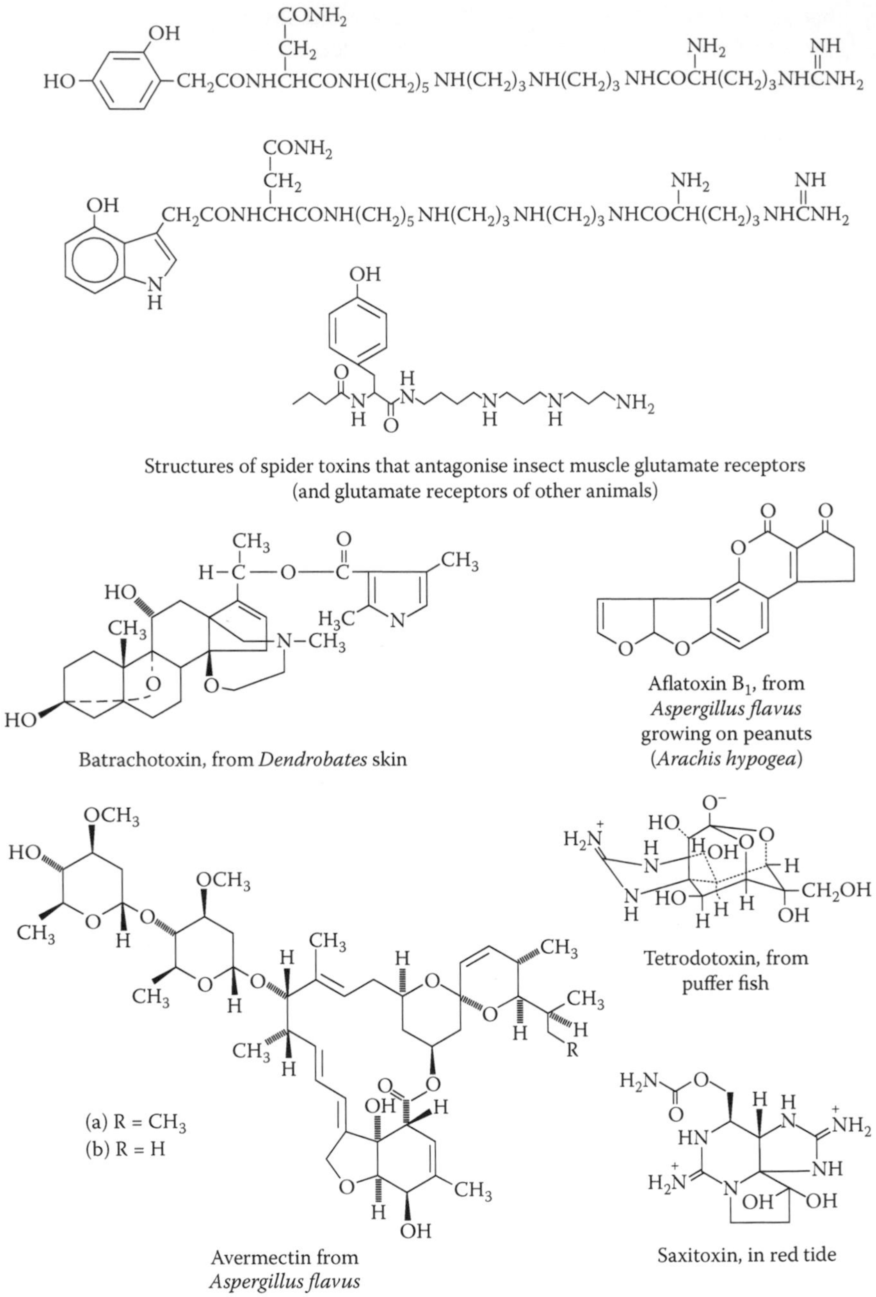

Structures of spider toxins that antagonise insect muscle glutamate receptors
(and glutamate receptors of other animals)

Batrachotoxin, from *Dendrobates* skin

Aflatoxin B$_1$, from
Aspergillus flavus
growing on peanuts
(*Arachis hypogea*)

(a) R = CH$_3$
(b) R = H

Avermectin from
Aspergillus flavus

Tetrodotoxin, from
puffer fish

Saxitoxin, in red tide

FIGURE 1.3   Some toxins from animals and microorganisms.

remain in the feces and effectively control the insects that inhabit dung pats (cow pie). Ecologists have argued that their large-scale use could have serious effects upon insect populations in grasslands where there are cattle.

Preparations of the bacterium *Bacillus thuringiensis* (BT) are applied as sprays to control insect pests on agricultural crops. The bacterium produces endotoxins that are highly toxic to insects.

A considerable number of mycotoxins that show high toxicity to vertebrates and/ or invertebrates are produced by organisms associated with crop plants (Flannigan 1991). There are many known cases of human poisoning caused by such compounds. There are three broad categories of mycotoxins represented here, based on the structures of the intermediates from which these secondary metabolites are derived. They are (1) compounds derived from polyketides, (2) terpenes derived from mevalonic acid, and (3) cyclic peptides and derivatives thereof.

Well-known members of group 1 are the aflatoxins, of which aflatoxin B1 will be taken as a particular example (Figure 1.3). Aflatoxin B1 is a toxic compound that acts as a hepatocarcinogen and can cause liver cirrhosis. It is converted into a highly reactive epoxide by P450-based monooxygenase attack. The epoxide forms adducts with guanine residues of DNA. It is synthesized by certain strains of the molds *Aspergillus flavus* and *A. parasiticus*, which grow on badly stored food products, predominately in tropical areas. Aflatoxins are most often associated with peanuts (groundnuts), maize, rice, and cottonseed, and Brazil, pistachio, and other nuts. The toxicity of aflatoxins came to be widely appreciated when large numbers of turkeys died after feeding upon contaminated groundnuts. There is some evidence of an association between levels of aflatoxins in food and the incidence of primary hepatocellular carcinoma in humans living in areas of central and southern Africa, Thailand, and Indonesia (Flannigan 1991).

Group 2 includes some 80 sesquiterpene trichothecenes, which are particularly associated with fungi belonging to the group *Fusarium*. *Fusarium* species are widely known both as plant pathogens and contaminants of stored foods such as maize. Trichothecenes are strong inhibitors of protein synthesis in mammalian cells. There have been many incidents of poisoning of farm animals caused by contamination of their food by these compounds.

Group 3 includes the ergot alkaloids. These are derivatives of lysergic acid, isolysergic acid, or ergoline. They are found in the ergots (sclerotia) of *Claviceps purpurea*, a fungus that infects grasses and cereals. There have been many examples of poisoning of both humans and animals by ergot alkaloids present in contaminated grain. Historically, there are well-recorded cases of humans becoming intoxicated after consuming rye contaminated by ergot, causing a demented state sometimes referred to as St. Anthony's fire. The development of fungicides in modern times has led to the effective control of this fungus, and cases of poisoning are now very uncommon.

1.4   HUMAN-MADE CHEMICAL WEAPONS

The synthesis of toxic organic compounds by humans, and their release into the natural environment began to assume significant proportions during the 20th century, especially after the Second World War. Prior to 1900, the chemical industry was relatively small, and the largest chemical impact of humans on the environment was probably due to the release of hydrocarbons, especially polycyclic aromatic hydrocarbons (PAHs), with the combustion of coal and other fuels.

Chemical warfare agents came to be used on the battlefield during the 1914–1918 war. These included mustard gas (yperite) and lewisite. More toxic compounds, the organophosphorous nerve gases, were produced later, during the Second World War (1939–1945), although they were not actually used in combat. Organophosphorous anticholinesterases were synthesized under the direction of G. Schrader at the I.G. Farbenindustrie in Germany from the mid-1930s, and some of these, for example, tetraethylpyrophosphate (TEPP) and parathion, came to be used as insecticides. Others were taken over by the government for development as potential chemical warfare agents (Holmstedt 1963; Maynard and Beswick 1992; Marrs, Maynard, and Sidell 2007).

From 1942–1944 the OP chemical warfare agents tabun (ethyl N,N-dimethyl phosphoroamido cyanidate) and sarin (isopropyl methylphosphonylfluoridate) were synthesized on a large scale at a factory at Duhernfurt near Breslau (now Wroclaw, Poland). In Great Britain, the related compounds dimethyl and diethyl phosphofluoridate and diisopropyl phosphofluoridate (DFP) were synthesized during 1940–1941. Subsequent to the war, further work was done on the development of chemical weapons of this type, the United States and Great Britain being active in this field. Fortunately, there has not yet been any large-scale use of these weapons, although it is alleged that Saddam Hussein used a nerve gas against Kurdish villagers in Iraq.

By contrast, many organic pesticides have been developed and widely used since the Second World War. Insecticides, herbicides, rodenticides, fungicides, nematicides, acaricides, and molluscicides have all been employed in the war against crop pests, diseases, and weeds, often with considerable success, but not without longer-term problems. In particular, their regular and intensive use has led to the development of resistance in target species. Similarly, insecticides have been widely employed in vector control, for example, the malaria mosquito and tsetse fly, and rodenticides have been used to control rats and mice, and again there have been problems of resistance. Pesticides of different kinds feature prominently among the organic pollutants mentioned in the this book. In the ensuing account, particular attention is given to the effects of pesticides on populations—both in terms of changes in population numbers and changes in population genetics (the development of resistance).

The pollutants to be described in this book are predominantly the products of human activity, although a few, such as PAHs and methyl mercury, are also naturally occurring. The harmful effects of these pollutants on individual organisms, on the numbers and genetic composition of populations, and on the structure and function of communities and ecosystems represent the basic material of this book. The main purpose of this introductory chapter is to put the matter into perspective. The effects to be described should always be seen against the background of the long and continuing history of chemical warfare on earth. Following from this, effects on populations may be caused not only by anthropogenic chemicals but by natural ones as well. Indeed, some population changes may be the consequence of the simultaneous action of chemicals originating from both sources; for example, effects of PAHs and organomercury on populations may be due to chemicals coming from both human and natural sources at the same time; natural estrogens may enhance the endocrine disruption caused by artificial estrogens. In the end, it is vital to adopt a holistic approach when attempting to understand the effects of chemicals in the field.

1.5 SUMMARY

In this introductory chapter, a broad overview is given of the history of chemical warfare on earth, and the compounds, species, and mechanisms involved. The impact of human-made compounds on the environment, which is the subject of this book, is an extremely recent event in evolutionary terms. It is important to take a holistic view, and to see the effects of human-made pollutants on the environment against the background of chemical warfare in nature.

FURTHER READING

Agosta, W. (1996). *Bombardier Beetles and Fever Trees*—A highly readable account of chemical warfare, without much detail on chemical or biochemical aspects.

Copping, L.G. and Menn, J.J. (2000). *Biopesticides: A Review of Their Action, Applications, and Efficacy*—A valuable account of naturally occurring compounds that act as pesticides.

Copping, L.G. and Duke, S.O. (2007). *Natural Products Used Commercially as Crop Protection Agents*—An updating and expansion of the previous title.

Harborne, J.R. (1993). *Introduction to Ecological Biochemistry,* 4th edition—A very popular text that contains chapters on the toxic compounds of plants and the coevolutionary arms race.

Harborne, J.R., Baxter, H., and Moss, G.P. (1996). *Dictionary of Plant Toxins,* 2nd edition—A dictionary that gives details of many toxins produced by plants.

Lewis, D.F.V. (1996). *Cytochromes P450*—Contains a very useful chapter on the evolution of forms of cytochrome P450.

2 Factors Determining the Toxicity of Organic Pollutants to Animals and Plants

2.1 INTRODUCTION

This chapter will consider the processes that determine the toxicity of organic pollutants to living organisms. The term *toxicity* will encompass harmful effects in general and will not be restricted to lethality. With the rapid advances of mechanistic toxicology in recent years, it is increasingly possible to understand the underlying sequence of changes that lead to the appearance of symptoms of intoxication, and how differences in the operation of these processes between species, strains, sexes, and age groups can account for selective toxicity. Thus, in a text of this kind, it is important to deal with these. Understanding why chemicals have toxic effects and why they are selective is of interest both scientifically and for more practical and commercial reasons. An understanding of mechanism can provide the basis for the development of new biomarker assays, the design of more effective and more environmentally friendly pesticides, and the development of new chemicals and strategies to control resistant pests.

Although many of the standard ecotoxicity tests use lethality as the endpoint, it is now widely recognized that sublethal effects may be at least as important as lethal ones in ecotoxicology. Pollutants that affect reproductive success can cause populations to decline. The persistent DDT metabolite *p,p'*-DDE caused the decline of certain predatory birds in North America through eggshell thinning and consequent reduction in breeding success (see Chapter 5). The antifouling agent tributyl tin (TBT) caused population decline in the dog whelk (*Nucella lapillus*) through making the females infertile (see Chapter 8).

Neurotoxic compounds can have behavioral effects in the field (see Chapters 5, 9, and 15), and these may reduce the breeding or feeding success of animals and their ability to avoid predation. A number of the examples that follow are of sublethal effects of pollutants. The occurrence of sublethal effects in natural populations is intimately connected with the question of persistence. Chemicals with long biological half-lives present a particular risk. The maintenance of substantial levels in individuals, and along food chains, over long periods of time maximizes the risk of sublethal effects. Risks are less with less persistent compounds, which are rapidly eliminated by living organisms. As will be discussed later, biomarker assays are already making an important contribution to the recognition and quantification of sublethal effects in ecotoxicology (see Chapter 4, Section 4.7).

In ecotoxicology, the primary concern is about effects seen at the level of population or above, and these can be the consequence of the indirect as well as the direct action of pollutants. Herbicides, for example, can indirectly cause the decline of animal populations by reducing or eliminating the plants they feed on. A well-documented example of this on agricultural land is the decline of insect populations and the grey partridges that feed on them, due to the removal of key weed species by herbicides (see Chapter 13). Thus, the toxicity of pollutants to plants can be critical in determining the fate of animal populations. When interpreting ecotoxicity data during the course of environmental risk assessment, it is very important to have an ecological perspective.

Toxicity is the outcome of interaction between a chemical and a living organism. The toxicity of any chemical depends on its own properties and on the operation of certain physiological and biochemical processes within the animal or plant that is exposed to it. These processes are the subject of the present chapter. They can operate in different ways and at different rates in different species—the main reasons for the selective toxicity of chemicals between species. On the same grounds, chemicals show selective toxicity (henceforward simply "selectivity") between groups of organisms (e.g., animals versus plants and invertebrates versus vertebrates) and also between sexes, strains, and age groups of the same species.

The concept of selectivity is a fundamental one in ecotoxicology. When considering the effects that a pollutant may have in the natural environment, one of the first questions is which of the exposed species/life stages will be most sensitive to it. Usually this is not known, because only a small number of species can ever be used for toxicity testing in the laboratory in comparison with a very large number at risk in the field. As with the assessment of risks of chemicals to humans, environmental risk assessment depends upon the interpretation of toxicity data obtained with surrogate species. The problem comes in extrapolating between species. In ecotoxicology, such extrapolations are particularly difficult because the surrogate species is seldom closely related to the species of environmental concern. Predicting toxicity to predatory birds from toxicity data obtained with feral pigeons (*Columba livia*) or Japanese quail (*Coturnix coturnix japonica*) is not a straightforward matter. The great diversity of wild animals and plants, and the striking differences between groups and species in their susceptibility to toxic chemicals cannot be overemphasized. For this reason, large safety factors are often used when estimating environmental toxicity from the very sparse ecotoxicity data.

Understanding the mechanistic basis of selectivity can improve confidence in making interspecies comparisons in risk assessment. Knowing more about the operation of the processes that determine toxicity in different species can give some insight into the question of how comparable different species are, when interpreting toxicity data. The presence of the same sights of action, or of similar levels of key detoxifying enzymes, may strengthen confidence when extrapolating from one species to another in the interpretation of toxicity data. Conversely, large differences in these factors between species discourage the use of one species as a surrogate for another.

Apart from the wider question of effects on natural environment, selectivity is a vital consideration in relation to the efficacy of pesticides and the risks that they pose to workers using them and to farm and domestic animals that may be exposed to them. In designing new pesticides, manufacturers seek to maximize toxicity to the target organism, which may be an insect pest, vertebrate pest, weed, or plant pathogen, while minimizing toxicity toward farm animals, domestic animals, and beneficial organisms. Beneficial organisms include beneficial insects such as pollinators and parasites and predators of pests. Understanding mechanisms of toxicity can lead manufacturers toward the design of safer pesticides. Physiological and biochemical differences between pest species and beneficial organisms can be exploited in the design of new, safer, and more selective pesticides. Examples of this will be given in the following text. On the question of efficacy, the development of resistance is an inevitable consequence of the heavy and continuous use of pesticides. Understanding the factors responsible for resistance (e.g., enhanced detoxication or insensitivity of the site of action in a resistant strain) can point to ways of overcoming it. For example, alternative pesticides not susceptible to the resistance mechanism may be used. In general, a better understanding of the mechanisms responsible for selectivity can facilitate the safer and more effective use of pesticides.

2.2   FACTORS THAT DETERMINE TOXICITY AND PERSISTENCE

The fate of a xenobiotic in a living organism, seen from a toxicological point of view, is summarized in Figure 2.1. This highly simplified diagram draws attention to the main processes that determine toxicity. Three main categories of site are shown in the diagram, each representing a different type of interaction with a chemical. These are

1. Sites of action. When a chemical interacts with one or more of these, there will be a toxic effect on the organism if the concentration exceeds a certain threshold. The chemical has an effect on the organism.
2. Sites of metabolism. When a chemical reaches one of these, it is metabolized. Usually this means detoxication, but sometimes (most importantly) the consequence is activation. The organism acts upon the chemical.
3. Sites of storage. When located in one of these, the chemical has no toxic effect, is not metabolized, and is not available for excretion. However, after release from storage, it may travel to sites of action and sites of metabolism.

In reality, things are more complex than this. For some chemicals, there may be more than one type of site in any of these categories. Some chemicals have more than one site of action. The organophosphorous (OP) insecticide mipafox, for example, can produce toxic effects by interacting with either acetylcholinesterase or neuropathy target esterase. Also, many chemicals undergo metabolism by two or more types of enzyme. Pyrethroid insecticides, for example, are metabolized by both monooxygenases and esterases. Also, lipophilic compounds can be stored in various hydrophobic domains within the body, including fat depots and in association with "inert" proteins (i.e., proteins that do not metabolize them or represent a site of action).

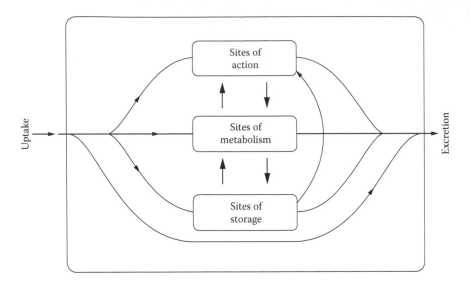

FIGURE 2.1   Toxicokinetic model.

Furthermore, any particular type of site belonging to any one of these categories may exist in a number of different cellular or tissue locations. For example, acetyl-cholinesterase is located in a number of different mammalian tissues (e.g., brain, peripheral nervous system, and red blood cells), and all of these may be inhibited by OP insecticides.

Despite these complicating factors, the model shown in Figure 2.1 identifies the main events that determine toxicity in general and selective toxicity in particular. More sophisticated versions of it can be used to explain or predict toxicity and selectivity. At this early stage of the discussion, it is important to distinguish between the forest and the trees. For many lipophilic compounds, rapid conversion into more polar metabolites and conjugates leads to efficient excretion, and thus efficient detoxication. This is emphasized by the use of a broad arrow running through the middle of the diagram. Inhibition of this process can cause large increases in toxicity (see later discussion of synergism).

For convenience, the processes identified in Figure 2.1 can be separated into two distinct categories: toxicokinetics and toxicodynamics. Toxicokinetics covers uptake, distribution, metabolism, and excretion processes that determine how much of the toxic form of the chemical (parent compound or active metabolite) will reach the site of action. Toxicodynamics is concerned with the interaction with the sites of action, leading to the expression of toxic effects. The interplay of the processes of toxicokinetics and toxicodynamics determines toxicity. The more the toxic form of the chemical that reaches the site of action, and the greater the sensitivity of the site of action to the chemical, the more toxic it will be. In the following text, toxicokinetics and toxicodynamics will be dealt with separately.

2.3   TOXICOKINETICS

From a toxicological point of view, the critical issue is how much of the toxic form of the chemical reaches the site of action. This will be determined by the interplay of the processes of uptake, distribution, metabolism, storage, and excretion. These processes will now be discussed in a little more detail.

2.3.1   UPTAKE AND DISTRIBUTION

The major routes of uptake of xenobiotics by animals and plants are discussed in Chapter 4, Section 4.1. With animals, there is an important distinction between terrestrial species, on the one hand, and aquatic invertebrates and fish on the other. The latter readily absorb many xenobiotics directly from ambient water or sediment across permeable respiratory surfaces (e.g., gills). Some amphibia (e.g., frogs) readily absorb such compounds across permeable skin. By contrast, many aquatic vertebrates, such as whales and seabirds, absorb little by this route. In lung-breathing organisms, direct absorption from water across exposed respiratory membranes is not an important route of uptake.

Once compounds have entered organisms, they are transported in blood and lymph (vertebrates), in hemolymph (invertebrates), and in the phloem or xylem of plants, eventually moving into organs and tissues. During transport, polar compounds will be dissolved in water or associated with charged groups on proteins such as albumin, whereas nonpolar lipophilic compounds tend to be associated with lipoprotein complexes or fat droplets. Eventually, the ingested pollutants will move into cells and tissues, to be distributed between the various subcellular compartments (endoplasmic reticulum, mitochondria, nucleus, etc.). In vertebrates, movement from circulating blood into tissues may be due to simple diffusion across membranes, or to transportation by macromolecules, which are absorbed into cells. This latter process occurs when, for example, lipoprotein fragments are absorbed intact into liver cells (hepatocytes). The processes of distribution are less well understood in invertebrates and plants than they are in vertebrates.

An important factor in determining the course of uptake, transport, and distribution of xenobiotics is their polarity. Compounds of low polarity tend to be lipophilic and of low water solubility. Compounds of high polarity tend to be hydrophilic and of low fat solubility. The balance between the lipophilicity and hydrophilicity of any compound is indicated by its octanol–water partition coefficient (K_{ow}), a value determined when equilibrium is reached between the two adjoining phases:

$$K_{ow} = \frac{\text{Concentration of compound in octanol}}{\text{Concentration of compound in water}}$$

Compounds with high K_{ow} values are of low polarity and are described as being lipophilic and hydrophobic. Compounds with high K_{ow} values are of high polarity and are hydrophilic. Although the partition coefficient between octanol and water is the one most frequently encountered, partition coefficients between other nonpolar liquids (e.g., hexane, olive oil) and water also give a measure of the balance between lipophilicity and hydrophilicity. K_{ow} values for highly lipophilic compounds are very large and are commonly expressed as log values to the base 10 (log K_{ow}).

K_{ow} values determine how compounds will distribute themselves across polar–nonpolar interfaces. Thus, in the case of biological membranes, lipophilic compounds of high K_{ow} below a certain molecular weight move from ambient water to the hydrophobic regions of the membrane, where they associate with lipids and hydrophobic proteins. Such compounds will show little tendency to diffuse out of membranes; that is, they readily move into membranes but show little tendency to cross into the compartment on the opposite side. Above a certain molecular size (about 800 kDa), lipophilic molecules are not able to diffuse into biological membranes. That said, the great majority of lipophilic pollutants described in the present text have molecular weights below 450 and are able to diffuse into membranes. By contrast, polar compounds with low K_{ow} values tend to stay in the aqueous phase and not move into membranes. The same arguments apply to other polar–nonpolar interfaces within living organisms, for example, those of lipoproteins in blood or fat droplets in adipose tissue. The compounds that diffuse most readily across membranous barriers are those with a balance between lipophilicity and hydrophilicity, having K_{ow} values of the order 0.1–1. Some examples of log K_{ow} values of organic pollutants are given in Table 2.1.

The compounds listed in the left-hand column are more polar than those in the right-hand column. They show less tendency to move into fat depots, and bioaccumulate than compounds of higher K_{ow}. That said, the herbicide atrazine, which has the highest K_{ow} in the first group, has quite low water solubility (about 5 ppm) and is relatively persistent in soil. Turning to the second group, these tend to move into fat depots and bioaccumulate. Those that are resistant to metabolic detoxication have particularly long biological half-lives (e.g., dieldrin, p,p'-DDT, and TCDD). Some of them (e.g., dieldrin, p,p'-DDT) have extremely long half-lives in soils (see Chapter 4, Section 4.2).

TABLE 2.1
Log K_{ow} Values of Organic Pollutants

Low K_{ow}		High K_{ow}	
Hydrogen cyanide	0.25	Malathion	2.89
Vinyl chloride	0.60	Lindane	3.78
Methyl bromide	1.19	Parathion	3.81
Phenol	1.45	2-chlorobiphenyl	4.53
Chloroform	1.97	4,4 dichlorobiphenyl	5.33
Trichlorofluoro methane	2.16	Dieldrin	5.48
Carbaryl	2.36	p,p'-DDT	6.36
Dichlorofluoro methane	2.53	benzo[a]pyrene	6.50
Atrazine	2.56	TCDD (dioxin)	6.64

Before leaving the subject of polarity and K_{ow} in relation to uptake and distribution, mention should be made of weak acids and bases. The complicating factor here is that they exist in solution in different forms, the balance between which is dependent on pH. The different forms have different polarities, and thus different K_{ow} values. In other words, the K_{ow} values measured are pH-dependent. Take, for example, the plant growth regulator herbicide 2,4-D. This is often formulated as the sodium or potassium salt, which has high water solubility. When dissolved in water, however, the following equilibrium is established:

$$R\text{--}COOH \rightarrow RCOO^- + H^+$$

where R = alkyl or aryl group.

If the pH is reduced by adding an acid, the equilibrium moves from right to left, generating more of the undissociated acid. This has a higher K_{ow} than the anion from which it is formed. Consequently, it can move readily by diffusion into and through hydrophobic barriers, which the anion cannot. If the herbicide is applied to plant leaf surfaces, absorption across the lipophilic cuticle into the plant occurs more rapidly at lower pH (e.g., in the presence of NH_4^+). The same argument applies to the uptake of weak acids such as aspirin (acetylsalicylic acid) across the wall of the vertebrate stomach. At the very low pH of the stomach contents, much of the aspirin exists in the form of the lipophilic undissociated acid, which readily diffuses across the membranes of the stomach wall and into the bloodstream. A similar argument applies to weak bases, except that these tend to pass into the undissociated state at high rather than low pH. Substituted amides, for example, show the following equilibrium:

$$R\text{--}CO\ NH_3^+ \rightarrow RNH_2 + H^+$$

As pH increases, the concentration of OH^- also goes up. H^+ ions are removed to form water, the equilibrium shifts from left to right, and more relatively nonpolar RNH_2 is generated.

Returning to the more general question of the movement of organic molecules through biological membranes during uptake and distribution, a major consideration, then, is movement through the underlying structure of the phospholipid bilayer. It should also be mentioned, however, that there are pores through membranes that are hydrophilic in character, through which ions and small polar organic molecules (e.g., methanol, acetone) may pass by diffusion. The diameter and characteristics of these pores varies between different types of membranes. Many of them have a critical role in regulating the movement of endogenous ions and molecules across membranes. Movement may be by diffusion, primary or secondary active transport, or facilitated diffusion. A more detailed consideration of pores would be inappropriate in the present context. Readers are referred to basic texts on biochemical toxicology (e.g., Timbrell 1999) for a more extensive treatment. The main points to be emphasized here are that certain small, relatively polar, organic molecules can diffuse through hydrophilic pores, and that the nature of these pores varies between membranes of different tissues and different cellular locations. Examples will be given, where appropriate, in the later text.

Considering again movement across phospholipid bilayers, where only passive diffusion is involved, compounds below a certain molecular weight (about 800 kDa) with very high K_{ow} values tend to move into membranes but show little tendency to move out again. In other words, they do not move across membranes to any important extent, by passive diffusion alone. On the other hand, they may be cotransported across membranes by endogenous hydrophobic molecules with which they are associated (e.g., lipids or lipoproteins). There are transport mechanisms, for example, phagocytosis (solids) and pinocytosis (liquids), which can move macromolecules across membranes. The particle or droplet is engulfed by the cell membrane, and then extruded to the opposite side, carrying associated xenobiotics with it. The lipids associated with membranes are turned over, so lipophilic compounds taken into membranes and associated with them may be cotransported with the lipids to other cellular locations. Compounds of low K_{ow} do not tend to diffuse into lipid bilayers at all, and consequently, do not cross membranous barriers unless they are sufficiently small and polar to diffuse through pores (see the preceding text). The blood–brain barrier of vertebrates is an example of a nonpolar barrier between an organ and surrounding plasma, which prevents the transit of ionized compounds in the absence of any specific uptake mechanism. The relatively low permeability of the capillaries of the central nervous system to ionized compounds is the consequence of two conditions: (1) the coverage of the basement membranes of the capillary endothelium by the processes of glial cells (astrocytes) and (2) the tight junctions that exist between capillaries, leaving few pores. Lipophilic compounds (organochlorine insecticides, organophosphorous insecticides, organomercury compounds, and organolead compounds) readily move into the brain to produce toxic effects, whereas many ionized compounds are excluded by this barrier.

2.3.2 METABOLISM

2.3.2.1 General Considerations

After uptake, lipophilic pollutants tend to move into hydrophobic domains within animals or plants (membranes, lipoproteins, depot fat, etc.), unless they are biotransformed into more polar and water soluble with compounds having low K_{ow}. Metabolism of lipophilic compounds proceeds in two stages:

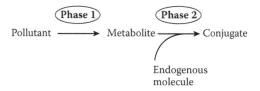

In phase 1, the pollutant is converted into a more water-soluble metabolites, by oxidation, hydrolysis, hydration, or reduction. Usually, phase 1 metabolism introduces one or more hydroxyl groups. In phase 2, a water-soluble endogenous species (usually an anion) is attached to the metabolite—very commonly through a hydroxyl group introduced during phase 1. Although this scheme describes the course of most biotransformations of lipophilic xenobiotics, there can be departures from it.

Sometimes, the pollutant is directly conjugated, for example, by interacting with the hydroxyl groups of phenols or alcohols. Phase 1 can involve more than one step, and sometimes it yields an active metabolite that binds to cellular macromolecules without undergoing conjugation (as in the activation of benzo[*a*]pyrene and other carcinogens). A diagrammatic representation of metabolic changes, linking them to detoxication and toxicity, is shown in Figure 2.2. The description so far is based on data for animals. Plants possess enzyme systems similar to those of animals, albeit at lower activities, but they have been little studied. The ensuing account is based on what is known of the enzymes of animals, especially mammals.

Many of the phase 1 enzymes are located in hydrophobic membrane environments. In vertebrates, they are particularly associated with the endoplasmic reticulum of the liver, in keeping with their role in detoxication. Lipophilic xenobiotics are moved to the liver after absorption from the gut, notably in the hepatic portal system of mammals. Once absorbed into hepatocytes, they will diffuse, or be transported, to the hydrophobic endoplasmic reticulum. Within the endoplasmic reticulum, enzymes convert them to more polar metabolites, which tend to diffuse out of the membrane and into the cytosol. Either in the membrane, or more extensively in the cytosol, conjugases convert them into water-soluble conjugates that are ready for excretion. Phase 1 enzymes are located mainly in the endoplasmic reticulum, and phase 2 enzymes mainly in the cytosol.

The enzymes involved in the biotransformation of pollutants and other xenobiotics will now be described in more detail, starting with phase 1 enzymes and then moving on to phase 2 enzymes.

For an account of the main types of enzymes involved in xenobiotic metabolism, see Jakoby (1980).

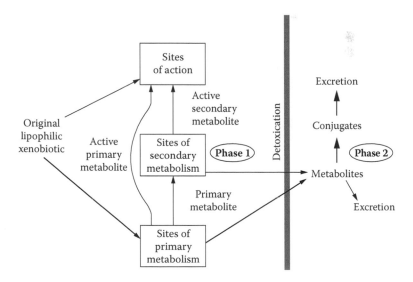

FIGURE 2.2   Metabolism and toxicity.

2.3.2.2 Monooxygenases

Monooxygenases exist in a great variety of forms, with contrasting yet overlapping substrate specificities. Substrates include a very wide range of lipophilic compounds, both xenobiotics and endogenous molecules. They are located in membranes, most importantly in the endoplasmic reticulum of different animal tissues. In vertebrates, liver is a particularly rich source, whereas in insects, microsomes prepared from midgut or fat body contain substantial amounts of these enzymes. When lipophilic pollutants move into the endoplasmic reticulum, they are converted through monooxygenase attack into more polar metabolites which partition out of the membrane into cytosol. Very often, metabolism leads to the introduction of one or more hydroxyl groups, and these are available for conjugation with glucuronide or sulfate. Monooxygenases are the most important group of enzymes carrying out phase 1 biotransformation, and very few lipophilic xenobiotics are resistant to metabolic attack by them, the main exceptions being highly halogenated compounds such as dioxin, p,p'-DDE, and higher chlorinated PCBs.

Monooxygenases owe their catalytic properties to the hemeprotein cytochrome P450 (Figure 2.3). Within the membrane of the endoplasmic reticulum (microsomal

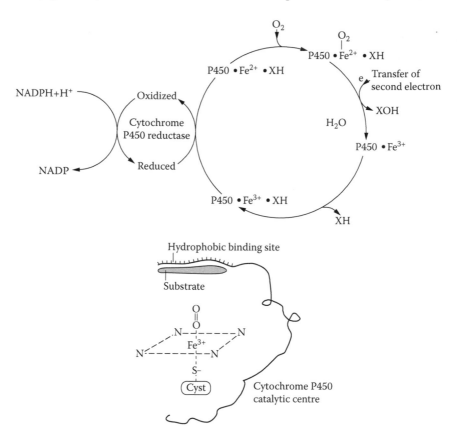

FIGURE 2.3   Oxidation by microsomal monooxygenases.

membrane), cytochrome P450 macromolecules are associated with another protein, NADPH/cytochrome P450 reductase. The latter enzyme is converted to its reduced form by the action of NADPH (reduced form of nicotine adenine dinucleotide phosphate). Electrons are passed from the reduced reductase to cytochrome P450, converting it to the Fe^{2+} state.

Xenobiotic substrates attach themselves to the hydrophobic binding site of P450, when the iron of the hemeprotein is in the Fe^{3+} state. After a single electron has been passed from the reductase to P450, the hemeprotein moves into the Fe^{2+} state, and molecular oxygen can now bind to the enzyme:substrate complex. It binds to the free sixth ligand position of the iron, where it is now in close proximity to the bound lipophilic substrate (Figure 2.3). A further electron is then passed to P450, and this leads to the activation of the bound oxygen. This second electron may come from the same source as the first, or it may originate from another microsomal hemeprotein, cytochrome b5, which is reduced by NADH rather than NADPH. After this, molecular oxygen is split—one atom being incorporated into the xenobiotic metabolite, and the other into water. The exact mechanism involved in these changes is still controversial. However, a widely accepted version of the main events is shown in Figure 2.4. The uptake of the second electron leads to the formation of a highly reactive superoxide anion, O_2^-, after which the splitting of molecular oxygen and "mixed function oxidation" immediately follow. The P450 returns to the Fe^{3+} state, and the whole cycle can begin again.

"Active" oxygen generated at the catalytic center of cytochrome P450 can attack the great majority of organic molecules that become attached to the neighboring substrate-binding site (Figure 2.3). When substrates are bound, the position of the molecule that is attacked ("regioselectivity") will depend on the spatial relationship between the bound molecule and the activated oxygen. Active oxygen forms are most likely to attack the accessible positions on the xenobiotic which are nearest to them. Differences in substrate specificity between the many different P450 forms are due,

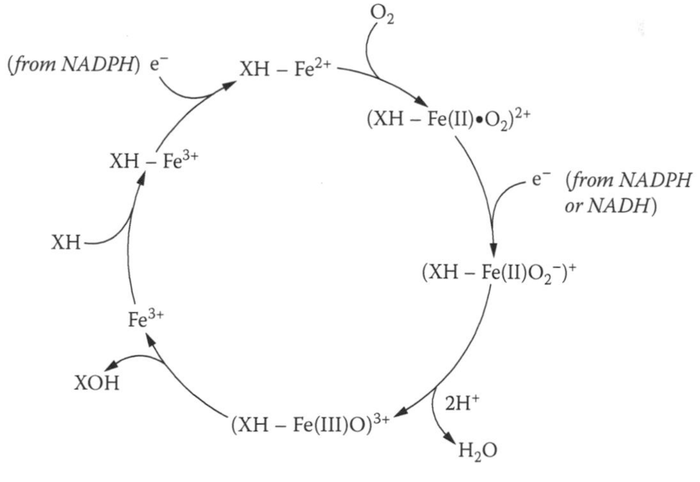

FIGURE 2.4   Proposed mechanism for monooxygenation by cytochrome P450.

very largely if not entirely, to differences in the structure and position of the binding site within the hemeprotein. The mechanism of oxidation appears to be the same in the different forms of the enzyme, so could hardly provide the basis for substrate specificity (see Trager 1988). This explains regiospecific metabolism, where different forms of P450 attack the same substrate but in different molecular positions. Regioselectivity is sometimes critical in the activation of polycyclic aromatic hydrocarbons that act as carcinogens or mutagens (see Chapter 9). Cytochrome P450 1A1, for example, tends to hydroxylate benzo[a]pyrene in the so called *bay region*, yielding bay-region epoxides that are highly mutagenic (Chapter 9). Other P450 forms attack different regions of the molecule, yielding less hazardous metabolites. The production of active forms of oxygen is, in itself, potentially hazardous, and it is very important that such reactive species do not escape from the catalytic zone of P450 to other parts of the membrane, where they could cause oxidative damage. There is evidence that, under certain circumstances, superoxide anion may escape in this way. This may occur when highly refractory substrates (e.g., higher chlorinated PCBs) are bound to P450, but resist metabolic attack (see Chapter 13, Section 13.3).

The wide range of oxidations catalyzed by cytochrome P450 is illustrated by the examples given in Figure 2.5. Aromatic rings are hydroxylated, as in the case of 2,6'-dichlorobiphenyl. The initial product is usually an epoxide, but this rearranges

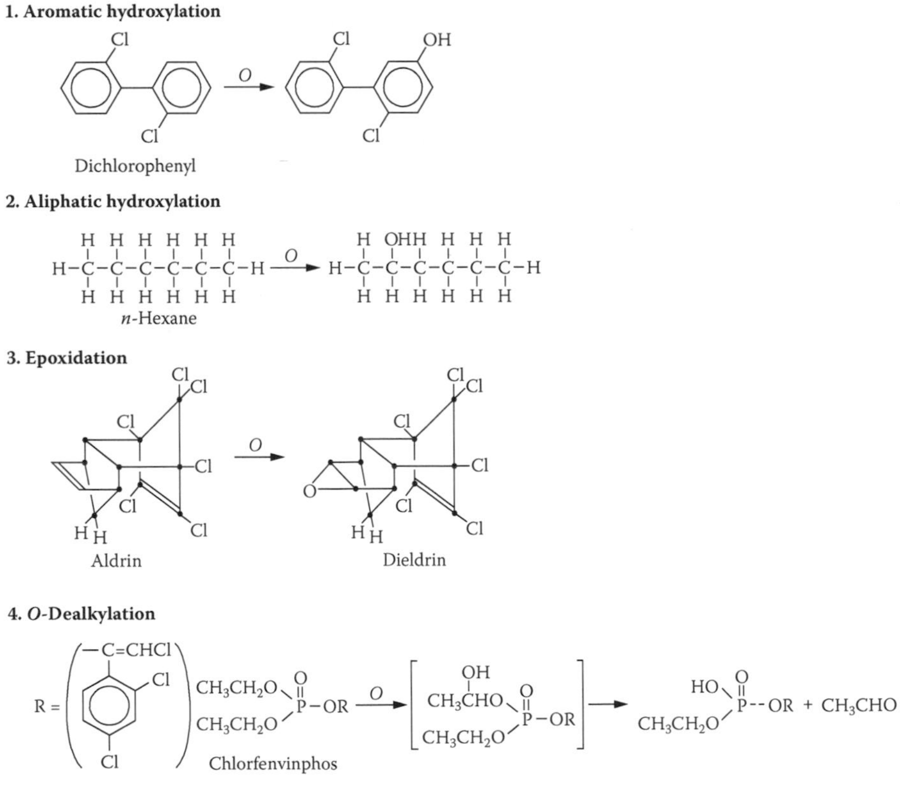

1. Aromatic hydroxylation

Dichlorophenyl

2. Aliphatic hydroxylation

n-Hexane

3. Epoxidation

Aldrin   Dieldrin

4. *O*-Dealkylation

Chlorfenvinphos

FIGURE 2.5 Biotransformations by cytochrome P450.

5. N-Dealkylation

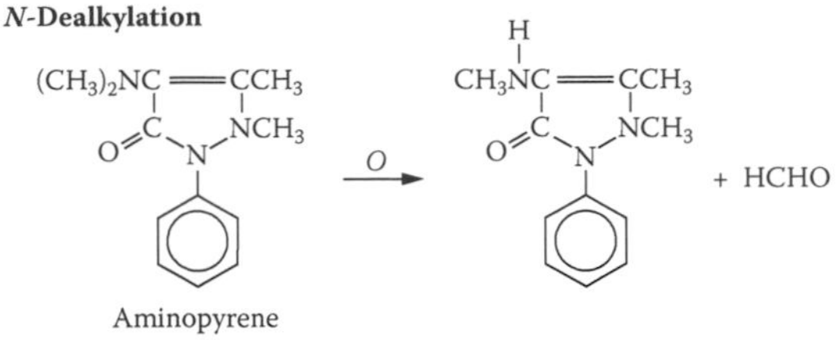

$(CH_3)_2NC$══CCH_3

Aminopyrene

CH_3NC══CCH_3 + HCHO

6. Oxidative desulphuration

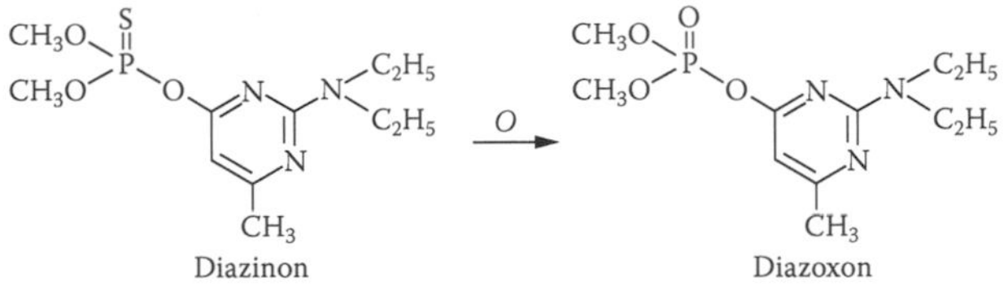

Diazinon

Diazoxon

7. Sulphur oxidation

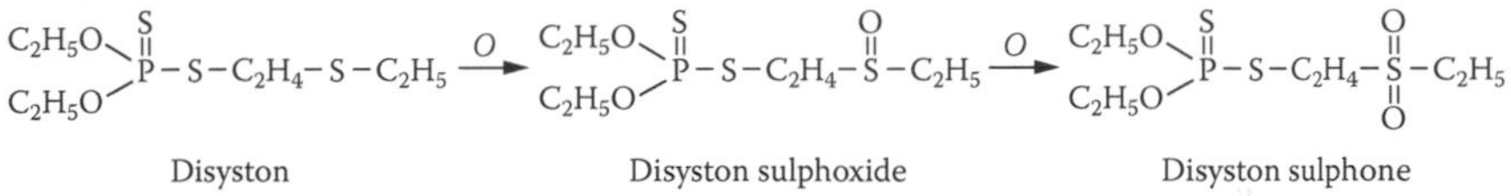

Disyston

Disyston sulphoxide

Disyston sulphone

8. N-Hydroxylation

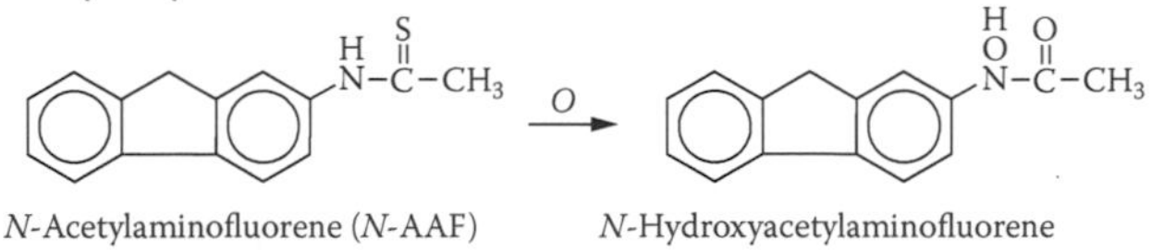

N-Acetylaminofluorene (N-AAF)

N-Hydroxyacetylaminofluorene

FIGURE 2.5 (CONTINUED)   Biotransformations by cytochrome P450.

to give a phenol. Alkyl groups can also be hydroxylated, as in the conversion of hexane to hexan-2-ol. If an alkyl group is linked to nitrogen or oxygen, hydroxylation may yield an unstable product, which rearranges. An aldehyde is released, leaving behind a proton attached to N or to O (N-dealkylation or O-dealkylation, respectively). Thus, with the OP insecticide chlorfenvinphos, one of the ethoxy groups is hydroxylated, and the unstable metabolite so formed cleaves to release acetaldehyde and desethyl chlorfenvinphos. In the case of the drug aminopyrene, a methyl group attached to N is hydroxylated, and the primary metabolite splits up to release formaldehyde and an amine. Sometimes the oxidation of C:C double bonds can generate stable epoxides, as in the conversion of aldrin to dieldrin, or heptachlor to heptachlor epoxide. Cytochrome P450s can also catalyze oxidative desulfuration. The example given is the OP insecticide diazinon, which is transformed into the active oxon, diazoxon. P=S is converted into P=O. With thioethers such as the OP insecticide disyston, P450 can catalyze the addition of oxygen to the sulfur bridge, generating sulfoxides and sulfones. P450s can also catalyze the N-hydroxylation of amines such as N-acetylaminofluorene (N-AAF).

This series of examples is by no means exhaustive, and others will be encountered in the later text. Although it is true that the great majority of oxidations catalyzed by cytochrome P450 represent detoxication, in a small yet very important number of cases, oxidation leads to activation. Activations are given prominence in the examples shown here, because of their toxicological importance. Thus, among the examples given earlier, the oxidative desulfuration of diazinon and many other OP insecticides causes activation; oxons are much more potent anticholinesterases than are thions. Some aromatic oxidations (e.g., of benzo[a]pyrene) yield highly reactive epoxides that are mutagenic. N-hydroxylation of certain amines (e.g., N-AAF) can also yield mutagenic metabolites. Finally, the epoxidation of aldrin or heptachlor yields highly toxic metabolites, while sulfoxides and sulfones of OP insecticides are sometimes more toxic than their parent compounds. Oxidation tends to increase polarity. Where this simply aids excretion, the result is detoxication. On the other hand, some metabolic products are much more reactive than the parent compounds, and this can lead to interaction with cellular macromolecules such as enzymes or DNA, with consequent toxicity.

Cytochrome P450 exists in a bewildering variety of forms, which have been assigned to 74 different gene families (Nelson et al. 1996). In one review (Nelson 1998), 37 families are described for metazoa alone. Although many of these appear to be primarily concerned with the metabolism of endogenous compounds, four families are strongly implicated in the metabolism of xenobiotics in animals. These are gene families CYP1, CYP2, CYP3, and CYP4 (see Table 2.2), which will shortly be described. A wider view of the different P450 forms and families was given earlier in Chapter 1, when considering evolutionary aspects of detoxifying enzymes. Differences in the form and function of P450s between the phyla will be discussed later in relation to the question of selectivity (Section 2.5).

To consider now P450 families of vertebrates that have an important role in xenobiotic metabolism, CYP1A1 and CYP1A2 are P450 forms that metabolize, and are

TABLE 2.2
Some Inhibitors of Cytochrome P450

Compound	Inhibitory Action
Carbon monoxide	Inhibits all forms of P450
	Competes with oxygen for heme-binding site
Methylene dioxyphenyls	Carbene forms generated, and these bind to heme
	Selective inhibitors
Imidazoles, triazoles, and pyridines	Contain ring N, which binds to heme
	Selective inhibitors
Phosphorothionates	Oxidative desulfuration releases active sulfur that binds to, and deactivates, P450
	Selective inhibitors
1-Ethynyl pyrene	Specific inhibitor of 1A1
Furafylline	Specific inhibitor of 1A2
Diethyldithiocarbamate	Specific inhibitor of 2A6
Sulfenazole	Specific inhibitor of 2C9
Quinine	Specific inhibitor of 2D1
Disulfiram	Specific inhibitor of 2E1

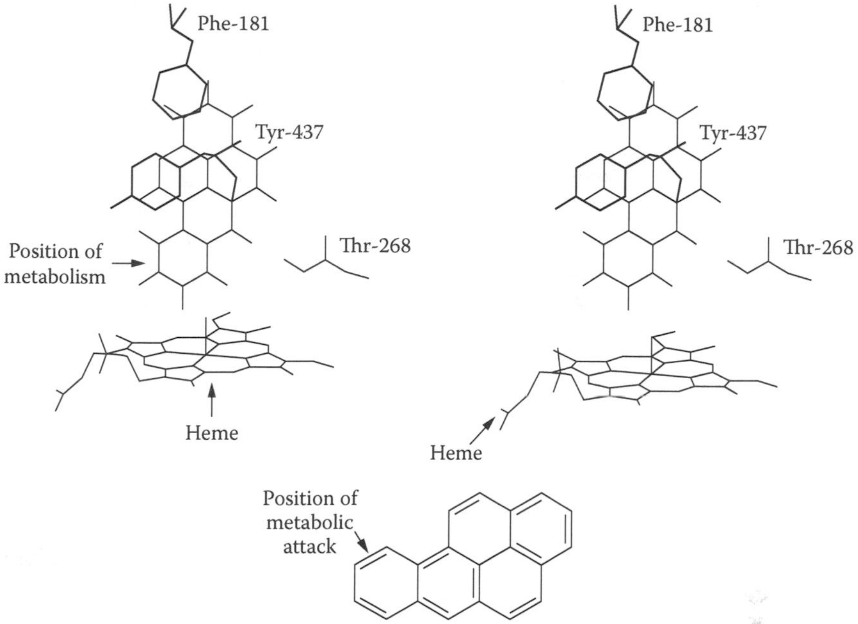

FIGURE 2.6   The procarcinogen benzo[*a*]pyrene oriented in the CYP1A1 active site (stereo view) via π– π stacking between aromatic rings on the substrate and those of the complementary amino acid side chains, such that 7,8-epoxidation can occur. The substrate is shown with pale lines in the upper structures. The position of metabolism is indicated by an arrow in the lower structure (after Lewis 1996).

inhibited by, planar molecules (e.g., planar PAHs and coplanar PCBs). This can be explained in terms of the deduced structure of the active site of these CYP1A enzymes (Figure 2.6; Lewis 1996, and Lewis and Lake 1996). This takes the form of a rectangular slot, composed of several aromatic side chains, including the coplanar rings of phenylalanine 181 and tyrosine 437, which restrict the size of the cavity such that only planar structures of a certain rectangular dimension will be able to take up the binding position. Small differences in structure between the active sites of CYP1A1 and CYP1A2 may explain their differences in substrate preference (Table 2.3), for example, phenylalanine 259 (CYP1A1) versus anserine 259 (1A2). CYP1A1 metabolizes especially heterocyclic molecules, whereas CYP1A1 is more concerned with PAHs. By contrast, the active sites of families CYP2 and 3 have more open structures and are capable of binding a wide variety of different compounds, some planar but many of more globular shape. CYP2 is a particularly diverse family, whose rapid evolution coincides with the movement of animals from water to land (for discussion, see Chapter 1). Very many lipophilic xenobiotics are metabolized by enzymes belonging to this family. Of particular interest from an ecotoxicological point of view, CYP2B is involved in the metabolism of organochlorine insecticides such as aldrin and endrin and some OP insecticides including parathion, CYP2C with warfarin metabolism, and CYP2E with solvents of low-molecular weight, including acetone and ethanol. CYP3 is noteworthy for the great diversity of substrates that it can metabolize, both

TABLE 2.3

Types of Carboxylesterase Isolated from Rat Liver Microsomes

PI Value	Genetic Classification	Substrates	Comments
5.6	ES3	Simple aromatic esters, acetanilide, lysophospholipids, monoglycerides, long-chain acyl carnitines	Sometimes called lysophospholipase to distinguish it from other esterases of this kind
6.2/6.4	ES4	Aspirin, malathion, pyrethroids, palmitoyl CoA, monoacylglycerol, cholesterol esters	May correspond to EC 3.1.2.2 and EC 3.1.1.23
6.0	ES8/ES10	Short-chain aliphatic esters, medium-chain acylglycerols, clofibrate, procaine	ES8 may be a monomer, ES10 a dimer
5.0/5.2	ES15	Mono- and diacylglycerols, acetyl carnitine, phorbol diesters	Corresponds to acetyl carnitine hydrolase

Source: Data from Mentlein et al. 1987.

endogenous and exogenous. Structural models indicate a highly unrestricted active site, in keeping with this characteristic (Lewis 1996). This is in marked contrast to the highly restricted active sites proposed for family CYP1A. Although CYP4 is especially involved in the endogenous metabolism of fatty acids, it does have a key role in the metabolism of a few xenobiotics, including phthalate esters.

Cytochrome P450 metabolism of xenobiotics has been less well studied in invertebrates compared to vertebrates. The importance of this subject in human toxicology has been a powerful stimulus for work on vertebrates, but there has been no comparable driving force in the case of invertebrate toxicology. Also, in the earlier stages of this work, there were considerable technical problems in isolating and characterizing the P450s of invertebrates, associated in part with the small size of many of them and also the instability of subcellular preparations made from them. Insects, however, have received more attention than other invertebrate groups, partly because of the importance of the use of insecticides for the control of major pest species and vectors of disease (e.g., malarial mosquito and tse-tse flies). In insects, P450s belonging to gene family CYP6 have been shown to have an important role in xenobiotic metabolism.

CYP6D1 of the housefly (*Musca domestica*) has been found to hydroxylate cypermethrin and thereby provide a resistance mechanism to this compound and other pyrethroids in this species (Scott et al. 1998; see also Chapter 12). Also, this insect P450 can metabolize plant toxins such as the linear furanocoumarins xanthotoxin and bergapten (Ma et al. 1994). This metabolic capability has been found in the lepidopteran *Papilio polyxenes* (black swallowtail), a species that feeds almost exclusively on plants containing furanocoumarins.

The classification of P450s, which is based on amino acid sequencing, bears some relationship to metabolic function. That said, some xenobiotic molecules, especially where they are large and complex, are metabolized by several different P450 forms. Different forms of P450 tend to show regioselectivity, for example, in the metabolism of PAHs such as benzo[a]pyrene and of steroids such as testosterone.

Oxidations catalyzed by cytochrome P450 can be inhibited by many compounds. Some of the more important examples are given in Table 2.2. Carbon monoxide inhibits all known forms of P450 by competing with oxygen for its binding position on heme. Indeed, this interaction was the original basis for the term *cytochrome P450*. Interaction of CO with P450 in the Fe^{2+} state yields a complex that has an absorption maximum of ~450 nm. Many organic molecules act as inhibitors, but they are, in general, selective for particular forms of the hemeprotein. Selectivity depends on the structural features of the molecules, how well they fit into the active sites of particular forms, and the position in the molecule of functional groups that can interact with heme or with the substrate-binding sites. A group of important inhibitors—methylene dioxyphenyl compounds such as piperonyl butoxide—that act as suicide substrates is described briefly here. The removal of two protons leads to the formation of carbenes, which bind strongly to heme, thereby preventing the binding of oxygen (Figure 2.7). Compounds of this type have been used to synergize insecticides such as pyrethroids and carbamates, which are subject to oxidative detoxication. A considerable number of compounds containing heterocyclic nitrogen are potent inhibitors (Figure 2.7). Included here are certain compounds containing heterocyclic groupings, such as imidazole, triazole, and pyridine. Some compounds of this type have been successfully developed as antifungal agents due to their strong inhibition of CYP51, which has a critical role in ergosterol biosynthesis (see Chapter 1). Their inhibitory potency depends on the ability of the ring N to ligate to the iron of heme, thus preventing the activation of oxygen. One

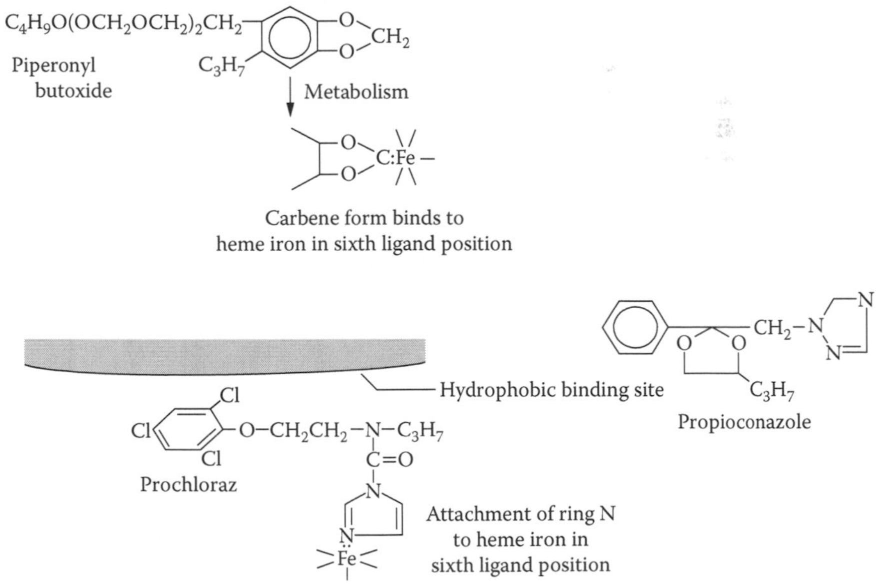

FIGURE 2.7   Cytochrome P450 inhibitors.

type of inhibition that is important in ecotoxicology is the deactivation of heme caused by the oxidative desulfuration of phosphorothionates (see Section 2.3.2.2). Sulfur atoms detached from phosphorothionates are bound in some form to cytochrome P450, destroying its catalytic activity. The exact mechanism for this is, at present, unknown. Apart from these broad classes of inhibitors, certain individual compounds are very selective for particular P450 forms, and are thus valuable for the purposes of identification and characterization. Some examples are given in Table 2.2.

There are marked differences in hepatic microsomal monooxygenase (HMO) activities between different species and groups of vertebrates. Figure 2.8 summarizes results from many studies reported in the general literature (Walker 1980, Ronis and Walker 1989). Mean activities for each species across a range of lipophilic xenobiotics are expressed relative to those of the male rat, making a correction for relative liver weight. Males and females of each species are represented by a single point wherever possible. For some species, there is just a single point because no distinction had been made between the sexes. The log relative activity is plotted against the log body weight.

The mammals, which are nearly all omnivorous, show a negative correlation between log relative HMO activity and log body weight. Thus, small mammals have much higher HMO per unit body weight than large mammals. This is explicable in terms of the detoxifying function of P450, much of the metabolism of these substrates being carried out by isoforms of CYP2. Small mammals have much larger surface area/body volume ratios than large mammals, and thus they take in food and associated xenobiotics more rapidly in order to acquire sufficient metabolic energy to maintain their body temperatures.

The birds studied differed widely in their type of food, ranging from omnivores and herbivores to specialized predators. Omnivorous and herbivorous birds had rather lower HMO activities than mammals of comparable body size, with galliform birds showing similar activities to mammals. Fish-eating birds and raptors, however, showed lower HMO activities than other birds and much lower activities than omnivorous mammals. This is explicable on the grounds that they have had little requirement for detoxication by P450 (e.g., isoforms of CYP2) during the course of evolution, in contrast to herbivores and omnivores that have had to detoxify plant toxins. Fish-eating birds, similar to omnivorous mammals, show a negative correlation between log HMO activity and log body weight. The slopes are very similar in the two cases. The bird-eating sparrow hawk shows a very low value for HMO activity, comparable to that of fish of similar body weight. This low detoxifying capability may well have been a critical factor determining the marked bioaccumulation of p,p'-DDE, dieldrin, and heptachlor epoxide by this species (see Chapter 5).

Fish show generally low HMO activities that are not strongly related to body weight. This may reflect a limited requirement of fish for metabolic detoxication; they are able to efficiently excrete many compounds by diffusion across the gills. The weak relationship of HMO activity to body weight is probably because fish are poikilotherms and should not, therefore, have an energy requirement for the maintenance of body temperature that is a function of body size. In other words, the rate of intake of xenobiotics with food is unlikely to be strongly related to body size.

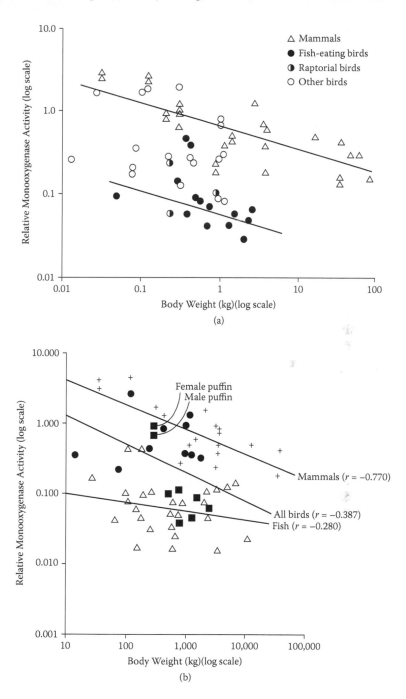

FIGURE 2.8 Monooxygenase activities of mammals, birds, and fish. (a) Mammals and birds. (b) Mammals, birds, and fish. Activities are of hepatic microsomal monooxygenases to a range of substrates expressed in relation to body weight. Each point represents one species (males and females are sometimes entered separately) (from Walker et al. 2000).

2.3.2.3 Esterases and Other Hydrolases

Many xenobiotics, both synthetic and naturally occuring, are lipophilic esters. They can be degraded to water-soluble acids and bases by hydrolytic attack. Two important examples of esteratic hydrolysis in ecotoxicology now follow:

$$
\underset{\text{Carboxyl ester}}{R-\overset{\overset{\textstyle O}{\|}}{C}-OX} \; + \; H_2O \; \Longrightarrow \; \underset{\text{Carboxylic acid}}{R-\overset{\overset{\textstyle O}{\|}}{C}-OH} \; + \; \underset{\text{Alcohol}}{XOH}
$$

$$
\underset{\text{Organophosphate triester}}{\overset{\textstyle RO}{\underset{\textstyle RO}{\diagdown}} \overset{\overset{\textstyle O}{\|}}{P}-OX + H_2O} \; \Longrightarrow \; \underset{\text{Organophosphate diester}}{\overset{\textstyle RO}{\underset{\textstyle RO}{\diagdown}} \overset{\overset{\textstyle O}{\|}}{P}-OH + XOH}
$$

Enzymes catalyzing the hydrolysis of esters are termed *esterases*. They belong to a larger group of enzymes termed *hydrolases*, which can cleave a variety of chemical bonds by hydrolytic attack. In the classification of hydrolases of the International Union of Biochemistry (IUB), the following categories are recognized:

3.1 Acting on ester bonds (esterases)
3.2 Acting on glyoacyl compounds
3.3 Acting on ether bonds
3.4 Acting on peptide bonds (peptidases)
3.5 Acting on C–N bonds other than peptide bonds
3.6 Acting on acid anhydrides (acid anhydrolases)
3.7 Acting on C–C bonds
3.8 Acting on halide bonds
3.9 Acting on P–N bonds
3.10 Acting on S–N bonds
3.11 Acting on C–P bonds

Although it is convenient to define hydrolases according to their enzymatic function, there is one serious underlying problem. Some hydrolases are capable of performing two or more of the preceding kinds of hydrolytic attack, and so do not fall simply into just one category. There are esterases, for example, that can also hydrolyze peptides, amides, and halide bonds. The shortcomings of the early IUB classification, which was originally based on the measurement of activities in crude tissue preparations, have become apparent with the purification and characterization of hydrolases. As yet, however, only limited progress has been made, and a comprehensive classification is still some distance away. In what follows, a simple and pragmatic classification will be described for esterases that hydrolyze xenobiotic esters (Figure 2.9). It should be emphasized that this is a classification seen from a toxicological point of view. Esterases are important both for their detoxifying function and as sites of action for toxic molecules. Thus, in Figure 2.9, esterases that degrade organophosphates serve

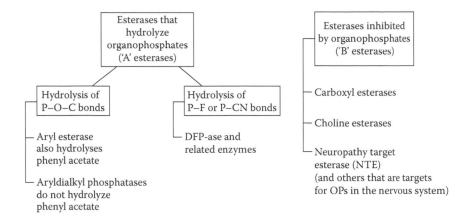

FIGURE 2.9   Esterases that are important in ecotoxicology.

a detoxifying function, whereas those inhibited by organophosphates often represent sites of action. The paradox of the latter is that esteratic hydrolysis leads to toxicity. Organophosphates behave as suicide substrates; during the course of hydrolysis, the enzymes become irreversibly inhibited, or nearly so. The inhibitory action of organophosphates on esterases will be discussed in Section 2.4.

Looking at the classification shown in Figure 2.9, esterases that effectively detoxify organophosphorous compounds by continuing hydrolysis are termed *A-esterases*, following the early definition of Aldridge (1953). They fall into two broad categories: those that hydrolyze POC bonds (the oxon forms of many organophosphorous insecticides are represented here), and those that hydrolyze P–F or P–CN bonds (a number of chemical warfare agents are represented here). Within the first category of A-esterase, two main types have been recognized. First, arylesterase (EC 3.1.1.2) can hydrolyze phenylacetate as well as organophosphate esters. It occurs in a number of mammalian tissues, including liver and blood, and has been purified and characterized. It is found associated with the high-density lipoprotein (HDL) of blood, and in the endoplasmic reticulum of liver. Other esterases that hydrolyze organophosphates but not phenylacetate have been partially purified and are termed *aryldialkylphosphatases* (EC 3.1.8.1) in recent versions of the IUB classification. These are also found in HDL of mammalian blood and in the hepatic endoplasmic reticulum of vertebrates. Within the second category of A-esterases are the diisopropylfluorophosphatases (EC 3.1.8.2) that catalyze the hydrolysis of chemical warfare agents ("nerve gases") such as diisopropyl phosphofluoridate (DFP), soman, and tabun.

There are marked species differences in A-esterase activity. Birds have very low, often undetectable, levels of activity in plasma toward paraoxon, diazoxon, pirimiphos-methyl oxon, and chlorpyrifos oxon (Brealey et al. 1980, Mackness et al. 1987, Walker et al. 1991; Figure 2.10). Mammals have much higher plasma A-esterase activities to all of these substrates. The toxicological implications of this are discussed in Chapter 10. Some species of insects have no measurable A-esterase activity, even in strains that have resistance to OPs (Mackness et al. 1982, Walker 1994). These include the peach potato aphid (*Myzus persicae*; Devonshire 1991) and the

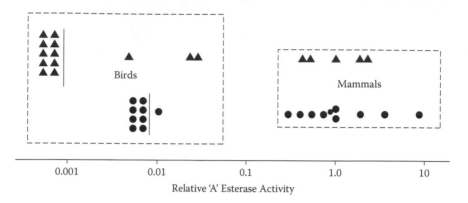

FIGURE 2.10 Plasma A-esterase activities of birds and mammals. Activities were originally measured as nanomoles product per milliliter of serum per minute, but they have been converted to relative activities (male rat = 1) and plotted on a log scale. Each point represents a mean value for a single species. Substrates: ●, paraoxon; ▲, pirimiphos-methyl oxon. Vertical lines indicate limits of detection, and all points plotted to the left of them are for species in which no activity was detected. (Activities in the male rat were 61 ± 4 and 2020 ± 130 for paraoxon and pirimiphos-methyl oxon, respectively.) (From Walker 1994a in Hodgson and Levi 1994.)

rust red flour beetle (*Tribolium castaneum*). Indeed, it has been questioned whether insects have A-esterase at all; some studies claiming to have detected it failed to distinguish between activities attributable to this enzyme and activities due to high levels of B-esterase (Walker 1994).

Dealing now with the B-esterases, the carboxylesterases (EC 3.1.1.1) represent a large group of enzymes that can hydrolyze both exogenous and endogenous esters. More than 12 different forms have been identified in rodents, and four of these have been purified from rat liver microsomes (Table 2.3; Mentlein et al. 1987). The four forms shown have been characterized on the basis of their substrate specificities and their genetic classification. They have molecular weights of about 60 kDa when in the monomeric state. They are separable by isoelectric focusing, and the PI value for each is shown in the first column. In the second column is the number assigned to each in the genetic classification. As can be seen, they all show distinct ranges of substrate specificity with a certain degree of overlap. All four can hydrolyze both exogenous and endogenous esters. ES4 and ES15 have activities previously associated with earlier entries in the IUB classification; entries were made on the basis of limited evidence. It may well be that some of these earlier entries can now be removed from the classification, the activities being due solely to members of EC 3.1.1.1. It is noteworthy that ES4 catalyzes the hydrolysis of pyrethroid insecticides and malathion. In mice, the carboxylesterases are tissue specific with a range of 10 different forms identified in the liver and kidney but only a few in other tissues. Only three forms have been found in mouse serum. As with other enzymes that metabolize xenobiotics, the liver is a particularly rich source.

Cholinesterases are another group of B-esterases. The two main types are acetylcholinesterase (EC 3.1.1.7) and "unspecific" or butyrylcholinesterase (EC 3.1.1.8). Acetylcholinesterase (AChE) is found in the postsynaptic membrane of cholinergic synapses of both the central and peripheral nervous systems. It is the site of action of OP and carbamate insecticides, and will be described in more detail in Section 2.4. Butyrylcholinesterase (BuChE) occurs in many vertebrate tissues, including blood and smooth muscle. Unlike AChE, it does not appear to represent a site of action for OP or carbamate insecticides. However, the inhibition of BuChE in blood has been used as a biomarker assay for exposure to OPs (see Thompson and Walker 1994). Neuropathy target esterase (NTE) is another B-esterase located in the nervous system. Inhibition of NTE can cause delayed neuropathy (see Section 2.4). Finally, other hydrolases of the nervous system that are sensitive to OP inhibition have been identified (Chapter 10, Section 10.2.).

The distinction between A- and B-esterases is based on the difference in their interaction with OPs. Cholinesterases have been more closely studied than other B-esterases and are taken as models for the whole group. They contain serine at the active center, and organophosphates phosphorylate this as the first stage in hydrolysis (Figure. 2.11). This is a rapid reaction that involves the splitting of the ester bond and the acylation of serine hydroxyl. The leaving group XO– combines with a proton from the serine hydroxyl group to form an alcohol, XOH. The next stage in the process, the release of the phosphoryl moiety, the restoration of the serine hydroxyl, and the reactivation of the enzyme, is usually very slow. The OP has acted as a suicide substrate, inhibiting the enzyme during the course of hydrolytic attack. A further complication may be the "aging" of the bound phosphoryl moiety. The "R" group is lost, leaving behind a charged PO– group. If this happens, the inhibition becomes irreversible, and the enzyme will not spontaneously reactivate.

This process of aging is believed to be critical in the development of delayed neuropathy, after NTE has been phosphorylated by an OP (see Chapter 10, Section 10.2.4). It is believed that most, if not all, of the B-esterases are sensitive to inhibition by OPs because they, too, have reactive serine at their active sites. It is important to emphasize that the interaction shown in Figure 2.11 occurs with OPs that contain an oxon group. Phosphorothionates, which contain instead a thion group, do not readily interact in this way. Many OP insecticides are phosphorothionates, but these need to be converted to phosphate (oxon) forms by oxidative desulfuration before inhibition of acetylcholinesterase can proceed to any significant extent (see Section 2.3.2.2).

The reason for the contrasting behavior of A-esterases is not yet clearly established. It has been suggested that the critical difference from B-esterases is the

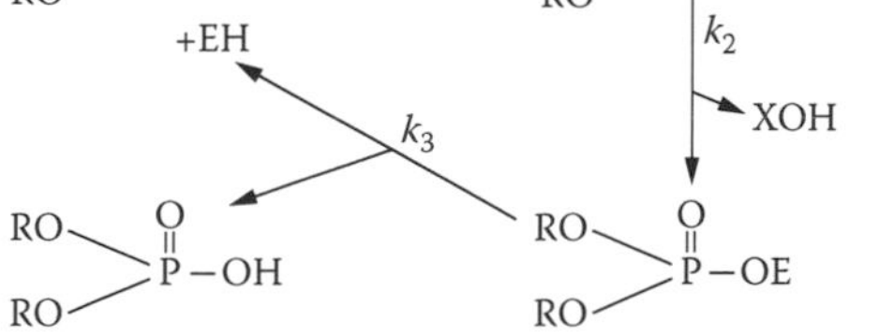

FIGURE 2.11   Interaction between organophosphates and B-esterases. R, alkyl group; E, enzyme.

presence of cysteine rather than serine at the active site. It is known that arylesterase, which hydrolyzes OPs such as parathion, does contain cysteine, and that A-esterase activity can be inhibited by agents that attack sulfhydryl groups (e.g., certain mercurial compounds). It may be that acylation of cysteine rather than serine would be followed by rapid reactivation of the enzyme (compare with Figure 2.11). In other words, if (RO)$_2$ P(O)SE is formed, it may be less stable than (RO)$_2$ P(O)O E, readily breaking down to release the reactivated enzyme.

Additional to the hydrolases identified earlier, there are others that have been less well studied and are accordingly difficult to classify. Examples will be encountered later in the text, when considering the ecotoxicology of various organic pollutants. In considering esterases, it is important to emphasize that we are only concerned with enzymes that split bonds by a hydrolytic mechanism. In early work on the biotransformation of xenobiotics, there was sometimes confusion between true hydrolases and other enzymes that can split ester bonds and yield the same products, but by different mechanisms. Thus, both monooxygenases and glutathione-S-transferases can break POC bonds of OPs and yield the same metabolites as esterases. The removal of alkyl groups from OPs can be accomplished by O-dealkylation or by their transfer to the S group of glutathione. For further details, see relevant sections of Chapter 2. In early studies, biotransformations were observed in vivo or in crude in vitro preparations such as homogenates, that is, under circumstances where it was not possible to establish the mechanisms by which biotransformations were being catalyzed. What appeared to be hydrolysis was sometimes oxidation or group transfer. This complication needs to be borne in mind when looking at certain papers in the older literature.

2.3.2.4 Epoxide Hydrolase (EC 4.2.1.63)

Epoxide hydrolases hydrate epoxides to yield transdihydrodiols without any requirement for cofactors. Examples are given in Figure 2.12. Epoxide hydrolases are

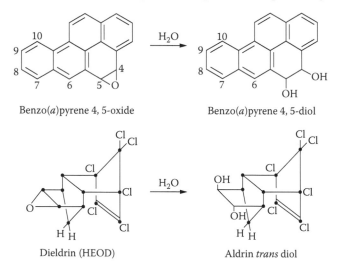

FIGURE 2.12 Epoxide hydration.

hydrophobic proteins of molecular weight ~50 kD and are found, principally, in the endoplasmic reticulum of a variety of cell types. Vertebrate liver is a particularly rich source; appreciable levels are also found in kidney, testis, and ovary. A soluble epoxide hydrolase is found in some insects, where it has the role of hydrating epoxides of juvenile hormones. The microsomal epoxide hydrolases of vertebrate liver can degrade a wide range of epoxides, including those of PAHs, PCBs, cyclodiene epoxides (including dieldrin and analogues thereof), as well as certain endogenous steroids. Epoxide hydrolase can detoxify potentially mutagenic epoxides formed by the action of cytochrome P450 on, for example, PAHs. Benzo[a]pyrene 4,5 oxide is an example. Its rapid hydration within the endoplasmic reticulum before it can migrate elsewhere is important for the protection of the cell. In general, the conversion of epoxides into more polar transdihydrodiols serves a detoxifying function, although there are a few exceptions to this rule.

2.3.2.5   Reductases

A range of reductions of xenobiotics are known to occur both in the endoplasmic reticulum and cytosol of a number of cell types. However, the enzymes (or other reductive agencies) responsible are seldom known in particular cases. Some reductions only occur at very low oxygen levels. Thus, they do not occur under normal cellular conditions, where there is a plentiful supply of oxygen.

Two important examples of reductive metabolism of xenobiotics are the reductive dehalogenation of organohalogen compounds, and the reduction of nitroaromatic compounds. Examples of each are shown in Figure 2.13. Both types of reaction can take place in hepatic microsomal preparations at low oxygen tensions. Cytochrome P450 can catalyze both types of reduction. If a substrate is bound to P450 in the

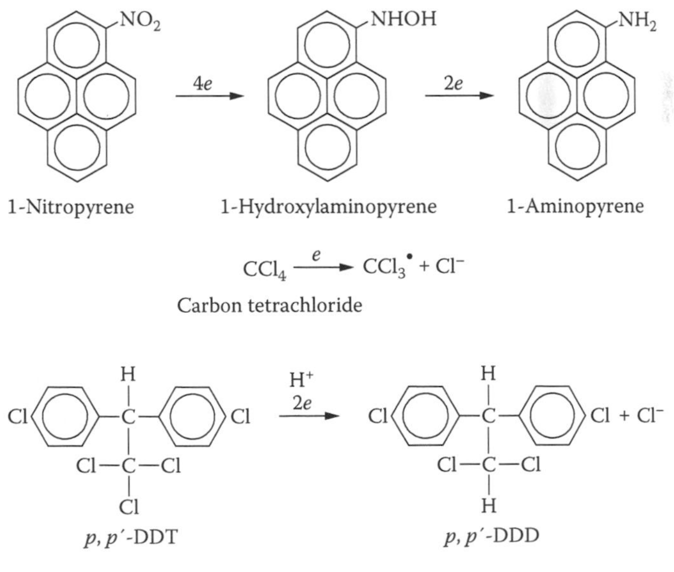

$$CCl_4 \xrightarrow{\ e\ } CCl_3^{\bullet} + Cl^-$$

Carbon tetrachloride

FIGURE 2.13   Reductase metabolism.

absence of oxygen, electrons can be passed from the iron atom of heme to the substrate. In the case of organohalogen compounds such as *p,p'*-DDT, carbon tetrachloride, and halothane, this leads to the loss of Cl⁻ and its replacement by hydrogen. If oxygen had been present, the electron would have passed to this (see Figure 2.4) and not directly to the substrate. With nitroaromatic compounds, reductions occur via an intermediate hydroxylamine stage to yield an amine (Figure 2.13). Often, the second stage is rapid, and the intermediate form is not detectable.

Although the role of P450 in this type of reductive metabolism has been well established in in vitro studies, there are uncertainties about the course of events in vivo. In the first place, as mentioned earlier, oxygen levels may be high enough to prevent the occurrence of this type of reaction (see discussion about the metabolism of *p,p'*-DDT in Chapter 5). Also, porphyrins other than P450 can catalyze reduction. So, too, can flavoprotein reductases such as NADPH or cytochrome P450 reductase. Even FAD can catalyze some reductions. Microorganisms in anaerobic soils and sediments can be very effective in degrading organohalogen compounds such as organochlorine insecticides, PCBs, and dioxins. The metabolic degradation of polyhalogenated compounds is often difficult and slow under aerobic conditions. Effective aerobic detoxication enzymes have yet to evolve for many compounds of this type. On the other hand, reductive dehalogenation is often an effective mechanism for biodegradation and has been exploited in the development of genetically manipulated microorganisms for bioremediation.

The transfer of electrons in xenobiotic reactions is tied up with the problem of the generation of active radicals, including those of oxygen. CCl_4, for example, is reduced to the highly reactive $CCl_3^•$ radical. Some organonitrocompounds, like the herbicide paraquat, can undergo redox cycling. One electron reduction of the paraquat yields an unstable radical. This radical passes an electron on to molecular oxygen, thereby generating the reactive superoxide ion and regenerating paraquat (see Chapter 13).

2.3.2.6   Conjugases

Conjugases catalyze phase 2 biotransformations, the coupling of xenobiotic metabolites (and sometimes original xenobiotics), with polar endogenous molecules, which are usually in the form of anions. Although phase 1 biotransformations of lipophilic compounds occur predominantly in the endoplasmic reticulum ("microsomal membrane"), phase 2 biotransformations often occur in the cytosol. Many different endogenous molecules are utilized for conjugation, and there can be large differences between groups and between species, in the preferred metabolic pathway. The critical thing is that polar conjugates are produced that can be rapidly excreted. The following account will be mainly concerned with three groups of enzymes that are responsible for most of the conjugations in vertebrates: glucuronyl transferases, sulfotransferases, and glutathione-*S*-transferases. It should be emphasized that less is known about the conjugases of invertebrates and plants. Although conjugations are seen to be detoxifying, and in general protective toward the organism, in some instances conjugates are broken down to release potentially toxic compounds. For example, some glutathione conjugates can break down in the kidney, with toxic effects.

UDP-Glucuronyl transferases (henceforward simply *glucuronyl transferases*) exist in a number of different forms with contrasting, yet overlapping, substrate specificities. Both exogenous and endogenous substrates are metabolized. They are associated with the endoplasmic reticulum of many vertebrate tissues, notably liver, and have molecular weights of 50–58 kD. They catalyze the transfer of glucuronate from UDP glucuronate (UDPGA) to a variety of substrates that possess functional groups with labile protons—principally –OH, but also –SH and –NH (see Figure 2.14). As noted earlier, these functional groups are often introduced during phase 1 metabolism. Sometimes, they are present in original xenobiotics, in which case conjugation can proceed without any preliminary phase 1 metabolism. The products are beta-D-glucuronides.

For the reaction to proceed, glucuronate must be delivered by UDPGA, which is generated in the cytosol, not in the endoplasmic reticulum. It is synthesized in a two-step process:

1. Glucose-6-phosphate interacts with UTP to form UDPD-glucose and pyrophosphate.
2. UDPD glucose is oxidized to UDPGA by the action of UDPG dehydrogenase, with NAD^+ as cofactor. UDPGA then migrates to the membrane-bound glucuronyl transferase, where conjugation of xenobiotics can proceed (Figure 2.14). Glucuronyl transferases can be activated by N-acetylhexosamine. Although well represented in mammals, birds, reptiles, and amphibians, glucuronyl transferases are at low levels in fish. In insects, by contrast, conjugation tends to be with glucose rather than glucuronate.

Sulfotransferases represent an important group of conjugases in cytosol. They are present in many tissues, the liver once again being a rich source. As with the glucuronyl transferases, the conjugating group is transferred from a complex molecule to the substrate (Figure 2.14). Here, the sulfate is donated by 3-phosphoadenine 5-phosphosulfate (PAPS). It is generated in the cytosol by a two-stage process:

1. ATP interacts with sulfate ion (SO_4^{2-}) to form adenosine 5-phosphosulfate (APS) in a reaction mediated by the enzyme ATP sulfate adenylyl transferase.
2. ATP interacts with APS to form PAPS, catalyzed by ATP adenylyl sulfate 3-phosphotransferase.

Sulfotransferases catalyze the transfer of sulfate from PAPS to wide-range xenobiotics that possess hydroxyl groups. Steroid alcohols are among the endogenous substrates. The sulfotransferases exist in different forms.

Glutathione-S-transferases represent another group of enzymes located primarily in the cytosol, although one form occurs in microsomes. Glutathione conjugations depend on the capacity of glutathione to act as a nucleophile (GS-), against xenobiotics, which are electrophiles. Although many glutathione conjugations can proceed spontaneously, without the participation of the enzyme, reaction tends to be slow. The glutathione-S-transferases are able to bind reduced glutathione (GSH) in close proximity to substrates held at a hydrophobic binding site and, by this means, increase the rate of conjugation.

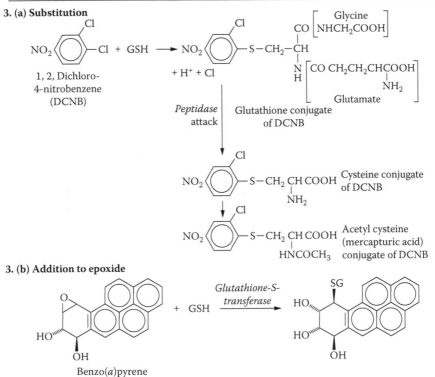

FIGURE 2.14 Phase 2 biotransformation—conjugation. (1) Glucuronide formation. (2) Sulfate formation. (3) Glutathione conjugation.

Many xenobiotics and xenobiotic metabolites undergo glutathione conjugation. Indeed, it is one of the most important phase 1 transformations in both vertebrates and invertebrates. Two contrasting types of reaction are known (Figure 2.14). Nucleophilic attack can lead to the displacement of halide from organohalogen compounds, to be replaced by a sulfur linkage to glutathione. In the example shown, the initial glutathione conjugate 1,2, dichloro-4-nitrobenzene undergoes further biotransformation, involving the removal of the glycine and glutamine moieties by peptidase attack to yield a cysteine conjugate, which is then acetylated to

---

BOX 2.1   STUDYING IN VITRO METABOLISM

Metabolism in vivo is studied by dosing animals with xenobiotics—often in radiolabeled form—and then extracting metabolites from urine, feces, bile, blood, and sometimes other tissues. In this way, a holistic picture of metabolism can be obtained, but there is little opportunity for the characterization of individual enzymes and their kinetic properties. Characterization of enzymic processes can be accomplished by in vitro studies in relatively simple systems under the close control of the experimenter. Examples of in vitro systems, in order of decreasing complexity, are: tissue slices, tissue homogenates, subcellular fractions, and purified enzymes. Following this sequence, the experimenter moves progressively away from the most complex systems, which are closest to in vivo conditions, to the simplest and furthest removed from the complex operation of the whole organism; this also represents progression from the least to the greatest experimental control. Thus, with the purified enzyme, there can be regulation not only of temperature, pH, ionic composition of medium, and levels of cofactors and inhibitors but also of cooperating enzymes and the environment as well (e.g., enzymes that have been extracted from membranes can be returned to reconstituted membranes with regulated phospholipid composition).

Vertebrate liver is a very rich source of enzymes that metabolize lipophilic xenobiotics, and subcellular fractions are prepared to study metabolism. Sometimes, other tissues such as brain, kidney, testis, and ovary are also treated in this way. A typical subcellular fractionation of liver might be as follows:

1. Homogenization of liver in buffer solution to give Crude homogenate
2. Low-speed centrifugation of crude homogenate at 11,000 g to give

11,000 g Supernatant + 11,000 g Precipitate

3. High-speed centrifugation of 11,000 g supernatant at 105,000 g to give

105,000 g Supernatant + Microsomal fraction

The microsomal fraction consists mainly of vesicles (microsomes) derived from the endoplasmic reticulum (smooth and rough). It contains cytochrome P450 and NADPH/cytochrome P450 reductase (collectively the microsomal monooxygenase system), carboxylesterases, A-esterases, epoxide hydrolases, glucuronyl transferases, and other enzymes that metabolize xenobiotics. The 105,000 g supernatant contains soluble enzymes such as glutathione-S-transferases, sulfotransferases, and certain esterases. The 11,000 g supernatant contains all of the types of enzyme listed earlier.

Microsomes are widely used to study the metabolism of xenobiotics. Enzymes can be chararacterized on the basis of their requirement for cofactors (e.g., NADPH, UDPGA), and their response to inhibitors. Kinetic studies can be carried out, and kinetic constants determined. They are very useful in studies of comparative metabolism, where many species not available for in vivo experiment can be compared with widely investigated laboratory species such as rats, mice, feral pigeon, Japanese quail, and rainbow trout.

form an acetyl cysteine conjugate (also referred to as a mercapturic acid conjugate). Mercapturic acid conjugates are often the main excreted forms following glutathione conjugation in mammals. However, cysteine conjugates are also excreted and, in insects, unchanged glutathione conjugates have been reported as the principal excreted forms. A contrasting type of glutathione conjugation involves attack upon epoxides. The epoxide ring is opened, a GS link is attached to one carbon, and OH is attached to the other. The net effect is addition across the epoxide bond without substitution. This latter type of conjugation can be critical in removing newly generated mutagenic epoxides such as benzo[a]pyrene 7,8-diol 9,10 epoxide (Figure 2.14) before they can cause cellular damage by binding to DNA.

Glutathione-S-transferases are known to exist in a number of isoforms. These are homo- or heterodimers, built from subunits of 22–28 kD. In rat, three classes of isoforms are known, built on subunits numbered 1–7.

Class	Constitution
Alpha	1:1, 1:2, and 2:2
Mu	3:3, 3:4, 4:4, and 6:6
Pi	7:7

There is less than 30% sequence homology between the different classes. Although glutathione-S-transferases are predominantly cytosolic enzymes, one microsomal form is known, which exists as a trimer of molecular mass 51 kD. Some glutathione conjugates are unstable and are better regarded as intermediates in the process of biotransformation than as stable conjugates for excretion. Thus, with certain organohalogen compounds, conjugation with glutathione involves dehydrohalogenation before the conjugate breaks down to release a dehalogenated metabolite. One example is the conjugation of dichloromethane. The conjugate is hydrolyzed, and formaldehyde is released (Figure 2.15). The dehydrochlorinations of the organochlorine insecticides *p,p'*-DDT and Y HCH are mediated by a glutathione-*S*-transferase and are thought to proceed via a glutathione conjugate as intermediate (Figure 2.15). The metabolite *p,p'*-DDE itself is epoxidized before undergoing glutathione conjugation (some PCB congeners are metabolized in the same way, see Bakke et al. 1982). The glutathione conjugate is then hydrolyzed, releasing a mercapto derivative of *p,p'*-DDE. The mercapto compound is methylated, and the resulting thiomethyl metabolite is oxidized to a sulfone (Figure 2.15). This metabolite has been found at relatively high levels in marine mammals, notably in lung tissue.

Apart from their catalytic function, at least one form of glutathione-*S*-transferases has the function of simply binding xenobiotics and transporting them, without metabolism. In effect, this is an example of storage (see Section 2.3.3). The form in question is termed *ligandin*, and binding is associated with one particular subunit. Binding is not associated with catalytic activity.

In addition to the foregoing, a number of other conjugations have been identified. Peptide conjugations are very common throughout the animal kingdom, the preferred conjugating peptide varying greatly between species. It appears to be a question of ready availability. Glycine is very commonly involved, whereas ornithine is preferred in some species of birds. Some insects use mainly arginine. As with earlier examples, the conjugates are charged, water soluble, and readily excretable.

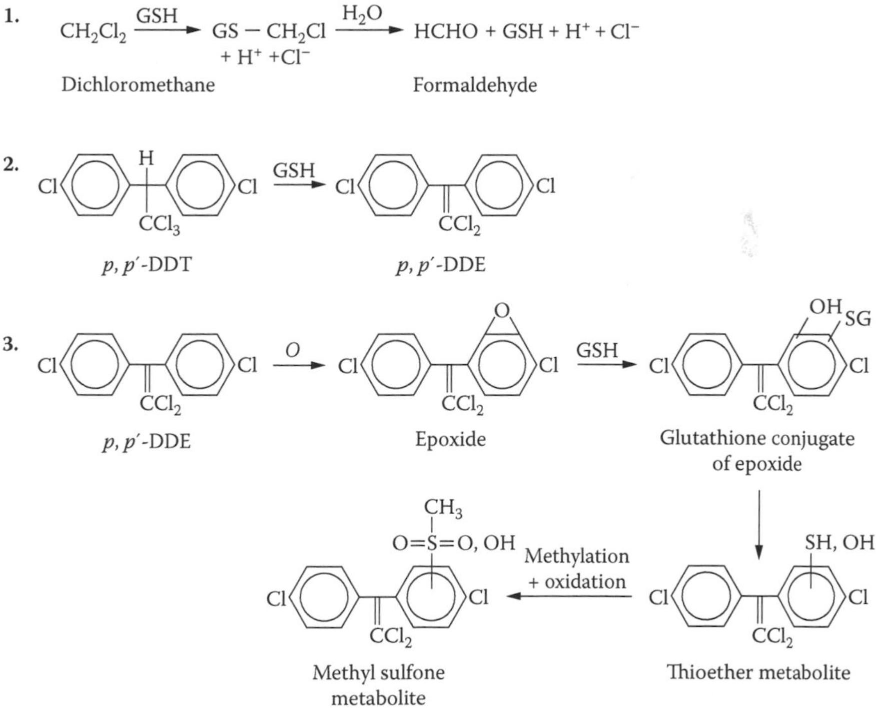

FIGURE 2.15   Glutathione-*S*-transferase attack on organochlorine compounds.

Acetylation is another conjugation reaction, but it differs from the foregoing examples in that the products tend to be less polar than the substrates.

Conjugates of xenobiotics can be hydrolytically degraded by enzymic attack. Very often the products are the original metabolites formed during phase 1 biotransformation. Thus, beta glucuronidases catalyze the hydrolysis of beta glucuronides, and sulfatases catalyze the hydrolysis of sulfate conjugates. If these biotransformations occur in the gut of vertebrates, reabsorption of the released metabolites (original xenobiotics) may follow. This may promote enterohepatic circulation of metabolites that are excreted in bile (see later discussion in Section 2.3.4.2) and so delay final elimination from the body. Endogenous molecules, for example certain hydroxy metabolites of steroids, also undergo enterohepatic circulation.

2.3.2.7   Enzyme Induction

The levels of enzymes existing in different tissues depend on gene expression. Some enzymes that metabolize xenobiotics are just present at what are regarded as "constitutive" levels and are not known to significantly increase in quantity when the organism is exposed to chemicals. On the other hand, certain other enzymes that have a role in xenobiotic metabolism increase in concentration, and consequently in activity, with exposure to chemicals. Such induction can be seen as advantageous to the organism. The quantity of enzyme synthesized is related to the requirement for detoxication, and energy is not wasted in the maintenance of unnecessarily high levels of enzyme. These are only maintained so long as there is exposure to significant levels of the chemical; once the chemical disappears, the enzyme levels fall again. Some P450s are readily inducible, as might be expected because of their critical role in detoxication, and have been studied in some detail. The following account will be principally concerned with them, making brief reference to some other enzymes that have been less well investigated.

Details of some inducible P450 forms that play key roles in the metabolism of xenobiotics are shown in Table 2.4. P450s belonging to family 1A are induced by various lipophilic planar compounds including PAHs, coplanar PCBs, TCDD and other dioxins, and beta naphthoflavone (Monod 1997). As noted earlier, such planar compounds are also substrates for P450 1A. In many cases, the compounds induce the enzymes that will catalyze their own metabolism. Exceptions are refractory compounds such as 2,3,7,8-TCDD, which is a powerful inducer for P450 1A but a poor substrate.

P450s belonging to family 2 are particularly important in the metabolism of a very wide range of nonplanar lipophilic compounds, and a number of them are inducible. The CYP forms 2B1, 2B2, and 2C1–2C4 inclusive are all inducible by phenobarbital. DDT and dieldrin are inducers of CYP2B isozymes. CYP2E1 is inducible by ethanol, acetone, benzene, and other small organic molecules. CYP3A isozymes are inducible by endogenous and synthetic steroids, phenobarbital, and the antifungal agents clotrimazole and ketoconazole. Finally, CYP4A forms are inducible by clofibrate, di-2-ethylhexylphthalate, and mono-2-ethylhexyl phthalate. Induction of CYP4A isoforms is associated with peroxisome proliferation.

The induction of P450s belonging to family CYP2 by phenobarbital and other inducing agents is accompanied by proliferation of the endoplasmic reticulum and

TABLE 2.4
Induction of Cytochrome P450s

Vertebrates			
Family	Individual Forms	Induced by	Typical Substrates
CYP1	1A1	PAHs (e.g., 3MC)	PAHs
	1A2	TCDD (dioxin)	Ethoxyresorufin
		Coplanar PCBs	
CYP2	11B1, 11B2	Phenobarbital	Wide range
	11C1–11C4		
	11E1	Ethanol	Ethanol
CYP3	111A1, 111A2	Steroids	Testosterone
CYP4	1VA1	Clofibrate	
Insects			
CYP6	CYP 6B1	Xanthotoxin	Linear furanocoumarins pyrethroids

hypertrophy of hepatocytes. The total P450 content of microsomal membranes can increase by as much as twofold, and there can be very large increases in the specific content of particular P450 isoforms and associated enzyme activities. By contrast, induction of CYP1A forms is not associated with a proliferation of the endoplasmic reticulum, and induction of CYP4A is usually accompanied by peroxisome proliferation.

The induction process can operate at different levels. The most important mechanisms for particular isoforms are summarized as follows:

Gene transcription—CYPs 1A1, 1A2, 2B1, 2B2, 2C7, 2C11, 2C12, 2D9, 2E1,
    2H1, 2H2, 3A1, 3A2, 3A6, 4A1
mRNA stabilization—CYPs 1A1, 2B1, 2B2, 2C12, 2E1, 2H1, 2H2, 3A1,
    3A2, 3A6
Enzyme stabilization—CYPs 2E1, 3A1, 3A2, 3A6

The mechanism of induction for CYP1A isoforms operates through the Ah receptor, which is located in the cytosol. The inducing agent (e.g., TCDD, coplanar PCB) binds to the Ah receptor, and then the complex so formed moves into the nucleus. Transcriptional activation of the CYP1A gene follows. Interaction of polyhalogenated compounds with the Ah receptor is associated with a variety of toxic responses (Ah-receptor-mediated toxicity), which will be discussed later in Chapter 7, Section 7.2.4. The regulatory elements on the P450 genes are discussed elsewhere (see Lewis 1996).

Apart from monooxygenases, other enzymes concerned wih xenobiotic metabolism may also be induced. Some examples are given in Table 2.5. Induction of glucuronyl transferases is a common response and is associated with phenobarbital-type induction of CYP family 2. Glutathione transferase induction is also associated with this. A variety of compounds, including epoxides such as stilbene oxide and

TABLE 2.5

Induction of Enzymes Other Than Monooxygenases

Type of Enzyme	Induced by	Typical Substrates
Carboxylesterases	Phenobarbital, clofibrate, aminopyrene	Carboxylesters
Epoxide hydrolases	Phenobarbital, trans stilbene oxide, 2-acetyl aminofluorene	Epoxides
Glucuronyl transferases	Various, including phenobarbital	Many compounds with –OH groups (and some with –SH groups)
Glutathione-*S*-transferases	Various, including phenobarbital	Wide range of electrophiles

dieldrin, can induce epoxide hydrolases. Finally, carboxyl esterases can be induced by aminopyrene, phenobarbital, and clofibrate, but only to a limited extent (Hosokawa et al. 1987). Generally speaking, these other inductions are of lower magnitude than many inductions of the P450 system.

The inductions described here are believed to have evolved to protect animals against plant and other naturally occurring toxins. However, when animals are exposed to human-made xenobiotics, induction can have the opposite effect. Some P450 forms can activate certain mutagens and insecticides, and induction of them can lead to enhanced toxicity. Some of the pesticides developed by industry (e.g., phosphorothionate insecticides, see Chapter 10) depend on oxidative activation for their efficacy. An interesting question is, to what extent, over the relatively short period during which such compounds have been in use, have systems to counter the production of active metabolites been evolved by species exposed to them (e.g., by loss or downscaling of inducible P450 forms that are responsible for activation)? This is relevant to the question of development of resistance by pests to pesticides, which will be addressed in Chapter 4 and in Part 2 of this book.

2.3.3 STORAGE

A xenobiotic is said to be stored when it is not available to sites of metabolism or action and is not available for excretion. In other words, it is held in an "inert" position from a toxicological point of view, where it is not able to express toxic action or to be acted upon by enzymes. A xenobiotic is stored when it is located in a fat depot (adipose tissue), bound to an inert protein or other cellular macromolecule, or simply held in a membrane that does not have any toxicological function (i.e., it does not contain or represent a site of toxic action, neither does it contain enzymes that can degrade the xenobiotic).

Highly lipophilic compounds such as organochlorine insecticides, PCBs, and PCDDs tend to be stored in fat depots, where they can reach concentrations 10–50 times higher than in brain, liver, muscle, or other metabolically active tissues. Although storage in fat can protect the organism in the short term, it may prove damaging in the long term. Rapid mobilization of fat depots as a consequence of starvation or disease can lead to rapid release of the stored xenobiotic and to delayed toxic effects. In one well-documented case in the Netherlands (see Chapter 5), wild female eider ducks (*Somateria mollissima*) experienced delayed neurotoxicity caused by dieldrin. The ducks had laid down large reserves of depot fat before breeding, and these reserves were run down during the course of egg laying. Dieldrin concentrations quickly rose to lethal levels in the brain. Male eider ducks did not lay down and mobilize body fat in this way and did not show delayed neurotoxicity due to dieldrin.

Binding to proteins can also represent storage. In the first place, highly lipophilic compounds, such as organochlorine insecticides, associate with lipoproteins and are circulated in blood in this state. Indeed, their water solubility is so low that only an extremely small proportion of the total concentration present in body fluids is in true solution. Their association with lipoproteins is due, at least in part, to the hydrophobic effect of water. They are excluded from water by the mutual attraction of water molecules and are pushed into association with the lipoproteins and other hydrophobic domains of the body, including lipoproteins. Apart from "dissolving" in mobile lipid depots, they can bind to hydrophobic proteins as a consequence of van der Waals interactions between the compounds and the surfaces of the proteins. More polar compounds (including ionic compounds) also interact with proteins, but in different ways. The formation of ionic bonds or hydrogen bonds leads to the binding of more polar xenobiotics to functional groups of certain proteins. Albumin, for example, is abundant in mammalian plasma and can bind a number of relatively polar xenobiotics. Hydroxy-metabolites of PCBs can bind to certain plasma proteins. One particular case, the binding of 3.3′,4,4′ tetrachloro-biphenyl to transthyretin, has been closely studied because it is associated with toxicity rather than storage (see Chapter 6, Section 6.2 and Brouwer et al. 1990).

Storage of lipophilic pollutants in the eggs of invertebrates, birds, amphibians, and reptiles is of importance in ecotoxicology. Organochlorine insecticides are transported by lipoproteins from females to eggs. At first, they are stored within the yolk of birds eggs. When the eggs develop, they are mobilized and can cause delayed toxicity in the developing embryo. Such effects have been observed with dieldrin and DDT.

2.3.4 EXCRETION

As explained in Chapter 1, there is strong evidence for the rapid evolution of enzyme systems concerned with the metabolism of xenobiotics coinciding in time with the movement of animals from water to land. Thus, radiation of the CYP2 family of P450s corresponds closely with the colonization of land at the start of the Devonian period circa 400 million years ago. The CYP2 family is particularly concerned with the metabolism of xenobiotics and is represented by a considerable number of different forms in terrestrial mammals (see Section 2.3.2.2). On the other hand, this family of P450s, and indeed xenobiotic metabolizing enzymes more generally, are less well developed in fish, as will be explained later. Fish can "excrete" many lipophilic xenobiotics by diffusion across the gills into ambient water. This excretion mechanism is not, however, available to terrestrial animals. They depend on the conversion of lipophilic compounds into water-soluble metabolites and conjugates, which are excreted in urine and bile. In the following account, aquatic animals and terrestrial animals will be treated separately.

2.3.4.1 Excretion by Aquatic Animals

Lipophilic xenobiotics are subject to exchange diffusion across the respiratory surfaces that aquatic animals present to ambient water. Thus, excretion can occur across the gills of fish and the permeable skins of certain amphibia (e.g., frogs). Exchange diffusion tends toward an equilibrium or steady state where the ratio of the concentration of the xenobiotic inside the organism to the concentration in ambient water represents the bioconcentration factor (BCF, see Chapter 4). Thus, loss by diffusion is limited because it is a two-way process. Uptake from the ambient medium works against loss by outward diffusion. However, the dilution volume of the ambient water in relation to that of the organism is very large indeed. Where a pollutant is at low concentration in water and is absorbed mainly from food or directly from sediment, the system should be very effective. On the other hand, if the pollutant is absorbed mainly from water, effective elimination by the diffusion process will depend upon the concentration in the water falling. A fall in pollutant concentration in ambient water may be due either to its removal by degradation, uptake by other organisms, adsorption to sediments, etc., or to the aquatic animal moving into cleaner water.

Loss by diffusion tends to be very effective for compounds of K_{ow} close to 1, but less so for compounds of very high K_{ow} (1×10^5 and above). Thus, many anthropogenic compounds of high K_{ow} (e.g., higher chlorinated PCBs, most organochlorine insecticides) can be strongly bioaccumulated by aquatic animals. The relatively weak development of enzyme systems in fish which can metabolize xenobiotics, suggests that diffusion has provided reasonably effective protection against naturally occurring xenobiotics during the course of evolution, with only a limited requirement for back up by a metabolic detoxication system. Some anthropogenic pollutants, however, appear to be asking new questions, and there is some evidence that strains of fish with greater metabolic capability have evolved or will evolve in some polluted areas (see Chapter 5, Section 5.3.5).

Excretion has also been studied in marine invertebrates, such as the bivalve *Mytilus edulis* (edible mussel). The principal excretion mechanism appears again to be passive diffusion, although there is some evidence for the active excretion of certain organic nitrogen compounds by *Mytilus* species utilizing a "multiple drug resistance" mechanism (Kurelec et al. 2000). Metabolism of xenobiotics is generally slower in marine invertebrates than in fish, with crustaceans (e.g., crabs) showing higher metabolic capability than mollusks.

2.3.4.2 Excretion by Terrestrial Animals

Excretory processes for xenobiotics are best understood for mammals, with far less work having been done on birds, reptiles, and amphibians. Highly lipophilic compounds show little tendency to be excreted unchanged. In the absence of effective metabolism, they tend to have very long biological half-lives in depot fat. Thus, half-lives of about 1 year have been reported for *p,p'*-DDE in birds, whereas higher chlorinated PCBs and higher brominated PBBs have half-lives running into years in certain mammals including humans. Related compounds that are easily metabolized (e.g., lower chlorinated PCBs or "biodegradable" cyclodienes such as the dieldrin analogues, endrin and HCE) have half-lives of a few hours or days. The effective elimination of strongly lipophilic compounds depends on their conversion to water-soluble conjugates and metabolites, leading to their excretion in bile and urine. A small qualification to this generalization is that some lipophilic compounds are excreted to a limited extent, with lipoproteins. Thus, organochlorine insecticides such as dieldrin and DDT are transported from female birds, reptiles, and insects into eggs. Also, female mammals excrete such compounds in milk.

The conversion of lipophilic xenobiotics into water-soluble metabolites and conjugates by vertebrates occurs principally in the liver, and less in the kidney and certain other metabolically active tissues. Most excretion occurs in the form of conjugates. When conjugates are formed in the liver, they may be excreted in bile or in urine. The excretory route depends upon molecular weight. As mentioned earlier, most conjugates are anions. Anionic conjugates below 300 kDa molecular weight tend to be excreted in urine, whereas above 600 kDa they tend to be excreted in bile. Between these two extremes, the preferred excretory route depends on the species. It has been proposed that there are "threshold" molecular weights for different species, above which 10% or more of the anionic conjugate is excreted in bile (Figure 2.16). The following threshold values are proposed for the named species—the rat 325 (±50), the guinea pig (±50), and the rabbit 475 (±50). Thus, rats show a greater tendency to excrete anionic conjugates in bile than do the other two species, when considering xenobiotics over a range of molecular weights (see Walker 1975).

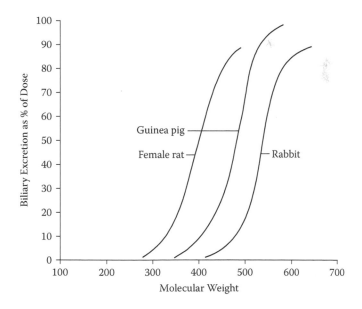

FIGURE 2.16 Excretion routes of xenobiotic anionic conjugates.

The reason for this pattern has yet to be definitely established. The most likely explanation is that there is selective reabsorption of the conjugates from bile into plasma. Conjugates are moved from hepatocytes into biliary canaliculi during the formation of primary bile. With some conjugates, for example, glucuronides, this movement appears to be driven by active transport. During subsequent passage through the biliary system, lower-molecular-weight conjugates are absorbed into blood, whereas larger ones remain in bile, eventually to be discharged into the alimentary tract. Such selective absorption may occur through pores. This could explain the reason for the differing thresholds between species. It is possible that rats have tighter gap junctions between hepatocytes than rabbits, so that larger conjugates are retained in the biliary system, whereas in the latter species they can escape into plasma. The species differences in preferred excretory route have implications for the later course of excretion and for toxicity. Conjugates that remain within the biliary system may be broken down in the gut to release the molecules that had been conjugated (often phase 1 metabolites). These molecules, being less polar than the conjugate, can then be reabsorbed into the blood and recirculated to the liver. The process is termed *enterohepatic circulation*, and some compounds may cycle 30 times or more before being finally voided with feces.

Enterohepatic circulation can lead to toxic effects. For example, the drug chloramphenicol is metabolized to a conjugate that is excreted in bile by the rat. Once in the gut, the conjugate is broken down to release a phase 1 metabolite that undergoes further metabolism to yield toxic products. When these are reabsorbed, they can cause toxicity. The rabbit, by contrast, excretes chloramphenicol conjugates in urine, and there are no toxic effects at the dose rates in question.

Excretion via the kidney can be a straightforward question of glomerular filtration, followed by passage down the kidney tubules into the bladder. However, there can also be excretion and reabsorption across the tubular wall. This may happen if an ionized form within the tubule is converted into its nonpolar nonionized form because of a change in pH. The nonionized form can then diffuse across the tubular wall into plasma. Additionally, there are active transport systems for the excretion of lipophilic acids and bases across the wall of the proximal tubule. The antibiotic penicillin can be excreted in this way.

Far less is known about excretion by terrestrial insects than by terrestrial mammals. Metabolism can take place in the midgut and fat body. Excretion can occur via the malpighian tubules.

2.4 TOXICODYNAMICS

The discussion until now has been concerned with toxicokinetics—the uptake, distribution, metabolism, storage, and excretion of organic pollutants within living organisms. These processes determine how much of a toxic molecule—original pollutant or active metabolite—reaches the sites of action. At this stage, we enter the field of *toxicodynamics*, which is concerned with the interaction of toxic molecules with their sites of action and the consequences thereof. Conceptually, it is useful to make a distinction between toxicokinetics and toxicodynamics because these two factors determine, in contrasting and largely independent ways, how toxic any particular chemical will be to living organisms. Selectivity is due to the differential operation of toxicokinetic and toxicodynamic processes between different species, strains, sexes, or age groups (see Section 2.5). In the study of the basis of selective toxicity (e.g., resistance of insects to insecticides), it is advantageous to design experiments that permit the distinction between toxicokinetic and toxicodynamic factors (for further discussion, see Walker 1994).

Most of the organic pollutants described in the present text act at relatively low concentrations because they, or their active metabolites, have high affinity for their sites of action. If there is interaction with more than a critical proportion of active sites, disturbances will be caused to cellular processes, which will eventually be manifest as overt toxic symptoms in the animal or plant. Differences between species or strains in the affinity of a toxic molecule for the site of action are a common reason for selective toxicity.

It should also be mentioned that some compounds of relatively low toxicity act as physical poisons, although such pollutants are seldom important in ecotoxicology. They have no known specific mode of action, but if they reach relatively high concentrations in cellular structures, for example, membranes, they can disturb cellular processes. Examples include certain ethers and esters, and other simple organic compounds.

Tables 2.6 and 2.7 give examples of the modes of action of pollutants in animals and in plants/fungi, respectively. It is noteworthy that many of the chemicals represented are pesticides. Pesticides are designed to be toxic to target species. On the other hand, manufacturers seek to minimize toxicity to humans, beneficial organisms and, more generally, nontarget species. Selective toxicity is an important issue. Regardful of the potential risks associated with the release of bioactive compounds into the environment, regulatory authorities usually require evidence of the mode of toxic action before pesticides can be marketed. Other industrial chemicals are not subject to such strict regulatory requirements, and their mode of action is frequently unknown.

The examples given in Tables 2.6 and 2.7 illustrate the wide range of different mechanisms by which pollutants cause toxic effects. The following account will focus on certain broad issues concerning mode of action. A more detailed description of individual examples will be given in later chapters devoted to particular types of pollutants.

Many of the pollutants expressing high toxicity to animals are lipophilic in character. This appears to be a consequence of toxicokinetic factors. After absorption, lipophilic compounds tend to remain within the organism, and effective excretion depends on their enzymic conversion to water-soluble and readily excretable products (see Section 2.3.2.1). They tend to associate with membranes and hydrophobic macromolecules within the body, thus facilitating interaction with enzymes, receptors for chemical messengers, and pore channels. Water-soluble organic compounds, on the other hand, tend to be rapidly excreted unchanged and do not tend to associate with lipophilic structures within the organism.

Broadly speaking, toxic interactions between chemicals and cellular sites of action are of two kinds:

1. The pollutant (xenobiotic) forms a stable covalent bond with its target. Examples include the phosphorylation of cholinesterases by the oxon forms of OPs, the formation of DNA adducts by the reactive epoxides of benzo[a] pyrene and other PAHs, and the binding of organomercury compounds to

TABLE 2.6
Mode of Action of Pollutants in Animals

Pollutants	Type of Toxic Action	Site of Action	Consequence	Comments	Reference
OP and carbamate insecticides	Neurotoxicity	Acetylcholinesterase	Disruption of synaptic transmission	OPs include insecticides and chemical warfare agents	Eto (1974), Ballantyne and Marrs (1992), Kuhr and Dorough (1977)
DDT and related compounds	Neurotoxicity	Na$^+$ channels of axonal membrane	Disturbance of nerve action potential	Persistent insecticides	Kuhr and Dorough (1977), Brooks (1974), Eldefrawi and Eldefrawi (1990)
Dieldrin and other cyclodiene insecticides	Neurotoxicity	GABA receptors	Can cause CNS disturbances	Dieldrin is persistent and highly toxic to animals	Brooks (1974), Eldefrawi and Eldefrawi (1990), Brooks (1992), Salgado (1999)
Pyrethroids	Neurotoxicity	Na$^+$ channels of axonal membrane	Similar action to DDT	Selective insecticides	Eldefrawi and Eldefrawi (1990), Salgado (1999), Leahey (1985)
Benzo[a]pyrene and other PAHs	Mutagens and carcinogens	DNA	Mutagenesis can lead to carcinogenesis	Little known about consequences of mutation other than carcinogenesis	Hodgson and Levi (1994)

Compound	Mechanism	Site of action	Effect	Comments	Reference
p,p'-DDE	Affects Ca^{++} transport in avian shell gland	Probably Ca^{++} ATPase, may also have effect on prostaglandin levels	Thinning of avian eggshells	Persistent metabolite of p,p'-DDT implicated in the decline of certain predatory birds	Lundholm (1987, 1997), Peakall (1992)
Dioxins and coplanar PCBs	"Ah-receptor-mediated toxicity"	Ah receptor	A range of toxic effects including endocrine disturbances	Mechanism by which binding to Ah receptor causes toxic effects still unclear	Safe (1990)
4-OH-3,3'4,5'-Cl-biphenyl	Thyroxine antagonism	Thyroxine-binding site of transthyretin	Vitamin A deficiency	Effect caused by this and certain other PCB metabolites	Brouwer et al. (1990)
Anticoagulant rodenticides	Vitamin K antagonism	Vitamin K binding sites of hepatic carboxylase	Hemorrhaging	Warfarin, diphenacoum, bromodiolone, flocoumafen, and brodiphacoum are examples	Thijssen (1995)
Organo-mercury fungicides	Neurotoxicity	Sulfhydryl groups of proteins	Damage to CNS	Both alkyl and aryl mercury compounds can bind to sulfhydryl groups	Environmental Health Criteria (EHC) 86 (1989)
Tributyl tin fungicides	Endocrine disruptors in mollusks	Thought to disturb testosterone metabolism by binding to P450	Cause "imposex" in dog whelks and a variety of other effects on other mollusks	Have caused population declines in certain mollusks	Matthiessen and Gibbs (1998)

TABLE 2.7
Mechanism of Action of Pollutants in Plants and Fungi

Pollutants	Type of Toxic Action	Site of Action	Consequence	Comments	Reference
Triazine and substituted urea herbicides	Photosynthesis inhibitors	Q_B binding site of the D1 protein in photosystem 11	Inhibition of electron transport in chloroplast and consequent reduction of photosynthesis	These herbicides have generally very low toxicity to animals	Hassall (1990), Sjut (1997)
Phenoxyakanoic herbicides	Disruption of growth regulation of plants	Unknown, but apparently the same as for the natural auxins whose action they mimic	Distorted growth patterns including malformed leaves and epinasty of stems	2,4-D, MCPA, CMPP, and MCPB are examples, generally of low toxicity to animals	Hassall (1990)
Dinitroaniline herbicides	Mitotic disrupters	Polymerase responsible for microtubule formation	Disruption of cell division	Examples include trifluralin and oryzalin	Hassall (1990)
Paraquat	Photosynthesis inhibitor	Electron transport system of photosystem 1	Paraquat diverts electrons from the electron transport system to oxygen to form O_2, which causes cellular damage	Paraquat and related herbicides are toxic to animals because they form reactive oxyradicals (see Chapter 14)	Hassall (1990), Sjut (1997)
DNOC	Mitochondrial poison	Inner membrane of mitochondrion	Uncouples oxidative phosphorylation by running down the proton gradient across inner mitochondrial membrane	DNOC and certain other dinitrophenols are general biocides because the mitochondrion is present in most living organisms	Hassall (1990), Nicholls (1982)
Ergosterol biosynthesis inhibitors (EBIs)	Inhibitors of ergosterol biosynthesis in fungi	Fungal cytochrome P450 involved in sterol metabolism	Destabilizing of fungal membranes	Some EBIs are inducers and or inhibitors of P450 forms of animals (see Section 2.6)	Kato (1986)

the sulfhydryl groups of proteins. It should be added that the reactive PAH epoxides and most of the reactive oxons of OPs are not the original pollutants absorbed by animals; they are unstable metabolites generated by monooxygenase attack (see Section 2.3.2.2).

2. The pollutant binds to the target without forming a covalent bond. Such interactions are usually reversible, and examples include the binding of the stable organochlorine insecticides p,p'-DDT and dieldrin to Na^+ channels and GABA receptors, respectively; the binding of planar PCBs and dioxins to the Ah receptor the binding of 4 OH`-, 3,3′,4,5′-tetrachlorobiphenyl to the thyroxine-binding site of transthyretin; and the binding of ergosterol biosynthesis inhibitor (EBI) fungicides to a fungal form of cytochrome P450. The reduction of DDT toxicity to insects with increasing temperature has been attributed to desorption of DDT from the sodium channel. (The same phenomenon has been reported with pyrethroid insecticides, which also interact with the Na^+ channel.)

The interaction of toxic molecules with their sites of action leads to physiological effects at the cellular and whole-organism level. The nature of these effects, and their significance in ecotoxicology, depend upon the site of action. It is noteworthy that the four most important groups of insecticides are all neurotoxins, albeit working through different modes of action (Table 2.5). Disturbances of the nervous system can have far-reaching consequences, especially when they arise in the central nervous system (e.g., the effects of dieldrin and HCH on GABA receptors of vertebrates). At an early stage of intoxication, there can be behavioral effects (evidence of behavioral effects will be discussed in Chapter 16). In another example, the action of endocrine disruptors may lead to deleterious effects on reproduction, as in the case of imposex caused by TBT in the dog whelk. Also, the reproduction of some raptors can be adversely affected by eggshell thinning caused by p,p'-DDE, apparently due to inhibition of Ca^{2+} ATPase. In a further example, anticoagulant rodenticides such as warfarin, brodifacoum, and flocoumafen act as antagonists of vitamin K, which has a critical role in the synthesis of clotting proteins in the liver of vertebrates. Following exposure to these compounds, synthesis of clotting proteins in the liver is incomplete, and precursors of them are released into blood. After a period of days, the level of clotting proteins falls in the blood, and hemorrhaging occurs. Thus, different modes of action lead to different toxic effects with differing ecotoxicological consequences.

Mechanism of action can be an important factor determining selectivity. In the extreme case, one group of organisms has a site of action that is not present in another group. Thus, most of the insecticides that are neurotoxic have very little phytotoxicity; indeed, some of them (e.g., the OPs dimethoate, disyston, and demeton-S-methyl) are good systemic insecticides. Most herbicides that act upon photosynthesis (e.g., triazines and substituted ureas) have very low toxicity to animals (Table 2.7). The resistance of certain strains of insects to insecticides is due to their possessing a mutant form of the site of action, which is insensitive to the pesticide. Examples include certain strains of housefly with knockdown resistance (mutant form of Na^+ channel that is insensitive to DDT and pyrethroids) and strains of several species of insects that are resistant to OPs because they have mutant forms of acetylcholinesterase. These

BOX 2.2   THE CONCEPT OF BIOMARKERS

A biomarker is here defined as a biological response to an environmental chemical at the individual level or below, which demonstrates a departure from normality. Responses at higher levels of biological organization are not, according to this definition, termed biomarkers. Where such biological responses can be readily measured, they may provide the basis for biomarker assays, which can be used to study the effects of chemicals in the laboratory or, most importantly, in the field. There is also interest in their employment as tools for the environmental risk assessment of chemicals.

Some biomarkers only provide a measure of exposure; others also provide a measure of toxic effect. Biomarkers of the latter kind are of particular interest and importance and will be referred to as "mechanistic biomarkers" in the present text. Some mechanistic biomarker assays directly measure effects at the site of action as described in Section 2.4 (see Chapter 4, Table 4.2, for examples). Inhibition of acetylcholinesterase is one example. Others measure secondary effects on the operation of nerves or the endocrine system (examples given in Table 4.2 and Chapters 15 and 16).

This concept will be discussed further in the context of effects on populations in Chapter 4, Section 4.4.

examples will be discussed in more detail when considering individual groups of pollutants. In contrast to the foregoing examples, some pollutants have wide-ranging toxicity to nearly all forms of life and earn the doubtful accolade of being termed *biocides*. Probably the best examples of these are uncouplers of oxidative phosphorylation such as dinitrophenol and the herbicide dinitroorthocresol (DNOC). These act upon the mitochondrial membrane of animals and plants, and run down the proton gradient across the membrane, which drives the process of ATP formation (see Chapter 13).

2.5   SELECTIVE TOXICITY

Selective toxicity (henceforward simply "selectivity") is of fundamental interest in ecotoxicology. For any pollutant that is, or may become, widely distributed in the environment, it is desirable to know which species or groups will be most sensitive to its toxic action. Sensitive species should be given particular attention in biological monitoring (e.g., in use of biotic indices) and, ideally, should have appropriate representatives in ecotoxicity testing protocols. Selectivity needs to be taken into account when estimating "environmental toxicity" (e.g., PNEC) from ecotoxicity data obtained with surrogates, where emphasis is placed on "the most sensitive species." A large safety factor needs to be incorporated in such calculations if the test organism is expected to be much less sensitive than key species in the natural environment. In the quest for new pesticides, a major objective is to design compounds that are selective between pest species and beneficial organisms, thereby ensuring compliance with the requirements for regulatory bodies for environmental safety.

Selective toxicity is also important in relation to the development of resistance or tolerance to pollutants from two distinct points of view. On the one hand, there is interest among scientists concerned with crop protection and disease control in mechanisms by which crop pests, vectors of disease, plant pathogens, and weeds develop resistance to pesticides. Understanding the mechanism should point to ways of overcoming resistance, for example, other compounds not affected by resistance mechanisms or synergists to inhibit enzymes that provide a resistance mechanism. On the other hand, the development of resistance can be a useful indication of the environmental impact of pollutants.

As discussed earlier, selectivity is the consequence of the interplay between toxicokinetic and toxicodynamic factors. Some examples are given in Table 2.8, which will now be briefly discussed (data from Walker and Oesch 1983, and Walker 1994a,b). These and other examples will be described in more detail under specific pollutants later in the text. In the table, comparisons are made between the median lethal doses or concentrations for different species or strains. Comparisons are made of data obtained in lethal toxicity tests where the same route of administration was used for species or strains that are compared. The degree of selectivity is expressed

TABLE 2.8
Selective Toxicity

Pollutant	Toxicity Test	Toxicity to Species or Strain A	Toxicity to Species or Strain B	Toxicity to A Toxicity to B
Dimethoate (OP)	Topical LD_{50}	Rat 925	Housefly 0.20 *	4.6×10^3
Diazinon (OP)	Topical LD_{50}	Rat 850	Housefly 1.9 *	447
Diazinon	Acute oral LD_{50}	Rat 450	Birds (mean 4 species) 4.5	100
Fenvalerate (pyrethroid)	Topical LD_{50}	Rat 4000	Honeybee 0.21 *	2.3×10^4
Dimethoate (OP)	Topical LD_{50}	*Myzus persicae* 'R' clone	*Myzus persicae* 'S' clone	500[a]
cis - cypermethrin (pyrethroid)	Topical LD_{50}	*Heliothis virescens* 'R' strain (PEG 87 d/3 3rd instar larva)	*H.virescens* 'S' Strain (BRCb/1c 3rd instar larva)	7×10^5 [b]

Note: LD_{50} values expressed as mg/kg (μg/g) except where marked with an asterisk. The latter are given as μg/insect.

[a]   From Devonshire and Sawicki (1979).
[b]   From McCaffery et al. (1991).

Source: Other data from Walker (1994a) and Walker and Oesch (1983).

as a selectivity ratio, LD_{50} (or LC_{50}) of species or strain A divided by LD_{50} (or LC_{50}) of species or strain B.

The organophosphorous insecticides dimethoate and diazinon are much more toxic to insects (e.g., housefly) than they are to the rat or other mammals. A major factor responsible for this is rapid detoxication of the active oxon forms of these insecticides by A-esterases of mammals. Insects in general appear to have no A-esterase activity or, at best, low A-esterase activity (some earlier studies confused A-esterase activity with B-esterase activity) (Walker 1994b). Diazinon also shows marked selectivity between birds and mammals, which has been explained on the grounds of rapid detoxication by A-esterase in mammals, an activity that is absent from the blood of most species of birds (see Section 2.3.2.3). The related OP insecticides pirimiphos methyl and pirimiphos ethyl show similar selectivity between birds and mammals. Pyrethroid insecticides are highly selective between insects and mammals, and this has been attributed to faster metabolic detoxication by mammals and greater sensitivity of target (Na^+ channel) in insects.

Some examples are also given of insect resistance. Here, the selectivity ratio is usually termed the *resistance factor*. Thus, a strain of housefly resistant to the OP malathion can detoxify the insecticide by B-esterase attack due to metabolism by a carboxylesterase (a form of B-esterase) not found in susceptible strains of this species. Some strains of housefly (*Musca domestica*) developed resistance to DDT as a consequence of possessing a mutant form of the target (Na^+ channel). Strains of the peach potato aphid (*Myzus persicae)* were resistant to OP insecticides generally because of very high levels of a carboxylesterase. The elevated levels of the enzyme were due to multiple copies of the gene coding for it. Resistant strains had 2, 4, 8, 16, or 32 copies of carboxylesterase gene, and the level of resistance was related to the number of copies in any particular aphid strain (Devonshire 1991). The strain showing a 500-fold resistance factor possessed 32 copies of the carboxylesterase gene. In another example, a resistant strain of the tobacco bud worm (*Heliothis virescens*) showed 70,000-fold resistance to the pyrethroid cypermethrin related to two major resistance factors: (1) enhanced oxidative detoxication by a P450-based monooxygenase and (2) insensitivity of a mutant form of target (Na^+ channel). Thus, resistance here was related to both toxicodynamic and toxicokinetic factors

2.6 POTENTIATION AND SYNERGISM

Ecotoxicity testing of industrial chemicals and pesticides during the course of environmental risk assessment is usually only performed on individual compounds. In the real world, however, living organisms are frequently exposed to complex mixtures of chemicals, inorganic as well as organic, and questions arise about the toxicity and ecological effects of mixtures (see Chapter 14, Walker et al. 1996). Examples include mixtures of hydrocarbons, and of polyhalogenated aromatic compounds such as PCBs and dioxins. The problem is, of course, that even testing single compounds is a difficult and expensive exercise; the resources do not exist to test more than a very small proportion of the combinations of chemicals found in the living environment. It is usually assumed that the toxicity of mixtures is roughly additive (in other words, the toxicity of the mixture is approximately the sum of the toxicities of the individual components). Although this assumption is usually borne out in practice, there are important exceptions, where there is *potentiation*, and the sum of the individual toxicities is considerably greater than additive. *Synergism* is a particular type of potentiation where the toxicity of one compound is enhanced by another compound (*synergist*), which does not express toxicity by itself at the levels of exposure in question. (For further discussion, see Moriarty 1999 and Walker et al. 2006.) Thus, potentiation is a broader term than synergism, and includes cases where two or more compounds, which themselves express toxicity, cause toxicity greater than additive toxicity when brought together in a mixture.

A basic problem, then, is recognizing where there is a serious risk of potentiation of toxicity when animals are exposed to mixtures of chemicals in the field. With regard to environmental risk assessment, it is clearly impossible to test more than a small proportion of the combinations that exist (or may exist in future as new chemicals are released into the environment). On the other hand, there is a case for testing certain combinations of chemicals that may arise in the field (e.g., as a result of mixing pesticides in spray tanks) if there are grounds for suspecting that there may be potentiation of toxicity. As knowledge grows of the mechanisms that determine toxicity, it should become increasingly possible to foresee where potentiation is likely to occur, and to limit any testing of the ecotoxicity of mixtures to a small number of combinations that give cause for concern.

From a mechanistic point of view, potentiation can be seen to arise from interactions occurring within the model given in Figure 2.1. Taking the example of two interacting compounds, potentiation may be due to one compound causing changes in either the toxicokinetics or the toxicodynamics of the other compound, or both of these. The first compound may alter the toxicokinetics of the other, so that more of the active form (original compound or metabolite) reaches the site of action. Alternatively, the first compound may alter the toxicodynamics of the other, so that it becomes more effective at the same tissue concentration. In practice, most of the well-studied cases of potentiation are due to one compound affecting the toxicokinetics of another, so that more of an active form reaches the site of action. Commonly, this has been due to inhibition of detoxication, but enhanced metabolic activation, increased rate of absorption, and reduced storage have all been implicated in certain cases.

Table 2.9 gives a number of examples of synergism due to one compound (the synergist) inhibiting the metabolic detoxication of the other. Strikingly, there can be an increase of toxicity as great as 400-fold when enzymic detoxication is inhibited. It is noteworthy that two of the aforementioned examples involve the inhibition of microsomal monooxygenases, enzymes that have a very wide-ranging detoxifying function in the animal kingdom. Understandably there is concern about the presence of monooxygenase inhibitors among environmental chemicals because of their potential to increase the toxicity of other compounds that are substrates of the enzyme. Inhibitors of monooxygenases, esterases, and other enzymes having a detoxifying function are sometimes useful for identifying the cause of resistance to pesticides. The inhibitor reduces the level of resistance by blocking an enzyme that is responsible for enhanced detoxication in the resistant strain.

Other cases of potentiation will be discussed in later chapters dealing with individual pollutants. These will include the synergism of pyrethroid toxicity to

TABLE 2.9
Synergism Due to Inhibition of Detoxication

Organism	Insecticide	Detoxifying Enzyme	Inhibitor (Synergist)	Synergistic Ratio
Insect strains resistant to pyrethroids (e.g., *H. virescens*)	Cypermethrin	Monooxygenase	Piperonyl butoxide (PBO)	<200
Insects	Carbaryl	Monooxygenase	PBO and other methylenedioxyphenyls	<400
Mammals and some malathion-resistant insects	Malathion	Carboxylesterase	Some OPs other than malathion	<200

Note: Synergistic ratio = toxicity without synergist/toxicity with synergist.
Sources: Kuhr and Dorough 1976, McCaffery et al. 1991, Walker and Oesch 1983.

honeybees by EBI fungicides, due to inhibition of detoxication by monooxygenases (Chapter 12); the increased toxicity of the carbamate insecticide carbaryl to red-legged partridges (*Alectoris rufa*) caused by the OP malathion due to inhibition of detoxication by monooxygenases (Johnston et al. 1994; Chapter 10); the enhanced toxicity of malathion to birds by EBI fungicides as a consequence of induction of monooxygenase and increased activation (Chapter 10); and the enhanced activation of mutagenic PAHs due to induction of monooxygenases by coplanar PCBs and dioxins (Chapter 9).

2.7 SUMMARY

The toxicity of an organic chemical depends on the operation of toxicokinetic and toxicodynamic processes within living organisms, and selectivity between species, strains, sexes, and age groups is the outcome of the differential operation of these processes between such contrasting groups. A model is presented that describes the fate of lipophilic organic compounds within living organisms, seen from a toxicological point of view. After uptake, the chemical is distributed to sites of action, metabolism, storage, and excretion. *Toxicokinetics* encompasses all of those processes that occur before the arrival of a toxic compound (original compound or active metabolite) at the site of action—processes that determine how much of the toxic compound reaches the site of action. Toxicodynamics is concerned with the interaction of the toxic compound with the site of action and consequent toxic effects.

Emphasis is given to the critical role of metabolism, both detoxication and activation, in determining toxicity. The principal enzymes involved are described, including monooxygenases, esterases, epoxide hydrolases, glutathione-*S*-transferases, and glucuronyl transferases. Attention is given to the influence of enzyme induction and enzyme inhibition on toxicity.

The mechanism of toxic action of some important organic pollutants is described and related, where possible, to ecotoxicological effects.

FURTHER READING

There are many texts dealing with the biochemical toxicology of organic compounds, but most of them are principally concerned with human toxicology. The following texts are suitable, in differing ways, as supplementary reading to the present chapter.

Hassall, K.A. (1990). A readable student text, particularly useful for the metabolism and mode of action of herbicides and fungicides.

Hodgson, E. and Kuhr, R.J. (1990). A multiauthor work with a wealth of information on the mechanism of action of insecticides.

Hodgson, E. and Levi, P. (1994). A multiauthor text dealing in depth with some aspects of biochemical toxicology highly relevant to the present topic. Examples of metabolism and mode of action of insecticides, and PAHs.

Jakoby, W.B. (1990). Although a little out of date, still a useful source on the enzymes concerned with xenobiotic metabolism.

Lewis, D.F.V. (1996). A detailed account of cytochrome P450, including evolutionary aspects.

3 Influence of the Properties of Chemicals on Their Environmental Fate

Chapter 2 was concerned with processes that determine the distribution of chemicals within living organisms and the relationship between distribution and toxicity. The importance of properties of chemicals in determining their fate within living organisms was given emphasis. Polarity, molecular size, the presence of functional groups, and molecular stability were all seen to influence toxicokinetic processes. The types of enzymes responsible for biotransformations were related to the structures of the chemicals undergoing biotransformation (e.g., esterases for esters, reductases for nitro compounds, monooxygenases for aromatic hydrocarbons, etc.). This chapter will consider the wider question of how the properties of chemicals determine their fate in the gross environment, and how these properties can be incorporated into descriptive and predictive models relating to the movement and distribution of environmental chemicals.

First, there will be a description of chemical and physical properties that determine the fate of organic pollutants in the gross environment, restricting the discussion to chemical and physical processes. Next, an account is given of how these data are incorporated into models that attempt to describe or predict environmental fate. Finally, the relationship between the properties of chemicals and the operation of biological processes that determine environmental fate will be discussed. Drawing on the background given in Chapter 2, emphasis will be given to the relationship between properties of environmental chemicals and their uptake excretion and metabolism by living organisms. The overall aim will be to lay a proper foundation for consideration of the more complex issues in Chapter 4, which will address the question of movement and distribution of chemicals in ecosystems where both biological and abiotic processes come into play. Key issues will be the fate of chemicals in soils and the movement of chemicals along terrestrial and aquatic food chains, situations in which fate is determined by both biological and chemical processes.

3.1   PROPERTIES OF CHEMICALS THAT INFLUENCE THEIR FATE IN THE GROSS ENVIRONMENT

Polarity is one of the most important determinants of the environmental fate of organic chemicals. In general, the more polar a compound, the higher its water solubility and the lower its octanol–water partition coefficient (K_{ow}). Conversely, the less polar a compound, the lower its water solubility and higher its K_{ow} (see earlier discussion in Chapter 2, Section 2.3.1). K_{ow} is a measure of "hydrophobicity," and is regarded as one of the most valuable indicators of the environmental behavior of organic chemicals. Many examples will be given in the chapters dealing with individual pollutants in the second part of this text. (For more detailed and critical discussion about the determination and utilization of K_{ow} values, readers are referred to the chapter by Connell in the work by Calow 1994.)

Hydrophobic compounds tend to be excluded from the aqueous phase of the environment owing to what has been termed the *hydrophobic effect* (Tanford 1980). Polar water molecules are drawn together because of the attraction of d+ and d– charges on adjacent molecules. Hydrophobic compounds present in water, which have little or no charge, are thereby "squeezed out." In soils and sediments, they tend to become adsorbed to the surface of colloidal material—organic matter and clay—where they are retained by van der Waals forces. In surface waters, they tend to move to the surface (into surface oil films, where present), into sediments, or into the waxy cuticle of aquatic macrophytes. The tendency of hydrophobic compounds to bind to colloid surfaces has two important consequences: (1) they are not very mobile and (2) they are often very persistent because they are not freely available to the enzyme systems (e.g., of microorganisms) that can degrade them. In soils, as in terrestrial animals, refractory hydrophobic molecules tend to be markedly persistent. Of the examples given in Table 2.1 (Chapter 2), dieldrin and *p,p′*-DDT with K_{ow} values of 5.48 and 6.36, respectively, have very long half-lives in soils and sediments.

Another very important factor determining the distribution of chemicals in the gross environment is *vapor pressure*. It is defined as the pressure exerted by a chemical in the vapor state on its own liquid or solid surface at equilibrium. It may be expressed as millimeters of mercury (torr), as a fraction of normal atmospheric pressure (760 mm Hg), or as pascals (Pa; 1 Pa = 0.0075 mm Hg). Vapor pressure is related to temperature, and liquids boil at temperatures that raise their own vapor pressures to 760 mm Hg. Water, for example, boils at 100°C. Some solids sublime—that is, they pass directly into the vapor state without liquefying. Many pesticides that exist in the solid state under normal temperature and pressure will sublime under field conditions. Volatilization represents a major source of loss of pesticides from the surface of crops and from the soil. With pollutants, high vapor pressure generally indicates that chemicals will tend to move into the atmosphere, with the attendant risk that they will be transported over large distances if they are stable enough. Chlorofluorocarbons (CFCs) present a striking example of the problem. They are highly volatile, and stable enough to reach the ozone layer within the stratosphere. Once there, however, they can reduce the concentration of ozone by interacting with it, an environmental process that is thought to have significantly reduced the extent of the ozone layer. The long-range movement of certain polychlorinated biphenyls

(PCBs) and organochlorine insecticides into polar regions has been attributed to aerial transport in the vapor state. In the second part of the book, examples will be given of pollutants that move readily into the air because of relatively high vapor pressure.

Much of the earth's surface is water, and an important aspect of the volatilization of pollutants is their movement from water to air. The movement of solutes between water and air is governed by Henry's law, which states that at equilibrium, the concentration of a chemical in the vapor state bears a constant relationship to the concentration in aqueous solution.

$$H' = Conc.air/Conc. \ water = [n/V][1/S] = P/RTS$$

where H' is a partition coefficient, n/V is the molar concentration of the vapor, S is the saturation solubility of the solute in water, P is vapor pressure, R is the universal gas constant, and T is the absolute temperature.

This equation can be simplified to give $H = P/S$, where H is now Henry's law constant, which has dimensions of atm m^3/mole. Values for H may be calculated or measured (Mackay et al. 1979), and are now widely used in *fugacity modeling* (see Section 3.2).

Another important determinant of the environmental fate of pollutants is their chemical stability. Environmental chemicals that are highly resistant to chemical degradation have been described as being "refractory," or "recalcitrant" (see Crosby 1998). Many such chemicals are not very photochemically stable. When exposed to solar radiation, they may be oxidized or may undergo molecular rearrangement. Some chemicals, for example, organophosphates (OPs), carbamates, and pyrethroids, are susceptible to hydrolysis, especially when exposed to water having high pH. By contrast, other compounds only undergo very slow degradation and can therefore be transported over large distances in air or water without substantial loss, unless biodegradation is significant (see Section 3.3). Typical examples are certain highly halogenated compounds such as *p,p'*-DDE, dieldrin, higher chlorinated PCBs, and TCDD (dioxin). Highly halogenated compounds tend to be resistant to oxidation and other mechanisms of chemical degradation.

In most cases, chemical instability of pollutants limits risk to the environment because it usually represents a reduction in toxicity. There are, however, exceptions and the devil may be found in the detail. When dieldrin residues are exposed to solar radiation, there is some conversion to the persistent and highly toxic photodieldrin. When the OP malathion is stored under hot conditions over long periods, it is converted to highly toxic isomalathion. When polycyclic aromatic hydrocarbons (PAHs) are exposed to radiation, they are converted into products that are highly toxic to fish. Such examples illustrate the dangers of hasty or superficial judgment in environmental risk assessment, and the importance of rigorous case-by-case analysis.

Another factor that can influence the environmental distribution of a chemical is the presence of charged groups. Some pollutants, such as the sodium or potassium salts of phenoxyalkanoic herbicides, dinitrophenols, and tetra- or penta-chlorophenol, exist as anions in solution. Others, such as the bipyridyl herbicides diquat and paraquat, are present as cations. In either case, the ions may become bound to organic macromolecules or minerals of soils or sediments that bear the opposite charge. Thus, the paraquat cation can be strongly bound (adsorbed) to negatively charged surfaces within clay minerals, in competition with exchangeable cations such as Ca^{2+}, K^+, and Na^+. It can also be bound to ionized carboxyl or phenolic groups of soil organic matter. Organic anions may bind to certain positively charged groups within soils and sediments. The influence of chemical properties on environmental fate is dealt with in much more detail in specialized texts on environmental chemistry, such as those by Schwarzenbach et al. (1993) and Crosby (1998).

3.2   MODELS OF ENVIRONMENTAL FATE

Confronted with the complexity of environmental pollutants and the great expense of actually measuring the levels of organic pollutants, the attractions of developing models for environmental fate are obvious. Because of the restraints of cost and time, the determination of actual levels of pollutants in different compartments of the environment can only be carried out to a very limited degree. Many different models have been developed that attempt to describe or predict the fate of chemicals in the gross environment. (For a detailed treatment of the subject, readers are referred to Mackay 1991, Mackay 1994, Jorgensen 1990, and Bacci 1994.) They range from limited single-media models, which are concerned with the fate of a chemical in a phase such as air, water, or soil, to wide-ranging multimedia models, which attempt to describe the movement of chemicals through different phases of the environment across a very large area over distances as great as thousands of kilometers across. The former may be expected to provide reasonably close estimates of environmental concentrations if the data base is good enough, whereas the latter can only give a broad and imprecise view of distribution through air, soil, water, sediment, and biota on what is approaching a global scale. Some multimedia models are only useful for ranking chemicals according to their tendency to move into particular environmental compartments (e.g., air), and cannot give reliable estimates of concentrations that will be found in the real world following defined releases of chemicals. There are too many uncontrolled variables such as temperature, wind speed, solar radiation, precipitation of rain and snow, etc. The term *evaluative models* is sometimes used to describe such systems (Bacci 1994). In certain cases, the multimedia approach has been applied to global pollution problems; for example, the geochemical cycling of CO_2 and the distribution of CFCs, which may affect the ozone layer.

Multimedia models can describe the distribution of a chemical between environmental compartments in a state of equilibrium. Equilibrium concentrations in different environmental compartments following the release of defined quantities of pollutant may be estimated by using distribution coefficients such as K_{ow} and H's (see Section 3.1). An alternative approach is to use *fugacity* (f) as a descriptor of chemical quantity (Mackay 1991). Fugacity has been defined as the tendency of a chemical to escape from one phase to another, and has the same units as pressure. When a chemical reaches equilibrium in a multimedia system, all phases should have the same fugacity. It is usually linearly related to concentration (C) as follows:

$$f = C/Z$$

where Z is a constant, sometimes termed the *fugacity capacity constant*. Values for Z depend on the chemical, nature of the absorbing medium, and temperature.

Examples of models of the environmental fate of chemicals, utilizing partition coefficients and fugacities, are given in the works cited earlier, and also in Chapter 3 in Walker et al. (2000, 2006).

3.3 INFLUENCE OF THE PROPERTIES OF CHEMICALS ON THEIR METABOLISM AND DISPOSITION

Up to this point, the discussion has been limited to the influence of chemical properties on the physical processes that influence distribution of the chemical through the environment without taking account of their effects on those biological processes that also affect movement. Although it is true that biological factors have little or no influence on the fate of chemicals transported by air, this is not the case with transport by surface waters or in sediments or soils. In the latter case, living organisms are involved in the uptake, metabolism, and transfer of organic pollutants. Stable pollutants are transported over considerable distances by migrating animals and birds, and move through terrestrial and aquatic food chains. Such complexities are not readily accommodated by the relatively simple "chemical" models described earlier.

In Chapter 2, the fate of organic pollutants in living organisms and the processes involved were described. In this section, the relationship between the properties of chemicals and the operation of those processes will be briefly reviewed. Many examples of the influence of properties of a chemical on its own metabolic fate will be encountered in Part 2 of this book.

Once again, the hydrophobicity of compounds is a critical determinant of their fate. Compounds with high K_{ow} values tend to undergo bioaccumulation by animals, because they move from the aqueous phase into the hydrophobic environment of fat depots and biological membranes. Water-soluble compounds do not have this tendency. Aquatic organisms do not tend to bioconcentrate them. When terrestrial animals absorb them from food or water, they are usually rapidly excreted in their unchanged forms. With lipophilic compounds, however, rapid elimination from the body depends on efficient metabolism to water-soluble and readily excretable metabolites and conjugates. This trend is particularly marked in terrestrial animals; aquatic species can lose lipophilic compounds to some extent by "exchange diffusion." Interestingly, in comparison with terrestrial animals, aquatic animals generally have relatively low levels of detoxifying enzymes such as cytochrome P450s of family 2, presumably because they have less need of them. Fish, for example, have considerably lower levels of such enzymes than omnivorous terrestrial animals and birds (Chapter 2, Walker 1978, 1980).

Many of the lipophilic pollutants described here are not persistent along food chains because they are readily biodegradable by terrestrial vertebrates. The main exceptions are polyhalogenated compounds such as some organochlorine (OC) insecticides, higher halogenated PCBs and PBBs, PCDDs, and PCDFs. As mentioned earlier, the main mechanism of primary metabolism of these compounds is P450-catalyzed attack. The problem is that highly halogenated compounds are resistant to such oxidative attack. Thus, polyhalogenated compounds of this type present a general problem of persistence and have long biological half-lives in most animals, terrestrial or aquatic. They are, therefore, liable to undergo biomagnification with movement along food chains, and also to survive transportation over long distances in migrating animals (e.g., fish or whales) and birds. Some organometallic compounds (e.g., methyl mercury) are also very persistent in terrestrial vertebrates, and are only slowly metabolized.

Some compounds that are readily and rapidly metabolized and excreted by terrestrial vertebrates degrade only very slowly in other species, for example, marine invertebrates such as bivalve mollusks. The general question of major differences between species and groups in metabolic capacity will be discussed in Chapter 4, Section 4.1. At this stage, the critical point is that there are compounds (such as PAHs) that tend to bioconcentrate/bioaccumulate in certain species low in aquatic food chains that are rapidly metabolized by fish, mammals, and birds occupying higher trophic levels. Thus, unlike many polyhalogenated compounds, they are not biomagnified as they move to the top of the food chain. This illustrates the limitations of using standard laboratory species such as rats, mice, or Japanese quail as metabolic models in ecotoxicology. There are many chemicals that are rapidly metabolized by these species, which are recalcitrant and consequently persistent in certain aquatic invertebrates (Livingstone 1991; Walker and Livingstone 1992).

The biotransformation of lipophilic pollutants into water-soluble and readily excretable products represents the main mechanism for their elimination by terrestrial animals. Its effectiveness, however, depends on the availability of the pollutants to the relevant enzymes. Where pollutants are stored in fat reserves, they usually become available to the liver in course of time, with the mobilization of lipids. Rates of elimination depend on rates of mobilization of fat reserves. The very strong binding of pollutants to proteins can severely limit the availability of chemicals to enzyme systems or for direct excretion, and can result in very long biological half-lives. Examples include the binding of superwarfarins to liver proteins (Chapter 11) and the binding of hydroxymetabolites of PCBs to plasma proteins (see Chapter 6, Section 6.2.4). Such long-term binding is limited by the number of available binding sites and, in the examples given, only relates to the persistence of low concentrations of chemical. Higher concentrations are freely available for metabolism or excretion.

The refractory nature of some pollutants, notably, persistent polyhalogenated compounds, has raised problems of bioremediation of contaminated sites (e.g., sediments and dumping sites). There has been interest in the identification, or the production by genetic manipulation, of strains of microorganisms that can metabolically degrade recalcitrant molecules. For example, there are bacterial strains that can reductively dechlorinate PCBs under anaerobic conditions.

3.4   SUMMARY

The environmental fate of chemicals is determined by both chemical/physical and biological processes; in turn, the operation of these processes is dependent on the properties of the environmental chemicals themselves. Polarity, vapor pressure, partition coefficients, and chemical stability are all determinants of movement and distribution in the physical environment. Here, constants such as vapor pressure, partition coefficients (e.g., K_{ow}), and fugacity values have been incorporated into models that describe or predict environmental fate. In these relatively simple situations, biological factors are largely ignored. For movement along food chains or fate in soils or sediments, biological factors such as uptake metabolism and excretion are critically important and need to be taken into account for developing models. These complications will be dealt with in Chapter 4.

FURTHER READING

Crosby, D.G. (1998). A very readable teaching text with many useful examples, which has the advantage of linking chemistry to toxicology.
Schwarzenbach, R.P., Gschwend, P.M., and Imboden, D.M. (1993). *Environmental Organic Chemistry.* An authoritative text on basic principles.

4 Distribution and Effects of Chemicals in Communities and Ecosystems

4.1 INTRODUCTION

In Chapter 3, the distribution of environmental chemicals through compartments of the gross environment was related to the chemical factors and processes involved, and models for describing or predicting environmental fate were considered. In the early sections of the present chapter, the discussion moves on to the more complex question of movement and distribution in the living environment—within individuals, communities, and ecosystems—where biological as well as physical and chemical factors come into play. The movement of chemicals along food chains and the fate of chemicals in the complex communities of sediments and soils are basic issues here.

Ecotoxicology deals with the study of the harmful effects of chemicals in ecosystems. This includes harmful effects upon individuals, although the ultimate concern is about how these are translated into changes at the levels of population, community, and ecosystem. Thus, in the concluding sections of the chapter, emphasis will move from the distribution and environmental concentrations of pollutants to consequent effects at the levels of the individual, population, community, and ecosystem. The relationship between environmental exposure (dose) and harmful effect (response) is fundamentally important here, and full consideration will be given to the concept of biomarkers, which is based on this relationship and which can provide the means of relating environmental levels of chemicals to consequent effects upon individuals, populations, communities, and ecosystems.

4.2 MOVEMENT OF POLLUTANTS ALONG FOOD CHAINS

The pollutants of particular interest here are persistent organic chemicals—compounds that have sufficiently long half-lives in living organisms for them to pass along food chains and to undergo biomagnification at higher trophic levels (see Box 4.1). Some compounds of lesser persistence, such as polycyclic aromatic hydrocarbons (PAHs) (Chapter 9), can be bioconcentrated/bioaccumulated at lower trophic levels but are rapidly metabolized by vertebrates at higher levels. These will not be discussed further here, where the issue is biomagnification with movement along the entire food chain. The best studied examples of this are the organochlorines (OCs) dieldrin and p,p'-DDE (see Chapter 5), and the PCBs (see Chapter 6), where concentrations in the tissues of predators of the highest trophic levels can be 10^4–10^5-fold higher than in organisms at the lowest trophic levels. Other examples include polychlorinated dibenzodioxin (PCDDs), polychlorinated dibenzofurans (PCDFs), and some organometallic compounds (e.g., methyl mercury).

Biomagnification along terrestrial food chains is principally due to bioaccumulation from food, the principal source of most pollutants (Walker 1990b). In a few instances, the major route of uptake may be from air, from contact with contaminated surfaces, or from drinking water. The bioaccumulation factor (BAF) of a chemical is given by the following equation:

$$\text{Concentration in organism/concentration in food} = \text{BAF}$$

Biomagnification along aquatic food chains may be the consequence of bioconcentration as well as bioaccumulation. Aquatic vertebrates and invertebrates can absorb pollutants from ambient water; bottom feeders can take up pollutants from sediments. The bioconcentration factor (BCF) of a chemical absorbed directly from water is defined as

$$\text{Concentration in organism/concentration in ambient water} = \text{BCF}$$

One of the challenges when studying biomagnification along aquatic food chains is establishing the relative importance of bioaccumulation versus bioconcentration. The processes that lead to biomagnification have been investigated with a view to developing predictive toxicokinetic models (Walker 1990b). When organisms are continuously exposed to pollutants maintained at a fairly constant level in food and/ or in ambient water/air, tissue concentrations will increase with time until either (1) a lethal concentration is reached and the organism dies or (2) a steady state is reached when the rate of uptake of the pollutant is balanced by the rate of loss. The BCF or BAF at the steady state is of particular interest and importance because (A) it represents the highest value that can be reached and therefore indicates the maximum risk, (B) it is not time dependent, and (C) the rates of uptake and loss are equal, thereby facilitating the calculation of the rate constants involved.

BCFs and BAFs measured before the steady state is reached have little value because they are dependent on the period of exposure of the organism to the chemical, and thus may greatly underestimate the degree of biomagnification that is possible. This statement should be qualified by the reservation that there may be situations in which the duration of exposure cannot be long enough for the steady state to be reached, for example, where the life span of an insect is very short. The principal processes of uptake and loss by different types of organisms are indicated in Table 4.1 (see also Box 4.2).

A rough indication of the relative importance of different mechanisms of uptake and loss is given by a scoring system on the scale +–> ++++. Within each category of organism there will be differences between compounds in the relative importance of different mechanisms, for example, due to differences in polarity and biodegradability.

BOX 4.1   PERSISTENT ORGANIC POLLUTANTS (POPS)

A list of hazardous environmental chemicals, sometimes referred to as the "dirty dozen," has been drawn up by the United Nations Environment Programme (UNEP). These are:

POP	Year of Introduction	Classification
Aldrin	1949	Insecticide
Chlordane	1945	Insecticide
DDT	1942	Insecticide
Dieldrin	1948	Insecticide
Endrin	1951	Insecticide and rodenticide
Heptachlor	1948	Insecticide
Hexachlorobenzene	1945	Fungicide
Mirex	1959	Insecticide
Toxaphene	1948	Insecticide and acaricide
PCBs	1929	Various industrial uses
Dioxins	1920s	By-products of combustion, for example, of plastics, PCBs
Furans	1920s	By-products of PCB manufacture

The selection of these compounds was made on the grounds of their toxicity, environmental stability, and tendency to undergo biomagnification; the intention was to move toward their removal from the natural environment. In the REACH proposals of the European Commission (EC; published in 2003), a similar list of 12 POPs was drawn up, the only differences being the inclusion of hexachlorobiphenyl and chlordecone, and the exclusion of the by-products, dioxins, and furans. The objective of the EC directive is to ban the manufacture or marketing of these substances. It is interesting that no fewer than eight of these compounds, which are featured on both lists, are insecticides.

TABLE 4.1
Principal Mechanisms of Uptake and Loss for Lipophilic Compounds

Habitat/Type of Organism	Mechanisms of Uptake			Mechanisms of Loss	
	Diffusion	From Food	From Ingested Water	Diffusion	Metabolism
Aquatic					
Mollusks	++++	+		++++	
Fish	++++	$+\rightarrow+++$		++++	$+\rightarrow+++$
Terrestrial					
Vertebrates		++++	< ++		++++

**BOX 4.2   MODELS FOR BIOCONCENTRATION
AND BIOACCUMULATION**

As indicated in Table 4.1, aquatic mollusks present a relatively simple picture because they have little capacity for biotransformation of organic pollutants, the principal mechanism of both uptake loss being diffusion. It is not surprising, therefore, that bioconcentration factors (BCFs) for diverse lipophilic compounds, measured at the steady state, are related linearly to log K_{ow} values (Figure 4.1). Thus, the more hydrophobic a compound is, the greater the tendency to partition from water into the lipids of the mollusk. The relationship shown in Figure 4.1 has been demonstrated in several species of aquatic mollusks, including the edible mussel (*Mytilus edulis*), the oyster (*Crassostrea virginica*), and soft clams (*Mya arenaria*) (Ernst 1977). A similar relationship has also been found with rainbow trout and other fish for some pollutants. On the other hand, some organic pollutants do not fit the model well (Connor 1983). It seems probable that some compounds that are metabolized relatively rapidly by fish will be eliminated faster than would be expected if diffusion were the only process involved (Walker 1987). Such compounds would not be expected to follow closely a model for BCF based on K_{ow} alone. This point aside, K_{ow} values can give a useful prediction of BCF values at the steady state for lipophilic pollutants in aquatic invertebrates. A great virtue of the approach is that K_{ow} values are easy and inexpensive to measure or predict (Connell 1994). Other more complex and sophisticated models have been developed for fish (see, for example, Norstrom et al. 1976) but are too time-consuming/expensive to be used widely in environmental risk assessment where cost-effectiveness is critically important. Modeling for bioaccumulation by terrestrial animals presents greater problems, and BAFs cannot be reliably predicted from K_{ow} values (Walker 1987). For example, benzo[*a*]pyrene and dieldrin have log K_{ow} values of 6.50 and 5.48, respectively, but their biological half-lives range from a few hours in the case of the former to 10 –369 days for the latter. Endrin is a stereoisomer of dieldrin with a similar K_{ow}, but has a half-life of only 1 day in humans, compared with 369 days in the case of dieldrin. These large differences in persistence have been attributed to differences in the rate of metabolism by P450-based monooxygenases (Walker 1981). Effective predictive models for bioaccumulation of strongly lipophilic compounds by terrestrial animals need to take account of rates of metabolic degradation. This is not a straightforward task and would require the sophisticated use of enzyme kinetics to be successful. In one model, it has been suggested that Lineweaver–Burke plots for microsomal metabolism might be used to predict BAF values in the steady state (Walker 1987) (Figure 4.2). In principle, when an animal ingests a lipophilic compound at a constant rate in its food, a steady state will eventually be reached where the rate of intake of the compound is balanced by the rate of its metabolism. It is assumed that the rate of loss of the unchanged compound by direct excretion is negligible. Primary metabolic attack upon many highly lipophilic compounds (e.g., polyhalogenated aromatic compounds and PAHs) takes place predominantly in the endoplasmic reticulum, particularly that of the liver in vertebrates. Thus, microsomes (especially hepatic microsomes of vertebrates) can serve as model systems for measuring rates of enzymic detoxication. Lineweaver–Burke and similar metabolic plots can relate concentrations of pollutants in microsomal membranes to rates of metabolism. In the steady state, rate of intake of chemical should equal rate of metabolism in the membranes of the endoplasmic reticulum. The concentration of the chemical required in the membranes to give this balancing metabolic rate can be estimated from the Lineweaver–Burke plot. The necessary balancing metabolic rate can be calculated from the defined rate of intake in food, and then the microsomal concentration that will give this rate can be read from the plot. Thus, the concentration in endoplasmic reticulum can be compared to the dietary concentration to give an estimate of BAF. Estimates can also be made of BAF for the liver or the whole body if approximate ratios of concentrations of chemical in different compartments of the body when at the steady state are known.

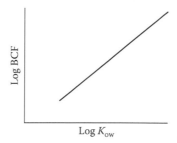

FIGURE 4.1   Relationship of BCF to log K_{ow} values.

The main points to bring out are as follows:

1. The uptake and loss by exchange diffusion is important for aquatic organisms but not for terrestrial ones.
2. Metabolism is the main mechanism of loss in terrestrial vertebrates, but is less important in fish, which can achieve excretion by diffusion into ambient water.
3. Most aquatic invertebrates have very little capacity for metabolism; this is particularly true of mollusks. Crustaceans (e.g., crabs and lobsters) appear to have greater metabolic capability than mollusks (see Livingstone and Stegeman 1998; Walker and Livingstone 1992).

The balance between competing mechanisms of loss in the same organism depends on the compound and the species in question. In fish, for example, some compounds

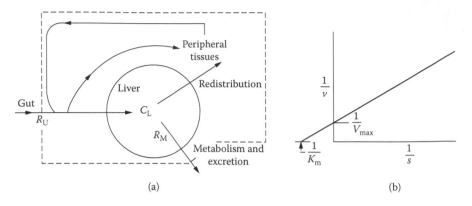

(a)   (b)

FIGURE 4.2 (a) A bioaccumulation model for terrestrial organisms. A kinetic model for liver. R_U, rate of uptake from the gut; R_M, rate of metabolism in liver; C_L, concentration of pollutant in liver. The arrows indicate the routes of transfer of pollutant within the animal. The rates of uptake and metabolism are expressed in terms of kilograms of body weight. The final elimination of water-soluble products (metabolites and conjugates) is in the urine. (b) Lineweaver–Burke plot to estimate the bioaccumulation factor; V_{max} and v are expressed as milligrams of pollutant metabolized per kilogram of body weight per day; S is expressed as the concentration of pollutant, ppm by weight (either in terms of grams of liver or milligrams of hepatic microsomal protein) (from Walker 1987).

that are good substrates for monooxygenases, hydrolases, etc., can be metabolized relatively rapidly even though they, as a group, have relatively low metabolic capacity (Chapter 2). So, in this case metabolism as well as diffusion is an important factor determining rate of loss. By contrast, many polyhalogenated compounds are only metabolized very slowly by fish, so metabolism does not make a significant contribution to detoxication, and loss by diffusion is the dominant mechanism of elimination.

Some further aspects of detoxication by fish need to be briefly mentioned. When fish inhabit polluted waters, exchange diffusion occurs until a steady state is reached, and no net loss will occur by this mechanism unless the concentration in water falls. When a recalcitrant pollutant is acquired from prey, digestion can lead to the tissue levels of that pollutant temporally exceeding those originally existing while in the steady state. Here, diffusion into the ambient water may provide an effective excretion mechanism in the absence of effective metabolic detoxication. Seen from an evolutionary point of view, the requirements of fish for metabolic detoxication would appear to have been limited on the grounds that loss by diffusion would often have prevented tissue levels becoming too high. The poor metabolic detoxication systems of fish relative to those of terrestrial omnivores and herbivores are explicable on these grounds (Chapter 2). However, the advent of refractory organic pollutants, which combine high toxicity with high lipophilicity, has exposed the limitations of existing detoxication systems of fish. The very high toxicity of compounds such as dieldrin and other cyclodiene insecticides to fish was soon apparent, with fish kills occurring at very low concentrations in water (see Chapter 5) and metabolically resistant strains of fish being reported in polluted rivers such as the Mississippi.

More rapid elimination was needed than could be provided by passive diffusion in order to prevent tissue concentrations reaching toxic levels.

Some models for predicting bioconcentration and biomagnification are presented in Box 4.1.

4.3 FATE OF POLLUTANTS IN SOILS AND SEDIMENTS

Regarding soils, a central issue is the persistence and movement of pesticides that are widely used in agriculture. Many different insecticides, fungicides, herbicides, and molluscicides are applied to agricultural soils, and there is concern not only about effects that they may have on nontarget species residing in soil, but also on the possibility of the chemicals finding their way into adjacent water courses.

Soils are complex associations between living organisms and mineral particles. Decomposition of organic residues by soil microorganisms generates complex organic polymers ("humic substances" or simply "soil organic matter") that bind together mineral particles to form aggregates that give the soil its structure. Soil organic matter and clay minerals constitute the colloidal fraction of soil; because of their small size, they present a large surface area in relation to their volume. Consequently, they have a large capacity to adsorb the organic pollutants that contaminate soil. Within a freely draining soil there are air channels and soil water, the latter being closely associated with solid surfaces. Depending on their physical properties, organic compounds become differentially distributed between the three phases of the soil, soil water, and soil air.

Hydrophobic compounds of high K_{ow} become very strongly adsorbed to soil colloids (Chapter 3, Section 3.1), and consequently tend to be immobile and persistent. OC insecticides such as DDT and dieldrin are good examples of hydrophobic compounds of rather low vapor pressure that have long half-lives, sometimes running into years, in temperate soils (Chapter 5). Because of their low water solubility and their refractory nature, the main mechanism of loss from most soils is by volatilization. Metabolism is limited by two factors: (1) being tightly bound, they are not freely available to enzymes of soil organisms, which can degrade them, and (2) they are, at best, only slowly metabolized by enzyme systems. Because of strong adsorption and low water solubility, there is little tendency for them to be leached down the soil profile by percolating water. The degree of adsorption, and consequently the persistence and mobility, is also dependent on soil type. Heavy soils, high in organic matter and/or clay, adsorb hydrophobic compounds more strongly than light sandy soils, which are low in organic matter. Strongly lipophilic compounds are most persistent in heavy soils. When OC insecticides are first incorporated into soil, they are lost relatively rapidly, mainly due to volatilization, before they become extensively adsorbed to soil colloids (Figure 4.3). With time, however, most residual OC insecticide becomes adsorbed, and subsequently there is a period of very slow exponential loss.

In marked contrast to hydrophobic compounds, more polar ones tend to be less adsorbed and to reach relatively high concentrations in soil water. Phenoxyalkanoic acids such as 2,4-D and MCPA are good examples (Figure 4.3). Their half-lives in soil are measured in weeks rather than years, and they are more mobile than OC insecticides in soils. When first applied they are lost only slowly. After a lag period of a

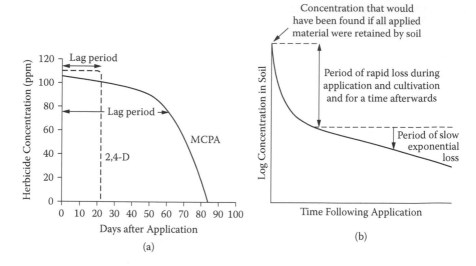

FIGURE 4.3 Loss of pesticides from soil. (a) Breakdown of herbicides in soil. (b) Disappearance of persistent organochlorine insecticides from soils (from Walker et al. 2000).

few days, however, they disappear very rapidly as a consequence of metabolism by soil microorganisms. This has been explained on the grounds that it takes time for a buildup in numbers of strains of microorganisms that can metabolize them; these microorganisms use the herbicides as an energy source. It has also been suggested that the lag period relates to the time it takes for enzyme induction to occur. Whatever the explanation, soils treated with these compounds stay enriched for a period, and further additions of the original compounds will be followed by rapid metabolism without a lag phase. If, however, the soils are untreated for a long period, they will revert to their original state and not show any enhanced capacity for degrading the herbicides. An important difference from the OC insecticides and related hydrophobic pollutants is that, because of their polarity and water solubility, they are freely available to the microorganisms that can degrade them. Interestingly, the phenoxyalkanoic acid 2,4,5-T is more persistent than either 2,4-D or MCPA. With three substituted chlorines in its phenyl ring, it is metabolized less rapidly than the other two compounds, and it would appear that metabolism is a rate-limiting factor determining rate of loss from soil.

It was long assumed that there is little tendency for most pesticides or other organic pollutants to move through soil into drainage water. Indeed, this is to be expected with intact soil profiles. Hydrophobic compounds will be held back by adsorption, whereas water soluble ones will be degraded by soil organisms. Some soils, however, depart from this simple model. Soils high in clay can crack and develop deep fissures during dry weather. If rain then follows, pesticides, in solution or adsorbed to mobile colloids, can be washed down through the fissures, to appear in neighboring drainage ditches and streams. This was found to happen with pesticides such as carbofuran, isoproturon, and chlorpyrifos in the Rosemaund experiment conducted in England during the period 1987–1993 (Williams et al. 1996).

The influence of polarity on movement of chemicals down through the soil profile has been exploited in the selective control of weeds using soil herbicides (Hassall 1990). In general, the more polar and water soluble the herbicide, the further it will be taken down into the soil by percolating water. Insoluble herbicides such as the triazine compound simazine (water solubility, 3.5 ppm), remain in the first few centimeters of soil when applied to the surface. More water-soluble compounds such as the urea herbicides diuron and monuron (water solubilities 42 ppm and 230 ppm, respectively) are more mobile, and can move farther down the soil profile. Selective weed control can be achieved in some deep-rooting crops by judicious selection from this range of herbicides, so that the herbicide will only percolate far enough down the soil profile to control surface rooting weeds without reaching the main part of the root system of the crop (depth selection). Thus, when applied to the soil surface, simazine should only be toxic to shallow rooting weeds and should not affect crops that root farther down. Other more water-soluble herbicides can give weed control to greater depths in situations where the rooting systems of the crops are sufficiently deep. When attempting depth selection in weed control, account needs to be taken of soil type. Herbicides will move farther down the profile in the case of light sandy soils than they will in heavy clays or organic soils.

Although the major concern about the fate of organic pollutants in soil has been about pesticides in agricultural soils, other scenarios are also important. The disposal of wastes on land (e.g., at landfill sites) has raised questions about movement of pollutants contained in them into the air or neighboring rivers or water courses. The presence of polychlorinated biphenyls (PCBs) or PAHs in such wastes can be a significant source of pollution. Likewise, the disposal of some industrial wastes in landfill sites (e.g., by the chemical industry) raises questions about movement into air or water and needs to be carefully controlled and monitored.

In certain respects, sediments resemble soils. Sediments also represent an association between mineral particles, organic matter, and resident organisms. The main difference is that they are situated underwater and are, in varying degrees, anaerobic. The oxygen level influences the type of organisms and the nature of biotransformations that occur in sediments. A feature with sediments, as with soils, is the limited availability of chemicals that are strongly adsorbed. Again, compounds with high K_{ow} tend to be strongly adsorbed, relatively unavailable, and highly persistent. There is much interest in the question of sediment toxicity and the availability to bottom-dwelling organisms of compounds adsorbed by sediments (Hill et al. 1993). One case in point is pyrethroid insecticides (see Chapter 12), which are strongly retained in sediments on account of their high K_{ow} values. Because of their ready biodegradability, they are not usually biomagnified in the higher trophic levels of aquatic food chains.

However, they are available to bottom-dwelling organisms low in the food chain. Questions are asked about the possible long-term buildup of pyrethroids in sediments and their effects on organisms in lower trophic levels.

4.4   EFFECTS OF CHEMICALS UPON INDIVIDUALS— THE BIOMARKER APPROACH

Until now this narrative has been concerned with questions about the movement and distribution of chemicals in the living environment, a topic that relates to the field of toxicokinetics in classical toxicology, although on a much larger scale. It is now time to move to a consideration of the effects that chemicals may have upon living organisms, which relates to the area of toxicodynamics in classical toxicology. Effects upon individuals will be discussed before dealing with consequent changes at the higher levels of biological organization—population, community, and ecosystem.

Measuring effects of chemicals upon free-living individuals in the natural environment is not an easy matter. Mobile animals need to be captured so that samples of tissues can be taken for analysis, but this is often difficult to do in a properly controlled way in the field. It is easier to obtain samples from sedentary species (e.g., mollusks or plants) or of eggs in the case of birds, reptiles, and some invertebrates. All too often sampling is destructive, which raises problems of experimental design and statistical evaluation of results. In principle, measuring behavioral effects of chemicals is an attractive option, but this can be hard to achieve in practice because of the difficulty of making reliable measurements in the field. The problems of sampling can, to some extent, be circumvented by deploying indicator species that have been maintained in a "clean" environment into the field. Thus, control fish from the laboratory can be held in cages in contaminated waters and samples taken from them after periods of exposure to pollutants. Uncontaminated birds' eggs can be introduced into the nests of birds of the same species that are breeding in a polluted area. In this way, changes caused by pollutants can be measured and evaluated.

Problems of sampling aside, the success of any strategy of this kind depends on the availability of reliable tests that can measure harmful effects of chemicals under field conditions. Reference has already been made to biomarkers (Chapter 2, Box 2.2). In the following account they are defined as "biological responses to environmental chemicals at the individual level or below, which demonstrate a departure from normal status." This definition includes biochemical, physiological, histological, morphological, and behavioral changes, but does not extend to changes at higher levels of organization. Changes at population, community, or ecosystem level are regarded instead as bioindicators.

The concept of biomarkers is illustrated in Figure 4.4. As the dose of a chemical increases, the organism moves from a state of homeostasis to a state of stress. With further increases in dose, the organism enters first the state of reversible disease, and eventually the state of irreversible disease, which will lead to death. In concept, all of these stages can be monitored by biomarker assays (lower part of conceptual diagram).

Some biomarker responses provide evidence only of exposure and do not give any reliable measure of toxic effect. Other biomarkers, however, provide a measure of toxic effects, and these will be referred to as mechanistic biomarkers. Ideally, biomarker assays of this latter type monitor the primary interaction between a chemical and its site of action. However, other biomarkers operating "down stream" from the original toxic lesion also provide a measure of toxic action (see Figure 14.3 in Chapter 14), as, for instance, in the case of changes in the transmission of action potential

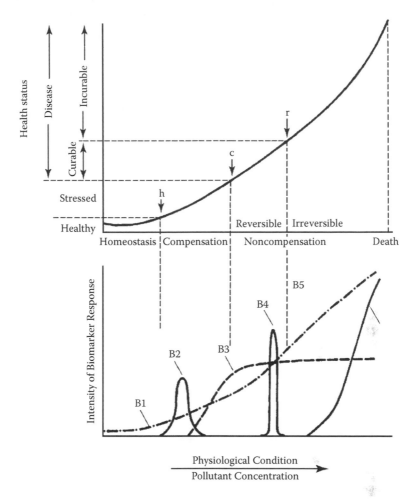

FIGURE 4.4 Relationship between exposure to pollutant, health status, and biomarker responses. Upper curve shows the progression of the health status of an individual as exposure to pollutant increases; h, the point at which departure from the normal homeostatic response range is initiated; c, the limit at which compensatory responses can prevent development of overt disease; r, the limit beyond which pathological damage is irreversible by repair mechanisms. The lower graph shows the response of five different hypothetical biomarkers used to assess the health of the individual. (Reproduced from Depledge et al. 1993. With permission from Springer-Verlag.)

following the interaction of DDT or pyrethroids with Na^+ channels, although these are changes that may also be caused by the operation of other toxic mechanisms. These will also be treated as mechanistic biomarkers in order to distinguish them from other responses that are only biomarkers of exposure (e.g., the induction of certain enzymes that can occur at low levels of exposure before any toxic effects are manifest). Mechanistic biomarkers (see Table 4.2 for examples) have potential for measuring adverse effects of chemicals in the field—effects that may be translated into changes at the population level and above.

TABLE 4.2
Some Mechanistic Biomarker Assays

Assay	Chemicals	Secondary Effects	Chapter in This Text
Brain acetylcholinesterase inhibition	OP and carbamate insecticides	Acetylcholine buildup in synapse and synaptic block	10, 16
Inhibition of neuropathy target esterase (NTE) in mammals	Mipafox, DFP, leptophos, TOCP, acetamidophos	Nerve degeneration	10, 16
Undercarboxylated Gla proteins in blood of vertebrates	Warfarins and superwarfarins	Hemorrhaging	11
Occupancy of retinol binding site of transthyretin in vertebrate blood	4-OH, 3,3',4,4' tetrachlobiphenyl and related PCB metabolites	Reduced levels of retinol and thyroxine (T4) in blood	6
Eggshell thinning in birds	p,p'-DDE	Egg breakage	5
Imposex in dog whelk	Tributyl tin	Infertility of females	8

Of the examples given, brain cholinesterase inhibition has been frequently measured in field studies involving OP insecticides and when investigating cases of poisoning on agricultural land (see Chapter 10). A number of studies on fish, rodents, and birds in the field and/or in the laboratory give evidence for a range of sublethal neurotoxic and behavioral effects of OP insecticides when brain acetylcholinesterase inhibition is in the range 40–50%—before the appearance of severe toxic manifestations and death (see Chapters 10 and 16). A few OP compounds cause delayed neuropathy in mammals and birds, and this has been related to inhibition of neuropathy target esterase (NTE). Symptoms of this type of poisoning appear after aging of the inhibited enzyme occurs (Chapter 10). Warfarin and related anticoagulant rodenticides act as competitive antagonists of vitamin K at binding sites for this cofactor in the liver and consequently inhibit the carboxylation of precursors of blood-clotting proteins (Chapter 11). Thus, undercarboxylated Gla proteins are released into the blood. After a period of time (usually 5 days or more) the blood becomes depleted of normal clotting proteins and loses its capacity to coagulate. A valuable biomarker assay involves the measurement of levels of undercarboxylated clotting protein in the blood by immunochemical determination. An increase of this nonfunctional protein in the blood provides a measure of the toxic process that leads to hemorrhaging and death—although it does not measure the primary interaction between rodenticide and the vitamin K binding site in the liver. Certain metabolites of coplanar PCBs such as 4-OH, 3,3',4,4'-TCB can compete with thyroxine (T4) for binding sites on the protein transthyretin in blood. This interaction leads to the breaking apart of the protein and the loss of thyroxine and retinol from the blood. Here, measurement of the reduction of binding of thyroxine to transthyretin provides the basis for a mechanistic biomarker assay that reports changes at the site of action caused by these metabolites of PCBs (Brouwer et al. 1990, 1998). *p,p'*-DDE causes eggshell thinning in birds by retarding the uptake of Ca^{++} across the wall of the shell gland (Chapter 5). There is still some uncertainty about the exact mechanism by which *p,p'*-DDE retards the movement of Ca^{++} into the shell gland, although both inhibition of calcium ATPase and changes in prostaglandin levels have been implicated (Lundholm 1997). Again, the secondary effect—thinning of the eggshell—provides a good monitor of the toxic process. Finally, imposex is a condition that can be caused by tributyl tin (Chapter 8). It seems probable that the primary effect is to inhibit the production of testosterone. This point aside, a convenient biomarker assay measures the development of a penis by the female dog whelk that, when sufficiently large, blocks the oviduct and causes infertility. Once again, a biomarker assay provides a convenient measure of the toxic process.

In all of these examples it should be noted that the toxic process occurs through different stages over a period of time. In some instances (e.g., delayed neuropathy caused by some OPs, and hemorrhaging caused by warfarin) there may be a delay of several weeks between initial exposure and the appearance of overt symptoms of poisoning, which raises two issues. First, it emphasizes the importance of having sensitive biomarkers that can provide early measures of intoxication before severe toxicity occurs. Second, if we are to understand toxicity at a deeper level, the different stages in the process need to be understood and be readily measurable. These issues will be returned to later in the text, especially in Chapters 15 and 16, when consideration will be given to the development of new biomarker strategies, incorporating the *omics*, which have the potential to address both questions (Box 4.3).

A number of attributes are sought when developing biomarker assays. At a practical level, assays should be robust, inexpensive, and relatively easy for nonspecialists to use. Unfortunately, some promising assays are really at the stage of being research tools, only usable by experts with specialized apparatus. There is a need for user-friendly kits (such as the diagnostic kits that are widely available to medical laboratories) for use in nonspecialist laboratories with relatively simple apparatus. Specificity is another desirable characteristic, in the sense of identifying a particular mechanism of toxicity and, therefore, a particular class of organic pollutant. When dealing with complex cases of pollution, it is extremely valuable to have biomarker assays that can provide evidence of causal links between levels of particular pollutants (or classes of pollutant) detected in the environment and associated harmful biological effects. Examples of such biomarkers include imposex in dog whelks related to tributyl tin, brain cholinesterase inhibition in vertebrates caused by organophosphorous insecticides, and eggshell thinning of some predatory birds caused by *p,p'*-DDE (Table 4.2). Sensitivity is another desirable characteristic that can facilitate the early detection of sublethal effects. The detection of later effects seen during the terminal stages of poisoning is of less value.

In practice, all of these attributes are unlikely ever to be found in a single bio-marker assay. However, the ultimate aim is to produce combinations of biomarker assays that collectively can give an in-depth picture of the toxic process (Peakall and Shugart 1993; Walker et al. 2006). In this way, the time-dependent changes follow-ing the interaction of a pollutant with its site of action can be traced through different levels of organization, progressing from the site of action itself through local cel-lular disturbances to effects expressed at the level of the whole organism, including behavioral effects—in other words, through the entire pathway of the toxic process.

Examples of biomarker assays operating at different levels are given in Table 4.2. The recent development of "omics" technology should provide strong support to this approach (Box 4.3). Microarray analysis, for instance, can give a time-related sequence of gene responses that relate to the cellular changes of toxicity.

BOX 4.3   THE OMICS

Recent rapid technological advances have spawned a host of new terms that have caused confusion not only to interested laypeople but even to members of the scientific community. Many recently defined *omics* represent a case in point. The term *genome* was proposed by Hans Winkler in 1920 to describe the complete set of genes for a particular species (Snape et al. 2004; van Straalen and Roelofs 2006). Much later the term *genomics* appeared in the title of a jour-nal (see, for example, McKusick 1997). Genomics has been defined as the study of how the genome translates into biological functions (Snape et al. 2004).

This definition of the term genomics is a broad one and includes both structural and functional aspects. Other terms relating to gene function are *transcriptomics* (study of the transcripts of DNA, principally mRNA), *pro-teomics* (study of all the proteins of the cell), and *metabolomics* (the study of metabolites) (van Straalen and Roelofs 2006). Other terms that have arisen are *toxicogenomics*, which applies genomics to mammalian toxicology (Nuwaysir et al. 1999), and *ecotoxicogenomics*, which applies genomics to ecotoxicology (Snape et al. 2004). New techniques in these fields are necessarily underpinned by sophisticated information technology.

In the present context, the techniques of genomics have great potential in ecotoxicology, particularly as the basis of biomarker assays. DNA microarray assays allow the measurement of changes in gene expression when organisms are exposed to chemicals and other stressors (Burczynski 2003). In essence, the products of gene expression (mRNA converted into cDNA) are measured by hybridizing them to complementary sequences of DNA printed onto a glass slide (microarray). After processing, genes that have increased their expres-sion will be colored green, those that have decreased their expression will be red, and the level of expression will be indicated by the intensity of the color. In principle, the time-dependent response to a chemical of certain genes of an entire genome can be measured. Such a sequence of gene responses can be compared with the corresponding sequence of cellular responses to a toxic chemical measured by mechanistic biomarker assays, as described in this section and in later parts of this text (note especially Chapters 15 and 16).

Unfortunately, as with many exciting new areas of science, ideas are much easier to conceive than to deliver. There are many difficulties both in the execution and—not least—the interpretation of results from toxicogenomic assays. There has been a serious problem with regard to controlling data quality. Indeed, some journals, for example, *Nature*, have adopted strict guidelines on data quality when assessing papers submitted to them for publication on this subject. Also, the value of genomic data in ecotoxicology needs to be seen in context. Genotoxic compounds represent a rather small proportion of the pollutants that are known to have had serious ecotoxicological effects, as will be apparent from the pages that follow. Most primary toxic interactions—or their immediate knock-on effects—are not at the level of the gene. Changes in gene expression bear testimony to toxic disturbances in the organism—to the disturbance of cellular processes, to the disruption of neurotransmission, to the ability of blood to clot, etc. They do not provide direct measures of the initial toxic interactions or the consequent cellular disturbances. Thus, genomic techniques have great potential when used in combination with mechanistic biomarker assays that measure the toxic process itself. They also have potential for screening where toxic effects are suspected in the environment. The pattern of genomic response may give critical evidence about the mode of toxic action that operates when pollution occurs.

4.5   BIOMARKERS IN A WIDER ECOLOGICAL CONTEXT

From an ecological point of view, chemicals can be seen as constituting one class of agents that stress biological systems ("stressors") (Van Straalen 2003). As we have seen, organic chemicals that have harmful effects upon living organisms may be naturally occurring as well as human made, and both may contribute to "chemical stress" (Chapter 1). Other stressors include extremes of temperature, acidity, humidity, and levels of inorganic ions not usually regarded as toxic (e.g., nitrate, phosphate, etc.). In an ideal world, the stress caused by all of these factors should be taken into account when considering the state—healthy or otherwise—of individuals, populations, communities, and ecosystems. It may even be argued that, in the ultimate analysis, ecotoxicology is part of stress ecology (Van Straalen 2003).

However, there are a number of issues here. In the first place, stress itself is a somewhat nebulous concept, and there is continuing debate about how it should be defined. Second, even with the benefit of multivariate statistics and the techniques of bioinformatics, measuring stress from all sources in a meaningful way is dauntingly complex and may not be realizable in practice.

So far as the discipline of ecotoxicology goes, there have been a number of cases where organic pollutants were shown to be the principal or sole cause of population declines in the natural environment, and these are described in the chapters that follow. These have included cases where the recovery of populations followed the reduction of pollutant levels in the environment. Examples include the declines of predatory birds caused by cyclodiene insecticides and p,p'-DDE, and the decline of dog whelks caused by tributyl tin. Here, other stress factors were evidently not implicated in either the declines or the recoveries. However, now that action has been taken to rectify problems such as these in most parts of the world, the effects of pollutants are less easy to detect and may increasingly need to be seen in relation to stress factors more generally.

From a toxicological point of view, stress needs to be seen in relation to the toxic process as a whole (Figure 4.4). As the dose of a toxicant increases, organisms move from a homeostatic state through a stressed state to a reversible disease state, before finally reaching an irreversible disease state. Thus, although the impact of stress needs to be taken into consideration in the early stages of intoxication, the more serious effects in terms of health tend to come later when moving beyond stress to fundamental disturbances of function that may lead to starvation, infertility, and death. Indeed, it was effects of the latter kind that led to the population declines referred to previously.

4.6   EFFECTS OF CHEMICALS AT THE POPULATION LEVEL

4.6.1   POPULATION DYNAMICS

In environmental risk assessment, the objective is to establish the likelihood of a chemical (or chemicals) expressing toxicity in the natural environment. Assessment is based on a comparison of ecotoxicity data from laboratory tests with estimated or measured exposure in the field. The question of effects at the level of population that may be the consequence of such toxicity is not addressed. This issue will now be discussed.

Toxic effects upon individuals in the field may be established and quantified in a number of assays. Lethal effects can be assessed by collecting and counting corpses found in the field following the application of a chemical, as in field trials with new pesticides. This is an imprecise technique because many individual casualties will escape detection, especially with mobile species such as birds. With very stable pollutants such as dieldrin, heptachlor epoxide, p,p'-DDT, and p,p'-DDE, the determination of residues in carcasses found in the field can provide evidence of lethal toxicity in the field. Such data may also be used to obtain estimates of the effects of chemicals upon mortality rates of field populations, which can then be incorporated into population models (see Chapter 5, Section 5.3.5.1). Mechanistic biomarker assays can also be used to quantify toxic effects in the field. Even when investigating cases of lethal poisoning, relatively stable biomarker assays such as cholinesterase inhibition may be used to establish the cause of death so long as carcasses are relatively fresh. Of particular interest is the use of biomarker assays to monitor the effects of chemicals on living organisms in the field. Here, biomarker assays can provide measures of sublethal toxic effects. Ideally, these should be nondestructive, to allow serial sampling of individuals.

The present section will be limited to the question of effects on populations that are a direct consequence of toxicity to individuals, whether lethal or sublethal. The problem of indirect effects have been touched upon when discussing effects at the level of a community or ecosystem (Section 4.5). The important point to emphasize at this early stage, though, is that a reduction in numbers of one species caused by a toxic chemical can have knock-on effects on populations of other species in a community or an ecosystem. The state of a population can be expressed in terms of numbers (population dynamics) or genetic composition (population genetics). This section will deal with questions relating to population dynamics; the next will deal with population genetics.

The population density of animals in the natural environment is far from constant. Seasonal fluctuations are normal, with an increase in numbers during and immediately after breeding, but a decline between this time and the next breeding occasion. In temperate climates, most breeding occurs during the warm season, when food is most readily available. Numbers fall during the cold season, when there is a shortage of food. Given these complications, it is understandable that ecologists and ecotoxicologists are particularly interested in the growth rate of populations (r). Population growth rate is defined as the population increase per unit time divided by the number of individuals in the population (see Chapter 12, Walker et al. 2000 and Gotelli 1998). Population growth rate may be positive, negative, or zero. When $r = 0$, the rates of recruitment and loss are equal, and the population density in the field represents the *carrying capacity*. In the field, population numbers are determined by, among other things, density-dependent factors such as availability of food, water, or breeding sites. Population density in the field cannot exceed, for any appreciable period, the carrying capacity. When investigating the effects of factors such as pollutants on population numbers, an important question is whether they bring population numbers below the carrying capacity. A pollutant may reduce survivorship or reproductive success, but this does not necessarily reduce population numbers below that which would normally be maintained by density-dependent factors.

In general, it can be stated that the population density of an animal depends on the balance between the rate of recruitment and the rate of mortality. In the context of ecotoxicology, the influence of pollutants upon either of these factors is of fundamental interest and importance. When a population is at or near its carrying capacity, these two factors are in balance, and the critical question about the effects of pollutants is whether they can adversely affect this balance and bring a population decline.

The population growth rate of an organism can be calculated from the Euler–Lotka equation

$$1 = 1/2n_1 \, l_1 e^{-rt}_1 + 1/2n_2 \, l_2 \, e^{-rt}_2 + 1/2 \, n_3 l_3 \, e^{-rt}_3, \text{ etc.}$$

where t_1 equals age of first breeding, t_2 equals age of second breeding, t_3 equals age of third breeding, etc.

l_1 is the probability of female surviving to the age of first breeding, l_2, the probability of reaching age of second breeding, etc.

n_1 is the number of offspring produced at the first breeding, n_2, the number of offspring produced at the second breeding, etc.

In predicting the effects of a pollutant on population growth rate, the effects of the chemical on the values of t, l, and n are of central interest. Chemical residue data and biomarker assays that provide measures of toxic effects are relevant here because they can, in concept, be used to relate the effects of a chemical upon the individual organism to a population parameter such as survivorship or fecundity (Figures 4.5 and 4.6). Examples of this are discussed in the second part of the text, including the reduction of survivorship of sparrow hawks caused by dieldrin (Chapter 5), the

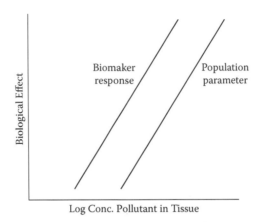

FIGURE 4.5   Schematic diagram of the relationship between biomarker response and change in population parameter.

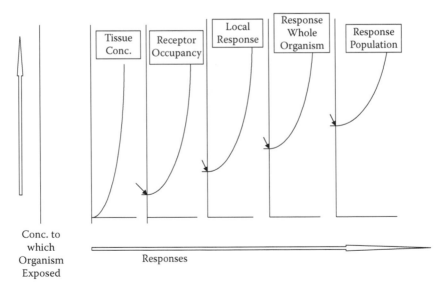

FIGURE 4.6   Schematic diagram illustrating response to pollutants at different levels of biological organization. The threshold tissue concentrations, for which no response is measurable, are indicated by arrows for each of the biological responses. These thresholds tend to increase with movement toward higher levels of biological organization.

reduction in breeding success of raptors caused by p,p'-DDE (Chapter 5), and the reduction in fecundity of dog whelks caused by TBT (Chapter 8). The measurement of responses of free-living organisms to pollutants in the field using appropriate biomarker assays can be a prelude to estimating the consequent effects upon t, l, or n from graphs of the type shown in Figure 4.5. The estimated values can then be incorporated into equations such as the one shown previously to predict the effects of the chemical on r. Such models can be tested in field studies or in mesocosm studies to establish their validity. It is also possible, in concept, to utilize biomarker responses determined in the laboratory to predict the effects at the population level of known or predicted exposures to particular chemicals in the field. In the simplest case, such an approach would assume that the relationship between the biomarker responses and the consequent change in the population parameter would be similar in the field to that in the laboratory. However, this approach would need to be rigorously tested in the field because the relationship might be substantially different between laboratory and field.

The development of models incorporating biomarker assays to predict the effects of chemicals upon parameters related to r has obvious attractions from a scientific point of view and is preferable, in theory, to the crude use of ecotoxicity data currently employed in procedures for environmental risk assessment. However, the development of this approach would involve considerable investment in research, and might prove too complex and costly to be widely employed in environmental risk assessment.

4.6.2 POPULATION GENETICS

As explained in Chapter 1, the toxicity of "natural" xenobiotics has exerted a selection pressure upon living organisms since very early in evolutionary history. There is abundant evidence of compounds produced by plants and animals that are toxic to species other than their own and which are used as chemical warfare agents (Chapter 1). Also, as we have seen, wild animals can develop resistance mechanisms to the toxic compounds produced by plants. In Australia, for example, some marsupials have developed resistance to naturally occurring toxins produced by the plants upon which they feed (see Chapter 1, Section 1.2.2).

It is only very recently that organic compounds synthesized by humans have begun to exert a selection pressure upon natural populations, with the consequent emergence of resistant strains. Pesticides are a prime example and will be the principal subject of the present section. It should be mentioned, however, that other types of biocides (e.g., antibiotics and disinfectants) can produce a similar response in microbial populations that are exposed to them.

The large-scale use of pesticides commenced in developed countries after the Second World War. In due course, they came to be more widely used in developing countries, notably for the large-scale control of important vectors of disease such as malarial mosquitoes and tsetse flies. With the continuing use of pesticides, problems of resistance began to emerge. The emergence of strains of pest species possessing genes that confer resistance was an inevitable consequence of the continuing selection pressure of the pesticides. Indeed, this can be seen to mirror the development of defense systems toward natural toxins much earlier in evolutionary history. It is highly probable that many of the defense systems now emerging in resistant strains were originally developed to combat natural toxins (e.g., P450 forms of some resistant strains of insects).

The development of resistant strains of pest species of insects has been intensively studied for sound economic reasons, and there are many good examples. For further information, see Brown (1971), Georghiou and Saito (1983), McCaffery (1998), and Oppenoorth and Welling (1976). Some examples of mechanisms of insect resistance are given in Table 4.3.

The levels of resistance (SR values) developed by insects toward insecticides in the field can be as high as several hundredfold relative to susceptible strains of the same species. Such high levels of resistance usually lead to loss of effective control of the pest. Broadly speaking, resistance mechanisms are of two kinds: (1) those depending on toxicokinetic factors such as reduced uptake, increased metabolism, or increased storage, and (2) those depending on toxicodynamic factors, principally making alterations in the site of action, which lead to decreased sensitivity to the insecticide. These two kinds of mechanism are considered in the following text.

Resistance mechanisms associated with changes in toxicokinetics are predominately cases of enhanced metabolic detoxication. With readily biodegradable insecticides such as pyrethroids and carbamates, enhanced detoxication by P450-based monooxygenase is a common resistance mechanism (see Table 4.3).

TABLE 4.3
Examples of Resistance of Insects to Insecticides

Insecticide	Species	Strain (RF)	Mechanism	Comment
Cypermethrin (cis-isomers)	*Heliothis virescens*	PEG 87 (70,000+)	P450 (major) Altered NaCh (minor)	Resistance sensitive to PBO
Cypermethrin (cis-isomers)	*H. virescens*	Field strains (85–315)	Altered NaCh (principal)	Little metabolic resistance
Parathion Methyl parathion	*H. virescens*	Field strains	Altered Ch-ase	
OPs	*Myzus persicae*	Resistant clones (85–315)	Enhanced B esterase	Multiple copies of gene
Cyclodienes	*H. virescens*	Field strains	Altered GABA	
DDT	*Musca domestica*	Many strains	Enhanced DDT-ase	DDTase inhibitors reduce resistance
Carbamates	Several	Various (< 200)	P450	PBO reduces resistance

Note: NaCh = sodium channel; RF = resistance factor, which is LD_{50} resistant strain/LD_{50} susceptible strain; GABA = gamma amino butyric acid receptor; PBO = piperonyl butoxide.

Sources: McCaffery (1998), Oppenoorth and Welling (1976), Devonshire (1991), Devonshire and Field (1991).

The existence of this type of resistance can often be established by toxicity studies with synergists. Inhibitors of P450 forms such as piperonyl butoxide (PBO) can substantially reduce the level of resistance shown by the resistant strains. OP resistance is sometimes due to enhanced esterase activity, as in the case of a series of clones of the peach potato aphid, which possess multiple copies of a gene encoding for a carboxyl esterase (see Devonshire 1991 and Section 10.2.2 of this text). A specialized example of metabolic resistance is that shown by houseflies (*Musca domestica*) and other insects to DDT. Resistant strains have elevated levels of DDT-dehydrochlorinase, a form of glutathione transferase, which is able to dehydrochlorinate some OCs. Metabolic resistance has been attributed to elevated levels of glutathione-*S*-transferases in the case of diazinon and some other OPs (Brooks 1972; Oppenoorth and Welling 1976).

Metabolic resistance may be the consequence of the appearance of a novel gene on the resistant strain, which is not present in the general population; it may also be due to the presence of multiple copies of a gene in different strains or clones as in the example of OP resistance in the peach potato aphid mentioned earlier.

Many cases of resistance are due to the existence of insensitive forms of the target site in the resistant strain (Table 4.3). The change of a single amino acid residue due to mutation can be enough to radically alter the affinity of an insecticide for its active site. For example, the replacement of a single leucine residue by phenylalanine in the sodium channel of the housefly can radically reduce the effectiveness of DDT or pyrethroids (Salgado 1999). The same substitution on the Na channel protein is also found in resistant strains of the diamondback moth (*Plutella xylostella*), the German cockroach (*Blatella germanica*), and peach potato aphid. Resistance to OPs and carbamates is sometimes due to the presence of altered forms of acetylcholinesterase in the resistant strains. Again, the substitution of a single amino acid residue by another can bring resistance. This type of resistance has been reported in several species of insects as well as in some red spider mites and cattle ticks (Oppenoorth and Welling 1976). The forms of acetylcholinesterase in resistant insects are discussed in Chapter 10, Section 10.2.5. Resistance to cyclodiene insecticides such as dieldrin and endosulfan has been related to the presence of an altered form of the GABA receptor in resistant strains of insects. Dieldrin, heptachlor epoxide, and other active forms of cyclodiene insecticides are refractory, and it may well be that metabolic resistance is unlikely to arise, leaving alteration of the target site as the main, if not the only, viable type of resistance mechanism.

In some resistant strains, both types of resistance mechanism have been shown to operate against the same insecticide. Thus, the PEG87 strain of the tobacco bud worm (*Heliothis virescens*) is resistant to pyrethroids on account of both a highly active form of cytochrome P450 and an insensitive form of the sodium channel (Table 4.3 and McCaffery 1998).

Apart from the resistance of insects to insecticides, resistance has been developed by plants to herbicides, fungi to fungicides, and rodents to rodenticides. Rodenticide resistance is discussed in Chapter 11, Section 11.2.5.

4.7   EFFECTS OF POLLUTANTS UPON COMMUNITIES AND ECOSYSTEMS—THE NATURAL WORLD AND MODEL SYSTEMS

The state of communities and ecosystems can be described according to both structural and functional parameters (see Chapter 14 in Walker et al. 2000). Functional analyses include the measurement of nutrient cycling, turnover of organic residues, energy flow, and niche metrics. Structural analyses include the assessment of species present, their population densities, and their genetic composition.

Changes in the composition of some communities and ecosystems are relatively easy to measure and to monitor using biotic indices (ecological profiling). One well-established system is the River Invertebrate Prediction and Classification System (RIVPACS) used in Britain to assess the quality of rivers (Wright 1995). Macroinvertebrate profiles have been established for normal unpolluted rivers of diverse kinds, which are used as standards. Pollution can cause departures from these profiles, for example, due to the removal of sensitive species by direct toxicity. Another system of this kind is the Invertebrate Community Index, now receiving much attention in the United States, which gives a quantitative index of the structure and function of aquatic invertebrate communities. A further system, used in some states as part of a regulatory mechanism, is the Index of Biotic Integrity (IBI) for aquatic communities (Karr 1981).

Biotic indices that are relatively simple and inexpensive to apply can be very useful for identifying environmental problems caused by pollutants. Serious effects of pollutants can cause departures from normal profiles. The problem is, however, identifying which pollutants—or which other environmental factors—are responsible for significant departures from normality. This dilemma illustrates well the importance of having both a top–down and a bottom–up approach to pollution problems in the field. Chemical analysis and biomarker assays can be used to identify chemicals responsible for adverse changes in communities detected by the use of biotic indices.

When new pesticides are developed, their effects upon soil communities are tested. Typically these tests use functional parameters (e.g., generation of CO_2 or nitrification) (Somerville and Greaves 1987). Many effects shown on soil communities are of short duration and are thought to lie within the range of normal fluctuations in soil processes.

In the quest for better methods of establishing the environmental safety (or otherwise) of chemicals, interest has grown in the use of microcosms and mesocosms—artificial systems in which the effects of chemicals on populations and communities can be tested in a controlled way, with replication of treatments. Mesocosms have been defined as "bounded and partially enclosed outdoor units that closely resemble the natural environment, especially the aquatic environment" (Crossland 1994). Microcosms are smaller and less complex multispecies systems. They are less comparable with the real world than are mesocosms. Experimental ponds and model streams are examples of mesocosms (for examples, see Caquet et al. 2000, Giddings et al. 2001, and Solomon et al. 2001). The effects of chemicals at the levels of population and community can be tested in mesocosms, although the extent to which such effects can be related to events in the natural environment is questionable. Although mesocosms have been developed by both industrial and government laboratories, there is uncertainty about the interpretation of the results obtained using them. Results coming from mesocosm tests are not yet of much use in environmental risk assessment. However, refinement of techniques used in mesocosm testing could make them more valuable for risk assessment in the future (see Section 4.8).

4.8   NEW APPROACHES TO PREDICTING ECOLOGICAL RISKS PRESENTED BY CHEMICALS

As discussed previously, current risk assessment practices do not deal with the fundamental question about possible effects at the level of population and above. In the foregoing sections, consideration was given to ways in which effects at these higher levels might be identified—and even predicted—by using data from field studies, laboratory studies, and mesocosms. At the fundamental level, the use of population models that can predict population growth rate (r) has obvious attractions. The incorporation of data from field and laboratory studies into such models should, in principal, allow reasonable predictions to be made of the effects of defined environmental levels of chemicals upon populations. The critical role of biomarker assays and/or residue data in establishing (1) the relationship between dose and toxic effect, and (2) the relationship between toxic effect and population change has already been emphasized here and in the wider literature (Peakall 1992; Peakall and Shugart 1993; Huggett et al. 1992; Fossi and Leonzio 1994; Walker et al. 1998). In the forthcoming chapters, examples will be given where this approach has already been successful in the retrospective investigation of pollution problems. The effects of TBT on dog whelks (Chapter 8, Section 8.3.4), dieldrin on sparrow hawks (Chapter 5, Section 5.3.5.1), and p,p'-DDE on peregrines and bald eagles (Chapter 5, Section 5.2.5.1) are all cases in point.

The more difficult thing is to develop models that can, with reasonable confidence, be used to predict ecological effects. A detailed discussion of ecological approaches to risk assessment lies outside the scope of the present text. For further information, readers are referred to Suter (1993); Landis, Moore, and Norton (1998); and Peakall and Fairbrother (1998). One important question, already touched upon in this account, is to what extent biomarker assays can contribute to the risk assessment of environmental chemicals. The possible use of biomarkers for the assessment of chronic pollution and in regulatory toxicology is discussed by Handy, Galloway, and Depledge (2003).

Another issue is the development and refinement of the testing protocols used in mesocosms. Mesocosms could have a more important role in environmental risk assessment if the data coming from them could be better interpreted. The use of biomarker assays to establish toxic effects and, where necessary, relate them to effects produced by chemicals in the field, might be a way forward. The issues raised in this section will be returned to in Chapter 17, after consideration of the individual examples given in Part 2.

4.9 SUMMARY

The movement of organic pollutants along food chains, and their fate in soils and sediments, is dependent upon biological as well as chemical factors. The chemical and biochemical properties of pollutants determine the rates at which they move between compartments of the environment, cross membranous barriers, or undergo chemical or biochemical degradation. Highly lipophilic compounds with high K_{ow} values are of particular concern because they tend to be immobile and persistent in soils and sediments. Where they are chemically stable and resist metabolic degradation, they tend to be biomagnified in food chains, reaching relatively high concentrations in top predators. Examples include persistent OC insecticides and PCBs, PCDDs, PCDFs, and methyl mercury.

In the field, effects of chemicals upon individuals may be measured by the use of mechanistic biomarkers. This approach has recently been strengthened by new technologies arising in the field of genomics. Free-living or deployed organisms may be sampled in order to measure responses to environmental chemicals.

In ecotoxicology, the largest concern is about effects of organic pollutants at the levels of population, community, and ecosystem. Population effects may be on numbers (population dynamics) or on gene frequencies (population genetics). Data on effects upon individuals obtained by using biomarker assays can provide vital evidence of causality at the population level and above when conducting field studies. In communities and in ecosystems there may be effects on either or both structure and function. The potential use of population models incorporating biomarker data for studying pollutant effects is discussed.

FURTHER READING

Burczynski, M. (Ed.) (2003). *An Introduction to Toxicogenomics*—Describes, with examples, the use of genomic techniques in toxicology.

Newman, M.C and Unger, M.A. (2003). *Fundamentals of Ecotoxicology*, 2nd edition—A valuable account of ecological effects of pollutants.

Peakall, D.B. (1992). *Animal Biomarkers as Pollution Indicators*—A wide-ranging account of biomarker assays in higher animals.

Peakall, D.B. and Shugart, L.R. (Eds.) (1993). *Biomarkers: Research and Application in the Assessment of Environmental Health*—Conclusions and statements of principle from a NATO Symposium.

Schuurmann, G. and Markert, B. (Eds.) (1998). *Ecotoxicology*—A multiauthor work giving a very detailed account of the environmental fate of certain chemicals.

Part 2

Major Organic Pollutants

The next eight chapters will be devoted to the ecotoxicology of groups of compounds that have caused concern on account of their real or perceived environmental effects and have been studied both in the laboratory and in the field. These are predominantly compounds produced by humans. However, a few of them, for example, methyl mercury, methyl arsenic, and polycyclic aromatic hydrocarbons (PAHs), are also naturally occurring. In this latter case, there can be difficulty in distinguishing between human and natural sources of harmful chemicals.

The compounds featured in Part 2 have been arranged into groups according to their chemical structures. In many cases, members of the same group show the same principal mechanism of toxic action. The chlorinated cyclodienes act upon gamma aminobutyric acid (GABA) receptors, the organophosphorous and carbamate insecticides are anticholinesterases, the pyrethroids act upon sodium channels, the anticoagulant rodenticides are vitamin K antagonists, and some of the PAHs are genotoxic. From an ecotoxicological point of view, there are advantages when groups of compounds can be identified that share the same mechanism of action. Here, it becomes possible to relate the toxicity of a mixture of similar compounds to a single event—for example, acetylcholinesterase inhibition or vitamin K antagonism—and so biomarker assays can be developed, which monitor the effects of combinations of chemicals (see Chapter 13).

Unfortunately, processes are not always so simple. The members of groups such as polychlorinated biphenyls (PCBs) and PAHs, for example, do not all operate through the same principal mechanism of action. Also, some individual pollutants such as p,p'-DDT or tributyl tin work through more than one mode of action.

Thus, it is often not possible to measure the combined effects of members of one group of pollutants with a single mechanistic biomarker assay. The situation becomes complex when dealing with mixtures of different types of pollutants operating through contrasting mechanisms of action, a problem that will be addressed in Part 3 of this text.

5 The Organochlorine Insecticides

5.1 BACKGROUND

The organochlorine insecticides (henceforward OCs) can be divided into three main groups, each of which will be discussed separately in the sections that follow. These are (1) DDT and related compounds, (2) the cyclodiene insecticides, and (3) isomers of hexachlorocyclohexane (HCH; Brooks 1974; Figure 5.1).

The first OC to become widely used was dichlorodiphenyl trichloroethane (DDT). Although first synthesized by Zeidler in 1874, its insecticidal properties were not discovered until 1939 by Paul Mueller of the Swiss company J.R. Geigy AG. DDT production commenced during the Second World War, in the course of which it was mainly used for the control of insects that are vectors of diseases, including malarial mosquitoes and the ectoparasites that transmit typhus (e.g., lice and fleas). DDT was used to control malaria and typhus both in military personnel and in the civilian population. After the war, it came to be used widely to control agricultural and forest pests. Following the introduction of DDT, related compounds rhothane (DDD) and methoxychlor were also marketed as insecticides, but they were only used to a very limited extent. Restrictions began to be placed on the use of DDT in the late 1960s, with the discovery of its persistence in the environment and with the growing evidence of its ability to cause harmful side effects.

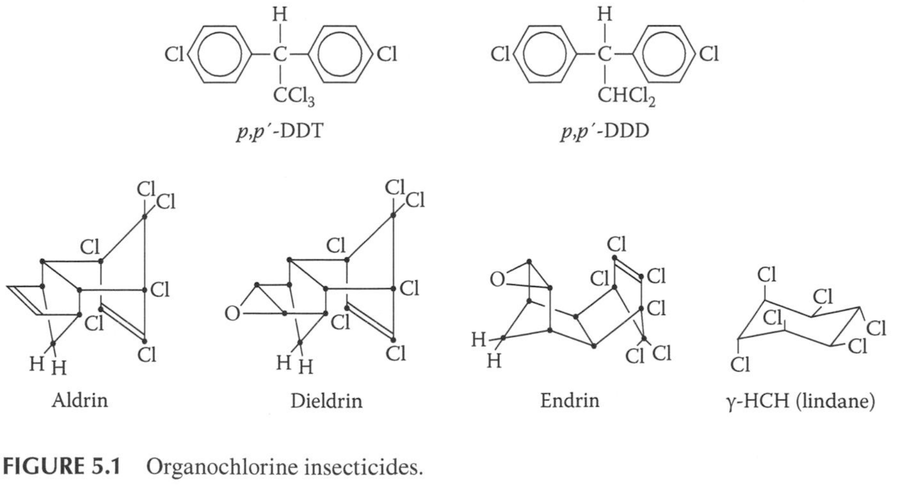

FIGURE 5.1 Organochlorine insecticides.

The cyclodiene insecticides aldrin, dieldrin, endrin, heptachlor, endosulfan, and others were introduced in the early 1950s. They were used to control a variety of pests, parasites, and, in developing countries, certain vectors of disease such as the tsetse fly. However, some of them (e.g., dieldrin) combined high toxicity to vertebrates with marked persistence and were soon found to have serious side effects in the field, notably in Western European countries where they were extensively used. During the 1960s, severe restrictions were placed on cyclodienes so that few uses remained by the 1980s.

HCH, sometimes misleadingly termed *benzene hexachloride* (BHC), exists in a number of different isomeric forms of which the gamma isomer has valuable insecticidal properties. These were discovered during the 1940s, and HCH came to be widely used as an insecticide to control crop pests and certain ectoparasites of farm animals after the Second World War. Crude technical BHC, a mixture of isomers, was the first form of HCH to be marketed. In time, it was largely replaced by a refined product called *lindane*, containing 99% or more of the insecticidal gamma isomer.

Those OCs that came to be widely marketed were stable solids that act as neurotoxins. Some OCs, or their stable metabolites, proved to have very long biological half-lives and marked persistence in the living environment. Where persistence was combined with high toxicity, as in the case of dieldrin and heptachlor epoxide (stable metabolite of heptachlor), there were sometimes serious environmental side effects. Because of these undesirable properties, no fewer than eight out of twelve chemicals or chemical groups identified by the United Nations Environment Programme (UNEP) as persistent organic pollutants (POPs or, more informally, "the dirty dozen") are OCs. These are aldrin, chlordane, DDT, dieldrin, endrin, heptachlor, mirex, and toxaphene. The intention is that high priority should be given by national and international environmental regulatory bodies to the eventual removal of POPs from the environment.

5.2 DDT [1,1,1,-TRICHLORO-2,2-BIS (P-CHLOROPHENYL) ETHANE]

5.2.1 CHEMICAL PROPERTIES

The principal insecticidal ingredient of technical DDT is *p,p′*-DDT (Table 5.1 and Figure 5.1). The composition of a typical sample of technical DDT is given in Table 5.2.

The composition of the technical insecticide varies somewhat between batches. However, the *pp′* isomer usually accounts for 70% or more of the total weight. The *o,p′* isomer is the other major constituent, accounting for some 20% of the technical product. *o,p′*-DDT is more readily degradable and less toxic to insects and vertebrates than the *p,p′* isomer. The presence of small quantities of *p,p′*-DDD deserves mention. Technical DDD has been marketed as an insecticide on its own (rhothane) and the *p,p′* isomer is a reductive metabolite of *p,p′*-DDT.

p,p′-DDT is a stable white crystalline solid with a melting point of 108°C. It has very low solubility in water and is highly lipophilic (log K_{ow} = 6.36); thus, there is a high potential for bioconcentration and bioaccumulation. It has a low vapor pressure,

TABLE 5.1
Chemical Properties of Organochlorine Insecticides

Chemical	Description	Water Sol. (mg/L)	log K_{OW}	Vapor Pressure mm Hg (at 25°C)
p,p'-DDT	Solid m.p. 108°C	<0.1	6.36	1.9×10^{-7} (20°C)
p,p'-DDD	Solid m.p. 109°C	<0.1		
Aldrin	Solid m.p. 104°C	<0.1		6.5×10^{-5}
Dieldrin	Solid m.p. 178°C	0.2	5.48	3.2×10^{-6}
Heptachlor	Solid m.p. 93°C	0.056	5.44	4×10^{-4}
Endrin	Solid m.p. 226–230°C	0.2	5.34	2×10^{-7}
Endosulfan	Solid m.p. 79–100°C	0.06–0.15		1×10^{-5}
Gamma HCH	Solid m.p. 112°C	7	3.78	9.4×10^{-6}

Note: m.p. = melting point.

TABLE 5.2
Composition of a Typical Sample of Technical DDT

Compound	Percentage of Technical Product
p,p'-DDT	72
o,p'-DDT	20
p,p'-DDD	3
o,o'-DDT	0.5
Other	4.5

and is consequently relatively slow to sublimate when applied to surfaces (e.g., leaves, walls, or surface waters).

p,p'-DDT is not very chemically reactive. However, one important chemical reaction is dehydrochlorination to form p,p'-DDE, which takes place in the presence of KOH, NaOH, and other strong alkalis. Dehydrochlorination is also a very important biotransformation and will be discussed further in Section 5.2.2. p,p'-DDT undergoes reductive dechlorination by reduced iron porphrins. In the presence of strong radiation, it undergoes slow photochemical decomposition.

5.2.2 Metabolism of DDT

p,p'-DDT is rather stable biochemically as well as chemically. Thus, it is markedly persistent in many species on account of its slow biotransformation. Metabolism of p,p'-DDT is complex, and there is still some controversy about its specifics. The most important metabolic pathways are shown in Figure 5.2.

A major route of biotransformation in animals is dehydrochlorination to the stable lipophilic and highly persistent metabolite p,p'-DDE. p,p'-DDE is far more persistent in animals than is p,p'-DDT. Therefore, dehydrochlorination does not promote excretion, although it usually results in detoxication because the metabolite is less acutely toxic than the parent compound. However, as will be seen, p,p'-DDE causes certain sublethal effects. Such metabolic conversion of parent compounds to persistent lipophilic metabolites also occurs with other OCs (see Section 5.3.2), and they may be regarded as a malfunction of detoxication systems that originally evolved to promote the elimination of naturally occurring lipophilic xenobiotics through the rapid excretion of their water-soluble metabolites and conjugates (Chapter 1). The dehydrochlorination of p,p'-DDT is catalyzed by a form of glutathione-S-transferase, and involves the formation of a glutathione conjugate as an intermediate.

Under anaerobic conditions, p,p'-DDT is converted to p,p'-DDD by reductive dechlorination, a biotransformation that occurs postmortem in vertebrate tissues such as liver and muscle and in certain anaerobic microorganisms (Walker and Jefferies 1978). Reductive dechlorination is carried out by reduced iron porphyrins. It is carried out by cytochrome P450 of vertebrate liver microsomes when supplied with NADPH in the absence of oxygen (Walker 1969; Walker and Jefferies 1978). Reductive dechlorination by hepatic microsomal cytochrome P450 can account for the relatively rapid conversion of p,p'-DDT to p,p'-DDD in avian liver immediately after death, and mirrors the reductive dechlorination of other organochlorine substrates (e.g., CCl_4 and halothane) under anaerobic conditions. It is uncertain to what extent, if at all, the reductive dechlorination of DDT occurs in vivo in vertebrates (Walker 1974).

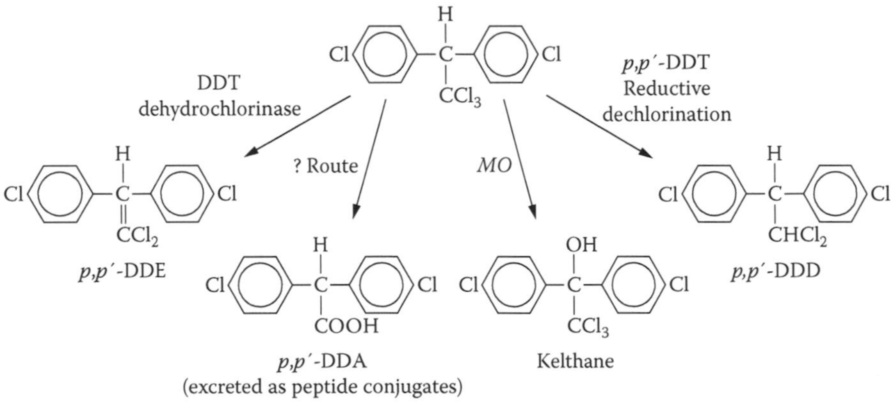

FIGURE 5.2 Metabolism of p,p'-DDT.

A major, albeit slow, route of detoxication in animals is conversion to the water-soluble acid p,p'-DDA, which is excreted unchanged, or as a conjugate. In one study, the major urinary metabolites of p,p'-DDT in two rodent species were p,p'-DDA-glycine, p,p'-DDA-alanine, and p,p'-DDA-glucuronic acid (Gingell 1976). The route by which p,p'-DDA is formed remains uncertain. Early studies suggested that conversion might be via p,p'-DDD, but the later observation that this is a postmortem process has cast some doubt on these findings. Some or all of the p,p'-DDD found in livers in these studies would have been generated postmortem because analysis was carried out after a period of storage. Another possibility is that this process, similar to dehydrochlorination, takes place via glutathione conjugation. After conjugation and consequent loss of HCl, the DDE moiety, which remains bound to glutathione, may undergo hydrolysis, leading eventually to deconjugation and formation of p,p'-DDA. A mechanism of this type has been proposed for the conversion of dichloromethane to HCHO (Schwarzenbach et al. 1993, p. 514; Chapter 2, Figure 2.15 of this book).

One other biotransformation deserving mention is the oxidation of p,p'-DDT to kelthane, a molecule that has been used as an acaricide. This biotransformation occurs in certain DDT-resistant arthropods, but does not appear to be important in vertebrates.

Unchanged p,p'-DDT tends to be lost only very slowly by land vertebrates. There can, however, be a certain amount of excretion by females into milk or across the placenta into the developing embryo (mammals) or into eggs (birds, reptiles, and insects).

5.2.3 Environmental Fate of DDT

In discussing the environmental fate of technical DDT, the main issue is the persistence of p,p'-DDT and its stable metabolites, although it should be born in mind that certain other compounds—notably, o,p'-DDT and p,p'-DDD—also occur in the technical material and are released into the environment when it is used. The o,p' isomer of DDT is neither very persistent nor very acutely toxic; it does, however, have estrogenic properties (see Section 5.2.4). A factor favoring more rapid metabolism of the o,p' isomer compared to the p,p' isomer is the presence, on one of the benzene rings, of an unchlorinated para position, which is available for oxidative attack. p,p'-DDD, the other major impurity of technical DDT, is the main component of technical DDD, which has been used as an insecticide in its own right (rhothane). p,p'-DDD is also generated in the environment as a metabolite of p,p'-DDT. In practice, the most abundant and widespread residues of DDT found in the environment have been p,p'-DDE, p,p'-DDT, and p,p'-DDD.

When DDT was widely used, it was released into the environment in a number of different ways. The spraying of crops, and the spraying of water surfaces and land to control insect vectors of diseases, were major sources of environmental contamination. Waterways were sometimes contaminated with effluents from factories where DDT was used. Sheep-dips containing DDT were discharged into water courses. Thus, it is not surprising that DDT residues became so widespread in the years after the war. It should also be remembered that, because of their stability, DDT residues can be circulated by air masses and ocean currents to reach remote parts of the globe. Very low levels have been detected even in Antarctic snow!

TABLE 5.3
Half-Lives of *p,p'*-DDT and Related Compounds

Compound	Material/Organism	t_{50} (Years)	Compound	Material/Organism	t_{50} (Days)
p,p'-DDT	Soil	2.8	*p,p'*-DDT	Feral pigeon (*Columba livia*)	28
p,p'-DDD	Soil	10+ (British soils)	*p,p'*-DDD	Feral pigeon	24
			p,p'-DDE	Feral pigeon	250
			p,p'-DDT	Bengalese finch (*Lonchura striata*)	10
			p,p'-DDT	Hens (*Gallus domesticus*)	36–56 (in fat)
			p,p'-DDT	Rat	57–107
			p,p'-DDT	Rhesus monkey	32 and 1520

Source: Data from Edwards (1973) and Moriarty (1975).

Some data on the half-lives of these three compounds are given in Table 5.3. All of them are highly persistent in soils, with half-lives running to years once they become adsorbed by soil colloids (especially organic matter—see Chapter 3). The degree of persistence varies considerably between soils, depending on soil type and temperature. The longest half-lives have been found in temperate soils with high levels of organic matter. (See, for example, Cooke and Stringer 1982.) Of particular significance is the very long half-lives for *p,p'*-DDE in terrestrial animals, approaching 1 year in some species, and greatly exceeding the comparable values for the other two compounds. This appears to be the main reason for the existence of much higher levels of *p,p'*-DDE than of the other two compounds in food chains even when technical DDT was widely used. Following the wide-ranging bans on the use of DDT in the 1960s and 1970s, *p,p'*-DDT residues have fallen to very low levels in biota, although significant residues of *p,p'*-DDE are still found, for example, in terrestrial food chains such as earthworms → thrushes → sparrow hawks in Britain (Newton 1986) and in aquatic food chains.

A nationwide investigation of OC residues in bird tissues and bird eggs was conducted in Great Britain in the early 1960s, a period during which DDT was widely used (Moore and Walker 1964). The most abundant residue was *p,p'*-DDE; levels of *p,p'*-DDT and *p,p'*-DDD were considerably lower. Levels in depot fat were some 10–30-fold higher than in tissues such as liver or muscle. The magnitude of residues was related to position in the food chain, with low levels in omnivores and herbivores and the highest levels in predators at the top of both terrestrial and aquatic food chains (see Walker et al. 1996, Chapter 4). Similar results were obtained with both bird tissues and eggs. The highest *p,p'*-DDE levels (9–12 ppm) were found in the eggs of sparrowhawks, which are bird eaters, and in herons (*Ardea cinerea*), which are fish eaters. Thus, when considering the fate of technical DDT in food chains generally, *p,p'*-DDE was found to be more stable and persistent (i.e., refractory) than either p,p'-DDT or p,p'-DDD and underwent strong biomagnification with transfer along food chains.

Studies on the marine ecosystem of the Farne Islands in 1962–1964 showed that p,p'-DDE reached concentrations over 1000-fold higher in fish-eating birds at the top of the food chain than those present in macrophytes at the bottom of the food chain (Figure 5.3). Fish-eating shag (*Phalocrocorax aristotelis*) contained residues some 50-fold higher than those in its main prey species, the sand eel (*Ammodytes lanceolatus*). The sand eel was evidently the principal source of p,p'-DDE for the shag, so there had apparently been very efficient bioaccumulation over a considerable period (Robinson et al. 1967a). However, as explained in Chapter 4, it should be borne in mind that the biomagnification of highly lipophilic chemicals along the entire aquatic food chain is a consequence not only of bioaccumulation through the different stages of the food chain, but also of bioconcentration of chemicals present in ambient water. For example, aquatic invertebrates of lower trophic levels acquire much of their residue burden of lipophilic compounds such as p,p'-DDE by direct uptake from ambient water (see Chapter 4).

In a study of marine food chains in the Pacific Ocean during the 1980s, bioconcentration factors of the order of 10,000-fold for total DDT residues (very largely p,p'-DDE) were reported when comparing levels in zooplankton with those in ambient water (Tanabe and Tatsukawa 1992). Striking levels of biomagnification were evident in the higher levels of the food chain. Thus, in comparison with residues in zooplankton, mycotophid (*Diaphus suborbitalis*) and squid (*Todarodes pacificus*) contained residues some tenfold greater, and striped dolphin (*Stenella coerolea alba*), several 100-fold greater. Total DDT residues of <50 mg/kg wet weight of blubber were reported for the striped dolphin. In a later study conducted in the Mediterranean in

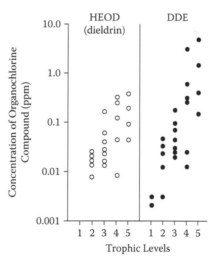

FIGURE 5.3 Organochlorine insecticides in the Farne Island ecosystem. From Walker et al. (2000). Trophic levels: (1) serrated wrack, oar weed; (2) sea urchin, mussel, limpet; (3) lobster, shore crab, herring, sand eel; (4) cod, whiting, shag, eider duck, herring gull; (5) cormorant, gannet, grey seal.

1990, total DDT residues of <230 mg/kg were reported for this species (Kannan et al. 1993; O'Shea and Aguilar 2001), producing further evidence of the continuing high level of pollution in this inland sea. In a study conducted in 1995, total DDT residues (very largely p,p'-DDE) were determined in marine organisms from the Barents Sea, near Svalbard in the Arctic (Borga, Gabrielsen, and Skaare 2001). Again, marked biomagnification was evident in residues, expressed as micrograms per gram in lipid with movement up the food chain, as indicated in the following table. Whole samples of fish and excised livers of birds were submitted for analysis.

Trophic level	Organisms	Biomagnification Factors for Total DDT (cf. trophic levels 2/3)
2/3	Crustaceans	1
4	Cod (*Gadus morhua*)	× 2
	Guillemots (*Uria lomvia* and *Cepphus grylle*)	× 47
	Kittiwake (*Rissa tridactyla*)	× 82
5	Glaucous gull (*Larus hyperboreus*)	× 2300

Once again, fish-eating species—the two guillemots and the kittiwake—in trophic level 4 show considerable biomagnification of residue in comparison with the invertebrates of levels 2 and 3. Most strikingly, though, the ultimate predator in trophic level 5, the glaucous gull, shows a biomagnification factor of 2300! It is suggested that this may be related to the fact that these predators feed upon fauna associated with ice. The mean value for total DDT levels in the livers of glaucous gulls, expressed as micrograms per gram in lipid, was 42. In a later study of organochlorine residues in arctic seabirds (Borga et al. 2007), p,p'-DDE remained a dominant residue in birds occupying positions in higher trophic levels, these species including little auk (*Alle alle*) as well as the two species of guillemot and kittiwake mentioned here. The differential biomagnification of organochlorine residues was examined in these species and related to factors such as diet, habitat, and metabolic capacity.

Finally, an investigation of total DDT levels in seal (*Phoca sibirica*) from Lake Baikal, Russia (the largest lake in the world), during the 1990s showed substantial levels with evidence of strong biomagnification in this aquatic food chain (Lebedev et al. 1998).

p,p'-DDE can also undergo bioaccumulation in terrestrial food chains. Studies with earthworms and slugs indicate that there can be a bioconcentration of total DDT residues (p,p'-DDT + p,p'-DDE + p,p'-DDD) relative to soil levels of one- to fourfold by earthworms, and above this by slugs (Bailey et al. 1974; Edwards 1973). When DDT was still widely used in orchards in Britain, blackbirds (*Turdus merula*) and song thrushes (*Turdus philomelis*) that had been found dead contained very high levels of DDT residues in comparison with those in the earthworms they ate. Some results from a study on one orchard sprayed with DDT are given in Table 5.4. Interpretation of field data involving such small numbers of individual specimens needs to be done with caution. However, the principal source of DDT residues for the two Turdus species appears to have been earthworms and other invertebrates (including slugs and snails). The birds found dead were probably poisoned by DDT

TABLE 5.4

DDT Residues from Samples Taken in Orchard near Norwich (1971–1972)

Species	Sampling Procedure	*p,p'*-DDT	*p,p'*-DDE	*p,p'*-DDD	Total DDT Residues/Sample
Soil	Random	1.2–3.5	0.5–1.1	0.22–0.72	2.1–5.3
Earthworm	Random	1.1–6.8	1.4–4.2	0.46–5.5	3.9–11.5
Blackbird	2 birds found dead	0/6.8	130/180	58/195	195/249
Blackbird	2 birds shot	0/2.4	24/33	14/30	49/53
Song thrush	2 birds found dead	0	164/192	81/128	273/292
Song thrush	1 bird shot	0	30	42	72

Source: From Bailey et al. (1974).

residues, and the levels found in them were some 20-fold higher than in the earthworms on which they were feeding, suggesting marked bioaccumulation. Birds that were shot also contained DDT residue levels well above those recorded for earthworms. It should be added that the relatively high levels of *p,p'*-DDE found in sparrowhawks in the 1980s from areas where DDT was once used may be due, in part, to transfer from soil sinks via soil invertebrates to the insectivorous birds upon which these raptors feed (Newton 1986). The virtual absence of *p,p'*-DDT coupled with the high levels of *p,p'*-DDD in liver and certain other tissues (but not fat) sampled from the dead birds strongly suggests postmortem conversion of the former to the latter by reductive dechlorination. By contrast, relatively high levels of *p,p'*-DDT were present in the earthworms upon which the birds were feeding and in eggs of both species sampled in the same area. It has been shown that little or no conversion of *p,p'*-DDT to *p,p'*-DDD occurs in birds' eggs, until embryo development commences (Walker and Jefferies 1978); thus the relative levels of the two compounds in eggs should reflect what is present in the birds' food, and in the tissues of the birds during life.

To summarize, *p,p'*-DDE is widespread in the natural environment—extending to polar ecosystems. Because of its lipophilicity and resistance to chemical and metabolic attack, it can undergo strong bioaccumulation to reach particularly high levels at the top of both aquatic and terrestrial food chains; this is true to a lesser extent with *p,p'*-DDT and *p,p'*-DDD, which are more readily biodegradable than *p,p'*-DDE. Although DDT has been banned in most countries for many years, residues of *p,p'*-DDE are still widely distributed through terrestrial and aquatic ecosystems, reflecting its environmental stability. The loss of *p,p'*-DDE from contaminated soils and sediments is so slow that they act as sinks, ensuring that there will be contamination of terrestrial and aquatic ecosystems for many decades to come.

5.2.4 TOXICITY OF DDT

The acute toxicity of *p,p'*-DDT to both vertebrates and invertebrates is attributed mainly to its action upon axonal Na^+ channels, which are voltage dependent (see Figure 5.4; Eldefrawi and Eldefrawi 1990). The molecule binds reversibly to a site

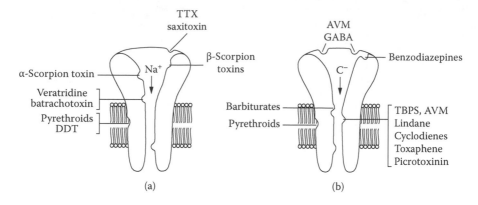

FIGURE 5.4   Sites of action of organochlorine insecticides: (a) sodium channel, (b) GABA receptor. (From Eldefrawi and Eldefrawi 1990. With permission.)

on the channel, thereby altering its function. Normally, when an Na^+ current is generated, the signal is rapidly terminated by the closure of the sodium channel. In DDT-poisoned nerves, the closure of the channel is delayed, an event that can cause disruption of the regulation of action potential and can lead to repetitive discharge. p,p'-DDT can also act upon the K^+ channel, which is concerned with the repolarization of the axonal membrane after passage of the action potential.

Apart from the action upon Na^+ channels, p,p'-DDT and its metabolites can have certain other toxic effects. It has been reported that p,p'-DDT can inhibit certain ATPases (see EHC 83). In fish, the inhibition of ATPases can affect osmoregulation.

The ability of p,p'-DDE to cause thinning of avian eggshells, even at very low concentrations in some species, has been a matter of considerable interest (see Ratcliffe 1967, 1993; Peakall 1993). The mechanism by which this is accomplished is still not fully established. It seems clear that the basic problem is the failure of Ca^{2+} transport across the wall of the eggshell gland (Lundholm 1997). Levels of p,p'-DDE that cause eggshell thinning in birds do not cause any reduction in plasma calcium levels. They do, however, bring an increase in concentration in the mucosa and a reduction in concentration in the lumen, which contains the developing egg. Thus, there appears to be a failure of the transport system into the lumen. It has been demonstrated that p,p'-DDE can inhibit the Ca^{2+} ATPase of the avian shell gland (Lundholm 1987); this has been proposed as a mechanism for the severe eggshell thinning caused by this compound in certain species of birds, including the American kestrel (*Falco sparverius*), sparrow hawk (*Accipiter nisus*), peregrine falcon (*Falco peregrinus*), and Gannet (*Sula bassana*; Wiemeyer and Porter 1970; Peakall 1993). However, there is also evidence that p,p'-DDE can affect prostaglandin levels in the eggshell gland, and this may be a contributory factor in eggshell thinning (Lundholm 1997). Dietary levels as low as 3 ppm have been shown to cause shell thinning in the American kestrel (Peakall et al. 1973; Wiemeyer and Porter 1970). The implications of this finding will be discussed in Section 5.2.5.

Finally, there is evidence that constituents of technical DDT can have a feminizing effect on avian embryos (for further discussion, see Chapter 15, Section 15.6). Of these, o,p'-DDT has been shown to have estrogenic activity in birds (Bitman et al. 1978; Holm et al. 2006). In a study with the California gull, o,p'-DDT was found to be a considerably more potent estrogen than p,p'-DDE (Fry and Toone 1981). It should be remembered that the foregoing effects involve interaction between an OC compound and protein targets that are located in lipophilic membrane. Relative to their concentrations in tissue fluids and blood, p,p'-DDT, p,p'-DDE, and other lipophilic OC compounds can reach very high concentrations at or near such hydrophobic domains.

From an ecotoxicological point of view, it has often been suspected that sublethal effects, such as those described here, can be more important than lethal ones. Both p,p'-DDT and p,p'-DDD are persistent neurotoxins, and may very well have caused behavioral effects in the field. This issue was not resolved when DDT was widely used, and remains a matter for speculation. More is known, however, about eggshell thinning caused by p,p'-DDE and its effects upon reproduction, which will be discussed in Section 5.2.5.1.

It can be seen from Table 5.5 that p,p'-DDT is toxic to a wide range of vertebrates and invertebrates. That said, it is considerably less toxic to most species than is dieldrin, heptachlor, or endrin. If applied topically, it is 180-fold more toxic to the housefly than to the rat, and appears to be reasonably selective between insects and mammals. Some aquatic invertebrates are very sensitive to p,p'-DDT, but there is a very wide range of susceptibility among freshwater invertebrates. The rather wide range of values for mammals is partly the consequence of the use of different vehicles. Oil solutions tend to be appreciably more toxic than solid formulations, presumably due to more rapid and/or efficient absorption from the gut. Thus, the lower part of the range (i.e., the values indicating greatest toxicity) should be more representative of the toxicity that will be shown when the compound is passed through the food chain, when it is dissolved in the fatty tissues of the prey species. In general, p,p'-DDE is less toxic than p,p'-DDT, especially in insects where dehydrochlorination of p,p'-DDT represents a detoxication mechanism despite the greater persistence of the metabolite compared to the parent compound.

TABLE 5.5
Ecotoxicity of p,p'-DDT and Related Compounds

Compound	Organism	Test	Median Lethal Dose or Concentration
p,p'-DDT	Marine invertebrates	LC_{50}	0.45–2.4 µg/L (48 or 96 h)
p,p'-DDE	Marine invertebrate (brown shrimp)	LC_{50}	28 µg/L (48 or 96 h)
p,p'-DDT	Freshwater invertebrates	LC_{50}	0.4–1800 µg/L (48 or 96 h)
p,p'-DDT	Fish (smaller fish most susceptible)	LC_{50} (96 h)	1.5–5.6 µg/L
p,p'-DDT	Mammals	LD_{50} (acute oral)	100–2500 mg/kg
p,p'-DDE	Rodents	LD_{50} (acute oral)	880–1240 mg/kg
p,p'-DDD	Rat	LD_{50} (acute oral)	400–3400 mg/kg
p,p'-DDT	Birds	LD_{50} (acute oral)	>500 mg/kg

Source: Data from ETC 9, ETC 83, and Edson et al. (1966).

5.2.5 ECOLOGICAL EFFECTS OF DDT

As explained in Section 5.2.3, p,p'-DDE is much more persistent in food chains than either p,p'-DDT or p,p'-DDD, and during the 1960s when DDT was still extensively used, it was often the most abundant of the three compounds in birds and mammals found or sampled in the field. Since the widespread banning of DDT, very little of the pesticides has been released into the environment, and p,p'-DDE is by far the most abundant DDT residue found in biota. While discussing the ecological effects of DDT and related compounds, effects on population numbers will be considered before those on population genetics (gene frequencies).

5.2.5.1 Effects on Population Numbers

An early indication of the damage that OC compounds can cause in the higher levels of the food chain came with a study on the East Lansing Campus of Michigan State University in 1961 and 1962 (Bernard 1966). Over several years, leading up to and including 1962, American robins (*Turdus migratorius*) and several other species of birds were virtually eliminated from a 75 ha study area in the spring, following the application of high levels of DDT (<25 lb/acre). The purpose of the exercise was to control Dutch elm disease. Subsequent investigation established that all American robins dying in this way contained more than 50 ppm of total DDT in the brain. Comparison with experimentally poisoned birds led to the conclusion that these levels were high enough to have caused lethal DDT poisoning. In these early days before the development of gas chromatography, it was difficult to distinguish between the different compounds derived from DDT, and a limitation of the study was that deductions were based on estimates of total DDT. As has been pointed out, the various impurities and metabolites arising from the technical material differ considerably in their toxicity, so an estimate of total DDT residues is only of limited usefulness when attempting to establish the cause of death. However, with the benefit of hindsight, it seems clear that many birds did die of DDT poisoning following these very high levels of application and that transfer through earthworms and other invertebrates made a major contribution to the level of residues in the birds. It also appeared that the effects were localized, seasonal, and transitory.

In another widely quoted earlier study, Hunt and Bischoff (1960) reported the decline of Western Grebe (*Aechmophorus occidentalis*) populations on Clear Lake, California, following the application of rhothane (DDD) over several years. There was evidence of a progressive buildup of p,p'-DDD residues in sediments over the period, and an analytical study of biota from the lake yielded the results shown in Table 5.6.

The levels of DDD found in dying or dead grebes were high enough to suggest acute lethal poisoning. As with the study on the American robin, there was strong evidence for the local decline of a species occupying a high trophic level of an ecosystem, a decline consequent upon the toxicity of a persistent OC compound obtained via its food. At first there was a tendency to explain the very large differences in DDD concentrations between the top and the bottom of the food chain in terms of progressive bioaccumulation with movement up the chain. On closer examination, however, much of this increase is explicable on the grounds of strong

TABLE 5.6
Results from an Analytical Study of Residues in Biota Sampled from Clear Lake, California

Species/Sample	*p,p'*-DDD Concentration (ppm wet weight)
Lake water	0.02
Plankton	5.0
Nonpredatory fish (fat)	40–1000
Predatory fish (fat)	80–2500
Predatory fish (flesh)	1–200
Western grebe (fat)	1600

Source: From Hunt and Bischoff (1960).

bioconcentration due to direct uptake from water, both by plankton and fish. The difference in concentration in body fat between predatory and nonpredatory fish is not very large, so there is no clear evidence of strong bioaccumulation of *p,p'*-DDD by the predatory fish from its food. A comparison of bioconcentration factors from water for nonpredatory and predatory fish would be necessary to establish how much bioaccumulation, if any, was achieved by the latter. The grebes, however, which were not expected to take up substantial quantities of insecticide directly from water, contained a mean level of *p,p'*-DDD in their depot fat well above the top of the range for nonpredatory fish and not much below the highest value found in predatory fish, suggesting some bioaccumulation in the last step of the food chain. Indeed, it seems very probable that the birds died from DDD poisoning while the tissue levels of the insecticide were still increasing (i.e., some time before a steady state was reached), so that the level of bioaccumulation found was below what might have been achieved at a lower level of exposure.

The two examples just given are of localized effects associated with the acute toxicity of DDT and DDD to organisms in higher trophic levels. A more wide-ranging toxic effect associated with population decline was eggshell thinning caused by the relatively high levels of *p,p'*-DDE in some predatory birds (see Table 5.7).

In North America, during the period late 1940s to late 1970s, the decline of several species of birds of prey was associated with eggshell thinning caused by *p,p'*-DDE. Peregrine populations declined or were extirpated when eggshell thinning of 18–25% occurred. This degree of eggshell thinning was associated with DDE residues in excess of 10 ppm (wet weight) in the eggs (Peakall 1993). The bald eagle showed a marked decline in many areas of North America, first reported in Florida in 1946 (Broley 1958). Shell thinning of 15% was associated with residues of 16 ppm *p,p'*-DDE in eggs of this species (Wiemeyer et al. 1993), and at the time of the initial decline (1946–1957), shell thinning of 15–19% was associated with diminished breeding success. In field studies carried out during 1969–1984, the picture was complicated by the fact that, although breeding success was negatively

TABLE 5.7

Population Declines Associated with Eggshell Thinning Caused by p,p'-DDE

Species/Area	Years	p,p'-DDE Residues (ppm)	Degree of Thinning (%)	Population Effect
Peregrine (New Jersey, Mass., S. California, Belgium)	1950–1973	Mostly > 20 in eggs	18–25	Extirpated
Bald eagle (United States)	1970	Mean 18 in carcasses		Declining
Gannet (Bonaventura Island, Quebec)	1969	19–30 in eggs	17–20	Declining; low reproductive success

Source: Data from Peakall (1993), Elliott et al. (1998), and Kaiser et al. (1980).

correlated with p,p'-DDE levels in the eggs, the correlation between breeding success and eggshell thinning was poor. A later study of bald eagles in the region of the Great Lakes (1987–1992) involved the measurement of p,p'-DDE and total PCB levels in the blood of nestlings (Bowerman et al. 2003). Geometric means of concentrations of these two parameters were negatively correlated with productivity and success rates of nesting within nine populations, the correlations being stronger for p,p'-DDE than for total PCBs.

Thus, as with studies on the double-crested cormorant in the Great Lakes (see Chapter 16 in Walker et al. 2006), there is evidence of a continuing (although reduced) effect of p,p'-DDE on reproductive success even after environmental levels had fallen and eggshell thinning was much less. This raises the possibility that p,p'-DDE may have had toxic effects other than eggshell thinning on these species (Nisbet 1989). There is the further complication that other OCs such as PCBs, dieldrin, and heptachlor epoxide were present in the same samples and may have had toxic effects.

Clearly, caution is needed when attempting to relate levels of eggshell thinning caused by p,p'-DDE to population effects. In Britain, eggshell thinning occurred in the peregrine falcon and sparrow hawk from 1946–1947 and was related to the presence of p,p'-DDE, but population declines did not occur until some 8 years later (Ratcliffe 1970, 1993). These declines coincided with the introduction of cyclodiene insecticides and will be discussed in Section 5.3.3. It is important to emphasize, however, that levels of DDT were higher and cyclodienes lower in North America in comparison with Britain and other Western European countries. The weight of evidence suggests that declines of the bald eagle, the peregrine, and the osprey (*Pandion haliaetus*) in the United States referred to earlier were mainly due to the effects of DDT, and especially due to eggshell thinning caused by p,p'-DDE.

In another study, on Bonaventura Island, Quebec, Canada, during the 1960s and early 1970s, gannets (*Sula bassanus)* showed a sharp population decline that was associated with poor breeding success. In 1969, there was clear evidence of severe shell thinning caused by p,p'-DDE residues of 19–30 ppm in eggs. Subsequently, pollution of the St. Lawrence river by DDT was reduced. The p,p'-DDE levels in the gannets fell, and by the mid- to late-1970s, shells became thicker, reproductive success increased, and the population recovered (Elliott et al. 1988). Taken overall, these findings illustrate very clearly the ecological risks associated with the wide dispersal of a highly persistent pollutant that can have sublethal effects.

Evidence for effects of p,p'-DDE on eggshell thickness and productivity has also come from studies on Golden eagles (*Aquila chrysaetos*) conducted up to the late 1990s in Western Norway (Nygard and Gjershaug 2001). Their evidence suggests that this species may be particularly sensitive to DDE-induced eggshell thinning, and the results are broadly comparable to those from earlier work on this species conducted in Scotland (Ratcliffe 1970).

Before leaving the question of effects of DDT and its derivatives upon populations, brief mention should be made of indirect effects. Sometimes insect populations increase in size because an insecticide reduces the numbers of a predator or parasite that keeps the insects' numbers in check. Such an effect was found in a controlled experiment where DDT was applied to a brassica crop infested with caterpillars of the cabbage white butterfly (*Pieris brassicae*; see Dempster in Moriarty 1975). Field applications of DDT severely reduced the population of carabid beetles, which prey upon and control the numbers of *Pieris brassicae* larvae. The infestation of the crop was initially controlled by DDT but, as the residues declined on the crop, the caterpillars eventually returned to reach much higher numbers than on control plots untreated by DDT, where natural predators maintained control of the pest. Thus the long-term indirect effect of DDT was to increase the numbers of the pest species. When DDT was used as an orchard spray, it was implicated, together with certain other insecticides, in the triggering of an epidemic of red spider mites (Mellanby 1967). The insecticides successfully controlled the capsid bugs (e.g., *Blepharidapterus angulatus*), which normally keep down the numbers of red spider mites, and this led to a population explosion of the latter—and to a new pest problem! These examples illustrate well a fundamental difference between ecotoxicology and normal medical toxicology. The well-established test procedures of the latter may tell us very little about what will happen when toxic chemicals are released into ecosystems.

5.2.5.2   Effects on Population Genetics (Gene Frequencies)

DDT had not been in general use for very long before there were reports of DDT resistance in insect populations that were being controlled by the insecticide. Examples included resistant strains of houseflies (*Musca domestica*) and mosquitoes (Georghiou and Saito 1983; Oppenoorth and Welling 1976). For further discussion, see Brown (1971). Two contrasting resistance mechanisms have been found in resistant strains of housefly. The first is metabolic resistance, usually due to enhanced levels of DDT dehydrochlorinase. In one resistant strain of housefly, enhanced monooxygenase activity was found, which might cause increased rates of detoxication to kelthane and other oxidative metabolites (Oppenoorth and Welling 1976). By contrast, some houseflies showed "knockdown" resistance ("kdr" or "super kdr"), due to nerve insensitivity. It now seems clear that this is the consequence of the appearance of a mutant form (or forms) of the Na^+ channel in resistant strains, which is insensitive to DDT (Salgado 1999). As explained earlier (Section 5.2.4), axonal Na⁺ channels represent the normal target for *p,p'*-DDT (and, incidentally, for pyrethroid insecticides). Interestingly, insects developing kdr or super kdr to DDT usually show cross-resistance to pyrethroids. In both cases, the resistant insects have insensitive forms of the target site. Knockdown resistance has also been reported in a number of other species exposed to DDT or pyrethroids or both, including *Heliothis virescens, Plutella xylostella, Blatella germanica, Anopheles gambiae*, and *Myzus persicae*. The appearance of resistant strains such as these can give valuable retrospective evidence of the environmental impact of pollutants. Assays for resistance mechanisms and the genes that operate them are valuable tools in ecotoxicology.

5.3   THE CYCLODIENE INSECTICIDES

Insecticides belonging to this group are derivatives of hexachlorocyclopentadiene, synthesized by the Diels–Alder reaction (see Brooks 1974). They did not come into use until the early 1950s and not to any important extent in Europe or North America before the mid-1950s. Thus, they did not begin to produce environmental side effects until at least 6 years after the onset of DDE-induced eggshell thinning in sparrow hawks, peregrines, and bald eagles, as described in Section 5.2.5.1. The cyclodienes include dieldrin, aldrin, heptachlor, endrin, chlordane, endosulfan, telodrin, chlordecone, and mirex. Some of them have only been used to a limited extent, and the following account will be restricted to dieldrin, aldrin, heptachlor, endrin, telodrin, and chlordane, compounds that have caused most concern about environmental side effects.

5.3.1   CHEMICAL PROPERTIES

The cyclodienes are stable solids having low water solubility and marked lipophilicity (Table 5.1), and their active ingredients have cage structures (Figure 5.5). The technical insecticides aldrin and dieldrin contain, respectively, the molecules HHDN and HEOD as their active ingredients. HHDN and HEOD are abbreviations of their formal chemical names (structures given in Figure 5.5). Here, the term "aldrin" will be used synonymously with HHDN, and the term "dieldrin" will be used synonymously with HEOD, unless otherwise indicated, thus following the common practice in the literature. Similarly, the common names of the other cyclodienes (e.g., heptachlor and endrin) will be used to refer to the chemical structures of the principal insecticidal ingredients of the technical products.

Of the examples given in Table 5.1, aldrin and heptachlor have the lowest water solubilities and the highest vapor pressures. They are readily oxidized, both chemically and biochemically, to their epoxides—dieldrin and heptachlor epoxide, respectively (Figure 5.5). (It should be noted that dieldrin has been marketed as an insecticide in its own right, whereas heptachlor epoxide has not.) The two epoxides have greater polarity, and consequently greater water solubility and lower volatility, than their precursors. Endrin is also an epoxide—in fact, it is a stereoisomer of dieldrin—and has greater water solubility and lower vapor pressure than aldrin or heptachlor. These relationships illustrate the importance of the electron-withdrawing power of oxygen atoms in determining the properties of organic compounds.

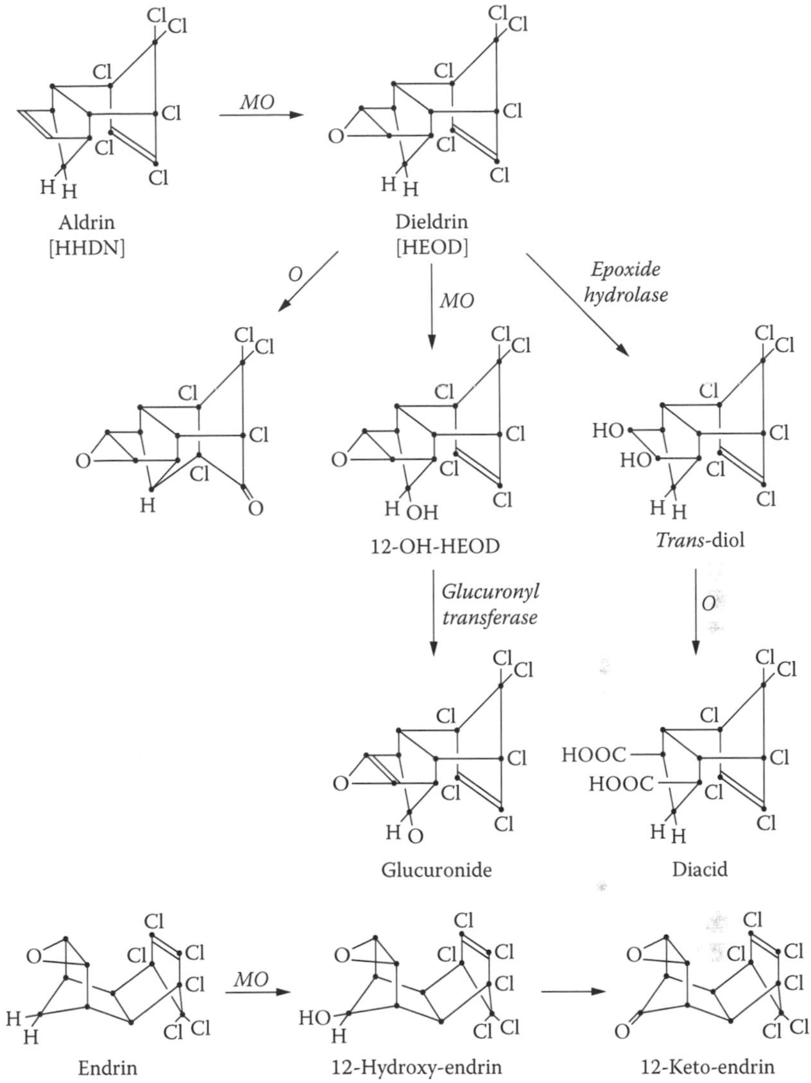

FIGURE 5.5 Metabolism of cyclodienes.

Apart from the oxidations just mentioned, cyclodienes are rather stable chemically. It should, however, be noted that dieldrin can undergo photochemical rearrangement under the influence of sunlight to the persistent and toxic molecule *photodieldrin*, which occurs as a residue following the application of this insecticide in the field.

5.3.2 THE METABOLISM OF CYCLODIENES

In terrestrial animals, cyclodienes such as dieldrin, like other refractive lipophilic pollutants, can be excreted in their unchanged forms, notably with lipoproteins, which are exported into milk (mammals), eggs (birds, reptiles, insects), or developing embryos (mammals). In a few cases (e.g., laying hens, which produce large numbers of eggs), this can represent a significant mechanism of loss. In general, however, it is not a sufficiently rapid mechanism to give much protection to the adult organism, although it constitutes a hazard to the next generation. Effective elimination depends on biotransformation into water-soluble and readily excretable metabolites and conjugates.

The metabolism of aldrin, dieldrin, endrin, and heptachlor in vertebrates is shown in Figure 5.5. As with p,p'-DDT and related compounds, a high level of chlorination greatly limits the possibility of metabolic attack by forming what is, in effect, a protective shield of halogen atoms. Monooxygenase attack might seem likely to be the most effective and rapid mechanism of biotransformation for compounds such as these which lack functional groups that are targets for more specialized enzymes (e.g., ester groups, which are attacked by esterases). However, monooxygenases do not readily attack C–Cl bonds—or, for that matter, C–Br or C–F bonds. Thus the most effective attack tends to be on other positions on the molecule—for example, on the C=C of the unchlorinated rings of aldrin and heptachlor, and on the endomethylene bridges across the same in the cases of dieldrin and endrin (Figure 5.5; Brooks 1974, Walker 1975, Chipman and Walker 1979). The first type of oxidation yields stable epoxides that are toxic and much more persistent than the parent compounds, and represents activation, not detoxication. The second line of attack is a typical phase 1 detoxication, yielding monohydroxy metabolites more polar than the parent compounds dieldrin and endrin; moreover, such monohydroxy metabolites readily undergo conjugation to form glucuronides and sulfates, which are usually rapidly excreted. The hydroxylation of endrin occurs relatively rapidly because the endomethylene bridge is in an exposed position for monooxygenase attack. It may be deduced that the molecule is bound to one or more forms of P450 belonging to gene family 2, and that the endomethylene group is thereby exposed to an activated form of oxygen generated from molecular oxygen bound to heme iron (see Chapter 2). The endomethylene group of dieldrin is less exposed than that of endrin, being screened by bulky neighboring chlorine atoms, and metabolic detoxication is consequently a good deal slower (Hutson 1976, Chipman and Walker 1979). Dieldrin is considerably more persistent in vertebrates than endrin despite the fact that the two compounds are stereoisomers with very similar physical properties, a logical consequence of the differential rates of metabolism. Thus, a stereochemical difference between two compounds having the same empirical formula may be reflected in large differences in toxicokinetics.

Cyclodiene epoxides such as dieldrin and heptachlor epoxide are also detoxified, albeit rather slowly, by epoxide hydrolase attack to form transdihydrodiols (Figure 5.5). The diols are relatively polar compounds that may be excreted unchanged, or as conjugates. There are very marked species differences in the ability to detoxify cyclodienes by epoxide hydrolase attack (Walker et al. 1978; Walker 1980). Using the readily biodegradable cyclodienes HEOM and HCE as substrates, mammals showed much higher microsomal epoxide hydrolase activities than birds or fish. Of the mammals, pigs and rabbits had particularly high epoxide hydrolase activity, and it is noteworthy that the trans diol has been shown to be an important in vivo metabolite of dieldrin in the rabbit, but not in the rat or the mouse, and not in birds (Korte and Arent 1965, Walker 1980, Chipman and Walker 1979). In general, the principal primary detoxication of dieldrin, endrin, and heptachlor epoxide is by monooxygenase attack.

In mammals, dieldrin and endrin are also converted into keto metabolites (Figure 5.5). In the rat, the keto metabolite is only a minor product, which, because of its lipophilicity, tends to be stored in fat. With endrin, a keto metabolite is formed by the dehydrogenation of the primary monohydroxy metabolite. In mammals, the trans diol of dieldrin is converted into a diacid in vivo (Oda and Muller 1972).

5.3.3 ENVIRONMENTAL FATE OF CYCLODIENES

During the 1950s, cyclodiene insecticides came to be widely used for a number of different purposes. They were used to control agricultural pests, insect vectors of diseases, rodents, and ectoparasites of farm animals, to treat wood against wood-boring insects, and to mothproof fabrics. Because of their very low solubility in water, such insecticides were usually formulated as emulsifiable concentrates or wettable powders. Following the discovery of their undesirable environmental side effects, the use of cyclodienes for many purposes was discontinued during the 1960s and early 1970s in Western Europe and North America. The following account will focus on the environmental fate of aldrin, dieldrin, and heptachlor—three insecticides that gave rise to persistent residues and were shown to cause serious and widespread environmental side effects. Other cyclodienes were less widely used, and some (e.g., endrin and endosulfan), although highly toxic, were far less persistent.

As mentioned earlier (Figure 5.5), aldrin and heptachlor are rapidly metabolized to their respective epoxides (i.e., dieldrin and heptachlor epoxide) by most vertebrate species. These two stable toxic compounds are the most important residues of the three insecticides found in terrestrial or aquatic food chains. In soils and sediments, aldrin and heptachlor are epoxidized relatively slowly and, in contrast to the situation in biota, may reach significant levels (note, however, the difference between aldrin and dieldrin half-lives in soil shown in Table 5.8). The important point is that, after entering the food chain, they are quickly converted to their epoxides, which become the dominant residues.

TABLE 5.8
Half-Lives of Cyclodienes

Compound	Material/Organism	Half-Life
Dieldrin	Soil	2.5 yr
Aldrin	Soil	0.3 yr
Heptachlor	Soil	0.8 yr
Dieldrin	Male rat	12–15 d
Dieldrin	Pigeon	47 d (mean)
Dieldrin	Dog	28–32 d
Dieldrin	Man	369 d (mean)

Sources: Soil data from Edwards (1973). Data for pigeon from Robinson et al. (1967b). Other data from Environmental Health Criteria 91.

Table 5.8 gives some examples of cyclodiene half-lives measured in (1) soils, and (2) experimentally dosed vertebrates. They should be taken as only rough indications of the environmental persistence of these compounds because there is great variability in this data: on the one hand, between different soil types in contrasting climates (see Chapter 4, Section 4.3), and on the other, between groups, species, strains, sexes, and age groups when considering persistence in animals (see Chapter 2, Section 2.3.2). With animals, such factors as manner of dosing, diet, and the method of analyzing data can all influence the values obtained (see Moriarty 1975 for further discussion). Looking at the estimated half-lives shown in Table 5.8, dieldrin, like *p,p'*-DDT, is markedly persistent in both soils and animals but less so than *p,p'*-DDE (Table 5.3). There is little data available for heptachlor epoxide, but on the grounds of its physical and chemical properties it is likely to have a similar persistence to dieldrin. Dieldrin is also highly persistent in sediments, which became clear in a study of British rivers conducted many years after the banning of the insecticide. Eels from all rivers investigated contained substantial dieldrin residues. In vertebrates, there are considerable interspecies differences in dieldrin half-lives. Male rats eliminate dieldrin more rapidly than female rats, pigeons, or dogs. For humans, the estimated dieldrin half-life is 369 days. By contrast, the half-life of endrin in humans is only 1–2 days (Environmental Health Criteria 130).

The rate of oxidative detoxication appears to be a critical factor in determining cyclodiene half-lives in animals. As noted earlier, endrin is rapidly detoxified by monooxygenase attack, whereas dieldrin is not. Also, male rats tend to have substantially higher monooxygenase activity toward cyclodiene substrates than pigeons or humans. The half-lives reported here seem to reflect these differences in rates of metabolism by monooxygenases; the higher the metabolic rate, the shorter the half-life. Toxicokinetic studies with cyclodienes support this interpretation. The rate of elimination of cyclodienes by terrestrial animals has been shown to be related to the rate at which they are converted into water-soluble and readily excretable metabolites (Chipman and Walker 1979, Walker 1981). These studies, which employed radiolabeled substrates, also provided further evidence for a point made earlier—that strongly lipophilic molecules of this type show little tendency to be excreted unchanged in urine or feces of terrestrial vertebrates.

Dieldrin, like *p,p'*-DDE, *p,p'*-DDD, *p,p'*-DDT, and many other organohalogenated compounds with high K_{ow} values, can undergo very marked bioconcentration by aquatic organisms (see Walker and Livingstone 1992). Bioconcentration factors (BCFs) in the steady state exceeding 1000 are usual. Thus Ernst (1977) gives a BCF value 1570 for *Mytilus edulis,* and Holden (1973) a value of 3700 for rainbow trout. With aquatic organisms, the rate of dieldrin metabolism is generally very low (especially in mollusks), and these high BCFs are a reflection of the passive exchange equilibrium between the organism and the ambient water (see Chapter 4).

With terrestrial organisms, metabolism is generally faster than in aquatic organisms but, as mentioned earlier, there are large species differences, and species that metabolize dieldrin slowly may strongly bioaccumulate the insecticide over long periods of exposure (Chipman and Walker 1979; Walker 1987, 1990a). In an incident at the London Zoo, 22 owls of diverse species died as the result of dieldrin poisoning (Jones et al. 1978). The source of dieldrin was the mice with which they had been fed. Comparison of the dieldrin liver concentrations in mice (geometric mean approximately 2.6 ppm, over two batches) with those in the 22 owls (geometric mean 28 ppm) suggested a bioaccumulation factor (BAF) of about 11. This can only be an approximation because the levels in mice were not monitored during the actual exposure of the owls. It should be recalled, however, that specialized predatory birds tend to have very low monooxygenase activities (see Chapter 2, Section 2.3.2; Walker 1980, Ronis and Walker 1989, Walker 1998a). This data, together with the evidence given earlier, strongly indicate that low metabolic capability is associated with a tendency for terrestrial organisms to strongly bioaccumulate dieldrin.

In the natural environment, as we have seen, dieldrin and heptachlor epoxide undergo bioconcentration and bioaccumulation with movement along food chains (see Section 5.2.3), reaching their highest concentrations in predators at the apex of food pyramids. Thus, dieldrin, like p,p'-DDE, was shown to exist at highest concentration in the fish-eating birds of the Farne Islands ecosystem in a study conducted during 1962–1964 (Figure 5.3; Robinson et al. 1967a). The mean concentration of dieldrin in carcasses of the shag (*Phalocrocorax aristotelis*; n = 8) was 1.0 ppm, whereas the mean concentration in the whole bodies of its principal prey in this area, the sand eel (*Ammodytes lanceolatus*; n = 16), was 0.016 ppm; this indicated an extraordinarily high BAF of 63. It should be emphasized that fish-eating species such as the shag and the cormorant evidently obtained most of their organochlorine residue from their food and not directly from the sea. They do not have permeable membranes, such as the gills of fish, across which pollutants can be readily absorbed from ambient water. Rather, they have relatively impermeable feathers and skin. Thus, they are not typical marine organisms like fish or aquatic invertebrates, which obtain much of their residue burden by bioconcentration from water. Further, they spend a good deal of time on dry land. Notwithstanding a necessary caution in the interpretation of field data, this provides further evidence for the marked bioaccumulation of dieldrin from prey by specialized predators at the apex of food pyramids. It is interesting to note that the gradient of increasing concentration with movement from trophic level 1 to trophic level 5 is steeper for p,p'-DDE than it is for dieldrin. Existing data suggests that dieldrin is more biodegradable and is eliminated more rapidly than p,p'-DDE, which may be the main reason for this difference in the degree of biomagnification.

Dieldrin also proved to be markedly persistent in terrestrial ecosystems. When it was widely used in the agricultural environment, however, very high concentrations sometimes existed at the beginning of the food chain, which is in marked contrast to the situation in the marine environment. Dieldrin and heptachlor were once commonly employed for dressing cereal and other crop seeds to give protection against soil pests. Newly treated seed had cyclodiene residues of about 800 ppm by weight associated with it (Turtle et al. 1963). Here, vertebrates feeding on the treated grain could soon acquire a lethal dose. Higher levels were often present in the grain than in the poisoned birds and other vertebrates. Large kills of granivorous birds on fields sown with cereals were a feature of this time. In the countrywide study of organochlorine residues in wild birds referred to earlier (Section 5.2.3), dieldrin showed much the same distribution pattern as p,p'-DDE, although the levels were usually lower. Once again, the highest average levels were found in predators, in both terrestrial and aquatic ecosystems (Moore and Walker 1964), but this broad survey did not reveal local patterns of residue accumulation in which grain-eating birds as well as predatory ones sometimes contained lethal concentrations of cyclodienes.

5.3.4 TOXICITY OF CYCLODIENES

There is strong evidence that the primary target for dieldrin, heptachlor epoxide, endrin, and other cyclodienes in the mammalian brain is the gamma aminobutyric acid (GABA) receptor, against which they act as inhibitors (Eldefrawi and Eldefrawi 1990). In insects, too, cyclodiene toxicity is attributed, largely or entirely, to the interaction with GABA receptors of the nervous system. Toxaphene and gamma HCH also act on this receptor. GABA receptors are found in the brains of both vertebrates and invertebrates, as well as in insect muscle; they possess chloride channels that, when open, permit the flow of Cl⁻ with consequent repolarization of nerves and reduction of excitability. They are particularly associated with inhibitory synapses. In vertebrates, the action of cyclodienes can lead to convulsions.

Given this mode of action upon the central nervous system (CNS), it is not surprising that cyclodienes can have a range of sublethal effects. These have been observed in humans occupationally exposed to aldrin or dieldrin (Environmental Health Criteria 91, Jaeger 1970). The symptoms observed included headache, dizziness, drowsiness, hyperirritability, general malaise, nausea, and anorexia. Sublethal effects included characteristic changes in electroencephalogram (EEG) patterns, and were observed over a wide range of blood concentrations. At the early stage of intoxication, muscle twitching and convulsions sometimes occurred. According to various authors, patients showing these symptoms had blood dieldrin levels in the range 8–530 µg/L (Environmental Health Criteria 91). The relationship between blood levels and the toxic effects of dieldrin in humans is shown in Figure 5.6. With increasing tissue levels of dieldrin, severe convulsions occurred, leading eventually

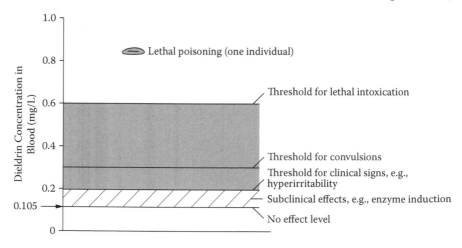

FIGURE 5.6 Dieldrin intoxication in humans and its relationship to blood levels. The hatched area represents the sublethal effects seen at 15–30% of lethal threshold concentration in blood (after Jager 1970).

to death. In cases of lethal poisoning, blood levels were found to exceed 600 µg/L (850 µg/L in one suicide case). Individuals showing sublethal effects made complete recoveries after discontinuation of exposure to the insecticide.

Experimental animals exposed to sublethal doses of cyclodienes show a similar picture, with changes in EEG patterns, disorientation, loss of muscular coordination and vomiting, as well as convulsions, the latter becoming more severe with increasing doses (Hayes and Laws 1991). It is clear from these wide-ranging studies that a number of neurotoxic effects can be caused by cyclodienes at levels well below those that are lethal. In the human studies described here, subclinical symptoms were frequently reported when dieldrin blood levels were in the range 50–100 µg/L, an order of magnitude below those associated with lethal intoxication.

Sublethal neurotoxic effects such as these have been associated with changes in behavior. In one study with dieldrin, squirrel monkeys were reported to show changes in learning ability and in EEG pattern after receiving doses of 0.01 or 0.1 mg/kg over 54 days (van Gelder and Cunningham 1975). Toxaphene, a chlorinated terpene, acts upon GABA receptors in a similar manner to dieldrin, and will, for convenience, be discussed here. In a study with goldfish (*Carassias auratus*), 96-hour exposure to 0.44 µg/L of toxaphene caused alterations in a number of behavioral parameters (Warner et al. 1966). The changes in response were more marked after 264 hours, although the fish remained outwardly healthy. These and other studies have established that sublethal effects arising from the interaction of neurotoxic chemicals with GABA receptors can cause behavioral changes, and these changes can be sensitively monitored using appropriately designed behavioral assays. The role of behavioral assays in biomarker strategies will be discussed in more detail in Chapter 16. The question of possible behavioral effects is an important one when considering the impact of cyclodienes in the field (see Section 5.3.5).

Some data on cyclodiene toxicity is presented in Table 5.9. Aldrin and dieldrin have similar levels of acute toxicity; indeed, the toxicity of aldrin has been largely attributed to its stable metabolite, dieldrin. Dieldrin is highly toxic to fish, mammals,

TABLE 5.9
Toxicity of Cyclodienes

Compound	Species	Toxicity
Dieldrin	*Daphnia magna*	96h LC_{50} 330 µg/L
Dieldrin	Fathead minnow	96h LC_{50} 4–18 µg/L
Dieldrin	Rainbow trout	96h LC_{50} 1.2–9.9 µg/L
Heptachlor	Rainbow trout	96h LC_{50} 7 µg/L
Dieldrin	Rat	Acute oral LD_{50} 37–87 mg/kg
Dieldrin	Rabbit	Acute oral LD_{50} 45–50 mg/kg
Heptachlor	Rat	Acute oral LD_{50} 40–162 mg/kg
Dieldrin	Pigeon (*Columba livia*)	Acute oral LD_{50} 67 mg/kg
Endrin	Rat	Acute oral LD_{50} 4–43 mg/kg

Sources: Environmental Health Criteria 38, 91, and 130.

and birds. Heptachlor has a similar order of toxicity. Endrin is rather more toxic than aldrin or dieldrin to rodents and rabbits (data not shown), despite the fact that it is more readily metabolized and therefore less persistent than dieldrin. Thus, in general, the principal cyclodiene insecticides are more toxic than DDT to land vertebrates but of a similar order of toxicity to fish. Lethal poisoning of vertebrates by cyclodiene insecticides in the field was frequently reported when they were widely used, and will be discussed in Section 5.3.5.1.

Because of its lipophilicity and refractory character, the toxic effects of dieldrin may be carried through to the next generation. In one example, dosing of female small tortoiseshell butterflies (*Aglais urticae*) with dieldrin led to an increased number of deformed adults emerging from pupae (Moriarty 1968).

A further feature of dieldrin toxicity, shared with other persistent OCs, is the ability to cause delayed neurotoxicity. In one example from a field in the Netherlands, female but not male eider ducks (*Somateria mollissima*) died with neurotoxic symptoms during the breeding season. Subsequent investigation established that they had stored dieldrin residues in depot fat during the period before going into lay. These fat reserves, together with the insecticide stored in them, were rapidly mobilized during egg laying, and blood dieldrin levels rose sharply as a consequence. Eventually, lethal concentrations were reached in the nervous system of ducks, causing convulsions and other symptoms of dieldrin toxicity (Koeman and van Genderen 1970, Koeman 1972).

5.3.5 Ecological Effects of Cyclodienes

The following account will be largely restricted to the effects of aldrin, dieldrin, and heptachlor—compounds that were once widely used, and for which there is the clearest record of ecological effects.

5.3.5.1 Effects on Population Numbers

Within a short time following the introduction of aldrin, dieldrin, and heptachlor seed dressings into Western Europe in the mid-1950s, there were reports of large-scale mortality of birds on farmland, including predatory birds. The carcasses contained sufficiently high residues of dieldrin or heptachlor epoxide or both in their tissues to have caused death (little or no aldrin or heptachlor was found in tissues because of rapid biotransformation to their respective epoxides). Also, characteristic symptoms of cyclodiene intoxication (e.g., convulsions) were observed in certain individuals before they died, and subsequent analysis revealed lethal levels of dieldrin or heptachlor epoxide or both in their tissues. Thus, there was early evidence of secondary poisoning due to transfer of the compounds to predators via their prey. Foxes were also victims of secondary cyclodiene poisoning. The toxic effects were largely attributed to dieldrin residues, although heptachlor epoxide was also implicated in certain areas. As mentioned earlier, dressed seed would typically contain 800 ppm of cyclodiene. Thus, grain-eating birds or mammals did not need to consume very large amounts of grain to acquire lethal doses. The question remained, however, whether lethal poisoning was on a large enough scale to cause the decline of vertebrate populations.

One of the earliest reports of cyclodiene poisoning in the field came from a coastal area of the Netherlands (Koeman et al. 1967, Koeman and van Genderen 1970, Koeman 1972), where sandwich terns (*Sterna sandvicensis*) died showing symptoms of cyclodiene poisoning. Both adults and chicks were affected. Upon analysis, they were found to contain residues of dieldrin, endrin, and telodrin at high-enough levels, singly or in combination, to cause lethal toxicity. The source turned out to be a neighboring factory that was synthesizing the compounds. These mortalities were linked to a local decline in the sandwich tern population that began around 1962. During the period 1965–1966, a decline of the buzzard (*Buteo buteo*) occurred in the Netherlands, which brought the species close to extinction in certain agricultural areas. Following investigation, this decline was attributed very largely to lethal poisoning by dieldrin (Fuchs 1967). Analysis of birds found in the field that had shown symptoms of cyclodiene poisoning before dying contained a mean dieldrin concentration of 18 ppm in their livers. From Britain came the evidence of widespread decline of the sparrowhawk (*Accipiter nisus*) and the peregrine (*Falco peregrinus*). The declines coincided in time with the introduction of aldrin, dieldrin, and heptachlor into the country in 1956, and occurred in areas where the chemicals were widely used (Ratcliffe 1993, Newton 1986; see Figures 5.7 and 5.8). Both species are bird-eating raptors, which were exposed to high levels of dieldrin and heptachlor epoxide in the tissues of the grain-eating birds on which they preyed. Cyclodiene poisoning was confirmed in individual cases on the grounds of lethal tissue levels of dieldrin or heptachlor epoxide or both, and toxic symptoms shown before death.

Sparrowhawks declined very sharply in the agricultural areas of eastern England where cyclodienes were widely used, becoming virtually extinct in parts of East Anglia where they had once been common. Such declines were not seen, however,

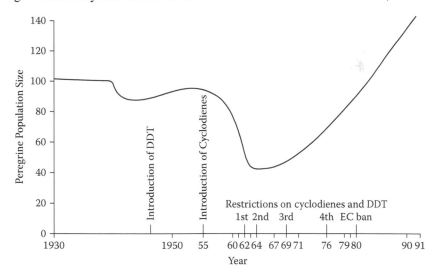

FIGURE 5.7 Peregrine population size in Britain (1930–1939 = 100) showing the 1961 population decline and subsequent recovery, together with an outline of pesticide usage (from Ratcliffe 1993).

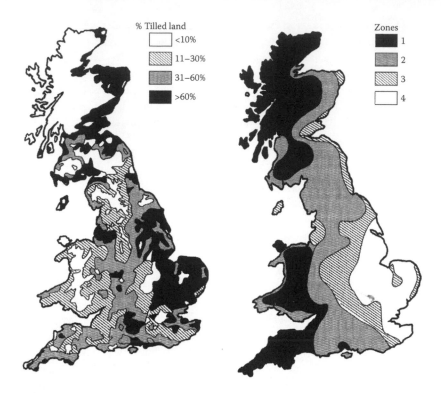

FIGURE 5.8 Changes in the status of sparrowhawks in relation to agricultural land use and organochlorine use. The agricultural map (left) indicates the proportion of tilled land where most pesticide is used. The sparrowhawk map (right) shows the status of the species in different regions and time periods. Zone 1: sparrowhawks survived in greater numbers through the height of the "organochlorine era" around 1960; the population decline is judged to be less than 50%, and recovery is effectively complete before 1970. Zone 2: the population decline is more marked than in zone 1, but it had recovered to more than 50% by 1970. Zone 3: the population decline is more marked than in zone 2, but it had recovered to more than 50% by 1980. In general, the population decline was more marked and the recovery later in areas with a greater proportion of tilled land (based on agricultural statistics for 1966) (from Newton and Haas 1984, reproduced in Newton 1986).

in more western and northern areas where there was much less use of cyclodienes (Newton 1986, Newton and Haas 1984). Reasonable numbers were also maintained in the New Forest area in Southern England where there was little use of these insecticides. Peregrines declined sharply in coastal areas of Southern England and Wales, disappearing completely from the southeast coast. They were less affected in Northern England and Scotland, maintaining good numbers in the Scottish Highlands (Ratcliffe 1993). It is important to emphasize that these declines did not occur until the mid-1950s, long after marked eggshell thinning was caused by p,p'-DDE that became clearly manifest by 1947 (Ratcliffe 1967). Sparrowhawks commence breeding at 1–2 years and peregrines at 2–3 years; thus, any population effects caused by a reduction in breeding success due to eggshell thinning would have been evident long before the population crashes of the mid- to late-1950s, which are associated with the introduction of cyclodienes.

Later evidence suggested that merlins (*Falco columbarius*) also declined in parts of Britain during the late 1950s or early 1960s as the result of exposure to cyclodienes (Newton, Meek, and Little 1978). The merlin, like the peregrine and sparrow hawk, is a bird-eating predator.

There are detailed records of the residues of dieldrin, heptachlor epoxide, and OCs in British sparrowhawks and kestrels (*Falco tinnunculus*) during the period 1963–1990s (Newton 1986, Newton and Wyllie 1992), thus permitting a detailed retrospective analysis of the relationship between levels of exposure and population changes in these two species (Walker and Newton 1998, 1999; Walker 2004). (The kestrel is another predatory bird that showed a sharp decline in agricultural areas over the period under consideration, although less marked than that of the sparrow hawk.) Residues in livers of sparrowhawks and kestrels are compared in Figure 5.9 (Walker and Newton 1998, 1999). Looking at the distribution of liver dieldrin levels

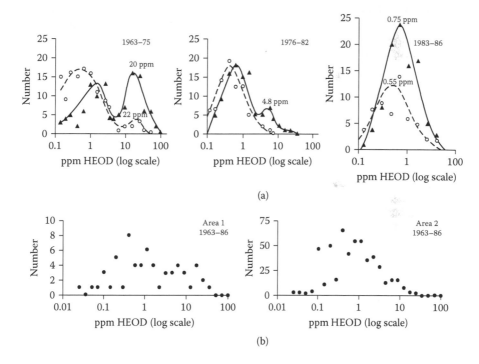

FIGURE 5.9 (a) Distribution of dieldrin (HEOD) residues in the livers of kestrels from two different areas of Britain. The HEOD residues are expressed as ppm wet weight. They are plotted on a log scale. The numbers of individuals with dieldrin residues falling within the ranges of concentrations represented by 0.15 log units are given on the vertical axis. The concentrations plotted represent the midpoints of each log range: area 1 (highest cyclodiene use), ▲; area 2 (lowest cyclodiene use), ○. (b) Distribution of dieldrin (HEOD) residues in the livers of sparrowhawks. Area 1 is zone 4, and area 2 is zones 1 and 2 in Figure 5.8 (from Walker and Newton 1999).

in birds found dead from different areas, both species show peaks centering on about 20 ppm dieldrin, which account for a high proportion of the total sample coming from the high-cyclodiene area of eastern England, collected during the period 1963–1975. This value is very close to the mean dieldrin concentration of 18 ppm found in the livers of buzzards poisoned by this insecticide in the Netherlands (Fuchs 1967; see also preceding text). On the basis of laboratory studies, a lethal threshold for acute dieldrin poisoning of about 10 ppm has been proposed for birds, and a high percentage of the birds in these high-residue peaks contained residues in excess of this. Further, certain individuals with high liver dieldrin levels were seen to show symptoms of cyclodiene poisoning before they died. Thus, on the basis of these lines of evidence, it was proposed that the peaks centering on approximately 20 ppm represented individuals who had been acutely poisoned by cyclodienes (Newton and Walker 1999, Walker 2004). In the samples taken during the period 1963–1975 from an area of low cyclodiene use, the peaks centering on about 20 ppm are relatively very small, representing no more than a few percentage individuals investigated.

These results provide strong evidence that dieldrin and heptachlor epoxide were primarily responsible for the initial population declines and the continuing low population of both the previously mentioned species in eastern England over the period 1963–1975. After 1975 came bans on the use of cyclodienes as seed dressings for autumn-sown cereal. Dieldrin levels began to fall sharply in sparrowhawks and kestrels after these bans came into force. However, populations did not begin to recover in either species until the geometric means of the dieldrin concentration in liver fell below 1 ppm (Newton and Wyllie 1992). As lethal toxicity of the insecticide in experimentally poisoned birds is associated with dieldrin levels of 10 ppm or more, this is a surprising observation. Very few (not more than 5%) individuals in samples with geometric means of 1 ppm could have residue levels as high as 10 ppm.

Inspection of the distribution of liver dieldrin levels in birds collected between 1976 and 1982 reveals the virtual disappearance of the lethal toxicity peak centering on about 20 ppm. On the other hand, another peak appears centering on 4.6 ppm in kestrel (Figure 5.9). As mentioned earlier (Section 5.3.4), dieldrin and other cyclodienes can produce a variety of sublethal neurotoxic effects, including disorientation, lack of coordination, and tremors at tissue concentrations well below those that cause death. It may therefore be suggested on the basis of the distributions shown in Figure 5.9 that those individuals with liver dieldrin levels between about 3 and 9 ppm experienced sublethal neurotoxic effects before death, and that these sublethal effects eventually caused death by impairing some function (e.g., hunting skills). Such a hypothesis could solve the dilemma mentioned earlier: why was recovery delayed until liver levels fell below 1 ppm? The delayed recovery is easier to understand if the earlier estimate of dieldrin casualties (<5% of sample) is too low, and a substantially larger proportion of the birds found dead were victims of dieldrin toxicity, lethal or sublethal. Loss of coordination and other sublethal neurotoxic effects could surely impair the hunting skills of these raptors, and failure to catch prey would lead to starvation (Walker and Newton 1998, 1999). Following this line of reasoning, a higher proportion of the recorded deaths of both species in eastern England during 1976–1982 would be attributed to dieldrin than if simply basing the estimate on the proportion of birds containing 9 ppm or more of dieldrin in the liver.

Further examination of Figure 5.9 reveals a difference between sparrow hawks and kestrels. Considering dieldrin levels of 3 ppm and above, the kestrel has proportionally more individuals in the category of 9 ppm and above than has the sparrow hawk. If the foregoing theory is correct, sparrowhawks, then, showed a higher incidence of deaths due to sublethal toxicity than kestrels and, conversely, kestrels showed a higher proportion of deaths due to direct acute poisoning than sparrowhawks. This suggests that sparrowhawks are more susceptible to sublethal effects than kestrels. This could be explained on the grounds that the sparrowhawk depends on greater maneuverability to catch its prey in the air than the kestrel, which simply drops on its prey from a hovering or perching position. The former species, therefore, may be more susceptible to neurotoxic disturbances than the latter.

These data for the sparrow hawk have been incorporated, together with other data for the species, into a population model (Sibly et al. 2000). Population changes in the sparrow hawk that occurred in Eastern England during 1963–1975 were predicted reasonably well by a model that assumed that all birds with dieldrin levels of 3 ppm or more died due to poisoning by the insecticide. The substantially smaller proportion of dead birds having 9 ppm or more, which were presumed to have died from direct poisoning, was not large enough to predict the rate or scale of the population decline that actually occurred or, for that matter, the delay in recovery until the residue level in livers fell below 1 ppm (Walker 2004).

In comparison to the situation in birds, there is far less evidence of dieldrin having had harmful effects in the field in mammals. It has been suggested that this is a reflection of the fact that mammals tend to be more reclusive and therefore more difficult to observe, catch, or count (Shore and Rattner 2001). That said, at the time when cyclodienes were widely used in Western Europe and North America, there were a fair number of reports of mammals being poisoned by them on agricultural land. Such animals included predators such as the fox (*Vulpes vulpes*) and badger (*Meles meles*), which had evidently acquired lethal doses from their prey.

Retrospectively, there has been considerable interest in the decline of the otter (*Lutra lutra*) in Britain. Chanin and Jefferies (1978) examined hunting records and concluded that a very marked and widespread decline of this species coincided in time and space with the introduction of dieldrin. Unfortunately, there is very little residue data for otters during that time, reflecting the point made earlier about the reclusiveness of many mammals. That said, there is strong circumstantial evidence to support the suggestion of these authors that dieldrin was the single most important cause of this decline. Apart from the coincidence in time and space, there is plenty of evidence for the existence of high dieldrin levels in British rivers at this time. Substantial levels of dieldrin were reported in freshwater fish and considerably higher concentrations in piscivorous birds such as the heron (*Ardea cinerea*) at this time. In a study at a heronry in Lincolnshire (1965–1968), young herons contained mean dieldrin residues in liver of 4.5 ppm and 3.05 ppm at two colonies, individual values ranging from 0.1 to 28.2 ppm (Prestt 1970). Fish sampled at the same heronries contained mean residue levels in liver of 0.70 ppm (roach), 0.90 ppm (eel), and 1.11 ppm (bream). These figures indicate that the young herons bioaccumulated residues in their livers some four times the levels in their food. Other data from that time showed that adult herons found dead over the whole of Britain contained a mean level of 9.3 ppm dieldrin in their livers. More generally, fish-eating birds, similar to other vertebrate predators, have been shown to strongly bioaccumulate dieldrin and other cyclodienes from their food (see Section 5.3.3). Indeed, it was argued earlier (see discussion about the evolution of cytochrome P450 in Chapter 2, Section 2.3.2) that predators have only low levels of detoxifying enzymes such as hepatic microsomal monooxygenases because there has been little need of them during the course of evolution. Coming back to the otter in Britain during the late 1950s–1960s, as a fish-eating predator it might reasonably have been expected to bioaccumulate similar levels of dieldrin as fish-eating birds of similar body size and, like them, to have suffered the lethal and sublethal toxic effects of dieldrin. This adds strength to the proposal of Chanin and Jefferies (1978).

Dieldrin and endrin were once used for the control of locusts and tse-tse flies in Africa (Koeman and Pennings 1970). Large numbers of birds and mammals were poisoned as a consequence, including many insectivores. The latter point is of interest because insects are usually poor "vectors" of insecticides on account of high sensitivity to their toxic effects. However, in the field program in Africa, relatively large quantities of these two highly toxic compounds were sprayed over substantial areas and were freely made available for uptake (cf. the limited availability of cyclodienes used as seed dressings in Western Europe and North America). Under these conditions, individual insects would have picked up by direct contact doses of insecticide far in excess of those needed to kill them, and passed on high doses to the birds feeding on them. With the general spraying of the area, insecticide would have also been absorbed by birds directly through inhalation, drinking water, and from leaf surfaces. It is not clear what effects these casualties had on the population numbers.

5.3.5.2 Development of Resistance to Cyclodienes

There have been many examples of insects developing resistance to aldrin or dieldrin, sometimes associated with cross-resistance to lindane (Brooks 1974, Salgado 1999). Examples include resistant strains of the fruit fly (*Drosophila melanogaster*), the housefly (*Musca domestica*), the German cockroach (*Blatella germanica*), the cabbage root fly (*Erioischia brassicae*), and the onion maggot (*Hylemya antiqua*). There is now good evidence that resistance is due to the presence of mutant forms of the $GABA_A$ receptor (Brooks 1992), which are relatively insensitive to cyclodienes and gamma HCH. A single gene conferring 4000-fold resistance to dieldrin has been cloned from a field strain of *D. melanogaster* and found to code for a subunit of the GABA receptor (Ffrench Constant et al. 1993). Resistance in this case was associated with a single-point mutation of a conserved alanine residue to serine in the second transmembrane domain of the protein, which is thought to line the pore of the chloride channel of the GABA receptor. Similar mutations in strains of the housefly, the German cockroach, and the mosquito *Anopheles arabiensis* (Du et al. 2005) also confer very high levels of resistance to dieldrin. An alanine-to-glycine substitution in another mosquito, *Anopheles gambiae*, has also been shown to confer resistance to dieldrin (Du et al. 2005). The levels of resistance of these strains to HCH and certain other cyclodienes are far less than those shown to dieldrin (Salgado 1999).

Some field studies have produced evidence suggesting that resistance is developed by wild vertebrates that have been regularly exposed to cyclodienes. Examples include resistance of pine mice (*Microtus pinetorum*) to endrin (Webb et al. 1972) and the resistance of mosquito fish (*Gambusia affinis*) in the Mississippi River to aldrin (Wells et al. 1973). In both cases, resistance was attributed to enhanced metabolic detoxication by the resistant strain.

5.4 HEXACHLOROCYCLOHEXANES

Hexachlorocyclohexanes (HCHs) have not caused the problems of bioconcentration or bioaccumulation associated with DDT or the cyclodienes, and they have not been implicated in large-scale environmental problems. They will only be discussed briefly here, drawing attention to certain differences from the foregoing groups.

The first commercially available HCH insecticide sometimes misleadingly called benzene hexachloride (BHC) was a mixture of isomers, principally alpha HCH (65–70%), beta HCH (7–10%), and gamma HCH (14–15%). Most of the insecticidal activity was due to the gamma isomer (Figure 5.1), a purified preparation of which (>99% pure) was marketed as lindane. In Western countries, technical HCH was quickly replaced by lindane, but in some other countries (e.g., China) the technical product, which is cheaper and easier to produce, has continued to be used. HCH has been used as a seed dressing, a crop spray, and a dip to control ectoparasites of farm animals. It has also been used to treat timber against wood-boring insects.

Gamma HCH (Figure 5.1) is more polar and more water soluble than most other OCs (Table 5.1), and is metabolized relatively rapidly to water-soluble products. The metabolism of gamma HCH is complex, and involves both dehydrochlorination reactions mediated by reduced glutathione (GSH) and hydroxylations mediated by cytochrome P450. The main excreted metabolites are various chlorophenols in free or conjugated forms, prominent among them are trichlorophenols (Environmental Health Criteria 124). Gamma HCH appears to be rapidly eliminated by vertebrate species, and residues in free-living vertebrates and invertebrates were found to be low when the compound was widely used in agriculture (e.g., levels in livers of birds of prey were 0.01–0.1 ppm).

Gamma HCH acts upon GABA receptors in a similar fashion to cyclodienes (Eldefrawi and Eldefrawi 1990), and cross-resistance of insects between the two types of insecticide is sometimes due to a mutant form of the receptor, as discussed in Section 5.3.5.2. Acute oral LD_{50}s of gamma HCH to rodents range from 55–250 mg/kg. Thus, it is of a similar order of toxicity as DDT but, on the whole, less toxic than aldrin, dieldrin, or endrin. It has not been implicated in field mortalities of birds, in contrast to cyclodienes. On the other hand, bats can be lethally poisoned when exposed to lindane-treated wood (Boyd et al. 1988). This poses a risk to bats roosting in lofts containing treated timber. Gamma HCH is highly toxic to fish, LC_{50} values for fish falling into the range 0.02–0.09 mg/L. When used as a sheep dip, there was a hazard to freshwater fish when sheep farmers discharged sheep dips into neighboring water courses.

Of the other HCH isomers, the alpha and beta forms are less toxic than the gamma form. However, the beta form is more persistent than the gamma form, and unacceptably high residues have sometimes been reported in foods originating from countries where technical HCH is still used (Environmental Health Criteria 123).

5.5 SUMMARY

OC insecticides such as aldrin, dieldrin, heptachlor, and DDT illustrate well the environmental problems associated with persistent lipophilic compounds of high toxicity, be that toxicity lethal or sublethal. They serve as models for environmental pollutants of this kind, having been studied in much greater depth and detail than most other pollutants. The original insecticides, or their stable metabolites, can undergo biomagnification of several orders of magnitude with movement along food chains. Attention has been drawn to the structural features of these polyhalogenated compounds that make them so resistant to biodegradation.

OCs have been shown to cause widespread population declines of certain raptorial birds, such as the peregrine falcon, the sparrow hawk, and the bald eagle. DDT has caused population declines of predatory birds in North America due to eggshell thinning, brought about by its highly persistent metabolite p,p'-DDE. Aldrin, dieldrin, and heptachlor were responsible for the population crashes of peregrines and sparrow hawks in Britain and certain other European countries during the late 1950s. Although acute lethal toxicity was important here, there is some evidence to suggest that sublethal neurotoxic effects were also involved.

Resistance to DDT has been developed in many insect species. Although there are some cases of metabolic resistance (e.g., strains high in DDT dehydrochlorinase activity), particular interest has been focused on kdr and super kdr mechanisms based upon aberrant forms of the sodium channel—the principal target for DDT. There are many examples of insects developing resistance to dieldrin. The best-known mechanism is the production of mutant forms of the target site (GABA receptor), which are insensitive to the insecticide.

FURTHER READING

Brooks, G.T. (1974). *The Chlorinated Insecticides,* Vols 1 and 2—A detailed and authoritative standard reference work on the chemistry, biochemistry, and toxicology of organochlorine insecticides.

Environmental Health Criteria 91 (1989)—A valuable source of information on the environmental toxicology of aldrin and dieldrin.

Jorgenson, J.L. (2001). *Aldrin and Dieldrin: A Review of Research on Their Production, Environmental Deposition and Fate, Bioaccumulation, Toxicology, and Epidemiology in the United States*—A relatively recent, wide-ranging review of two of the most important cyclodienes.

Moriarty, F. (Ed.) (1975). *Organochlorine Insecticides: Persistent Organic Pollutants*—A collection of focused chapters on ecotoxicological aspects of the organochlorine insecticides.

Shore, R.F. and Rattner, B.A. (2001). *Ecotoxicology of Wild Mammals*—A very detailed, well-structured, and well-referenced text in which organochlorine insecticides and their effects are fully represented. A standard reference work.

6 Polychlorinated Biphenyls and Polybrominated Biphenyls

6.1 BACKGROUND

The polychlorinated biphenyls (PCBs) and polybrominated biphenyls (PBBs) are industrial chemicals that do not occur naturally in the environment. The properties, uses, and toxicology of the PCBs are described in detail in Safe (1984), Robertson and Hansen (2001), and Environmental Health Criteria 140. PBBs are described in Safe (1984) and Environmental Health Criteria 152.

PCBs were first produced commercially around 1930. The commercial products are complex mixtures of congeners, generated by the chlorination of biphenyl. Most of them are very stable viscous liquids, of low electrical conductivity and low vapor pressure. Their principal commercial applications have been

1. As dielectrics in transformers and large capacitors
2. In heat transfer and hydraulic systems
3. In the formulation of lubricating and cutting oils
4. As plasticizers in paints, and as ink solvents in carbonless copy paper

With such a diversity of uses, they entered the natural environment by many different routes before they were subject to bans and restrictions. The level of chlorination determines the composition and properties and, ultimately, the commercial use of PCB mixtures. Depending on reaction conditions, levels of chlorination ranging from 21 to 68% (percentage by weight) have been achieved. The commercial products are known by names such as Aroclor, Clophen, and Kanechlor, usually superseded by a code number that indicates the quality of the product. Thus, in one series of products, Aroclor 1242 and Aroclor 1260 contain about 42% chlorine and about 60% chlorine respectively. The first two numbers of the code indicate that the product is derived from biphenyl, and the second two indicate the approximate level of chlorination. Since the discovery of pollution problems in the 1960s, the production of PCBs has greatly declined, and there are few remaining uses at the time of writing. Further details of the regulation of PCBs internationally are given in Robertson and Hansen (2001).

PBBs have also been marketed as mixtures of congeners, produced in this case by the bromination of biphenyl. Their main commercial use has been as fire retardants, for which purpose they were introduced in the early 1970s. The most widely known commercial PBB mixture was Firemaster, first produced in 1970 in the United States, with production discontinued in 1974 following the recognition of pollution problems.

Many of the components of PCB and PBB mixtures are both lipophilic and stable, chemically and biochemically. Similar to the persistent organochlorine insecticides and their stable metabolites, they can undergo strong bioconcentration and bioaccumulation to reach relatively high concentrations in predators.

6.2 POLYCHLORINATED BIPHENYLS

6.2.1 CHEMICAL PROPERTIES

In theory, there are 209 possible congeners of PCB. In practice, only about 130 of these are likely to be found in commercial products. The structures of some congeners are shown in Figure 6.1. The more highly chlorinated a PCB mixture is, the greater the proportion of higher chlorinated congeners in it. Thus, in Aroclor 1242 (42% chlorine), some 60% of the mass is in the form of tri- or tetrachlorobiphenyls, whereas in Aroclor 1260 (60% chlorine), some 80% of the mass is as hexa- and heptabiphenyls. Small amounts of PCDFs are found in commercial products (see Chapter 7).

Individual PCB congeners are often crystalline, but most commercial mixtures exist as viscous liquids, turning into resins with cooling. Highly chlorinated mixtures, such as Aroclor 1260, are resins at room temperature. In general, PCBs are very stable compounds of low chemical reactivity; they have rather high density, and are fire resistant. They have low electrical resistance that, in combination with their heat stability, makes them very suitable as cooling liquids in electrical equipment. They have low water solubility and high lipophilicity, and low vapor pressures (see Table 6.1). With increasing levels of chlorination, vapor pressure and water solubility tend to decrease, and log K_{ow} to increase. (Note: The values for vapor pressure and water solubility in Table 6.1 are expressed as negative.)

Some PCB congeners have coplanar structures (see, e.g., 3,4,3′,4′-tetrachlorobiphenyl in Figure 6.1). The coplanar conformation is taken up when there is no chlorine substitution in ortho positions. If there is substitution of chlorine in only one ortho position, the molecule may still be close to coplanarity, because of only limited interaction between Cl and H on adjoining rings. Substitution of chlorines in

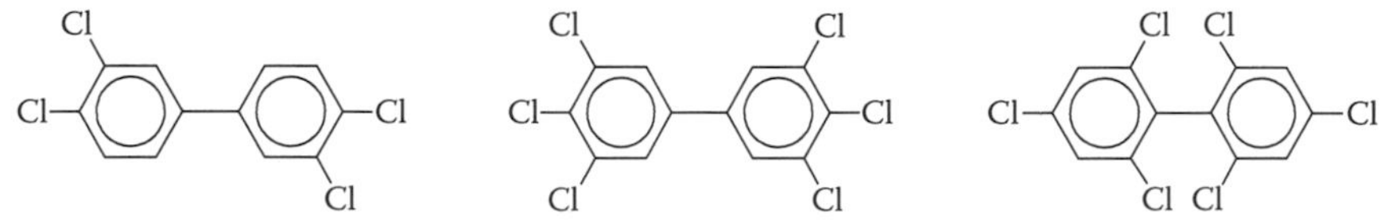

3, 3′, 4, 4′-Tetrachlorobiphenyl (coplanar)   3, 3′, 4, 4′, 5, 5′-Hexachlorobiphenyl (coplanar)   2, 2′, 4, 4′, 6, 6′-Hexachlorobiphenyl (not coplanar)

FIGURE 6.1 Some PCB congeners.

TABLE 6.1

Properties of PCB Congeners

Compound	Structure	Vapor Pressure [atmospheres] $-\log P_o$	Water Solubility [moles/liter] $-\log C$	$\log K_{ow}$
4,4'-DCB	Coplanar	7.32	6.53	5.33
2',3,4-TCB	Coplanar	6.88	6.52	5.78
2,2',5,5'-TCB	Nonplanar	7.60	7.06	6.18
2,2',4,5,5'-PCB	Nonplanar	8.02	7.40	6.36
2,2',3,3',4,4'-HCB	Nonplanar	9.65	8.72	6.97

Note: All values estimated at 25°C.
Source: Schwarzenbach et al. (1993).

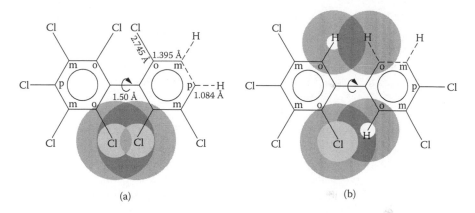

(a)   (b)

FIGURE 6.2 Planar and coplanar PCBs. Structural features of CB congeners influencing enzymatic metabolism. Areas where the principal enzymatic reaction occurs are given by broken lines. For atoms in the ortho position, the outer circle represents the area within the van der Waals radius of an atom; the dotted inner circle represents the part of this area that is also within the single bond covalent radius. The van der Waals radius indicates the maximum distance for any possible influence of an atom. The covalent radius represents the minimum distance that atoms can approach each other. (a) Vicinal atoms in the meta and para positions. Overlapping covalent radii for two ortho Cl show that a planar configuration is highly improbable when three or four ortho Cl are present. (b) Vicinal atoms in the ortho and meta positions. Nonoverlapping covalent radii for ortho Cl and ortho H show that a planar configuration causes a much lower energy barrier when chlorine atoms do not oppose each other. (Reproduced from Boon et al. 1992. With permission.)

adjacent ortho positions leads to movement of rings away from planarity to accommodate the overlap of the orbitals of the bulky halogen atoms (Figure 6.2). Most PCBs have nonplanar structures because of chlorine substitutions in ortho positions. There are important biochemical differences between coplanar and nonplanar PCB congeners that will be described in later sections.

6.2.2 METABOLISM OF PCBs

In terrestrial vertebrates, the elimination of PCBs, similar to that of OC insecticides, is largely dependent on metabolism. The rate of excretion of the unchanged congeners is generally very slow, although it should be noted that small amounts are "excreted" into milk (mammals) or eggs (birds, amphibians, reptiles, and insects), presumably transported by lipoproteins (see Chapter 2). In mammals there can also be transport across the placenta into the developing embryo. Although such "excretions" do not usually account for a very large proportion of the body burden of PCBs, the translocated congeners may still be in sufficient quantity to cause embryo toxicity.

In animals, primary metabolism of PCBs is predominantly by ring hydroxylation, mediated by different forms of cytochrome P450, to yield chlorophenols. The position of attack is influenced by the location of substitutions by chlorine. As with other lipophilic polychlorinated compounds, oxidative attack does not usually occur directly on C-Cl positions; it tends to occur where there are adjacent unsubstituted ortho-meta or meta-para positions on the aromatic ring. Unchlorinated para positions are particularly favored for hydroxylation, a mode of metabolism associated with P450s of family 2 rather than P4501A1/1A2. In the case of aromatic hydroxylations, it has been suggested that primary attack is by an active form of oxygen generated by the heme nucleus of P450 (see Chapter 2) to form an unstable epoxide, which then rearranges to a phenol (for further discussion of mechanism, see Trager 1988 and Crosby 1998). Two examples of hydroxylations of PCBs are shown in Figure 6.3: one PCB is planar, the other coplanar.

Monooxygenase attack upon the coplanar PCB 3,3′,4,4′-tetrachlorobiphenyl (3,3′4,4′-TCB) is believed to occur at unsubstituted ortho-meta (2′,3′) or meta-para (3′,4′) positions, yielding one or other of the unstable epoxides (arene oxides) shown in the figure. Rearrangement leads to the formation of monohydroxy metabolites. In one case, a chlorine atom migrates from the para to the meta position during this rearrangement (NIH shift), thus producing 4′ OH, 3,3′,4,5′-tetrachloro biphenyl. The mechanism of formation of 2′OH, 3,4,3′,4′-TCB is unclear (Klasson-Wehler 1989).

In the rabbit, the nonplanar PCB 2,2′,5,5′-tetrachlorobiphenyl (2,2′,5,5′-TCB) is converted into the 3′,4′-epoxide by monooxygenase attack on the meta-para position, and rearrangement yields two monohydroxymetabolites with substitution in the meta and para positions (Sundstrom et al. 1976). The epoxide is also transformed into a dihydrodiol by epoxide hydrolase attack (see Chapter 2, Section 2.3.2.4). This latter conversion is inhibited by 3,3,3-trichloropropene-1,2–oxide (TCPO), thus providing strong confirmatory evidence for the formation of an unstable epoxide in the primary oxidative attack (Forgue et al. 1980).

In the examples given, there is good evidence for the formation of an unstable epoxide intermediate in the production of monohydroxymetabolites. However, there is an ongoing debate about the possible operation of other mechanisms of primary oxidative attack that do not involve epoxide formation, for example, in the production of 2′OH 3,3′,4,4′-TCB (Figure 6.3). As mentioned earlier, P450s of gene family 1 (CYP 1) tend to be specific for planar substrates, including coplanar PCBs; they do not appear to be involved in the metabolism of nonplanar PCBs. On the other hand,

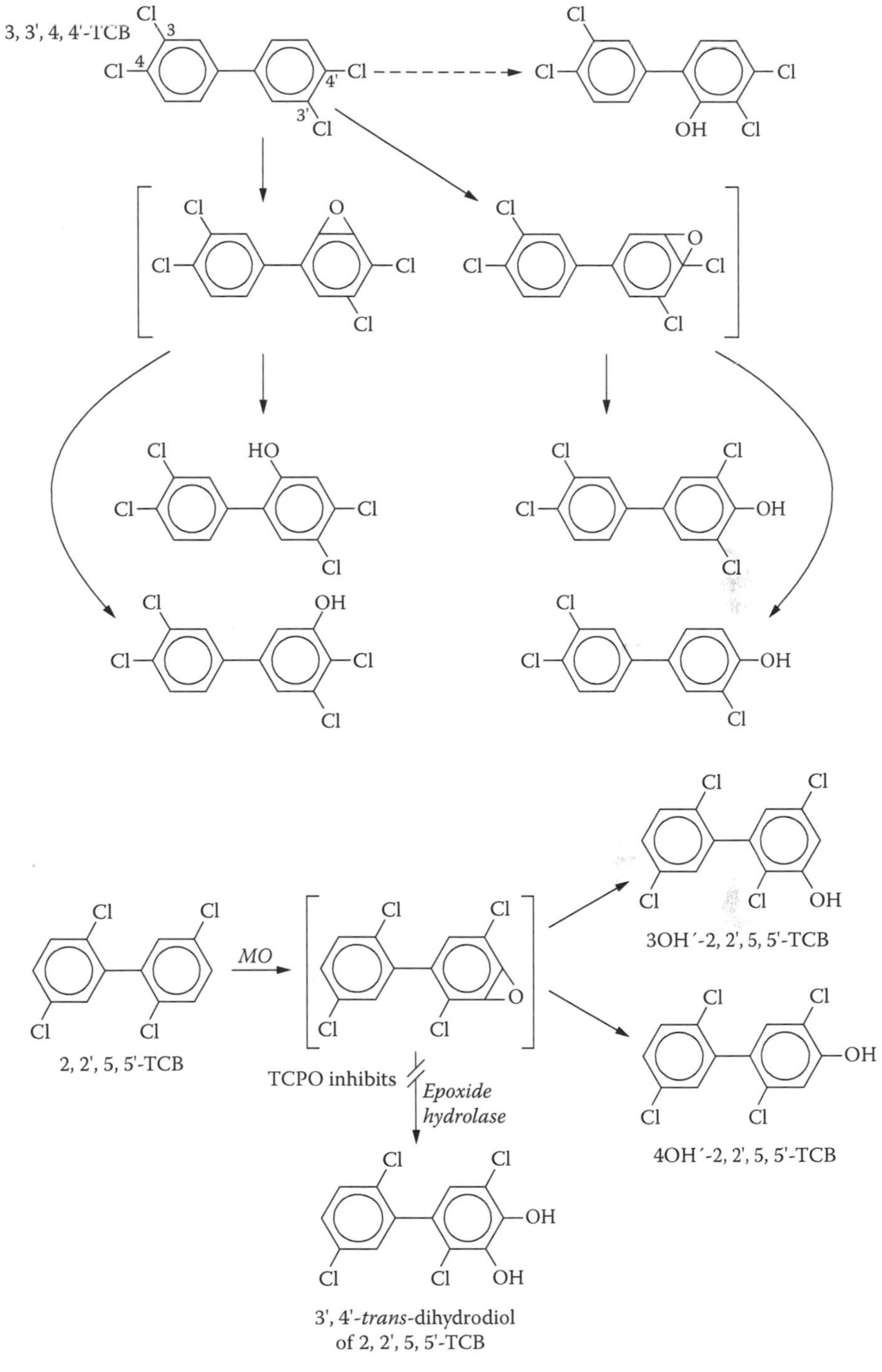

FIGURE 6.3 Metabolism of PCBs.

P450s of gene family 2 (CYP 2) are more catholic, and can metabolize both planar and nonplanar PCBs.

Having more unsubstituted ring positions available for metabolic attack, lower chlorinated PCBs are usually more rapidly metabolized than higher chlorinated PCBs. Reflecting this, the pattern of PCB residues changes with movement along food chains (Figure 6.4). Lower chlorinated PCBs decline in relative abundance or disappear altogether at higher trophic levels. The more highly chlorinated compounds, which are refractory to metabolic attack, become dominant in predators (Boon et al. 1992; Norstrom 1988), which tend to have smaller ranges of PCB congeners as residues than do the species below them in the food chain. These trends are readily recognized by comparing capillary GC analyses of tissues from organisms representing different trophic levels (Figure 6.4). The early fast-running peaks representing lower chlorinated congeners give way to the slower-moving peaks representing more highly chlorinated compounds with movement along the food chain. Some predatory species such as cetaceans and fish-eating birds metabolize PCBs relatively slowly (Walker and Livingstone 1992), in keeping with their very low microsomal monooxygenase activities toward lipophilic xenobiotics (Walker 1980).

Several studies have related the structures of PCBs to their rates of elimination by mammals. In one study (Mizutani et al. 1977), the elimination of tetrachlorobiphenyl

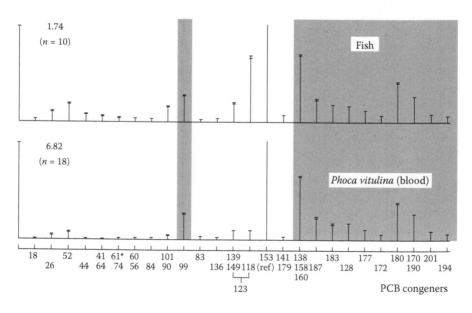

FIGURE 6.4   Mean concentration of CB153 in μg/g pentane-extractable liquid (PEL) in whole fish from the Dutch Wadden Sea and the cellular fraction of the blood of harbor seals. Numbers of CBs are given in order of elution from the GC column by their systematic numbers according to IUPAC rules as proposed by Ballschmitter and Zell (1980). All concentrations are proportional to the height of the bar. (Reproduced from Boon et al. 1992. With permission.)

congeners was studied in mice that had been fed diets containing a single isomer for 20 days. The estimated half-lives were as follows:

2,2′,3,3′-TCB	0.9 days
2,2′,4,4′-TCB	9.2 days
2,2′,5,5′-TCB	3.4 days
3,3′,4,4′-TCB	0.9 days
3,3′,5,5′-TCB	2.1 days

In another study (Gage and Holm 1976), the influence of molecular structure was studied on the rate of excretion by mice for 14 different congeners. The results were as follows:

Most rapidly eliminated
 4,4′-DCB, 3,3′,4′,6′-TCB, 2,2′,3,4′,6′-PCB and 2,2′3,4,4′,5′-HCB
Most slowly eliminated
 2,2′,4,4′,5,5′-HCB and 2,2′, 3,4,4′,5′-HCB

In the latter example, the most slowly eliminated compounds were nonplanar, and lacked vicinal carbons in either the ortho-meta or the meta-para positions that were without any chlorine substitution (i.e., there were no vicinal ortho-meta or meta-para positions that were substituted solely with hydrogen). The more rapidly eliminated compounds all possessed vicinal ortho-meta positions that were without chlorine substitution. In the former example, the most persistent compound was nonplanar, and lacked carbons unsubstituted by chlorine in the meta-para positions. Interestingly, both of the coplanar compounds were eliminated rapidly, even though one of them (3,3′,5,5′-TCB) lacked unsubstituted vicinal carbons in either position. This suggests that P4501A1/1A2 was able to hydroxylate the molecule reasonably rapidly without any vicinal unsubstituted carbons, presumably without the formation of an epoxide intermediate.

In a study with captive male American kestrels (Drouillard et al. 2001), birds were dosed with Aroclor-contaminated diet and the toxicokinetics of 42 PCB congeners contained therein was studied. Those congeners that were most rapidly cleared contained vicinal meta-para hydrogen substituents on at least one phenyl ring. This provides further evidence for the importance of "open" (i.e., not substituted by chlorine) meta-para positions for metabolic attack, an issue that will be returned to in the next section (Section 6.2.3).

Working with rats, Lutz et al. (1977) compared the rates of loss from blood of 4,-CB (rapidly metabolized) with that of 2,2′,4,4′,5′-HCB (slowly metabolized). Both showed biphasic elimination, with the former disappearing much more rapidly than the latter. Estimations were made of the rates of hepatic metabolism in vitro, which were then incorporated into toxicokinetic models to predict rates of loss. The predictions for HCB were very close to actual rates of loss for the entire period of elimination. For 4,-CB, prediction was good for the initial rate of loss, but loss was overestimated in the later stages of the experiment.

Looking at the foregoing results overall, the rates of loss in vivo are related to the rates of metabolism in vitro, measured or estimated. As with the OC insecticides, problems of persistence are associated with compounds that are not readily metabolized, for example, 2,2′,4,4′,5,5′-HCB in the foregoing examples. For further discussion of the dependence of elimination of lipophilic xenobiotics on metabolism, see Walker (1981).

Residues of PCBs in animal tissues include not only the original congeners themselves, but also hydroxy metabolites that bind to cellular proteins, for example, transthyretin (TTR; Klasson-Wehler et al. 1992; Brouwer et al. 1990; Lans et al. 1993). Small residues are also found of methyl-sulfonyl metabolites of certain PCBs (Bakke et al. 1982, 1983). These appear to originate from the formation of glutathione conjugates of primary epoxide metabolites, thus providing further evidence of the existence of epoxide intermediates. Further biotransformation, including methylation, yields methyl-sulfonyl products that are relatively nonpolar and persistent.

PCBs can act as inducers of P450, and consequently accelerate the rate of their own metabolism. Coplanar PCBs bind to the Ah receptor and thereby induce P450s 1A1/1A2. Inductions of P450 1A1/1A2 by organohalogen compounds are associated with a number of toxic effects (Ah-receptor-mediated toxicity), which will be discussed in Section 6.2.4. It should also be noted that induction of these P450s can increase the rate of activation of a number of carcinogens and mutagens, for example, certain PAHs. Nonplanar PCBs can induce cytochrome P450s belonging to family 2. The induction of P450 forms by PCBs and other pollutants provides the basis for valuable biomarker assays that are coming to be widely used in field studies (Rattner et al. 1993; Walker 1998d).

Certain anaerobic bacteria can reductively dechlorinate PCBs in sediments (EHC 140). Higher chlorinated PCBs are degraded more rapidly than lower chlorinated ones, which is in contrast to the trend for oxidative metabolism described earlier. Genetically engineered strains of bacteria have been developed to degrade PCBs in bioremediation programs.

6.2.3   ENVIRONMENTAL FATE OF PCBS

PCBs, similar to persistent OC insecticide residues, have become widely distributed around the globe, including in snow and biota of polar regions (Muir et al. 1992). Long-range aerial transport and subsequent deposition has been the major factor here (Mackay 1991). At the time of writing, little PCB is being released into the environment, but redistribution is evidently still occurring from "sinks" such as contaminated sediments and landfall sites, from which the persistent congeners are only slowly being lost. The levels of higher chlorinated PCBs are still undesirably high in predators—notably mammals and fish-eating birds at the top of marine food chains (Walker and Livingstone 1992; de Voogt 1996).

Although higher chlorinated PCBs are degraded more rapidly than lower chlorinated ones in anaerobic sediments, the reverse is true in terrestrial and aquatic food chains (see Section 6.2.2). As explained earlier, hydroxylations tend to be very slow in the absence of unchlorinated positions favorable for oxidative attack. Recalcitrant higher chlorinated PCBs tend to be strongly bioaccumulated and bioconcentrated with movement along food chains.

An early indication of the tendency for certain PCB congeners to be biomagnified came from studies on the Great Lakes of North America. The concentration of total PCBs in the food chain was found to be as follows:

Phytoplankton	0.0025 ppm
Zooplankton	0.123 ppm
Rainbow trout smelt	1.04 ppm
Lake trout	4.83 ppm
Herring gull eggs	124 ppm

There have been a number of estimates of bioconcentration factors for total PCBs in aquatic species following long-term exposure to PCB mixtures (EHC 140). Values for both invertebrates and fish have been extremely variable, ranging from values below 1 to many thousands. Bioaccumulation factors for birds and mammals for different Aroclors have indicated only limited degrees of bioaccumulation from food, for example, 6.6 and 14.8 for the whole carcasses of big brown bats (*Eptesicus fuscus*) and white pelican (*Pelecanus erythrorhynchos*), respectively (see Environmental Health Criteria 140).

As with OC insecticides (Chapter 5), residue data need to be interpreted with caution. However, it is clear that there can be biomagnification by several orders of magnitude with movement up the food chain. Moreover, the values for total PCBs underestimate the biomagnification of refractory higher chlorinated PCBs. The marked bioaccumulation of refractory PCB congeners is illustrated by the data for fish-eating birds given in Table 6.2 (Walker and Livingstone 1992; Norstrom 1988; Borlakoglu et al. 1988). In the Canadian study on the herring gull, a comparison was first made between the concentration of PCB congeners in eggs, with those present in a fish, the alewife (*Alosa pseudoharengus*), its principal food in the area of study, Lake Ontario. Some 80% of the total PCBs in the birds was accounted for by about 20 refractory congeners (Norstrom 1988). One congener, 2,2',4,4',5 (IUPAC code, PCB No. 153; see also Table 6.1), was among the most strongly bioaccumulated, and was used as a reference compound. It was assigned a "bioaccumulation index" of 1.0, and the bioaccumulation factors of other PCBs were expressed relative to this. In another study (Borlakoglu et al. 1992; Walker and Livingstone 1992), the pattern of PCB congeners found in fish-eating seabirds collected in British and Irish coastal waters were compared with the pattern in the PCB mixture Aroclor 1264. The species studied included the cormorant (*Phalacrocorax carbo*), shag (*Phalacrocorax aristotelis*), guillemot (*Uria aalge*), razorbill (*Alca torda*), and puffin (*Fratercula arctica*).

With both studies, the congeners that were strongly bioaccumulated had one feature in common: they lacked free adjacent meta-para positions on the rings. Also, they were predominantly nonplanar. This suggested that persistence was related to the failure of P450 forms (notably those belonging to family 2) to metabolize such nonplanar congeners. Interestingly, coplanar congeners, for example, 3,3',4,4'-TCB, were not among the most persistent compounds. Their relative abundance was considerably less than in original PCB mixtures. Presumably, they had been extensively

TABLE 6.2
Bioaccumulation of PCB Isomers by Seabirds

IUPAC number	Isomer Cl Substitution		Number of Unsubstituted Adjacent Carbons		Relative BF[a]	Enrichment Index Seabirds[b]
	Ring 1	Ring 2	om	mp		
74	4	2,4,5	2	0	0.86	
66	2,4	3,4	2	0	0.43	
99[c]	2,4	2,4,5	1	0	0.74	
114	4	2,3,4,5	2	0	<0.1	
118[c]	3,4	2,4,5	1	0	0.83	23.60
146	2,3,5	2,4,5	0	0	1.00	
153[c]	2,4,5	2,4,5	0	0	1.00	1.44
105	3,4	2,3,4	2	0	0.57	
138	2,3,4	2,4	0	0	0.93	3.10
178	2,3,5	2,3,4,6	0	0	0.52	
187[c]	2,4,5	2,3,5,6	0	0	0.94	
128[c]	2,3,4	2,3,4	2	0	0.77	
177	2,3,4	2,3,5,6	1	0	0.59	
180[c]	2,4,5	2,3,4,5	0	0	1.10	1.28
170[c]	2,3,4	2,3,4,5	1	0	1.05	1.35
201	2,3,5,6	2,3,4,5	0	0	0.98	
196[c]	2,3,4,5	2,3,4,6	0	0	1.07	
194[c]	2,3,4,4	2,3,4,5	0	0	0.90	
202	2,3,5,6	2,3,5,6	0	0		3.52
52	2,5	2,5	0	2	<0.1	
70	2.5	3,4	1	1	0.02	
101	2,5	2,4,5	0	1	0.24	0.40
97	2,3	2,4,5	1	1	<0.1	0.20
87	2,5	2,3,4	1	1	0.12	
110	3,4	2,3,6	1	1	0.21	
151	2,5	2,3,5,6	0	1	0.02	0.1
149	2,3,6	2,4,5	0	1	0.18	0.17
141	2,5	2,3,4,5	0	1	0.19	
174	2,3,6	2,3,4,5	0	1	0.07	

[a] Bioaccumulation factor (BF) of herring gull (*Larus argentatus*) eggs/Alewife (*Alosa pseudoharengus*) relative to PCB no 153 (Norstrom 1988).

[b] The enrichment index, which is PCB congener as a percentage of total PCB in seabird depot fat/PCB as a percentage of total PCB in Aroclor 1260.

[c] Among the 14 PCB congeners found at the highest concentration in eggs from the Mediterranean Sea and the Atlantic Ocean (Renzoni et al. 1986). These authors also found numbers 156, 172, and 183. Numbers 172 and 183 were also reported in Borlakoglu et al. (1988).

metabolized by the birds, or lower in the food chain by monooxygenase attack. The metabolism of planar compounds such as these is particularly associated with P4501A1/1A2, which are found in birds and in fish, although their characteristics are somewhat different from mammalian forms (Livingstone and Stegeman 1998; Walker 1998a). Cetaceans and seals also have a marked capacity to bioaccumulate a limited number of refractory highly chlorinated PCBs (Boon et al. 1992; Tanabe and Tatsukawa 1992). Seals appeared to metabolize PCBs lacking free meta-para ring positions more rapidly than did cetaceans, irrespective of whether the molecules were planar or nonplanar. In a study of total PCB residues in cetaceans from the Pacific Ocean, high levels were found in certain species of whale, with particularly high levels in killer whales (*Orcinus orca*), which contained about 800 ppm in blubber (Tanabe and Takatsukawa 1992). Killer whales were found to have very low levels of hepatic microsomal monooxygenase activity toward aldrin and aniline, that is, 0.041 and 0.22 nmol/mg protein/min. Comparable values for the rat were 1.58 and 2.24, respectively. Activities toward these substrates have been attributed mainly to P450s of family 2, and these data suggest very low levels of this type of detoxication in the killer whale. This fits in well with the general picture of poor P450-mediated oxidative metabolism of lipophilic xenobiotics by specialized predators (see Chapter 2).

6.2.4 THE TOXICITY OF PCBs

The acute toxicity of PCB mixtures to vertebrates tends to be low, typically, 1–10 g/kg to rats. The concern is about sublethal and chronic toxicity. Different PCB congeners show different modes of toxic action, so it is not surprising that their mixtures can produce a disconcerting range of toxic effects (Robertson and Hansen 2001).

Some early work on PCB toxicity in birds attempted to deal with this complexity, drawing attention to, among other things, effects on the thyroid (Jefferies 1975; Jefferies and Parslow 1976). Unfortunately, this work was largely overlooked at the time, and it was not until some time later that mechanisms of toxicity of PCBs began to be unraveled that could explain some of the effects reported in this study. Another much more recent study pointing to sublethal effects of PCBs on birds involved the dosing of Aroclor mixtures to captive American kestrels (Fernie et al. 2001). Birds so dosed laid eggs containing environmentally relevant levels of PCBs (34 μg/g whole egg wet weight). In a two-year study, reproductive effects were observed during but not after dosing. PCB-dosed pairs laid smaller clutches later in the season and laid more totally infertile clutches than did controls. It was interesting that these and other differences observed between treated birds and controls were only evident during dosing and not afterward. This pointed to effects being caused by readily degradable PCBs, that is, those containing one or more "open" meta-para positions, as discussed earlier (see Section 6.2.2).

It has come to be suggested that coplanar congeners as a group express toxicity through a common mechanism: interaction with the cytosolic Ah receptor (Safe 1990; Ahlborg et al. 1994) (Environmental Health Criteria 140). Although the full picture has yet to be elucidated, many toxic endpoints, including porphyria, indices of hepatotoxicity, and mortality of embryos of birds are correlated with the degree to which planar polychlorinated compounds bind to the Ah receptor. As described earlier (Chapter 2, and Section 6.2.2), interaction with the Ah receptor also leads to induction of P450 1A1/1A2, a response that is closely linked to what has been termed Ah-receptor-mediated toxicity (Fernandez-Salguero et al. 1996).

The complex formed between coplanar PCBs and the Ah receptor migrates to the nucleus, where it triggers the induction of the P450s and certain other enzymes, events associated with the initiation of toxic responses by mechanisms as yet unknown. It should be added that P4501A1/1A2 is involved in the activation of some mutagens/carcinogens. Thus, if organisms are exposed to coplanar PCBs in combination with, for example, mutagenic PAHs, there may be an increased rate of DNA adduct formation because the former induce P4501A1/2 (see Walker and Johnston 1989).

Ah-receptor-mediated toxicity is particularly associated with the highly toxic compound 2,3,7,8-tetrachloro-dibenzo-p-dioxin (TCDD), commonly referred to as *dioxin*. TCDD, and the concept of toxicity equivalency factors (TEFs) based on TCDDs, will be dealt with in Chapter 7. The main point to make at this juncture is that the toxicity of each individual coplanar congener in a mixture can be expressed in terms of a toxic equivalent calculated relative to the toxicity of dioxin. Summation of the toxic equivalents of the individual coplanar PCBs gives a measure of the toxicity of the whole mixture, as expressed through the Ah receptor mechanism.

In experimental mammals and certain other vertebrates, another mechanism of action shown by certain coplanar PCBs depends on metabolic activation: conversion of the PCBs to monohydroxy metabolites by cytochrome P4501A1/1A2 (Figure 6.3). 4′OH 3,3′,4,5′-TCB is a metabolite of 3,3′4,4′ TCB, which is closely related structurally to thyroxine (T4). 4′OH 3,3′,4,5′ competes very strongly with T4 for binding positions on the protein transthyretin (TTR) (Figure 6.3; Brouwer et al. 1990; Brouwer 1991; Brouwer et al. 1998). Some other hydroxy metabolites of PCBs do the same, although less strongly (Lans et al. 1993). When the metabolite binds to TTR, thereby excluding T4, the associated retinol (vitamin A) binding protein breaks away (Figure 6.5). Thus, the TCB metabolite can reduce the levels of bound thyroxine and retinol in blood with consequent physiological effects (e.g., vitamin A deficiency). These changes have provided the basis for the development of biomarker assays for the toxic action of PCBs (Brouwer 1991). 3,4,3′,4′-TCB is also an inducer of P4501A1/1A2, and can enhance the rate of its own activation if the levels of exposure are high enough (for further discussion of this question, see Walker and Johnston 1989). In a laboratory study with common tern (*Sterna hirundo*) chicks, dosing with coplanar 3,3′4,4′5-PCB either alone or in combination with the nonplanar 2,2′4,4′5,5′-HCB caused a reduction in plasma total thyroxine, which was negatively correlated with hepatic TEQ concentrations (Bosveld 2000). Cyto P4501A1 was also induced in these birds.

In mammals, there is evidence that hydroxy PCBs are transferred across the placenta into the fetus, where they accumulate (Morse et al. 1995). In an experiment with pregnant rats exposed to 3,3′,4,4′-TCB, substantial levels of 4′ OH,3,3′,4,5′-TCB accumulated in the fetus, with concomitant reduction in levels of T4. This was due to competitive binding of the PCB metabolite to TTR, thereby excluding T4. Similar

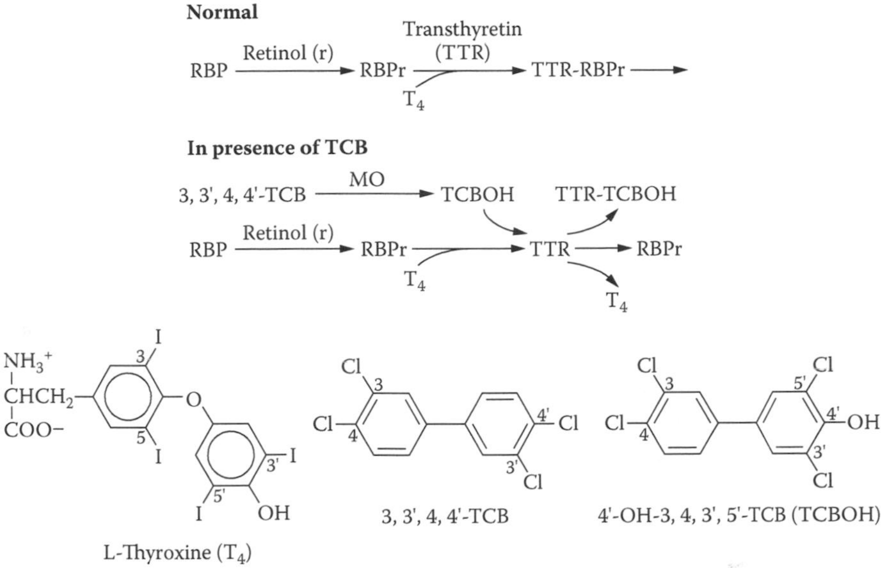

Normal

Transthyretin

RBP $\xrightarrow{\text{Retinol (r)}}$ RBPr $\xrightarrow[\text{T}_4]{\text{(TTR)}}$ TTR-RBPr $\longrightarrow$

In presence of TCB

3, 3', 4, 4'-TCB $\xrightarrow{\text{MO}}$ TCBOH   TTR-TCBOH

RBP $\xrightarrow{\text{Retinol (r)}}$ RBPr $\xrightarrow[\text{T}_4]{}$ TTR $\longrightarrow$ RBPr

T$_4$

L-Thyroxine (T$_4$)

3, 3', 4, 4'-TCB

4'-OH-3, 4, 3', 5'-TCB (TCBOH)

FIGURE 6.5 Thyroxine antagonism. Mechanism of toxicity of a polychlorinated biphenyl. Retinol (r) binds to retinol-binding protein (RBP), which is then attached to transthyretin (TTR). Thyroxine (T$_4$) binds to TTR and is transported via the blood in this form. The coplanar PCB, 3,3',4,4'-tetrachlorobiphenyl (3,3',4,4'-PCB), is converted into hydroxymetabolites by the inducible cytochrome P450 called P4501A1. The metabolite 4'-OH-3,3',4,5'-tetrachlorobiphenyl (TCBOH) is structurally similar to thyroxine and strongly competes for thyroxine binding sites. The consequences are loss of thyroxine from TTR, the fragmentation of the TTR–RBP complex, and loss of both thyroxine and retinol from blood (after Brouwer 1991).

effects were found when pregnant rats were exposed to Aroclor 1254, where relatively high levels of 4 OH, 2,3,3',4',5 pentachlorobiphenyl were detected in fetal plasma and brain (Morse et al. 1995). Similar effects have also been demonstrated in pregnant mice fed 3,3',4,4'-TCB (Darnerud et al. 1996). Changes in fetal and neonatal levels of thyroid hormones caused by PCBs may have a number of harmful consequences, including effects on brain development and behavior.

The functional form of thyroxine (T3) is generated by the deiodination of T4, and PCBs can influence the tissue levels of this form by disturbing metabolism, as well as by reducing the binding of T4. PCBs have been shown to inhibit the sulfation of thyroid hormones and the deiodination of T4 to T3. They can also induce the glucuronyl transferase that conjugates T4 (Brouwer et al. 1998).

There is considerable evidence that PCB mixtures and higher chlorinated congeners act as immunosuppressants in mammals (Environmental Health Criteria 140). Apart from a number of studies showing effects on gross measures of immunological function (e.g., spleen and thymus weights, lymphocyte counts, and histology of lymphoid tissue), functional alterations in humoral and cell-mediated immunity have also been demonstrated. Immunosuppression has been reported in seals fed fish diets rich in higher chlorinated PCBs (see de Voogt 1996). In a study with the harbor seal

(*Phoca vitulina*), pups were dosed PCBs and effects measured on mitogen-induced peripheral blood lymphocyte proliferation (Levin et al. 2004). A positive correlation was found between PCB concentrations in blubber and both T-cell mitogen- and B-cell mitogen-induced lymphocyte proliferation.

Before leaving the question of mechanisms of toxicity, two other issues will be briefly mentioned. The first, which is more relevant to human toxicology than to ecotoxicology, is the question of carcinogenicity. There is no clear evidence that PCB mixtures are primary carcinogens. They do, however, induce P450s that activate carcinogens and mutagens, and there is evidence that treatment with PCBs after exposure to certain carcinogens leads to an enhancement of carcinogenic action (Environmental Health Criteria 140). That said, there is still uncertainty about the importance of carcinogenicity in wild species. It has been questioned whether it can have a significant effect on population numbers in the wild. The other point deserving mention is the possibility that refractory PCB congeners may promote oxyradical toxicity. When they are bound as substrates to P450s, they are not susceptible to the normal process of oxidation (see Chapter 2), and highly reactive oxygen species, such as superoxide anion (O_2.-), may escape from the vicinity of the heme nucleus and cause damage to neighboring cellular macromolecules, instead of interacting with the PCB. More work needs to be done to test the validity of this theory.

6.2.5   Ecological Effects of PCBs

6.2.5.1   Physiological and Biochemical Effects in the Field

Before dealing with studies that have attempted to relate levels of PCBs in the environment with population changes, evidence will be discussed for physiological and biochemical effects caused by these compounds, effects that have been detected in wild vertebrates using biomarker assays. A number of field studies have linked the induction of P450s with the presence of PCBs in fish and birds. In one early study (Ronis et al. 1989), the levels of P450 IA1 measured by Western blotting were related to levels of total PCB in body fat in three species of fish-eating seabirds collected from the Irish Sea during 1978–1988. The species were cormorant, razorbill, and puffin. In a later study of pipping embryos of the black-crowned night (*Nycticorax nycticorax*), induction of P450s was related to total PCB (Rattner et al. 1993). There were close correlations between total PCB concentration in tissues and both ethoxy resorufin deethylase activity (EROD) and arylhydrocarbon hydroxylase (AHH) activities in hepatic microsomes, thus indicating P450 1A1 induction. In a study on osprey chicks in Canada and Northern United States, levels of P450 1A in liver were found to be related to total PCBs and dioxin equivalents in yolk sacs (Elliott et al. 2001). All of these studies give evidence for induction of this heme protein in birds in the field, thus raising questions about possible Ah-receptor-mediated toxicity caused by these compounds.

6.2.5.2   Population Effects

One widely reported incident was a large "seabird wreck" that occurred in the Irish Sea in the autumn of 1969 (Natural Environment Research Council Publication

1971). An estimated 17,000 dead seabirds were washed ashore along the coastline of the Irish Sea. Most of them were guillemots, but there were also razorbills, gannets, and herring gulls. In the following season, the breeding numbers of some of these species in the region of the Irish Sea were depressed, and this may have been due, at least in part, to this large-scale kill. Birds found dead contained, on average, some 56 ppm of PCB in liver. The significance of such a high concentration seems to have been overlooked in the report, which attempts to relate total body burdens rather than critical tissue concentrations to possible effects (Walker 1990a). Unfortunately, in the absence of analytical data on individual congeners and estimations of TEQ values, it is not possible to draw any strong conclusions about causality. However, certain other points deserve consideration. Some of the birds showed pathological changes in liver and kidney that were consistent with PCB poisoning. Also, some birds contained liver concentrations of total PCB that were high enough to suggest lethal poisoning, on the basis of the limited evidence available. Cormorants experimentally poisoned with Clophen 60 had the following tissue levels of total PCB: liver, 210–285 ppm; brain, 76–180 ppm (Koeman 1972). The problem is that, in the absence of analytical data for individual congeners, the TEQ values for coplanar PCBs in these samples cannot be estimated, and total PCB concentrations provide only a rough guide to likely toxic effects. Differences in susceptibility to PCBs between cormorants and other birds and differences in susceptibility between birds in the field and laboratory are unknown (there are likely to be situations in which birds are more susceptible to particular pollutants in the field than in the laboratory, because of the impact of other factors, e.g., disease, shortage of food, and bad weather). With the benefit of hindsight, it is clear that PCDDS and PCDF residues as well as PCB residues may have contributed to Ah-receptor-mediated toxicity. However, these were not determined because suitable methods of analysis were not available at the time of the seabird wreck. At the end of the day, the balance of evidence suggests that PCBs were an important contributory factor to this environmental disaster.

Some deaths of cormorants in the field in Holland were attributed to PCB poisoning on more substantial evidence than was available in the foregoing incident (Koeman 1972; Walker 1990a). Birds found dead had mean liver and brain levels of 319 and 190 ppm, respectively, which are not very different from those found in experimentally poisoned individuals.

PCBs have been implicated in a number of population declines reported from the field. Some of the best-documented examples are of certain predatory birds inhabiting the Great Lakes of North America. These observations were made between the early 1970s and the mid-1990s (Gilbertson et al. 1998). The herring gull, the Caspian tern (*Hydroprogne caspia*), and the double-crested cormorant (*Phalacrocorax auritus*), which are largely or entirely fish-eating species, showed population declines when PCB levels were high, a situation that was complicated by the presence of relatively high levels of other pollutants (e.g., *p,p'*-DDT and its metabolites). Recoveries have occurred with falling levels of PCBs, *p,p'*-DDE, and other pollutants. It is a situation in which interpretation of data is made more difficult because of the correlation between the temporal trends of different pollutants.

With the fall of p,p'-DDE levels, and associated eggshell thinning, populations of these affected species have recovered in many areas. However, reproductive failure, and physical deformities (such as crossed bills in double-crested cormorants) lasted into the mid-1990s in some areas where PCB levels remained high. With the Caspian tern, there was a strong correlation between TEQs (dioxin equivalents) in eggs, and embryonic mortality. With the double-crested cormorant, there was a negative correlation between TEQ values and reproductive success, and a positive correlation between TEQ and the incidence of crossed bills (Gilbertson et al. 1998). As with the Caspian tern, this species showed a negative correlation between TEQ values and embryonic mortality; however, the slopes of the regression lines were very different in the two species (Ludwig et al. 1996). There is, therefore, evidence linking the depressed state of certain populations of piscivorous birds in the Great Lakes with Ah-receptor-mediated toxicity caused by planar PCBs and other planar polyhalogenated aromatic compounds.

It has also been suggested that seal populations in and near the North Sea (Wadden and Baltic seas) have been adversely affected by PCBs (Brouwer 1991; Ross et al. 1995; Boon et al. 1992). Thyroid hormone antagonism, skeletal deformities, impaired reproduction, and immunosuppression have been reported either in free-living animals or in animals dosed with fish caught in the North Sea that contained high levels of PCBs. It was also suggested that the spread of a distemper virus that caused high mortality in seal populations was promoted by immunosuppression due to relatively high environmental levels of PCBs. As noted in Section 6.2.4, there is experimental evidence to lend some support to this theory (de Voogt et al. 1996; Levin et al. 2004). Population declines of Californian sea lions have also been linked to high tissue levels of PCBs (Environmental Health Criteria 140).

Despite wide-ranging restrictions and limitations on their release, levels of PCBs have been slow to come down in certain locations. It appears that redistribution of PCBs from sinks is still going on. It has frequently been suggested that they have had—and in some cases are still having—adverse effects on predators at the top of food chains, for example, fish-eating birds in some parts of the Great Lakes, and in marine mammals. The complexity of PCB pollution, with the possibility of interactive effects between different PCB congeners and/or between PCBs and other persistent pollutants, has made this a difficult point to prove or disprove. There is a need for the development and application of biomarker assays that can provide evidence of causality, to link levels of pollutants, taken singly or in combination, with consequent harmful effects, and then to relate the harmful effects to the state of populations using population dynamic models (see Chapters 12 and 15 in Walker et al. 2000).

Apart from direct evidence of population declines that may be due, largely or entirely, to PCBs, some studies indicate effects on reproductive success, which may be translated into population declines. In a study of osprey populations in the Delaware river and bay in 2002, Toschik et al. (2005), using a logistic regression model, report that the concentration of PCBs and other persistent organochlorine pollutants in eggs were predictive of hatching success. Arenal et al. (2004) compared the breeding success of starlings (*Sturnus vulgaris*) at a Superfund site (high PCB contamination) with that in a reference site. At the Superfund site, EROD levels were elevated in 15-day chicks, but not in adults; reduced nest provisioning behavior and decreased chick survival were also observed.

6.2.5.3 Population Genetics

Genetic resistance to PCBs has been reported in a nonmigratory estuarine fish (*Fundulus heteroclitus*) from New Bedford Harbor in Massachusetts, a body of water that is highly polluted by them (Nacci et al. 1999; Bello et al. 2001). Interestingly, resistant fish from the New Bedford Harbor showed less induction of P4501A1 when treated with coplanar PCB than did fish from control areas. In a general way, this fits with the concept of Ah-receptor-mediated toxicity discussed in Chapter 7, Section 7.2.4. Insensitivity to P4501A1 induction suggests that planar PCBs, dioxins, etc., are less likely, in resistant fish, to operate the toxic mechanisms normally associated with this event; in turn, this could mean poor affinity for the Ah receptor in the resistant fish. It should also be noted that P4501A has an activating function toward coplanar PCBs such as 3,3′,4,4′-tetrachlorobiphenyl (see Section 6.2.4). In other words, resistance may be the consequence of failure to induce the activation of one or more coplanar PCBs—and this again could come back to unresponsiveness of the Ah receptor to coplanar PCBs.

6.3 POLYBROMINATED BIPHENYLS

The principal source of pollution by polybrominated biphenyls (PBBs) has been the commercial mixture "Firemaster," which was produced in the United States between 1970 and 1974. Production was discontinued in 1974 following a severe pollution incident in Michigan, when Firemaster was accidentally mixed with cattle feed on a farm. In due course, PBBs entered the human food chain via contaminated animal products. Substantial residues were found in humans from the area, and were subsequently found to be highly persistent.

Firemaster is a stable solid, resembling a PCB mixture in its lipophilicity, chemical and thermal stability, and low vapor pressure. Firemaster contains some 80 out of a possible 209 PBB congeners, but just two of them—2,2′,4,4′,5,5-hexabromobiphenyl and 2,2′,3,4,4′,5,5′, heptabromobiphenyl—account for around 85% of the commercial product (Environmental Health Criteria 152). These two compounds were found to be very slowly eliminated by humans exposed to them during the Michigan incident. A half-life of about 69 weeks was estimated for 2,4,5,2′,4′,5′-HBB.

PBB residues became widespread in terrestrial and aquatic ecosystems of the United States (Sleight 1979) and Europe. On account of their lipophilicity and slow elimination by mammals, hexa- and heptabromobiphenyls may be expected to undergo biomagnification with movement along food chains in an analogous manner to higher chlorinated PCBs. Although they have been shown to have severe toxic effects on farm animals receiving high doses, nothing is known of any toxic effects that they may have had on wild vertebrates.

6.4 SUMMARY

PCB mixtures were once used for a variety of purposes, and came to cause widespread environmental pollution. Over 100 different congeners are present in commercial products such as Aroclor 1248 and Aroclor 1254. PCBs are lipophilic, stable, and of low vapor pressure. Many of the more highly chlorinated PCBs are refractory, showing very strong biomagnification with movement along food chains.

The toxicology of PCBs is complex and not fully understood. Coplanar PCBs interact with the Ah-receptor, with consequent induction of cytochrome P4501A1/2 and Ah-receptor-mediated toxicity. Induction of P4501A1 provides the basis of valuable biomarker assays, including bioassays such as CALUX. Certain PCBs, for example, 3,3′,4,4′-TCB, are converted to monohydroxymetabolites, which act as thyroxine antagonists. PCBs can also cause immunotoxicity (e.g., in seals).

PCBs have been implicated in the decline of certain populations of fish-eating birds, for example, in the Great Lakes of North America. Although their use is now banned in most countries and very little is released into the environment as a consequence of human activity, considerable quantities remain in sinks (e.g., contaminated sediments and landfill sites), from which they are slowly redistributed to other compartments of the environment. There continues to be evidence that PCB residues are still having environmental effects, for example, on birds and fish.

PBB mixtures have been used as fire retardants. Many of their constituent congeners are highly persistent, and there was a major environmental accident in the United States in which farm animals and humans became heavily contaminated by them.

FURTHER READING

Elliott, J.E., Wilson, L.K., and Henny, C.J. et al. (2001). Assessment of biological effects of chlorinated hydrocarbons in osprey chicks. *Environmental Toxicology and Chemistry* 20, 866–879.

Environmental Health Criteria 140 (1993). Polychlorinated Biphenyls and Terphenyls—A detailed reference work giving much information on the environmental toxicology of PCBs.

Fernandez-Salguero, P.M., Hilbert, D.M., Rudikoff, S. et al. (1996). Aryl-hydrocarbon receptor-deficient mice are resistant to 2,3,7,8-tetrachlorodibenzo-p-dioxin-induced toxicity. *Toxicology and Applied Pharmacology* 140, 173–179.

Levin, M., De Guise, S., and Ross, P. (2004). Association between lymphocyter proliferation and PCBs in free-ranging harbour seal (*Phoca vitulina*) pups from British Columbia. *Canada Environmental Toxicology and Chemistry* 24, 1247–1252.

Robertson, L.W. and Hansen, L.G. (Eds.) (2001). *PCBs: Recent Advances in Environmental Toxicology and Health Effects*—A wide ranging collection of articles arising from a meeting held at the University of Kentucky in 2000.

Safe, S. (1990). An authoritative account of the toxicology of PCBs, and the development of toxic equivalency factors.

Toschik, P.C., Rattner, B.A., McGowan, P.C. et al. (2005). Effects of contaminant exposure on reproductive success of Ospreys nesting in Delaware river and bay. *Environmental Toxicology and Chemistry* 24, 17–628.

Waid, J.S. (1985–1987). *PCBs in the Environment 1-111*. A useful source of information on environmental pollution by PCBs REFS.

7 Polychlorinated Dibenzodioxins and Polychlorinated Dibenzofurans

7.1 BACKGROUND

Neither of these groups of polychlorinated lipophilic compounds occurs naturally, and neither of them is synthesized intentionally. Both occur as by-products of chemical synthesis, industrial processes, and occasionally of interaction between other organic contaminants in the environment. Because of human exposure to them during the Vietnam War, industrial accidents, and the high mammalian toxicity of some of them, they have received much attention as hazards to human health. There has also been concern about some polychlorinated dibenzodioxins (PCDDs) and polychlorinated dibenzofurans (PCDFs) from an ecotoxicological point of view because they combine marked biological persistence with high toxicity. The following account will deal mainly with PCDDs, which have been studied in some detail and have been important in developing the concept of Ah receptor-mediated toxicity. Less is known about PCDFs, which will be described only briefly.

7.2 ORIGINS AND CHEMICAL PROPERTIES

PCDDs are polychlorinated planar molecules with an underlying structure of two benzene rings linked together by two bridging oxygens, thus creating a third "dioxin" ring (Figure 7.1). They are sometimes simply termed *dioxins* (Environmental Health Criteria 88). In theory, there are 75 different congeners, but only a few of them are regarded as being important from an ecotoxicological point of view. 2,3,7,8-tetra-chlorodibenzodioxin (TCDD) has received far more attention than the others due to its high toxicity and persistence and the detection of significant levels in the environment (structure given in Figure 7.1). They are formed when o-chlorophenols, or their alkali metal salts, are heated to a high temperature (see Crosby 1998). The formation of 2,3,7,8-TCDD from an orthotrichlorophenol is shown in Figure 7.1.

PCDDs have been released into the environment in a number of different ways. Sometimes this has been due to the use of a pesticide that is contaminated with them. 2,4,5-T and related phenoxyalkanoic herbicides have been contaminated with them as a consequence of the interaction of chlorophenols used in the manufacturing

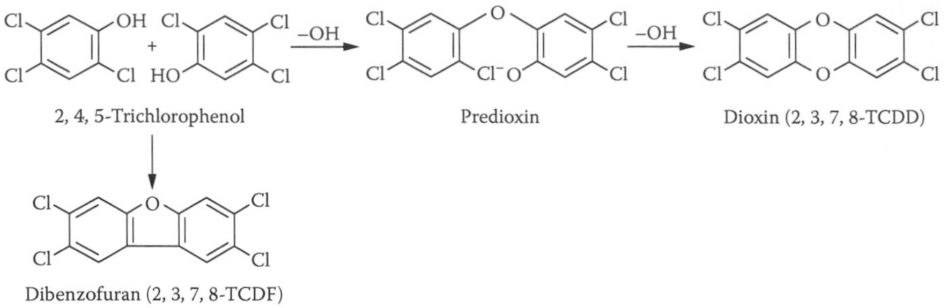

FIGURE 7.1 Formation of dioxin and dibenzofuran (from Crosby 1998).

process. Relatively high levels of 2,3,7,8-TCDD occurred in a herbicide formulation containing 2,4,5-T, which was known as "Agent Orange." It was sprayed as a defoliant on extensive areas of jungle during the Vietnam War. Consequently, many humans as well as wild animals and plants were exposed to dioxin. PCDDs are also present in pentachlorophenols, which are used as pesticides.

Industrial accidents have also led to environmental pollution by PCDDs. In 1976, there was an explosion at a factory in Seveso, Italy, which was concerned with the production of trichlorophenol antiseptic. A cloud containing chlorinated phenols and dioxins was released, which caused severe pollution of neighboring areas. People who had been exposed showed typical symptoms of early PCDD intoxication (chloracne). Another source of PCDDs is the effluent from paper mills where wood pulp is treated with chlorine (Sodergren 1991). This has been a problem in Northern Europe (including Russia) and in North America. Evidently, chlorine interacts with phenols derived from lignin to generate chlorophenols, which then interact to form dioxins. Finally, PCDDs and PCDFs can be generated during the disposal of PCB residues by combustion in specially designed furnaces. If combustion is incomplete in the furnaces, PCDDs and PCDFs can be formed and released into the air to pollute surrounding areas. Presumably, chlorinated phenols are first produced, which then interact to form PCDDs and PCDFs. Investigation of such cases of pollution has sometimes led to the closure of the commercial operations responsible for it.

2,3,7,8-TCDD has been more widely studied than other PCDDs, and will be taken as an example for the whole group of compounds. It is a stable solid with a melting point of 306°C. Its water solubility is very low, which has been estimated to be 0.01–0.2 μg/L; its log K_{ow} is 6.6. More highly chlorinated PCDDs are even less soluble in water.

PCDFs are similar in many respects to PCDDs but have been less well studied, and will be mentioned only briefly here. Their chemical structure is shown in Figure 7.1. Like PCDDs, they can be formed by the interaction of chlorophenols, and are found in commercial preparations of chlorinated phenols and in products derived from phenols (e.g., 2,4,5-T and related phenoxyalkanoic herbicides). They are also present in commercial polychlorinated biphenyl (PCB) mixtures, and can be formed during the combustion of PCBs. They have similar physical properties to PCDDs, and have low water solubilities and high K_{ow} values. The compound 2,3,7,8-TCDF has a log K_{ow} value of 5.82. Some PCDFs bind very strongly to the Aryl hydrocarbon (Ah) receptor.

From here on, these two related groups of pollutants will be treated together.

7.3   METABOLISM

The compound 2,3,7,8-TCDD has been much more widely studied than other PCDDs and will be taken as representative of the group. Metabolism is slow in mammals. Because of the high toxicity of the compound, only low doses can be given in in vivo experiments, making the quantification of metabolites difficult. However, there appear to be two distinct types of metabolite: (1) monohydroxylated PCDD deriva-tives, and (2) products of the cleavage of one of the ether bonds of the dioxin ring. Both types can be generated following epoxidation of the aromatic ring. Epoxidation in the meta-para position yields metabolites of category 1, and epoxidation in the ortho-para position yields metabolites of category 2 (see Environmental Health Criteria 88, and Poiger and Buser 1983). Hydroxymetabolites excreted in rat bile are present largely as glucuronide or sulfate conjugates.

7.4   ENVIRONMENTAL FATE

Higher chlorinated PCDDs, including 2,3,7,8-TCDD, are lipophilic and biologi-cally stable, and are distributed in the environment in a similar fashion to higher chlorinated PCBs, reaching relatively high levels at the top of food chains. Within vertebrates, however, they show a greater tendency to be stored in liver, and a lesser tendency to be stored in fat depots than most PCBs. Some biological half-lives for 2,3,7,8-TCDD are given in Table 7.1.

Because of its high toxicity, there is concern about very low levels of 2,3,7,8-TCDD in biota. This raises analytical problems, and high resolution capillary gas chromatog-raphy (GC) is needed to obtain reliable isomer-specific analyses at low concentrations. In the analysis of herring gull eggs collected from the Great Lakes, Hebert et al. (1994)

TABLE 7.1
Half-Lives of 2,3,7,8-TCDD

Species	Dose (route)	Half-Life (days)
Rat (different strains)	1–50 µg/kg (oral)	17–31
Mouse (different strains)	0.5–10 µg/kg (ip)	10–24
Hamster	650 µg/kg (oral)	15
Guinea pig	0.56 µg/kg (ip)	94

reported the following residues of 2,3,7,8-TCDD. Further details of this study are given in Norstrom et al. 1986.

Lake Ontario, early 1970s	2–5 µg/kg (i.e., 0.002–0.005 ppm by weight)
Lake Ontario, 1984/1985	0.08–0.1 µg/kg
Lake Michigan, 1971	0.25 µg/kg
Lake Michigan, 1972	0.07 µg/kg
Lake Michigan, 1984/1985	0.001–0.002 µg/kg

In another study conducted during 1983–85, fish from the Baltic Sea were found to contain 0.003–0.029 µg/kg of 2,3,7,8-TCDD (Rappe et al. 1987).

Recognizing the widespread occurrence of PHAHs in the natural environment, interest has grown in the development of simple rapid assays that can be used in a cost-effective way to identify "hot spots," where there are particularly high levels of them. Indeed, a wide range of tests are now available that are suitable for such biomonitoring (Persoone et al. 2000). In the next section, details will be given of the CALUX assay, which is based on a line of rat hepatoma cells that are responsive to "dioxin-like" compounds. This assay has been used for environmental monitoring. In one study, sediments were sampled from coastal and inland sites in the Netherlands and assayed for dioxin-like activity (Stronkhurst et al. 2002). The importance of this approach is that it measures the operation of a toxic mechanism in environmental samples and so brings to attention particular areas that require further investigation. Once hot spots have been identified, chemical analysis and biomarker assays can be used, which are expensive and time-consuming, to gain more detailed information about the nature and scale of the problem.

7.5 TOXICITY

2,3,7,8-TCDD is a compound of very high toxicity to certain mammals, and there has been great interest in the elucidation of its mode of action. The situation is complicated, at least on the surface, by the variety of symptoms associated with dioxin toxicity. Symptoms include dermal toxicity, immunotoxicity, reproductive effects, teratogenicity, and endocrine toxicity, which, together with induction of CYP1A, have been associated with the very strong binding of this molecule to the Ah receptor. These toxic effects have been referred to collectively as *Ah-receptor-mediated toxicity*. Interestingly, a strain of mice deficient in the Ah receptor do not respond in this way to 2,3,7,8-TCDD or to related compounds that behave in a similar way (Fernandez-Salguero et al. 1996). Although this observation supports the idea that toxicity is being mediated through the Ah receptor, it does not prove that this is the case. Until the mechanisms are known by which this toxicity is expressed, there will remain questions about whether this toxicity is truly mediated through this receptor, or whether this occurs through another receptor (or receptors) with similar binding properties to the Ah receptor, which is similarly affected by dioxin-like compounds. There is a further potential problem; it is well known that cytochrome P450 1A1 has a marked capacity to activate planar compounds such as PAHs or coplanar PCBs. To what extent, then, is Ah-receptor-mediated toxicity simply a consequence of the enhanced activation of compounds such as these (perhaps even endogenous planar compounds) when P450 1A1 is induced?

Apart from 2,3,7,8-TCDD, other PCDDs, PCDFs, and coplanar PCBs also interact with the Ah receptor causing induction of P450 1A1 with associated toxic effects (dioxin-like compounds). There are, however, large differences between individual compounds of this type both in their affinity for the receptor and in their toxic potency. Notwithstanding the theoretical uncertainties described in the preceding text, attempts have been made to develop a practical approach for risk assessment for these compounds more generally by estimating toxic equivalency factors (Safe 1990, 2001; Ahlborg et al. 1994).

Toxic equivalency factors (TEFs) are estimated relative to 2,3,7,8-TCDD, which is assigned a value of 1. They are measures of the toxicity of individual compounds relative to that of 2,3,7,8-TCDD. A variety of toxic indices, measured in vivo or in vitro, have been used to estimate TEFs, including reproductive effects (e.g., embryo toxicity in birds), immunotoxicity, and effects on organ weights. The degree of induction of P450 1A1 is another measure from which estimations of TEF values have been made. The usual approach is to compare a dose-response curve for a test compound with that of the reference compound, 2,3,7,8-TCDD, and thereby establish the concentrations (or doses) that are required to elicit a standard response. The ratio of concentration of 2,3,7,8-TCDD to concentration of test chemical when both compounds produce the same degree of response is the TEF. Once determined, a TEF can be used to convert a concentration of a dioxin-like chemical found in an environmental sample to a toxic equivalent (TEQ).

Thus, $[C] \times TEF = TEQ_{dioxin}$, where $[C]$ = environmental concentration of planar polychlorinated compound. The TEQ is an estimate of the concentration of TCDD that would produce the same effect as the given concentration of the dioxin-like chemical.

The criteria for including a compound in the aforementioned scheme and assigning it a TEF value, were set out in a WHO–European Centre for Environmental Health consultation in 1993 (Ahlborg et al. 1994). They are as follows:

1. The compound should show a structural relationship to PCDDs and PCDFs.
2. It should bind to the Ah receptor.
3. It should elicit biochemical and toxic responses that are characteristic of 2,3,7,8,-TCDD.
4. It should be persistent and accumulate in the food chain.

Some examples of TEF values used in an environmental study on polyhalogenated aromatic hydrocarbons (PHAHs) in fish are given in Table 7.2 (data from Giesy et al. 1997). The first point to notice is that values for PCDDs and PCDFs are generally much higher than those for PCBs. Even the most potent of the PCBs, 3,3′,4,4′,5-PCB, only has a TEF value of 2.2×10^{-2}, which is lower than nearly all the values for PCDDs and PCDFs. That said, PCBs tend to be at much higher concentrations in environmental samples than the other two groups, with the consequence that they have been found to contribute higher overall TEQ values than PCDDs or PCDFs in many environmental samples despite their low TEF values.

TABLE 7.2

TEF Values for Polyhalogenated Aromatic Hydrocarbons Used in a Study of Residues in Fish

Compound	TEF	Compound	TEF
PCDDs		**Nonortho PCBs**	
1,2,3,7,8-PCDD	4.2×10^{-1}	3,4,4',5-TCB	1.9×10^{-3}
1,2,3,4,7,8-HCDD	8.3×10^{-2}	3,3',4,4'-TCB	1.8×10^{-5}
1,2,3,4,6,7-HpCDD	2.3×10^{-2}	3,3',4,,4',5-PCB	2.2×10^{-2}
PCDFs		**Mono-ortho PCBs**	
2,3,7,8-TCDF	2.0×10^{-1}	2,3',4,3',4',5-PCB	3.5×10^{-7}
2,3,4,7,8-PCDF	2.8×10^{-1}	2,3,3',4,4'-PCB	8.0×10^{-6}
1,2,3,6,7,8-HCDF	6.0×10^{-1}	2,3,3',4,4',5-HCB	5.5×10^{-5}
1,2,3,4,7,8,9-HpCDF	2.0×10^{-2}	2,2',3,4,4',5'-HCB	1.5×10^{-5}

Source: Data from Giesy et al. (1997).

Up to this point, discussion of TEQs has been restricted to their estimation from concentrations of individual compounds determined chemically, employing TEFs as conversion factors. It is also possible to measure the total TEQ value directly by means of a bioassay. Rat and mouse hepatoma lines, which contain the Ah receptor, show P450 1A1 induction when exposed to planar PHAHs. The degree of induction can be measured in terms of the increase in ethoxyresorufin-O-deethylase (EROD) activity (see Chapter 2). One example of a cellular line of this type is the rat H4 IIE line, which has come to be widely used for environmental bioassays (see, for example, Giesy et al. 1997, Koistinen 1997, and Whyte et al. 1998). A development of this approach is the "chemically activated luciferase gene expression," or CALUX, assay (Aarts et al. 1993, Garrison et al. 1996). Here, a reporter gene for the enzyme luciferase is linked to the operation of the Ah receptor, so the degree of induction is indicated by the quantity of light that is emitted by the cells. This system has the advantage of not requiring an EROD assay to determine TEQ values. Values obtained by the direct measurement of TEQ (i.e., TCDD equivalents) using cellular systems can be directly compared to values estimated from chemical data using TEFs.

When using TEFs to estimate TEQs, it is assumed that all of the compounds are acting by a common mechanism through a common receptor, and that effects of individual components in a mixture are simply additive, without potentiation or antagonism. There are, however, reservations about adopting such a simple approach too generally. For one, the mechanism of toxicity has not been elucidated, and there may be toxic mechanisms operating, thus far unidentified, which do not involve the Ah receptor. It is true that early application of the approach in human toxicology produced relatively consistent results that tended to encourage this simple approach. In studies with experimental animals (mainly rodents), the summation of TEQ values for mixtures of compounds have often been found to relate reasonably well to toxic effects (see the following text). However, there is now evidence suggesting that PHAHs can express toxicity by other mechanisms, for example, in certain cases of developmental neurotoxicity and carcinogenicity (Brouwer 1996, Verhallen et al. 1997). Also, PCBs in mixtures have sometimes shown antagonistic effects, so additivity cannot be automatically assumed (Davis and Safe 1990, Giesy et al. 1997).

The use of TEQs in environmental risk assessment has great attractions. It offers a way of tackling the problem of determining the biological significance of levels of diverse PHAHs found in mixtures. There are, however, practical and theoretical problems that need to be resolved before it can be widely used with confidence. First, the issue—to what extent can the variety of toxic effects caused by PHAHs in vertebrates generally be explained in terms of Ah-receptor-mediated toxicity? Only limited work has been done on birds or fish, and practically nothing on amphibians or reptiles. So, this question extends far beyond the limited number of species of experimental animals so far studied. Second, even if it were possible to restrict the argument to effects that are directly associated with binding of PHAH to the Ah receptor, how comparable are the Ah receptors of unrelated species? It has been suggested that the Ah receptor is highly conserved and may not differ very much between groups of vertebrates. However, in field studies, there have sometimes been large differences between species in the relationship between TEQ values and toxic effects (see Ludwig et al. 1996, McCarty and Secord 1999, and further discussion in Chapter 6, Section 6.2.4 and Chapter 7, 7.2.5). Also, work on the estuarine fish, *Fundulus heteroclitus*, has revealed the existence of a strain inhabiting a polluted area that is resistant to dioxin-like compounds (see Chapter 6, Section 6.2.5). This species evidently possesses two forms of the Ah receptor, AHR1 and AHR2, which have differential tissue expression in susceptible strains as compared with the resistant strain (Powell et al. 2000). The upshot is that the resistant strain has a greater predominance of the form of the Ah receptor, which is insensitive to dioxin-like compounds, and this provides the basis of the resistance. Indeed, this type of resistance mechanism is apparently similar to that found in many strains of insects that have acquired resistance to insecticides, namely, target insensitivity. Examples include insensitive forms of the sodium channel, which confer resistance to DDT and pyrethroids (Chapter 5, Section 5.2.5.1, and Chapter 12, Section 12.6), insensitive forms of GABA receptors, which confer resistance to cyclodienes (Chapter 5, Section 5.3.5.2), and aberrant forms of acetylcholinesterase, which confer OP resistance. Thus, it seems that the application of the TEQ concept in ecotoxicology is complicated by the fact that there are different forms of the Ah receptor with differential sensitivities to dioxin-like compounds. Such differences may be expected to cause different responses to these compounds not only between strains but also between species, sexes, and age groups. A further likely consequence is that there will be differences in the comparative potency of individual dioxin-like compounds when interacting with contrasting forms of the Ah receptor. Thus, TEF values are likely to show species and strain variation, which may go some way to explaining some of the anomalies in field data, to be discussed in Section 7.6.

More work needs to be done to clarify this. At the moment, the TEF values in use have been obtained for a very limited number of species. There needs to be caution in using them for calculating TEQs in untested species.

Because of the concern over human health hazards associated with PCDDs, many toxicity tests have been performed on rodents. Some toxicity data are given for 2,3,7,8-TCDD as follows:

Acute oral LD_{50}/rat: 22–297 µg/kg (Different strains were tested)
Acute oral LD_{50}/mice: 114–2570 µg/kg (Different strains were tested)
Acute oral LD_{50}/guinea pig: 0.6–19 µg/kg

A variety of toxic symptoms were shown, the pattern differing between species. There were often long periods between commencement of dosing and death. There were also large species differences in toxicity, the guinea pig being extremely susceptible, the mouse far less so. However, the critical point is that 2,3,7,8-TCDD is an exceedingly toxic compound even to the mouse. With such differences in toxicity between closely related species, it seems probable that there will be even larger differences across the wide range of vertebrate species found in nature.

2,3,7,8-TCDF is not very toxic to the mouse (acute oral LD_{50} < 6000 µg/kg) but is highly toxic to the guinea pig (acute oral LD_{50} 5–10 µg/kg). Symptoms of toxicity in the guinea pig were similar to those found with 2,3,7,8-TCDD. Thus, the selectivity pattern was similar to that for 2,3,7,8-TCDD, but toxicity was considerably less (see Environmental Health Criteria 88).

7.6 ECOLOGICAL EFFECTS RELATED TO TEQS FOR 2,3,7,8-TCDD

Estimates of TCDD toxicity in field studies have depended on the estimates of TEQ values in an attempt to relate Ah-receptor-mediated toxicity caused by total PHAHs to effects on individuals and populations. Although some encouraging progress has been made, there have also been a number of problems. Not infrequently, TEQ values determined by bioassay have considerably exceeded values calculated from chemical data using TEFs. These discrepancies have not been explicable in terms of antagonistic effects, and the balance of evidence suggests that environmental compounds other than the PHAHs determined by analysis have contributed to the measured TEQ values. This has been observed in studies on fish (Giesy et al. 1997) and white-tailed sea eagles (*Haliaetus albicilla*) (Koistinen et al. 1997). In the study on fish, as much as 75% of the measured TEQ could not be accounted for using chemical data. The results of a number of studies are summarized in Table 7.3.

The results for the double-crested cormorant and the Caspian tern in the Great Lakes both show a relationship between TEQs and reproductive success. They were obtained during the late 1980s, at a time when DDE-related thinning of eggshells had fallen and TEQ values were based on PCBs alone. However, in certain areas PHAH levels remained high, and populations of these two species were still depressed. These investigations suggested that populations were still being adversely affected by PHAHs as a consequence of Ah-receptor-mediated toxicity. The data for white-tailed sea eagles from the Baltic coast also relate to an area where, at the time of the investigation, there was evidence of reduced breeding success in the local population—at a time when the species was increasing elsewhere in Scandinavia. The TEQs in the most highly contaminated individuals were high enough to support the suggestion that PHAHs were contributing to lack of breeding success.

TABLE 7.3
TEQ Values Found in Field Studies, and Ecological Effects Associated with Them

Area	Species	TEQ pg/g (method of determination)	Observation	Reference
Great Lakes	Double-crested cormorant (eggs)	100–300 (bioassay); only PCBs assayed	Egg mortalities of 8–39% correlated well with TEQ values Relationship to continued poor breeding success	Tillett et al. (1992)
	Caspian tern (eggs)	170–400 (chemical analysis); only PCBs assayed	Related to embryonic mortality	Ludwig et al. (1996)
		2700	Total reproductive failure	Ludwig et al. (1993)
Baltic Sea	White-tailed sea eagle (egg, muscle)	<1220 (bioassay) <1040 (chemical determination)	PCB fraction accounted for 75%+ of TEQ by either assay; reduced productivity of birds in this area	Koistinen et al. (1997)
Upper Hudson River	Tree swallows (*Tachycineta bicolor*) (nestlings)	410–25,400 (chemical determination)	TEQs mainly due to PCBs, especially 3,3′,4,4′-TCB	Secord et al. (1999)
			Reduced reproductive success, but less effect than expected from high TEQs	McCarty and Secord (1999)
Saginaw Bay, Great Lakes	Fish (whole body); sampled 1990	11–348 (bioassay); 14–70 (chemical determination)	PCDDs, PCDFs, and PCBs made variable, but on the whole similar contributions to TEQ values; probably not high enough to adversely affect fish populations	Giesy et al. (1997)
Woonasquatucket River	Tree swallows eggs	343–1281 (chemical determination)	TCDD mainly 2,3,7,8-TCDD Reduced hatching success	Custer et al. (2005)

The results for tree swallows (*Tachycineta bicolor*) in the area of the Upper Hudson River are surprising because the birds were still able to breed with TEQs far above the levels that had severe/fatal effects on other species of birds. However, there was evidence of reduced reproductive success in 1994, and of high rates of abandonment of nests and supernormal clutches in 1995 (McCarty and Secord 1999). It would appear that tree swallows are particularly insensitive to this type of toxic action. This finding raises questions about the validity and wider applicability of the use of TEQ values and brings to mind the large interspecific differences in TCDD toxicity found in some toxicity tests.

In another study of tree swallows at a Superfund site along Woonasquatucket River, Rhode Island, during 2000–2001, exceptionally high levels of 2,3,7,8-TCDD were found in eggs from the most polluted areas (300 to >1000 pg/g wet weight). Eggs sampled from a reference site contained only 12–29 pg/g. Concentrations in the food of the birds were more than 6–18 times higher than that regarded as safe for birds (10–12 pg/g). Hatching success was negatively correlated with TCDD concentrations in eggs, and only about 50% of eggs hatched in the most polluted area, compared to >77% in the reference area. In contrast to the study described in the area of the Hudson River, 2,3,7,8-TCDD was the dominant residue so far as TEQ estimations were concerned. PCB levels were below those known to affect avian reproduction (Custer et al. 2005).

A noteworthy finding of the investigations thus far is the considerable variation in the relative contributions of PCBs, PCDDs, and PCDFs to TEQ values determined for wild vertebrates. Thus, the TEQ values for white-tailed sea eagles and tree swallows on the Hudson River were accounted for very largely by coplanar PCBs, whereas 2,3,7,8-TCDD was dominant in tree swallows in the Rhode Island study area. Fish from Saginaw Bay showed relatively low TEQs. PCDDs, PCDFs, and PCBs all made similar contributions to TEQ. There was considerable variation between species and age groups of fish, with contributions to total TEQs in the following ranges:

PCDD: 5–38%
PCDF: 13–69%
PCB: 10–50%

7.7 SUMMARY

Both PCDDs and PCDFs are refractory lipophilic pollutants formed by the interaction of chlorophenols. They enter the environment as a consequence of their presence as impurities in pesticides, following certain industrial accidents, in effluents from pulp mills, and because of the incomplete combustion of PCB residues in furnaces. Although present at very low levels in the environment, some of them (e.g., 2,3,7,8-TCDD) are highly toxic and undergo biomagnification in food chains.

PCDDs and PCDFs, together with coplanar PCBs, can express Ah-receptor-mediated toxicity. TCDD (dioxin) is used as a reference compound in the determination of TEFs, which can be used to estimate TEQs (toxic equivalents) for residues of PHAHs found in wildlife samples. Biomarker assays for Ah-receptor-mediated toxicity have been based on the induction of P450 1A1. TEQs measured in field samples have sometimes been related to toxic effects upon individuals and associated ecological effects (e.g., reproductive success).

FURTHER READING

Ahlborg et al. (1996). Discusses the wider use of TEFs.

Environmental Health Criteria 88 (1989). Gives information on the environmental toxicology of PCDDs and PCDFs.

Safe, S. (1990). An authoritative account of the development of TEFs.

8 Organometallic Compounds

8.1 BACKGROUND

Metalloids such as arsenic and antimony, and metals such as mercury, lead, and tin—which occupy a similar location to metalloids in the periodic system—all tend to form stable covalent bonds with organic groups. Some authorities regard tin as a metalloid. By contrast, metals such as sodium, potassium, calcium, strontium, and barium, which belong to groups 1 and 2 of the periodic system, do not form covalent bonds with organic groups. The compounds used as examples here all possess covalent linkages between a metal and an organic group—most commonly an alkyl group. The elements in question are mercury, tin, lead, and arsenic, all of which are appreciably toxic in their inorganic forms as well as in their organometallic forms. The attachment of the organic group to the metal can bring fundamental changes in its chemical properties, and consequently in its environmental fate and toxic action. In particular, the attachment of alkyl or other nonpolar groups to metals increases lipophilicity and thereby enhances movement into and across biological membranes, storage in fat depots, and adsorption by the colloids of soils and sediments. Thus, the question of speciation is critical to understanding the ecotoxicology of these metals.

In the first place, organometallic compounds of mercury, tin, lead, and arsenic have been produced commercially, mainly for use as pesticides, biocides, or bactericides. Additionally, methyl mercury and methyl arsenic are generated from their inorganic forms in the environment, so residues of them may be both anthropogenic and natural in origin. Most of the following account will be devoted to organomercury and organotin compounds, which have been extensively studied. Organolead and organoarsenic compounds have received less attention from an ecotoxicological point of view, and will be dealt with only briefly.

8.2 ORGANOMERCURY COMPOUNDS

8.2.1 ORIGINS AND CHEMICAL PROPERTIES

A range of organomercury compounds have been produced commercially since early in the 20th century, principally for use as antifungal agents. Most of them have the general formula R–Hg–X, where R is an organic group and X is usually an inorganic group (occasionally a polar organic group such as acetate). The organic group is nonpolar (or relatively so) and gives the molecule a lipophilic character. The most common organic groups are alkyl, phenyl, and methoxyethyl (see Environmental Health Criteria 86).

163

General formula   RHgX

Where R = C_nH_{2n+1}, C_6H_5 or $CH_3OC_2H_5$

CH_3HgCl Methylmercuric chloride

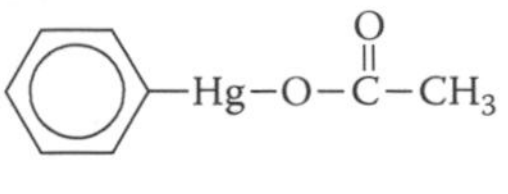

Phenylmercuric acetate

FIGURE 8.1   Organomercury compounds.

TABLE 8.1

Properties of Organomercury Compounds

Compound	Water Sol mg/L	Vapor Pressure mm Hg
Methyl mercuric chloride	1.4	8.5×10^{-3}
Phenylmercuric acetate	4400	—

The solubility of organomercury compounds depends primarily on the nature of the X group; nitrates and sulfates tend to be "salt-like" and relatively water-soluble, whereas chlorides are covalent, nonpolar compounds of low water solubility. Methyl mercury compounds tend to be more volatile than other organomercury compounds.

The structures of some organomercury compounds are shown in Figure 8.1, and some physical properties are given in Table 8.1

The R–Hg bond is chemically stable and is not split by water or weak acids or bases. This is a reflection of the low affinity of Hg for oxygen. It can, however, be readily broken biochemically. Organomercury, like other organometallic compounds, has a strong affinity for SH–groups of proteins and peptides.

$$R–Hg–X + \text{protein-SH} \rightarrow R–Hg–S\text{-protein} + X^- + H^+$$

This tendency to interact with –SH groups appears to be the fundamental chemical reaction behind most of the adverse biochemical effects of organomercury compounds; it is also the basis for one mechanism of detoxication.

Apart from the release of human-made organomercurial compounds, methyl mercury can also be generated from inorganic mercury in the environment as indicated in the following equation:

$$Hg \rightarrow Hg^{++} \rightarrow CH_3\,Hg^+ \rightarrow [CH_3]_2\,Hg$$

Thus, both elemental mercury and the mineral form cinnabar (HgS) can release Hg^{++}, the mercuric ion. Bacteria can then methylate it to form sequentially $CH_3\,Hg^+$, the methyl mercuric cation, and dimethyl mercury. The latter, like elemental mercury, is volatile and tends to pass into the atmosphere when formed. The methylation of mercury can be accomplished in the environment by bacteria, notably in sediments.

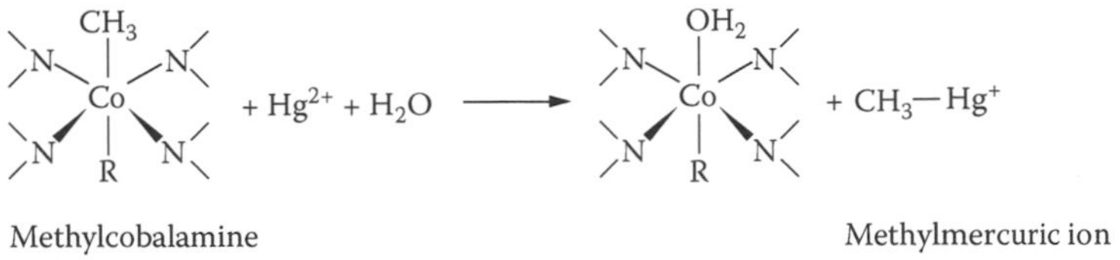

FIGURE 8.2   Methylation of inorganic mercury by methylcobalamine (from Crosby 1998).

A form of vitamin B12 can produce methyl carbanion, a reactive species that is responsible for methylation of Hg^{++} (see Figure 8.2, Craig 1986, and IAEA Technical Report 137). Methyl carbanion acts as a nucleophilic agent toward Hg ions.

It is difficult to establish to what extent methyl mercury residues found in the environment arise from natural as opposed to human sources. There is no doubt, however, that natural generation of methyl mercury makes a significant contribution to these residues. Samples of Tuna fish caught in the late 18th century, before the synthesis of organomercury compounds by humans, contain significant quantities of methyl mercury.

8.2.2   Metabolism of Organomercury Compounds

As mentioned earlier, methyl mercury compounds can undergo further methylation to generate highly volatile dimethyl mercury. Organomercury compounds can also be converted back into inorganic forms of mercury by enzymic action. Oxidative metabolism is important here, and has been reported in both microorganisms and invertebrates. Methyl mercury is slowly degraded by alpha oxidation, whereas other alkyl forms are subject to more rapid beta oxidation. This may explain why methyl mercury is degraded more slowly than other forms and is correspondingly more persistent. Phenyl mercury is degraded relatively rapidly to inorganic mercury by vertebrates and is generally less persistent than alkyl mercury.

$$R–Hg^+ \rightarrow Hg^{++}$$

Another type of detoxication involves the production of cysteine conjugates, which are readily excreted. (Again, organomercury compounds show their affinity for –SH groups). Methyl mercuric cysteine is an important biliary metabolite in the rat and is degraded within the gut (presumably by microorganisms) to release inorganic mercury (see IAEA Report 137, 1972).

The following ranges of half-lives have been reported for vertebrate species, which are presumably related to rates of biotransformation as the original lipophilic compounds show little tendency to be excreted unchanged.

Alkyl mercury   15–25 days
Phenyl mercury   2–5 days

8.2.3 Environmental Fate of Organomercury

As noted earlier, diverse forms of organomercury are released into the environment as a consequence of human activity. Methyl mercury presents a particular case. As a product of the chemical industry, it may be released directly into the environment, or it may be synthesized in the environment from inorganic mercury which, in turn, is released into the environment as a consequence of both natural processes (e.g., weathering of minerals) and human activity (mining, factory effluents, etc.).

The environmental cycling of methyl mercury is summarized in Figure 8.3. Dimethyl mercury, being highly volatile, tends to move into the atmosphere following its generation in sediments; once there, it can be converted back into elemental mercury by the action of UV light. Some dimethyl mercury is taken up by fish and transformed into a methyl cysteine conjugate, which is excreted. However, the most important species of methyl mercury in aquatic and terrestrial food chains is CH_3Hg^+, which exists in various states of combination with S– groups of proteins and peptides, and with inorganic ions such as chloride. Total methyl mercury of tissues, sediments, etc., is determined by chemical analysis, but the state of combination is not usually known. Some free forms of methyl mercury, for example, CH_3HgCl, are highly lipophilic and undergo bioaccumulation and bioconcentration with progression along food chains in similar fashion to lipophilic polychlorinated compounds.

In a report from the U.S. EPA (1980), fish contained between 10,000 and 100,000 times the concentration of methyl mercury present in ambient water. In a study of methyl mercury in fish from different oceans, higher levels were reported in predators than in nonpredators (see Table 8.2). Taken overall, these data suggest that predators have some four- to eightfold higher levels of methyl mercury than do nonpredators, and it appears that there is marked bioaccumulation with transfer from prey to predator.

In a laboratory study (Borg et al. 1970), bioaccumulation of methyl mercury was studied in the goshawk (*Accipiter gentilis*). The details are shown in Table 8.3 below.

Thus, chickens bioaccumulated methyl mercury to about twice the level in their food, whereas goshawks bioaccumulated methyl mercury to about four times the level present in the chicken upon which they were fed. The period of exposure was similar

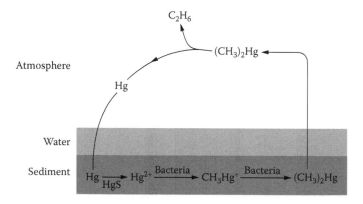

FIGURE 8.3   Environmental fate of methyl mercury (adapted from Crosby 1998).

TABLE 8.2
Methyl Mercury Residues in Fish (mg Hg/kg wet weight)

Type of Fish	Atlantic Ocean	Pacific Ocean	Indian Ocean	Mediterranean Sea
Nonpredators	0.03–0.27	0.03–0.25	0.005–0.16	0.1–0.24
Predators	0.3–1.3	0.3–1.6	0.004–1.5	1.2–1.8.

Source: Data from Environmental Health Criteria 101 Methylmercury.

TABLE 8.3
Bioaccumulation of Methylmercury

Material/Species	CH₃ Hg (ppm Hg)	Duration of Feeding (days)	Approximate Bioaccumulation Factor
Dressed grain	8		
Muscle of chickens fed dressed grain	10–40	40–44	2
Chicken tissue fed to goshawks	10–13		
Muscle from goshawks fed chicken tissue	40–50	30–47	4

in both cases. This provides further evidence for the slow elimination of methyl mercury by vertebrates and the relatively poor detoxifying capacity of predatory birds toward lipophilic xenobiotics compared to nonpredatory birds (see Chapters 2 and 5). In a related study with ferrets fed chicken contaminated with methyl mercury, a somewhat higher bioaccumulation factor was indicated (about sixfold), albeit over the somewhat longer exposure period of 35–58 days. This provided further evidence for strong bioaccumulation by predators.

Since the widespread banning of organomercury fungicides, significant levels of organomercury have continued to be found in certain areas—much of it, presumably, having been biosynthesized from inorganic mercury. Particular interest has come to be focused on methyl mercury pollution of the aquatic environment and on levels in fish and piscivorous birds. In North America, the common loon (*Gavia immer*) has been identified as a suitable indicator organism for this type of pollution (Evers et al. 2008). The half-life of methyl mercury in the blood of juvenile loons after moulting has been estimated to be 116 days (Fournier et al. 2002). In another study, the methyl mercury half-life in blood of another piscivorous bird, Cory's shearwater (*Calonectris diomedea*), was estimated to be 40–60 days (Monteiro and Furness 2001). The ecological effects of methyl mercury on common loons will be discussed later in Section 8.2.5.

Apart from CH₃ Hg⁺, other forms of R-Hg⁺ have been found in the natural environment, which originate from anthropogenic sources but are not known to be generated from inorganic mercury. These forms have been found in terrestrial and aquatic food chains. A major source has been fungicides, in which the R group is phenyl, alkoxyalkyl, or higher alkyl (ethyl, propyl, etc.). These forms behave in a similar manner

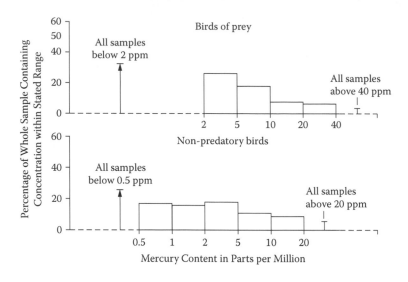

FIGURE 8.4   Mercury residues in the livers of Swedish birds (from Walker 1975).

to CH_3–Hg X and do tend to undergo biomagnification, but they are generally more easily biodegradable to inorganic mercury and tend to bioaccumulate less strongly.

At one time, a major source of organomercury pollution in Western countries was fungicidal seed dressings used on cereals and other agricultural products (see IAEA Technical Report 137 1972). Another important source was organomercury antifungal agents used in the wood pulp and paper industry. Most of these uses were discontinued by the 1970s, but certain practices have continued into the 1990s, including the use of phenylmercury fungicides as seed dressings in Britain and some other countries. In the 1950s and early 1960s, Sweden and other Scandinavian countries had serious pollution problems due to the use of methyl mercury compounds as seed dressings. Deaths of seed-eating birds and raptors preying upon them were attributed to methyl mercury poisoning (Borg et al. 1969). Thus, as with dieldrin, bioaccumulation led to secondary poisoning. Interestingly, seed-eating rodents contained lower mercury levels than seed-eating birds. Some data for total mercury levels found in Swedish birds during the mid-1960s are shown in Figure 8.4. Most of the mercury was in the methyl form. Findings such as these led to the banning of methyl mercury seed dressings in Scandinavia. Other forms of organomercury seed dressing (e.g., phenyl mercury) were not implicated in poisoning incidents and continued to be marketed in many Western countries after methyl mercury compounds were banned.

8.2.4   TOXICITY OF ORGANOMERCURY COMPOUNDS

The toxicity of organomercury, like that of certain other types of organometals, has been related to their strong affinity for functional –SH groups of proteins (Crosby 1998). Exposure of animals to organic mercury leads to a reduction in the number of free –SH groups in their tissues. Both mercury and methyl mercury bind strongly to these groups. This is consistent with the wide range of physiological and biochemical effects arising from mercury poisoning, for many proteins depend on free –SH groups for their normal function.

A prime target for methyl and other organic forms of mercury is the nervous system, especially the central nervous system (CNS). Here lies an important distinction between the toxicity of organic and inorganic mercury salts. Although inorganic forms of mercury can also bind to –SH groups, they cannot readily cross the blood–brain barrier, and so show less tendency than lipophilic organic mercury to reach the CNS and cause toxic effects there. Rather, inorganic mercury expresses its toxicity elsewhere (e.g., in the kidney and on cardiac function). Methyl mercury can cause extensive brain damage, including degeneration of small sensory neurons of the cerebral cortex. At the biochemical level, it binds to cysteine groups of acetylcholine receptors (Crosby 1998) and also inhibits Na^+/K^+ ATP-ase (Clarkson 1987).

With growing interest in the sublethal effects of methyl mercury, evidence has come to light of changes in the concentration of neurochemical receptors of the brain during the early stages of poisoning. Studies with mink dosed in captivity have shown that environmentally realistic levels of methyl mercury can cause (1) an increase in the concentration of brain muscarinic receptors for acetylcholine, and (2) a decrease in the concentration of N-methyl-D-aspartic acid glutamate receptors for glutamate (Basu et al. 2006, and Scheuhammer and Sandheinrich 2008). In the next section, field studies will be discussed, which have looked for evidence of effects of this kind in wild mammals and birds.

BOX 8.1   THE MINAMATA INCIDENT

The neurotoxicity of organomercury was graphically illustrated in an environmental disaster at Minamata Bay in Japan during the late 1950s and early 1960s. Release of both organic and inorganic mercury from a factory led to the appearance of high levels of methyl mercury in the neighboring marine ecosystem. Levels were high enough in fish to cause lethal intoxication of local people for whom fish was the main protein source. People died as a consequence of brain damage caused by methyl mercury. The victims had brain Hg levels in excess of 50 ppm.

In mammals, methyl mercury toxicity is mainly manifest as damage to the CNS with associated behavioral effects (Wolfe et al. 1998). Initially, animals become anorexic and lethargic and, with progression of toxicity, muscle ataxia and visual impairment are seen. Finally, convulsions occur, which lead to death. In dosing experiments with mink (*Mustela vison*), dietary levels of methyl mercury of 1.1 ppm fed over a period of 93 days produced subclinical neurological lesions (Wobeser et al. 1976), and this has been proposed as a *lowest observed adverse effect level* (LOAEL). In another study, otters (*Lutra canadensis*) were dosed with 2, 4, or 8 ppm methyl mercury in the diet (Connor and Nielsen 1981). Anorexia and ataxia were reported at 2 ppm in two-thirds of individuals; anorexia, ataxia, and neurological lesions at 4 ppm; and all the symptoms, leading to death at 8 ppm. The brain Hg concentrations (ppm per unit wet weight) at dose levels of 2, 4, and 8 ppm were

13.3, 21, and 23.7 ppm, respectively. Thus, symptoms of neurotoxicity were observed in individuals containing brain concentrations of Hg substantially lower than those associated with lethal toxicity. Interestingly, the proportion of the total mercury accounted for as organomercury declined with time, indicating that demethylation slowly occurred in the brain.

Captive goshawks dying from methyl mercury poisoning contained 30–40 ppm Hg in brain and 40–50 ppm Hg in muscle (Borg et al. 1970, see also Section 8.2.3). In this study and others (Wolfe et al. 1998), it became apparent that birds, like mammals, experience a range of sublethal effects before tissue levels became high enough to cause death. The first symptoms of methyl mercury poisoning in birds are reduced food consumption and weakness of the extremities. Muscular coordination is poor, there is ataxia, and birds can neither walk nor fly (See Rissanen and Mietinnen in IAEA Technical Report 137 1972). The severity of sublethal neurotoxic effects produced by methyl mercury would have reduced the likelihood of predatory birds acquiring lethal concentrations of methyl mercury when **chronically** exposed in the field. More likely they would have died from starvation due to sublethal effects before they could build up lethal concentrations. Predators would lose their ability to catch prey once muscular coordination was affected. These feeding skills are not tested in laboratory trials in which birds are presented with food and they may be expected to tolerate relatively high levels of methyl mercury in tissues before losing their ability to feed. This contrasts with **acute** exposures in the field where predators sometimes consumed high doses of methyl mercury in poisoned prey, and a single meal might have contained a lethal dose for the predator. More generally, impairment of ability to fly would have adversely affected herbivores and omnivores in their ability to feed or escape predation.

The acute toxicity of different types of organomercury compounds to mammals, expressed as mg/kg, fall into the following ranges:

Methyl mercury compounds	16–32
Ethyl mercury compounds	16–28
Phenyl mercury compounds	5–70

Thus, there is not a great deal of difference between the three classes in acute toxicity; all are highly toxic. However, methyl mercury is more persistent than the other two types, and so has the greater potential to cause chronic toxicity. The latter point is important when considering the possibility of sublethal effects.

8.2.5　Ecological Effects of Organomercury Compounds

Of the different forms of organomercury, methyl mercury is the one most clearly implicated in toxic effects in the field. When methyl mercury seed dressings were used in Sweden and other Northern European countries during the 1950s and 1960s, many deaths of seed-eating birds, and of predatory birds feeding upon them, were attributed to methyl mercury poisoning. There was evidence of birds experiencing sublethal effects such as inability to fly. There may well have been local declines of bird populations consequent upon these effects, but these were not clearly established at the time. Methyl mercury seed dressings were subsequently banned in Western countries, so the question is now rather an academic one.

Another major incident concerning methyl mercury was the severe pollution of Minamata bay in Japan (see Box 8.1). Here fish, fish-eating and scavenging birds, and humans feeding upon fish all died from organomercury poisoning. There may have been localized declines of marine species in this area due to methyl mercury, but there is no clear evidence of this.

Despite the banning of methyl mercury fungicides, methyl mercury continues to exist in some areas at levels high enough to cause adverse ecological effects. In a wide-ranging review, Wolfe et al. (1998) cite a number of studies that give evidence of sublethal effects of methyl mercury upon wild vertebrates in the time since severe restrictions were placed on the use of methyl mercury fungicides. A major reason for this is the continuing synthesis of methyl mercury in the environment from inorganic mercury, the latter originating from both natural and human sources. There has been particular concern about some aquatic habitats, for it appears that a very high percentage of total mercury in the higher trophic levels of aquatic food chains is in the form of methyl mercury. Indeed, more than 95% of the mercury found in the tissues of fish and marine invertebrates sampled from different oceans of the world was found to be in this form (Bloom 1992, Wolfe et al. 2007). Relatively high levels have been reported in lakes of Northern America, including the Great Lakes (Meyer et al. 1998; Evers et al. 1998, 2003, and 2008), and in the Mediterranean Sea (Renzoni in Walker and Livingstone 1992, Aguilar and Borrell 1994, Aguilar et al. 1996). Apart from the question of possible direct toxic effects caused by methyl mercury, there is also the possibility that there are adverse interactive effects (potentiation) with other pollutants such as PCBs, PCDDs, PCDFs, p,p'-DDE, metals, and selenium, which reach significant levels in aquatic organisms (Walker and Livingstone 1992, Heinz and Hoffman 1998).

The common loon (*Gavia immer*) is one fish-eating bird inhabiting lakes of North America that has been closely studied in connection with organomercury pollution (Evers et al. 2002, 2003, and 2008). Eggs collected between 1995 and 2001 in this area contained 0.07–4.42 µg/g Hg (wet weight). Although fertility was not related to the mercury content of eggs, there was an inverse relationship between Hg content and egg volume. Female loons with blood Hg concentrations of >3.0 µg/g laid eggs containing >1.3 µg/g and often had decreased reproductive success, laying fewer eggs than the less contaminated individuals (Evers et al. 2008). Adult common loons collected in Canada were analyzed for total and methyl mercury and for neurochemical receptors (Scheuhammer et al. 2008). Most of the mercury in brain was in methyl form (>78% in all cases). A positive correlation was found between total mercury in brain and the concentration of brain muscarinic receptors. On the other hand, a negative correlation was found between total mercury and *N*-methyl-D-aspartic acid receptors. This immediately raises questions about possible neurotoxic and behavioral effects due to methyl mercury. Similar correlations were also reported for bald eagles in the same study.

In a study conducted during the period 1998–2000 at North American sites, the relationship was studied between methyl mercury blood levels in common loons and behavioral parameters (Evers et al. 2008). Adult behaviors were divided into two categories: (1) high-energy and (2) low-energy. A negative correlation was found between high-energy behaviors and total mercury concentration in the blood of adult loons. High-energy behaviors included foraging for chicks, foraging for self, swimming and flying, preening, and agonistic behaviors. There was also a strong negative correlation between mercury levels in female loons and reproductive success (Burgess and Meyer 2008, Evers et al. 2008). In a laboratory investigation, adverse effects were observed when loon chicks were dosed with levels of methyl mercury comparable to the highest levels of exposure recorded in the preceding study (Kenow et al. 2007). There was evidence of demyelination of central nervous tissue and reduced immune function when the chicks were fed fish containing 0.4 µg/g of methyl mercury or more.

The mink (*Mustela vision*) is a piscivorous mammal that also has been exposed to relatively high dietary levels of methyl mercury in North America in recent times. In a Canadian study, mink trapped in Yukon territory, Ontario, and Nova Scotia were analyzed for levels of mercury and abundance of muscarinic, cholinergic and dopaminergic receptors in the brain (Basu et al. 2005). A correlation was found between total Hg levels and abundance of muscarinic receptors, but a negative correlation was found between total Hg and abundance of dopaminergic receptors. Thus, it was suggested that environmentally relevant concentrations of Hg (much of it in methyl form) may alter neurochemical function. The highest levels of mercury contamination were found in mink from Nova Scotia that had a mean concentration of total Hg of 5.7 µg/g in brain, 90% of which was methyl mercury.

8.3 ORGANOTIN COMPOUNDS

8.3.1 CHEMICAL PROPERTIES

Like mercury, tin is a metal that has a tendency to form covalent bonds with organic groups. The compounds to be discussed here are tributyl derivatives of tetravalent tin. The general formula for them is

$$[\text{n-C}_4 \text{H}_9]_3 \text{ Sn-X, where X is an anion.}$$

The most important of the compounds from an ecotoxicological point of view, and the one that will be used here as an example, is tributyltin oxide (TBTO). Its structure is shown in Figure 8.5.

TBTO is a colorless liquid of low water solubility and low polarity. Its water solubility varies between <1.0 and >100 mg/L, depending on the pH, temperature, and presence of other anions. These other anions determine the speciation of tributyltin in natural waters. Thus, in sea water, TBT exists largely as hydroxide, chloride, and carbonate, the structures of which are given in Figure 8.5. At pH values below 7.0, the predominant forms are the chloride and the protonated hydroxide; at pH8 they are the chloride, hydroxide, and carbonate; and at pH values above 10 they are the hydroxide and the carbonate (EHC 116).

The K_{ow} for TBTO expressed as log P_{ow} lies between 3.19 and 3.84 for distilled water, and is about 3.54 for sea water. TBTO is adsorbed strongly to particulate matter.

General formula

$$(C_4H_9)_3Sn - X \quad \text{where X is usually an anion}$$

Tributyltin oxide

$$(C_4H_9)_3Sn - O - Sn(C_4H_9)_3$$

Forms existing in water

$$(C_4H_9)_3 - SnOH \qquad \text{Hydroxide}$$

$$(C_4H_9)_3 - SnOH_2^+ \qquad \text{Protonated hydroxide}$$

$$(C_4H_9)_3 - SnCl \qquad \text{Chloride}$$

$$\left[(C_4H_9)_3 - Sn\right]_2 CO_3^- \qquad \text{Carbonate}$$

Metabolism

$$(C_4H_9)_3Sn^+ \xrightarrow[\text{(P450)}]{O} (C_4H_9)_2Sn^{2+}$$

FIGURE 8.5 Tributyltin.

Another type of organotin compound, triphenyltin (TPT), has been used as a fungicide.

8.3.2 METABOLISM OF TRIBUTYLTIN

In rats and mice, TBT compounds are hydroxylated by microsomal monooxygenase attack (see Environmental Health Criteria 116). Hydroxylation can occur on either the alpha or beta carbon of the butyl group, that is, alpha or beta in relation to the Sn atom (see Figure 8.5). After hydroxylation, the hydroxylated moiety breaks away to leave behind dibutyltin.

TBTs also cause inhibition of the P450 of monooxygenases. In fish and in the common whelk, TBT causes conversion of P450s to the inactive P420 form (Fent et al. 1998, Mensink 1997). In fish, inactivation was also found with TPT, and was related to the inhibition of ethoxy resorufin deethylase activity (EROD) activity. In these studies, organotin compounds were found both as substrates and deactivators of the hemeprotein (cf. the interaction of organophosphates with B-type esterases).

TBT, like most other organic pollutants, is metabolized more rapidly by vertebrates than by aquatic invertebrates such as gastropods.

8.3.3 ENVIRONMENTAL FATE OF TRIBUTYLTIN

The main uses of TBT compounds are related to their toxic properties. They have been used as antifoulants on boats, ships, buoys, and crabpots; as biocides for cooling systems, pulp and paper mills, breweries, leather processing plants, and textile mills; and as molluscicides. Of particular interest and importance is their incorporation into antifouling paints used on boats of many kinds ranging from small leisure craft to large oceangoing vessels. Release of TBT from antifouling paints has provided a small yet highly significant source of pollution to surface waters.

When TBTO is released into ambient water, a considerable proportion becomes adsorbed to sediments, as might be expected from its lipophilicity. Studies have shown that between 10 and 95% of TBTO added to surface waters becomes bound to sediment. In the water column it exists in several different forms, principally the hydroxide, the chloride, and the carbonate (Figure 8.5). Once TBT has been adsorbed, loss is almost entirely due to slow degradation, leading to desorption of diphenyltin (DPT). The distribution and state of speciation of TBT can vary considerably between aquatic systems, depending on pH, temperature, salinity, and other factors.

TBT levels have been monitored in coastal areas of Western Europe and North America. These have ranged upward to 5.34 µg/L in Western Europe, and 1.71 µg/L in North America (Environmental Health Criteria 116). The highest levels were recorded in the shallow waters of estuaries and harbors where there were large numbers of small boats.

TBT is taken up by aquatic organisms directly from water and from food. Comparison of concentrations in mollusks with concentrations in ambient water indicate very strong bioconcentration/bioaccumulation. When mollusks such as the edible mussel (*Mytilus edulis*) and the Pacific oyster (*Crassostrea gigas*) were exposed experimentally to TBTO in ambient water, bioconcentration factors (BCFs) ranging between 1,000-fold and 7,000-fold were found (see Environmental Health Criteria 116). With mussels exposed to relatively low levels of TBT, tissue levels were still increasing after 7 weeks' exposure, no plateau level having been reached. Exposure of *Mytilus edulis* under natural conditions indicated higher BCFs than this—5,000–60,000-fold (Cheng and Jensen 1989). Further investigation has shown that uptake of TBT from food can be greater than uptake directly from water. Thus, BCFs are a reflection more of bioaccumulation than of bioconcentration in this case.

8.3.4 TOXICITY OF TRIBUTYLTIN

Mechanistic studies have shown that TBT and certain other forms of trialkyltin have two distinct modes of toxic action in vertebrates. On the one hand they act as inhibitors of oxidative phosphorylation in mitochondria (Aldridge and Street 1964). Inhibition is associated with repression of ATP synthesis, disturbance of ion transport across the mitochondrial membrane, and swelling of the membrane. Oxidative phosphorylation is a vital process in animals and plants, and so trialkyltin compounds act as wide-ranging biocides. Another mode of action involves the inhibition of forms of cytochrome P450, which was referred to earlier in connection with metabolism. This has been demonstrated in mammals, aquatic invertebrates and fish (Morcillo et al. 2004, Oberdorster 2002). TBTO has been shown to inhibit P450 activity in cells from various tissues of mammals, including liver, kidney, and small intestine mucosa, both in vivo and in vitro (Rosenberg and Drummond 1983, Environmental Health Criteria 116).

Of particular interest in the present context is that TBT can inhibit cytochrome-P450-based aromatase activity in both vertebrates and aquatic invertebrates (Morcillo et al. 2004, Oberdorster and McClellan-Green 2002). The conversion of testosterone to estradiol is catalyzed by aromatase, and so inhibition of the enzyme can, in principle, lead to an increase in cellular levels of testosterone. The significance of this is that many mollusks experience endocrine disruption when exposed to TBTs, and there is evidence that this is the consequence of elevated levels of testosterone (Matthiessen and Gibbs 1998, Morcillo et al. 2004). In other words, inhibition of aromatase may be the reason why TBT compounds act as endocrine disruptors in mollusks. Apart from this widely proposed mechanism, there is some evidence to suggest that TBT may affect testosterone levels in mollusks by inhibiting conjugation with sulfate or glucuronide (Ronis and Mason 1996, Morcillo et al. 2004).

Returning to the question of endocrine disruption, it is now known that over 100 species of gastropods worldwide suffer from a condition described as "imposex," the development of male characteristics by females. It has already been established for some species (e.g., dog whelk) that TBT is the cause of this, and it is suspected that organotin compounds account for most cases of imposex worldwide (Matthiessen and Gibbs 1998). Some examples of masculinization of female gastropods caused by tributyltin are given in Table 8.4.

It can be seen that there are differences between species in the physiological changes that are caused. However, one common factor is the development of male characteristics, usually with an adverse effect on reproduction (Matthiessen and Gibbs 1998, Gibbs et al. 1988). As explained previously, the underlying cause appears to be elevated levels of testosterone following exposure to TBT, an effect that has now been observed in several species of gastropods, and the most widely held explanation of this is inhibition of the P450 isozyme, known as aromatase, by TBT, which fits in with the well-documented ability of TBT to deactivate P450s. Although it is true that inhibition of aromatase is known to lead to an increase in the cellular level of testosterone in certain species, there is still some debate over the importance of this mechanism in mollusks. Among others, there are a number of P450 forms involved in the metabolism of steroid hormones apart from aromatase, and TBT appears to act upon several different forms (Mensink et al. 1996,

TABLE 8.4
Masculinization of Gastropods Caused by TBT

Species	Effects	Comment
Dog whelk (*Nucella lapillus*)	TBT causes imposex Development of penis blocks oviduct	Cause of population declines along South Coast of England
Sting whelk (*Ocenebra erinacea*)	Similar effect to that observed in dog whelk	
Common whelk (*Buccinum undatum*)	Also causes imposex	Effect only seen in juvenile Effects reported from North Sea in the vicinity of shipping lanes
Periwinkle (*Littorina littoria*)	Gross malformation of oviduct usually termed "intersex"	Causes infertility and population decline in England, some without development of penis, but some recovery following 1987 ban on TBT use on small vessels

Source: From Matthiessen and Gibbs (1998).

Fent et al. 1998). Thus, inhibition of P450s other than aromatase might cause elevation of testosterone levels.

Apart from gastropods, harmful effects of TBT have also been demonstrated in oysters (Environmental Health Criteria 116, Thain and Waldock 1986). Early work established that adult Pacific oysters (*Crassostrea gigas*) showed shell thickening caused by the development of gel centers when exposed to 0.2 µg/L of TBT fluoride (Alzieu et al. 1982). Subsequent work established the no observable effect level (NOEL) for shell thickening in this, the most sensitive of the tested species, at about 20 ng/L. It has been suggested that shell thickening is a consequence of the effect of TBT on mitochondrial oxidative phosphorylation (Alzieu et al. 1982). Reduced ATP production may retard the function of Ca^{++} ATPase, which is responsible for the Ca^{++} transport that leads to $CaCO_3$ deposition during the course of shell formation. Abnormal calcification causes distortion of the shell layers.

Other disturbances have been shown to be caused by TBT in oysters. These include effects on gonad development and gender in adult oysters, and on settlement, growth and mortality of larval forms. In one experiment, European flat oysters (*Ostrea edulis*) were exposed to 0.24 or 2.6 µg/L TBT over a period of 75 days (Thain and Waldock 1986). There was a large production of larvae in a related control group, but none in either of the treated groups. On subsequent examination of the treated oysters, no females were found in either of the two treated groups (20% of the control group were females). In the group receiving the highest exposure to TBT, 72% were undifferentiated.

TBTO has appreciable toxicity to fish, LC_{50} values ranging from 1–30 µg/L for most of the species that have been tested. Acute oral LD_{50} values for the rat and the mouse are in the range 85–240 mg/kg. Recently, evidence has come forward that TBT can cause masculinization of fish (Shimasaki et al. 2003). Japanese flounder (*Paralichthys olivaceus*) fed a diet containing 0.1 µg/g TBTO showed a 26% increase in the ratio of sex-reversed males. Interestingly, several authors have reported that TBT can inhibit cytochrome P450 forms in fish (Fent and Bucheli 1994, Morcillo et al. 2004), so once again, elevation of testosterone levels following inhibition of aromatase is a possible explanation.

8.3.5   Ecological Effects of TBT

A striking example of the harmful effects of low levels of TBT was the decline of dog whelk populations around the shores of Southern England (completely from certain shallow waters such as marinas, estuaries, and harbors) where there were large numbers of small boats. It was found that females showed the first signs of imposex with levels of TBT as low as 1 ng/L in water, and that development of male characteristics increased progressively with dose. TBT concentrations exceeding 5 ng/L caused blockage of the oviduct because of the proliferation of the vas deferens and led to breeding failure. Before 1987, levels high enough to have this effect were common in shallow waters with large numbers of small craft. In 1987, a ban on the use of TBT as an antifoulant on small boats was introduced in Britain, after which recovery of the population began.

An interesting local phenomenon connected with TBT pollution in Southern England is the so-called *Dumpton Syndrome*. In the eponymous coastal area in Kent, a local population survived levels of TBT that caused extinction of the population elsewhere. This was found to be related to a local genetic deficiency of male dog whelks, which had an underdeveloped genitalia (small or absent penis and incomplete gonoducts). Some females in the area did not develop imposex, and bred successfully. The tentative interpretation was that there was a genetic deficiency in certain males and females from the Dumpton area that caused low levels of testosterone; thus, neither males nor females could properly develop male characteristics. It is thought that the Dumpton females were able to breed with males that were unaffected by the Dumpton Syndrome (Gibbs 1993).

In the late 1970s there were severe problems with oyster populations along certain stretches of the French coast, notably in the bay of Arcachon (Alzieu et al. 1982). Poor shell growth, shell malformations, and very poor spatfalls were all observed. Subsequent investigations attributed these harmful effects largely or entirely to TBT. As with imposex in the dog whelk, affected areas had relatively large numbers of small boats and associated (relatively) high levels of TBT in ambient water. As with the dog whelk, populations recovered in badly affected areas following a ban on use of TBT on small craft (<25 m) and a consequent reduction in environmental levels of TBT. In the bay of Arcachon, the percentages of the population showing deformities were 95–100%, 70–80%, and 45–50% in 1980/1981, 1982, and 1983, respectively. The spatfall was excellent by 1983, having failed completely in 1980 and 1981. Harmful effects of TBTs on oysters have been reported from many other locations, including other coastal areas of Western Europe, and coastal areas of the United States, and Japan.

8.4   ORGANOLEAD COMPOUNDS

Lead tetraalkyl compounds, of which lead tetraethyl is the best known, were once widely used as "antiknock" compounds, that is, they were added to petrol to control semiexplosive burning. In many countries, this use has been greatly curtailed, with the large-scale introduction of lead-free petrol. The reduction in use of tetraalkyl lead has been part of a wider aim of reducing human exposure to lead. In fact, lead tetraalkyls in petrol are broken down to a considerable degree during the operation of the internal combustion engine so that most of the lead in car exhausts is in the inorganic form.

The main concern regarding tetraalkyl lead has been about human health hazards, a concern that has resulted in the progressive replacement of leaded petrol by unleaded petrol in most countries (Environmental Health Criteria 85). There has been particular concern about possible brain damage to children in polluted urban areas. Little work has been done on the effects of organolead compounds on wildlife or ecosystems, so the following account will be brief.

Lead tetramethyl and lead tetraethyl are covalent lipophilic liquids of low water solubility. Certain inorganic forms of lead, for example, lead tetrachloride, have similar properties, but other forms such as lead nitrate and lead dichloride are ionic and water soluble. Covalent and lipophilic forms of lead, like lipophilic forms of organomercury and organotin, can readily cross membranous barriers such as the blood–brain barrier. Consequently, they readily enter the CNS of animals, where they cause damage by combining with sulfhydryl groups.

Lead tetraethyl is oxidized by the P450-based monooxygenase system to form the lead triethyl cation

$$Pb \ [C_2H_5]_4 \rightarrow Pb \ [C_2H_5]_3{}^+$$

There is only limited information on the ecotoxicity of organolead compounds. The toxicity of tetraethyl lead to fish ranged from 0.02–2.0 mg/L in tests on three different species (see Environmental Health Criteria 84). These and other data suggest that alkyl lead compounds are 10–100 times more toxic to fish than inorganic lead. Turning to birds, toxicity tests upon starlings (*Sturnus vulgaris*) were carried out with trimethyl and triethyl lead (Osborn et al. 1983). Birds were dosed with each compound at two different rates, 0.2 and 2.0 mg/day, for 11 days. The high dose corresponded to 28 mg/kg/day, and all birds died within 6 days of commencement of dosing in the case of both compounds. Symptoms of neurotoxicity and behavioral effects were seen before death, most noticeably with trimethyl lead.

In 1979, a major poisoning incident involving over 2000 birds of different species, the majority of them dunlin (*Calidris alpina*), was attributed to alkyl lead poisoning (Bull et al. 1983, Osborn et al. 1983, and Environmental Health Criteria 85). It occurred in the Mersey estuary, United Kingdom, where there were periodically high levels of alkyl lead originating from the effluent of a petrochemical works. Casualties included various species of gulls, waders, and duck. Smaller numbers of dead birds were found in incidents that occurred in the following 2 years. The total lead content of dead birds averaged about 11 ppm in liver, most of it in the form of alkyl lead. Surviving birds that showed symptoms of poisoning contained about 8.9 ppm of lead. One important food source for the waders was the mollusk *Macoma baltica*, which was found to contain about 1 ppm of lead at the time of the first incident, suggesting that there may have been strong bioaccumulation by the birds.

8.5   ORGANOARSENIC COMPOUNDS

Organoarsenic compounds have been of importance in human toxicology but have not as yet received much attention in regard to environmental effects. Like methyl mercury compounds, they are both synthesized in the environment from inorganic forms and released into the environment as a consequence of human activity (Environmental Health Criteria 18). They can cause neurotoxicity.

Concerning anthropogenic sources, methyl arsenic compounds such as methyl arsonic acid and dimethylarsinic acid have been used as herbicides, and were once a significant source of environmental residues. Dimethyl-arsinic acid (Agent Blue) was used as a defoliant during the Vietnam War.

Methyl arsenic, like methyl mercury, is generated from inorganic forms of the element by methylation reactions in soils and sediments. However, the mechanism is evidently different from that for mercury, depending on the attack by a methyl carbonium ion rather than a methyl carbanion (Craig 1986, Crosby 1998). Methylation

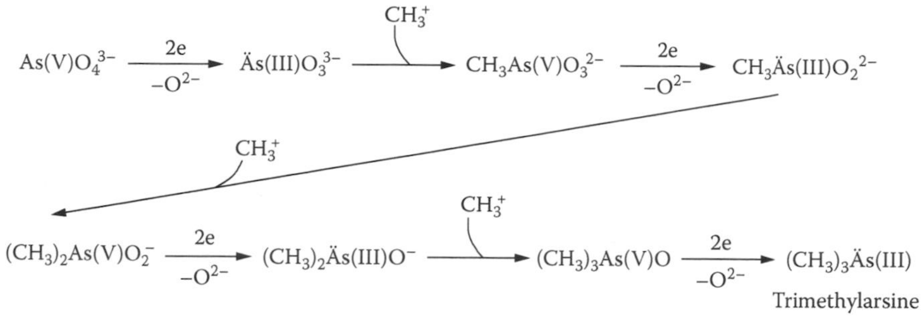

$$As(V)O_4^{3-} \xrightarrow[-O^{2-}]{2e} \ddot{A}s(III)O_3^{3-} \xrightarrow{CH_3^+} CH_3As(V)O_3^{2-} \xrightarrow[-O^{2-}]{2e} CH_3\ddot{A}s(III)O_2^{2-}$$

$$(CH_3)_2As(V)O_2^- \xrightarrow[-O^{2-}]{2e} (CH_3)_2\ddot{A}s(III)O^- \xrightarrow{CH_3^+} (CH_3)_3As(V)O \xrightarrow[-O^{2-}]{2e} (CH_3)_3\ddot{A}s(III)$$

Trimethylarsine

FIGURE 8.6 Methylation of arsenate (after Environmental Health Criteria 18).

of arsenic occurs with the trivalent rather than the pentavalent form, and up to three methyl groups can be bound to one As atom (Figure 8.6). The final product, trimethylarsine, is both volatile and highly toxic to mammals. It has been implicated in cases of human poisoning following its generation by microorganisms from inorganic arsenic (Paris Green) in old wallpaper.

Significant quantities of organoarsenic compounds have been found in marine organisms, and questions have been asked about the possible health risk to humans consuming sea food.

8.6 SUMMARY

Mercury, tin, lead, arsenic, and antimony form toxic lipophilic organometallic compounds, which have a potential for bioaccumulation/bioconcentration in food chains. Apart from anthropogenic organometallic compounds, methyl derivatives of mercury and arsenic are biosynthesized from inorganic precursors in the natural environment.

Methyl mercury compounds are neurotoxic and can cause behavioral effects in vertebrates. They bind strongly to sulfhydryl groups of proteins and can cause changes in the abundance of neurochemical receptors of the vertebrate brain. Cases of human poisoning (Minamata Bay incident) and poisoning of wildlife (Sweden in the 1950s and 1960s) have been caused by methyl mercury. Even after bans on use of methyl mercury, fungicide levels of methyl mercury are still high in some areas (e.g., of North America), and there is evidence of continuing side effects on wildlife.

Tributyltin compounds used as antifouling agents on boats have had serious toxic effects upon many mollusks, including populations of oysters and dog whelks. Females of the latter species developed a condition known as imposex, which rendered them infertile and caused local extinction of the population in shallow coastal waters. Imposex provides the basis for a valuable biomarker assay.

Alkyl lead compounds are also highly neurotoxic and were implicated in large-scale kills of wading birds.

FURTHER READING

The following issues of Environmental Health Criteria are valuable sources of information on the environmental toxicology of organometallic compounds:

Craig, P.J. (Ed.) (1986). *Organometallic Compounds in the Environment*—A collection of detailed chapters on the environmental chemistry and biochemistry of organometallic compounds.

Environmental Health Criteria 18 Arsenic

Environmental Health Criteria 85 Lead Environmental Aspects

Environmental Health Criteria 86 Mercury, Environmental Aspects

Environmental Health Criteria 101 Methylmercury

Environmental Health Criteria 116 Tributyltin

International Atomic Energy Agency (1972). *Mercury Contamination in Man and His Environment*, Technical Report Series 137—Contains some useful accounts of work done in Sweden on ecotoxicology of organomercury compounds that is difficult to find in the general literature.

Matthiessen, P. and Gibbs, P.E. (1998). *Critical Appraisal of the Evidence for TBT-Mediated Endocrine Disruption in Molluscs*—A concise review of effects of TBT on molluscs.

Scheuhammer, A.M and Sandheinrich, M.B (Eds.) (2008). Special issue of the journal *Ecotoxicology* devoted to effects of methyl mercury on wildlife, which gives recent results of field studies conducted in North America.

9 Polycyclic Aromatic Hydrocarbons

9.1 BACKGROUND

Hydrocarbons are compounds composed of carbon and hydrogen alone. They may be classified into two main groups:

1. Aromatic hydrocarbons that contain ring systems with delocalized electrons, for example, benzene.
2. Nonaromatic hydrocarbons that do not contain such a ring system. Included here are alkanes, which are fully saturated hydrocarbons; alkenes, which contain one or more double bonds; and alkynes, which contain one or more triple bonds.

Some examples of different types of hydrocarbons are given in Figure 9.1. Nonaromatic compounds without ring structure are termed *aliphatic*, whereas those with a ring structure (e.g., cyclohexane) are termed *alicyclic*. Aromatic hydrocarbons often consist of several fused rings, as in the case of benzo[*a*]pyrene.

Both classes of hydrocarbon occur naturally, notably in oil and coal deposits. Aromatic compounds are also products of incomplete combustion of organic compounds, and are released into the environment both by human activities, and by certain natural events, for example, forest fires and volcanic activity.

Aromatic hydrocarbons, the subject of this chapter, are of particular concern because of their mutagenic and carcinogenic properties. In the first place, this raises issues about human health risks. However, there are also questions about possible harmful effects on ecosystems that are exposed to high levels of aromatic hydrocarbons. Nonaromatic hydrocarbons are usually of low toxicity, have not received much attention in ecotoxicology, and will not be discussed further in the present text. It should, however, be remembered that crude oil consists mainly of alkanes, and large releases into the sea due to the wreckage of oil tankers have caused the death of many seabirds and other marine organisms because of their physical effects of *oiling*, or *smothering* (see Clark 1992). Also, released crude oil may act as a vehicle for other lipophilic pollutants that dissolve in it, for example, organotin compounds and PCBs, both of which may be present in or on wrecked vessels.

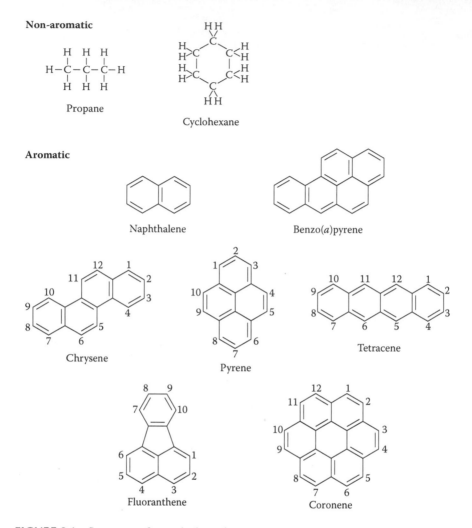

FIGURE 9.1   Structures of some hydrocarbons.

9.2 ORIGINS AND CHEMICAL PROPERTIES

The largest releases of polycyclic aromatic hydrocarbons (PAHs) are due to the incomplete combustion of organic compounds during the course of industrial processes and other human activities. Important sources include the combustion of coal, crude oil, and natural gas for both industrial and domestic purposes, the use of such materials in industrial processes (e.g., the smelting of iron ore), the operation of the internal combustion engine, and the combustion of refuse (see Environmental Health Criteria 202, 1998). The release of crude oil into the sea by the offshore oil industry and the wreckage of oil tankers are important sources of PAH in certain areas. Forest fires, which may or may not be the consequence of human activity, are a significant and usually unpredictable source of PAH. In general, environmental contamination is by complex mixtures of PAHs, not by single compounds.

The structures of some PAHs of environmental interest are given in Figure 9.1. Naphthalene is a widely distributed compound consisting of only two fused benzene rings. It is produced commercially for incorporation into mothballs. Many of the compounds with marked genotoxicity contain 3–7 fused aromatic rings. Benzo[a] pyrene is the most closely studied of them, and will be used as an example in the following account.

Because of the fusion of adjacent rings, PAHs tend to have a rigid planar structure. As a class, they are of low water solubility and marked lipophilicity and have high octanol–water coefficients (K_{ow}) (Table 9.1); the higher the molecular weight, the higher the lipophilicity, and the higher the log K_{ow}. Vapor pressure is also related to molecular weight; the higher the molecular weight, the lower the vapor pressure. PAHs have no functional groups and are chemically rather nonreactive. They can, however, be oxidized both in the natural environment and biochemically (see Figure 9.2). Photodecomposition can occur in air and sunlight to yield oxidative products, such as quinones and endoperoxides. Nitrogen oxides and nitric acid can convert PAHs to nitro derivatives, whereas sulfur oxides and sulfuric acids can produce sulfanilic and sulfonic acids.

9.3   METABOLISM

Because of the absence of functional groups, primary metabolic attack on PAHs is limited to oxidation, usually catalyzed by cytochrome P450. As with coplanar PCBs, oxidative attack involves P450 forms of more than one gene family, including members of gene families 1, 2, and 3. The position of oxidative attack on the ring system (regioselectivity) depends on the P450 form to which a PAH is bound. P450 1A1 is particularly implicated in the metabolic activation of carcinogens, such as benzo[a] pyrene, where oxygen atoms can be inserted into the critical bay region positions. The metabolism of benzo[a]pyrene has been studied in some depth and detail, and will be used as an example of the metabolism of PAHs more generally (Figure 9.2).

TABLE 9.1
Properties of Some Polycyclic Aromatic Hydrocarbons

Compound	Vapor Pressure (Pa @ 25°C)	log K_{ow}	Water Solubility at 25°C μg/L
Naphthalene	10.4	3.4	3.17×10^{-4}
Anthracene	8.4×10^{-4}	4.5	73
Pyrene	6.0×10^{-4}	5.18	135
Chrysene	8.4×10^{-5}	5.91	2.0
Benzo[a]pyrene	7.3×10^{-7}	6.50	3.8
Dibenz[a,h]anthracene	1.3×10^{-8}	6.50	0.5

Source: From Environmental Health Criteria 202.

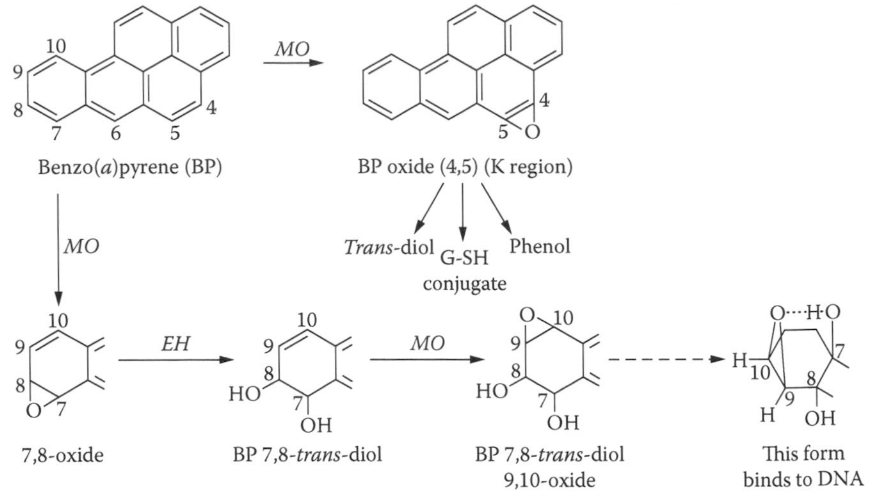

FIGURE 9.2   Metabolism of benzo[a]pyrene.

Initial metabolic attack can be upon one of a number of positions on the benzo[a] pyrene molecule to yield various epoxides. Epoxides tend to be unstable and can quickly rearrange to form phenols. They may also be converted to transdihydrodiols by epoxide hydrolase, or to glutathione conjugates by the action of glutathione-S-transferases. Hydroxymetabolites are more polar than the parent molecules, and can be converted into conjugates, such as glucuronides and sulfates, by the action of conjugases. Two important oxidations of benzo[a]pyrene are shown in Figure 9.2: formation of the 7,8 oxide and the 4,5 oxide. The 4,5 oxide is unstable under cellular conditions, undergoing rearrangement to form a phenol, and biotransformation to a transdihydrodiol or a glutathione conjugate. In vitro, it shows mutagenic properties, for example, in the Ames test. However, in vertebrates in vivo, it appears to be detoxified very effectively, thus preventing the formation of DNA adducts to any significant degree. The 7,8 oxide is converted to the 7,8 transdihydrodiol by the action of epoxide hydrolase. The 7,8 transdihydrodiol is a substrate for P450 1A1, and consequent oxidation yields the highly mutagenic 7,8 diol, 9,10 oxide, a metabolite that, under cellular conditions, interacts with guanine residues of DNA. Thus, a mutagenic diol epoxide generated in the endoplasmic reticulum is able to escape detoxication in situ, or elsewhere in the cell, as it migrates to the nucleus to interact with DNA. This ability to escape full detoxication may be related to the fact that the isomer of the 7,8 diol 9,10 oxide responsible for adduction (Figure 9.2) is a poor substrate for epoxide hydrolase. It should be added that benzo[a]pyrene is an inducer of P450 1A1, so it can increase the rate of its own activation! The toxicological significance of this type of interaction will be discussed in Section 9.5.

In terrestrial animals, the excreted products of PAHs are mainly conjugates formed from oxidative metabolites. These include glutathione conjugates of epoxides, and sulfate and glucuronide conjugates of phenols and diols.

PAHs, such as benzo[a]pyrene and 3-methyl cholanthrene, induce P450 1A1/2 (Chapter 2).

Induction of P450 1A1/2 provides the basis for biomarker assays for PAHs and other planar organic pollutants, such as coplanar PCBs, PCDDs, and PCDFs.

9.4   ENVIRONMENTAL FATE

Viewed globally, the largest emissions of PAH are into the atmosphere, and the main source is the products of incomplete combustion of organic compounds. As mentioned earlier, emissions are mainly the consequence of human activity, although certain natural events, for example, forest fires, are sometimes also important. Emissions into the air are of complex mixtures of different PAHs, including particulate matter, as in smoke. PAHs in the vapor phase can be adsorbed on to airborne particles. Airborne PAHs eventually enter surface waters owing to precipitation of particles or to diffusion. Once there, because of their high K_{ow} values, they tend to become adsorbed to the organic material of sediments, and are taken up by aquatic organisms. Similarly, airborne PAH can eventually reach soil to become adsorbed by soil colloids, and absorbed by soil organisms.

Apart from release into air, which is important globally, the direct transfer of PAH to water or land surfaces can be very important locally. Wreckages of oil tankers and discharges from oil terminals cause marine pollution by crude oil, which contains appreciable quantities of PAH. Disposal of waste containing PAH around industrial premises has caused serious pollution of land in some localities.

When crude oil is released into the sea, oil films (*slicks*) can spread over a large area, the extent and direction of movement being determined by wind and tide (see Clark 1992). The hydrocarbons of lowest molecular weight have the highest vapor pressures, and tend to volatilize, leaving behind the least volatile components of crude oil. Eventually, the residue of relatively involatile hydrocarbons will sink to become associated with sediment. Thus, long after the surface film of oil has disappeared, residues of PAH will exist in sediment, where they are available to bottom-dwelling organisms. To illustrate the range of PAHs found in sediments, some values follow for PAH residues detected in sediment from the highly polluted Duwamish waterway in the United States (see Varanasi et al. 1992). All concentrations are given as mean values expressed as ng/g dry weight.

Naphthalene	400
Fluorene	390
Phenanthrene	2400
Anthracene	610
Fluoranthene	3900
Benz[*a*]anthracene	2000
Chrysene	2900
Benzo[*a*]pyrene	2300
Pyrene	4800
Dibenz[*a,h*]anthracene	470
Perylene	900
Benzofluoranthenes	4900

PAHs can be bioconcentrated or bioaccumulated by certain aquatic invertebrates low in the food chain that lack the capacity for effective biotransformation (Walker and Livingstone 1992). Mollusks and *Daphnia* spp. are examples of organisms that readily bioconcentrate PAH. On the other hand, fish and other aquatic vertebrates readily biotransform PAH; so, biomagnification does not extend up the food chain as it does in the case of persistent polychlorinated compounds. As noted earlier, P450-based monooxygenases are not well represented in mollusks and many other aquatic invertebrates (see Chapter 4, Section 4.2); so, this observation is not surprising. Oxidation catalyzed by P450 is the principal (perhaps the only) effective mechanism of primary metabolism of PAH.

An example of PAH residues in polluted marine ecosystems is given in Table 9.2. They are from studies carried out at different coastal sites of the United States (Varanasi et al. 1992). The residues are categorized into "lower aromatic hydrocarbons (LAH)," composed of 1–3 rings, and "higher aromatic hydrocarbons (HAH)," composed of 4–7 rings. HAHs predominate in sediment, showing levels in the range 1,800–12,600 ng/g, that is, 1.8–12.6 ppm by weight. As mentioned earlier, some invertebrates can bioaccumulate PAH, which is the main reason for significant levels of PAH in the remains of food found within fish stomachs. Analysis of invertebrates from some of these areas showed levels of 300–3500 ng/g total aromatic hydrocarbon in their tissues, of which over 90% was accounted for as HAH. The concentrations of aromatic hydrocarbons are, however, substantially below those found in samples of sediment. Thus, HAH levels range from 135 to 2700 ng/g in stomach contents. Interestingly, no residues of aromatic hydrocarbons were detected in the livers of the fish analyzed in this study, illustrating their ability to rapidly metabolize both LAH and HAH.

It appears that organisms at the top of aquatic food chains are not exposed to substantial levels of PAH in food because of the detoxifying capacity of organisms beneath them in the food chain. On the other hand, fish, birds, and aquatic mammals feeding on mollusks and other invertebrates are in a different position. Their food may contain substantial levels of PAH. Although they can achieve rapid metabolism of dietary PAH, it should be remembered that oxidative metabolism causes

TABLE 9.2
Concentration of Aromatic Hydrocarbons (AHs) in Samples from U.S. Coastal Sites

Sample	Total AH (ng/g)	1–3 Rings (LAH) (ng/g)	4–7 Rings (HAH) (ng/g)
Sediments	700–14,000	200–1,400	1,800–12,600
Fish stomach	150–3,000	15–300	135–2,700
Contents	—	—	—
Fish liver	Nil	Nil	Nil

Source: Extracted from data presented by Varanasi et al. (1992).

activation as well as detoxication. The types of P450 involved determine the position of metabolic attack (see Section 9.3). Cytochrome P450 1A1, for example, activates benzo[a]pyrene by oxidizing the bay region to form a mutagenic diol epoxide. The tendency for PAHs to be activated as opposed to detoxified depends on the balance of P450 forms, and this balance is dependent on the state of induction. Many PAHs, PCDDs, PCDFs, coplanar PCBs, and other planar organic pollutants are inducers of P450s belonging to Family 1A. Thus, activation of PAH may be enhanced owing to the presence of pollutants that induce P450 1A1/1A2, forms that are particularly implicated in the process of activation (Walker and Johnston 1989).

9.5 TOXICITY

PAHs are rather nonreactive in themselves, and appear to express little toxicity. Toxicity is the consequence of their transformation into more reactive products, by chemical or biochemical processes. In particular, the incorporation of oxygen into the PAH ring structure has a polarizing effect; the electron-withdrawing properties of oxygen leading to the production of reactive species such as carbonium ions. This is evidently the reason why PAHs become more toxic to fish and *Daphnia* following exposure to ultraviolet (UV) radiation (Oris and Giesy 1986, 1987); photooxidation of PAH to reactive products increases toxicity.

Much research on the toxicity of PAH has been concerned with human health hazards, and has focused on their mutagenic and carcinogenic action. These two properties are to some extent related, because there is growing evidence that certain DNA adducts formed by metabolites of carcinogenic PAHs become fixed as mutations of oncogenes or tumor-suppressor genes that are found in chemically produced cancers (Purchase 1994). Typical mutations occur at specific codons in the ras, neu, or myc oncogenes or in P53, retinoblastoma or APC tumor suppressor genes. These genes code for proteins involved in growth regulation, with the consequence that mutated cells have altered growth control. One example of such a carcinogenic metabolite is the 7,8-diol-9,10 oxide of benzo[a]pyrene (Figure 9.2). More generally, many compounds found to be mutagenic in bacterial mutation assays (e.g., the Ames test) are also carcinogenic in long-term dosing tests with rodents. However, a substantial number of carcinogens act by nongenotoxic mechanisms (Purchase 1994).

Benzo[a]pyrene is converted to its 7,8-diol-9,10 oxide by the action of cytochrome P450 1A1 and epoxide hydrolase, as shown in Figure 9.2. In one of its enantiomeric forms, this metabolite can then form DNA adducts by alkylating certain guanine residues (Figure 9.3). The metabolite acts as an electrophile, due to strong carbonium ion formation on the 10 position of the epoxide ring, which is located in the bay region. The epoxide ring cleaves, and a bond is formed between C10 of the PAH ring and the free amino group of guanine. The oxygen atom of the cleaved epoxide ring acquires a proton, thus leaving a hydroxyl group attached to C9. This adduct, similar to the others formed between reactive metabolites of PAHs and DNA, is bulky and can be detected by P32 postlabeling and immunochemical techniques (e.g., Western blotting). It has been proposed that there is a particular tendency for strong carbonium ion formation to occur on the bay region of PAHs, and that such

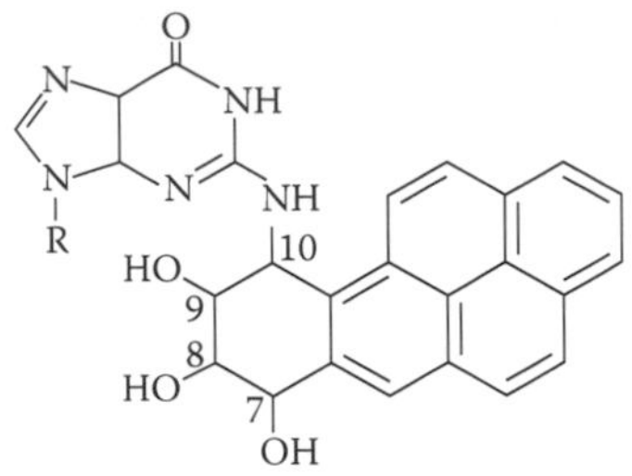

Attachment of BP 7,8-diol 9,10-oxide
to guanine residue of DNA

FIGURE 9.3   DNA adduct formation.

"bay region epoxides" have a strong tendency to form DNA adducts (see Hodgson and Levi 1994) .

There is strong evidence that DNA adduction by these bulky reactive metabolites of PAHs is far from random, and that there are certain "hot spots" that are preferentially attacked. Differential steric hindrance and the differential operation of DNA repair mechanisms ensure that particular sites on DNA are subject to stable adduct formation (Purchase 1994). DNA repair mechanisms clearly remove many PAH/ guanine adducts very quickly, but studies with P^{32} postlabeling have shown that certain adducts can be very persistent—certainly over many weeks. Evidence for this has been produced in studies on fish and *Xenopus* (an amphibian; Reichert et al. 1991; Waters et al. 1994).

Although genotoxicity is of central importance in human toxicology, its significance in ecotoxicology is controversial. However, PAH has been shown to cause tumor development in fish in response to, for example, oral, dermal, or intraperitoneal administration of benzo[a]pyrene and 3-methyl cholanthrene. Hepatic tumors have been reported in wild fish exposed to sediment containing about 250 mg/kg of PAH (Environmental Health Criteria 202). K-ras mutations occurred in pink salmon embryos (*Onchorhynchus gorbuscha*) following exposure to crude oil from the tanker *Exxon Valdez*, which caused extensive pollution of coastal regions of Alaska (Roy et al. 1999). However, it is not clear whether cancer is a significant factor in determining the survivorship or reproductive success of free-living vertebrates or invertebrates. Cancers usually take a long time to develop, and the life span of free-living animals is limited by factors such as food supply, disease, predation, etc. Do they live long enough for cancers to be a significant cause of population decline?

Apart from carcinogenicity, there is the wider question of other possible genotoxic effects in free-living animals, effects that may be heritable if the mutations are in germ cells. Studying aquatic invertebrates exposed to PAH, Kurelec (1991) noticed a number of longer-term physiological effects, which he termed collectively *genotoxic disease syndrome*. Although the basis for these effects has not yet been elucidated, the observations raise important questions that should be addressed. PAHs have been shown to form a variety of adducts in fish and amphibians, so there is a strong suspicion that some of these may lead to the production of mutations. If mutations occur in germ cells, there are inevitably questions about their effects on progeny. Most mutations are not beneficial!

There is evidence for immunosuppressive effects of PAHs in rodents (Davila et al. 1997). For example, strong immunosuppressive effects were reported in mice that had been dosed with benzo[a]pyrene and 3-methyl cholanthrene, effects that persisted for up to 18 months (Environmental Health Criteria 202). Multiple immunotoxic effects have been reported in rodents, and there is evidence that these result from disturbance of calcium homeostasis (Davila et al. 1997). PAHs can activate protein tyrosine kinases in T cells that initiate the activation of a form of phospholipase C. Consequently, release of inositol triphosphate—a molecule that immobilizes Ca^{2+} from storage pools in the endoplasmic reticulum—is enhanced.

Turning to the acute toxicity of PAH, terrestrial organisms will be dealt with before considering aquatic organisms, to which somewhat different considerations apply. The acute toxicity of PAHs to mammals is relatively low. Naphthalene, for example, has a mean oral LD50 of 2700 mg/kg to the rat. Similar values have been found with other PAHs. LC50 values of 150 mg/kg and 170–210 mg/kg have been reported, for phenanthrene and fluorene, respectively, in the earthworm. The NOEL level for survival and reproduction in the earthworm was estimated to be 180 mg/kg dry soil for benzo[a]pyrene, chrysene, and benzo[k]fluoranthene (Environmental Health Criteria 202).

Toxicity of PAH to aquatic organisms depends on the level of UV radiation to which the test system is exposed. PAHs can become considerably more toxic in the presence of radiation, apparently because photooxidation transforms them into toxic oxidative products. PAHs, such as benzo[a]pyrene, can have LC50s as low as a few micrograms per liter toward fish when there is exposure to UV (Oris and Giesy 1986, 1987). PAHs can also show appreciable toxicity to sediment-dwelling invertebrates. LC50 values of 0.5–10 mg/kg (concentration in sediment) have been reported for marine amphipods for benzo[a]pyrene, fluranthene, and phenanthrene, used singly or in mixtures. These values are much lower than the LC50 or NOEL concentrations for earthworms quoted earlier, which were exposed to contaminated soil. It has also been shown that exposure of adult fish to anthracene and artificial UV radiation can impair egg production (Hall and Uris 1991).

9.6 ECOLOGICAL EFFECTS

Serious ecological damage has been caused locally by severe oil pollution. The wreckages of the oil tankers—the *Torrey Canyon* (Cornwall, United Kingdom, 1967), the *Amoco Cadiz* (Brittany, France, 1978), the *Exxon Valdez* (Alaska, United States, 1989), and the *Sea Empress* (South Wales, United Kingdom, 1996)—all caused serious pollution locally. Less dramatically, leakage from offshore oil operations has also caused localized pollution problems. Most of the reported harmful effects of such marine pollution have been due to the physical action of the oil rather than the toxicity of PAHs. Fish, however, may have been poisoned by oil in situations where there was strong UV radiation (see Section 9.5). The oiling of seabirds and other marine organisms has been the cause of some local population declines. Sometimes the reduction of invertebrate herbivores on polluted beaches and rock pools has led to an upsurge of the plants upon which they feed. The flourishing of seaweeds and algal blooms has sometimes followed such environmental disasters, to be reversed when pollution is reduced and the herbivores recover. In the neighborhood of oil terminals the diversity of benthic fauna has been shown to decrease (see Clark 1992).

Although it has been relatively easy to demonstrate local short-term effects of oil pollution on seabird populations, establishing longer-term effects on marine ecosystems has proved more difficult, notwithstanding the persistence of PAH residues in sediments. In one study, the edible mussel (*Mytilus edulis*) was used as an indicator organism to investigate PAH effects along a pollution gradient in the neighborhood of an oil terminal at Sullom Voe, Shetlands, United Kingdom (Livingstone et al. 1988). The impact of PAH was assessed using a suite of biomarker assays. One of the assays was "scope for growth," an assay that seeks to measure the extra available energy of the organism that can be used for growth and reproduction: extra, that is, in relation to the basic requirement for normal metabolic processes. A strong negative relationship was shown between scope for growth and the tissue concentration of 2- and 3-ring PAHs (Figure 9.4). Although this observation might be criticized on the grounds that other pollutants could have followed the same pollution gradient and had similar effects upon scope for growth, there was some supporting evidence from a controlled mesocosm study, which showed a similar dose–response relationship over part of the range, and gave a similar regression line. Also, controlled laboratory studies with *M. edulis* showed that scope for growth could be reduced by dosing with diesel oil to give tissue levels of 2- and 3-ring PAH similar to those found in the field study. In addition to the reduction in scope for growth, some biomarker responses

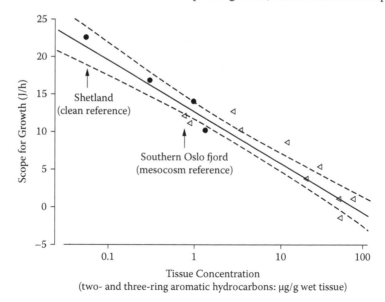

FIGURE 9.4 Relationship between scope for growth and whole tissue concentration of 2- and 3-ring aromatic hydrocarbons in *Mytilus edulis* (mean ±95% confidence limits). △, Data from Solbergstrand mesocosm experiment, Oslo Fjord, Norway. ●, Data from Sullom Voe, Shetland Islands (Moore et al. 1987).

were observed (Livingstone 1985). The lowest level of dosing with diesel oil (29 ppb for 4 months) caused a doubling of P450 levels, a reduction in lysosomal stability, and an increase in cytochrome P450 reductase. Thus, strong evidence was produced for the harmful effects of PAH on *M. edulis* near an oil terminal, but it was not established whether these were sufficient to cause a population decline.

As discussed earlier (see Chapter 6, Section 6.2.5), several studies have linked the presence of high PAH levels in the marine environment with a high incidence of tumors in fish. The ecological significance of these observations, however, is not known.

9.7 SUMMARY

On the global scale, most PAH release is the consequence of incomplete combustion of organic compounds. In the marine environment, however, there can be significant levels of PAH pollution locally, due to the large-scale release of crude oil—especially due to the wreckage of oil tankers, but also to leakage of crude oil during offshore oil operations. PAHs can be biomagnified by some aquatic invertebrates, but not in organisms higher in the food chain in which they are relatively rapidly metabolized.

PAHs do not have high acute toxicity to terrestrial animals. To fish, however, they can show considerable toxicity in the presence of UV light, as a consequence of their photooxidation. In human toxicology, the main concern has been about the mutagenic and carcinogenic properties of some PAHs. The metabolic activation of such compounds leads to the formation of DNA adducts, which can be stabilized as mutations of oncogenes or tumor suppressor genes, genes that have a role in growth regulation. There is growing evidence that PAHs can also form DNA adducts in wild vertebrates; however, there is controversy about the significance of this from an ecological point of view. PAHs cause the induction of P450 1A1/2, a response that has been utilized in the development of biomarker assays for these and other planar lipophilic organic pollutants.

FURTHER READING

Clark, R.B. (1992). *Marine Pollution*, 3rd edition, Clarendon Press, Oxford—Offers a readable account of marine pollution caused by crude oil, but does not deal with biochemical aspects.

Environmental Health Criteria 202 (1998). WHO, Geneva—Gives a very detailed account of the environmental toxicology of PAHs seen from the global point of view. However, it largely ignores marine pollution caused by oil spills.

10 Organophosphorus and Carbamate Insecticides

10.1 BACKGROUND

Organophosphorus insecticides (OPs) and carbamate insecticides are dealt with here in a single chapter because they share a common mode of action: cholinesterase (ChE) inhibition. Unlike DDT and most of the cyclodiene insecticides, they do not have long biological half-lives or present problems of biomagnification along food chains. When OCs such as DDT and dieldrin began to be phased out during the 1960s, they were often replaced by OPs or carbamates, which were seen to be more readily biodegradable and less persistent, although not necessarily as effective for controlling pests, parasites, or vectors of disease. They replaced OCs as the active ingredients of crop sprays, sheep dips, seed dressings, sprays used for vector control, and various other insecticidal preparations.

When OCs were phased out, the less persistent insecticides that replaced them were thought to be more "environment friendly." However, some of the insecticides that were used as replacements also presented problems because of very high acute toxicity. The insecticides to be discussed in this chapter illustrate well the ecotoxicological problems that can be associated with compounds that have low persistence but high neurotoxicity.

OPs were first developed during World War II, both as insecticides and chemical warfare agents. During this time, several new insecticides were synthesized by G. Schrader working in Germany, prominent among which was parathion, an insecticide that came to be widely used in agriculture after the war. In the postwar years, many new OPs were introduced and used for a wide range of applications. Early insecticides had only "contact" action when applied to crops in the field, but later ones, such as dimethoate, metasystox, disyston, and phorate, had systemic properties. Systemic compounds can enter the plant, to be circulated in the vascular system. Sap-feeding insects, such as aphids and whitefly, are then poisoned by insecticides (or their toxic metabolites) that circulate within the plant. Some OPs were developed that were highly selective between mammals and insects, and showed low mammalian toxicity (e.g., malathion and pirimiphos-methyl), making them suitable for certain veterinary uses, and protecting stored grain against insect pests.

The rapid growth in the use of OPs and the proliferation of new active ingredients and formulations was not without its problems. Some OPs proved to be too hazardous to operators because of very high acute toxicity. A few were found to cause delayed neurotoxicity, a condition not caused by ChE inhibition (e.g., mipafox, leptophos). There was also the problem of the development of resistance, for example, by cereal aphids. In due course, other insecticides, such as carbamates, were developed, and came to replace OPs for certain uses where there were problems. New carbamate insecticides were introduced and came to take a significant share of the market. Some had the advantage of being nematicides or molluskicides as well as being insecticides. Some had systemic action (e.g., aldicarb and carbofuran). Sometimes, they overcame problems of resistance that had arisen because of the intensive use of OPs in cereal aphids, such as *Myzus persicae*. Unfortunately, some carbamates also caused environmental problems because of high vertebrate toxicity.

In the following account, OPs will be discussed before considering carbamates.

10.2   ORGANOPHOSPHORUS INSECTICIDES

The chemical and biological properties of the OPs are described briefly in the next three sections. More detailed accounts are given by Eto (1974), Ballantyne and Marrs (1992), and Fest and Schmidt (1982).

10.2.1   CHEMICAL PROPERTIES

The OPs to be discussed here correspond to one or other of the two following structural formulas:

Compounds corresponding to structure [1] are referred to as *oxons*. R1, R2, and X are all linked to P through oxygen, and the compound is a triester of orthophosphoric acid that may be termed a phosphate. If only one or two of these links are through oxygen, then the compounds are termed *phosphinate* or *phosphonate*, respectively. Compounds corresponding to structure [2] are termed *thions*; R1, R2, and X are all linked to P through oxygen. Compounds of this type are triesters of phosphorothioic acid (phosphorothioates). If one of the links to P is through S, then the molecule is a phosphorodithioate. R1 and R2 are usually alkoxy groups, whereas X is usually a more complex group, linked to P through oxygen or sulfur. X is sometimes termed the *leaving group*, because it can be removed by hydrolytic attack, either chemically or biochemically.

Some properties of OPs are given in Table 10.1 and some structures in Figure 10.1. There is some variation in the values quoted for the aforementioned properties in the literature, reflecting purity of sample, accuracy of method, etc. The foregoing are representative values, and are not necessarily the most accurate ones for the purest samples.

Of the compounds listed in Table 10.1, all except dimethoate and azinphos-methyl exist as liquids at normal temperature and pressure. Looking through the table, it can be seen that there is considerable variation in both water solubility and vapor

TABLE 10.1

Properties of Some Organophosphorus Insecticides

Compound	Water Solubility (µg/mL @ 25°C)	log K_{ow}	Vapor Pressure (mmHg @ 25°C)
Parathion	11	3.83	6.7×10^{-6}
Diazinon	40	3.40	1.4×10^{-4}
Dimethoate	5000		8.5×10^{-6}
Azinphos-methyl	33		3.8×10^{-4}
Malathion	145	2.36	3.98×10^{-5}
Disyston	25		1.8×10^{-4}
Demeton-S-methyl	3300	1.32	3.6×10^{-4}
Chlorfenvinphos	145		3×10^{-6}

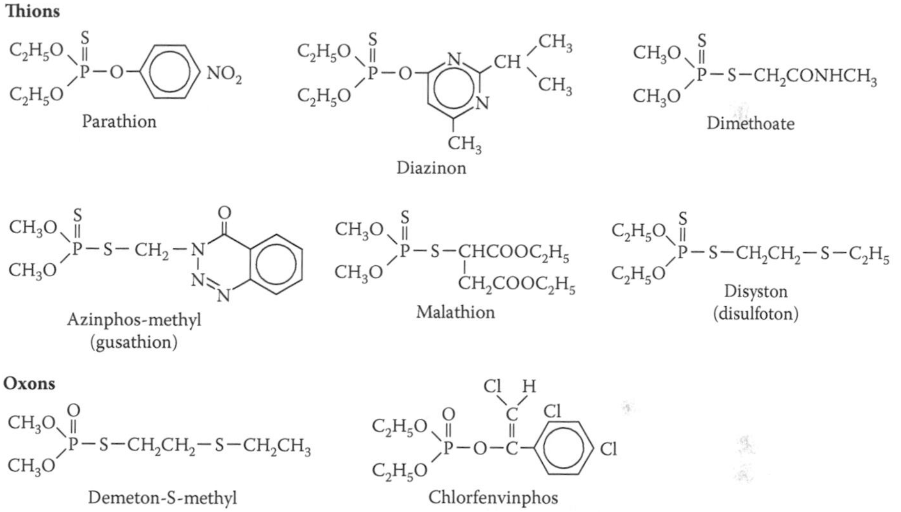

FIGURE 10.1 Some OPs.

pressure. Thus, dimethoate and demeton-S-methyl have appreciable water solubility and show marked systemic properties whereas parathion, chlorfenvinphos, and azinphos-methyl have low water solubility and are not systemic. Disulfoton, although of low water solubility in itself, undergoes biotransformation in plants to yield more polar metabolites, including sulfoxides and sulfones, which are systemic. In general, OPs are considerably more polar and water soluble than OCs.

The relatively high vapor pressure of most OPs limits their persistence when sprayed on to exposed surfaces (e.g., on crops, seeds, or farm animals). Some, such as chlorfenvinphos, have relatively low vapor pressure, and consequently tend to be more persistent than most OPs. Chlorfenvinphos has been used as a replacement for OC compounds both as an insecticidal seed dressing and as a sheep dip.

The environmental fate and behavior of compounds depends on their physical, chemical, and biochemical properties. Individual OPs differ considerably from one another in their properties and, consequently, in their environmental behavior and the way they are used as pesticides. Pesticide chemists and formulators have been able to exploit the properties of individual OPs in order to achieve more effective and more environment-friendly pest control, for example, in the development of compounds like chlorfenviphos, which has enough stability and a sufficiently low vapor pressure to be effective as an insecticidal seed dressing, but, like other OPs, is readily biodegradable; thus, it was introduced as a more environment-friendly alternative to persistent OCs as a seed dressing.

Of the compounds shown in Figure 10.1, six are thions and only two (demeton-S-methyl and chlorfenvinphos) are oxons. Four of the thions possess two sulfur linkages to P and are therefore phosphorodithionates. The oxons tend to be more unstable and reactive than the thions, and they are much better substrates for esterases, including acetylcholinesterase (AChE). Oxygen has stronger electron-withdrawing power than sulfur; so, oxons tend to be more polarized than thions. In fact, the thions are not effective anticholinesterases in themselves and need to be converted to oxons by monooxygenases before toxicity is expressed (see Chapter 10, Section 10.2.4). As technical products, thions have an advantage over most oxons in being more stable.

Organophosphorus insecticides as a class are chemically reactive and not very stable either chemically or biochemically. The leaving group (X in structural formula) can be removed hydrolytically, and OPs generally are readily hydrolyzed by strong alkali. Examples of enzymic hydrolysis are given in Figure 10.3. After OPs have been released into the environment, they undergo chemical hydrolysis in soils, sediments, and surface waters. The rate of hydrolysis depends on pH; in most cases, the higher the pH, the faster the hydrolysis of the OP. Demeton-S-methyl, for example, shows half-lives in aqueous solution of 63, 56, and 8 days at pH values of 4, 7, and 9, respectively (Environmental Health Criteria 197). Thus, most OPs are not very persistent in alkaline soils or waters.

Thions are prone to oxidation, and can be converted to oxons under environmental conditions. Also, some OPs can undergo isomerization under the influence of sunlight or high temperatures, a well-documented example being the conversion of malathion to isomalathion. Although malathion is a thion of low mammalian toxicity, isomalathion is an oxon of high mammalian toxicity. Cases of human poisoning have been the consequence of malathion undergoing this conversion in badly stored grain.

Another group of organophosphorus anticholinesterases deserving brief mention, which have not been employed as insecticides, are certain chemical warfare agents, often termed *nerve gases* (Box 10.1). Examples include soman, sarin, and tabun. These compounds have, as befits their intended purpose, very high mammalian toxicity and high vapor pressure. All the examples given are oxons, which tend to have greater mammalian toxicity than thions. Also, they are phosphinates rather than phosphates, having only one P linkage through oxygen or sulfur.

10.2.2 Metabolism

As examples of OP metabolism, the major metabolic pathways of malathion, diazinon, and disyston are shown in Figure 10.2, identifying the enzyme systems involved. OPs are highly susceptible to metabolic attack, and metabolism is relatively complex, involving a variety of enzyme systems. The interplay between activating transformations on the one hand, and detoxifying transformations on the other, determines toxicity in particular species and strains (see Walker 1991). Because of this complexity, knowledge of the metabolism of most OPs is limited. Further information on OP metabolism may be found in Eto (1974), Fest and Schmidt (1982), and Hutson and Roberts (1999).

All three insecticides shown in Figure 10.2 are thions, and all are activated by conversion to their respective oxons. Oxidation is carried out by the P450-based microsomal monooxygenase system, which is well represented in most land vertebrates and insects, but less well represented in plants, where activities are very low. Oxidative desulfuration of thions to oxons does occur slowly in plants, and may be due to monooxygenase attack and peroxidase attack (Drabek and Neumann 1985; Riviere and Cabanne 1987). Compounds, such as disyston, which have thioether bridges in their structure, can undergo sequential oxidation to sulfoxides and sulfones. Other examples are demeton-S-methyl (Figure 10.1) and phorate. The oxon forms of OP sulfoxides and sulfones can be potent anticholinesterases, and sometimes make an important contribution to the systemic toxicity of insecticides, such as demeton-S-methyl, disyston, and phorate.

The oxidation of OPs can bring detoxication as well as activation. Oxidative attack can lead to the removal of R groups (oxidative dealkylation), leaving behind P-OH, which ionizes to PO$^-$. Such a conversion looks superficially like a hydrolysis, and was sometimes confused with it before the great diversity of P450-catalyzed biotransformations became known. Oxidative deethylation yields polar ionizable metabolites and generally causes detoxication (Eto 1974; Batten and Hutson 1995). Oxidative demethylation (O-demethylation) has been demonstrated during the metabolism of malathion.

The bond between P and the "leaving group" (X) of oxons is susceptible to esterase attack, the cleavage of which represents a very important detoxication mechanism. Examples include the hydrolysis of malaoxon and diazoxon (see Figure 10.2). Such hydrolytic attack depends on the development of d^+ on P as a consequence of the electron-withdrawing effect of oxygen. By contrast, thions are less polarized and are not substrates for most esterases. Two types of esterase interact with oxons (see Chapter 2, Figure 2.9 and Section 2.3.2.3). A-esterases continuously hydrolyze them, yielding a substituted phosphoric acid and a base derived from the leaving group as metabolites. B-esterases, on the other hand, are inhibited by them, the oxons acting as "suicide substrates." With cleavage of the ester bond and release of the leaving group, the enzyme becomes phosphorylated and is reactivated only very slowly. If "aging" occurs it is not reactivated at all. Thus, continuing hydrolytic breakdown of oxons by B-esterases is, at best, slow and inefficient. Nevertheless, B-esterases produced in large quantities by resistant aphids can degrade or sequester OPs to a sufficient extent to substantially lower their toxicity and thereby provide a resistance

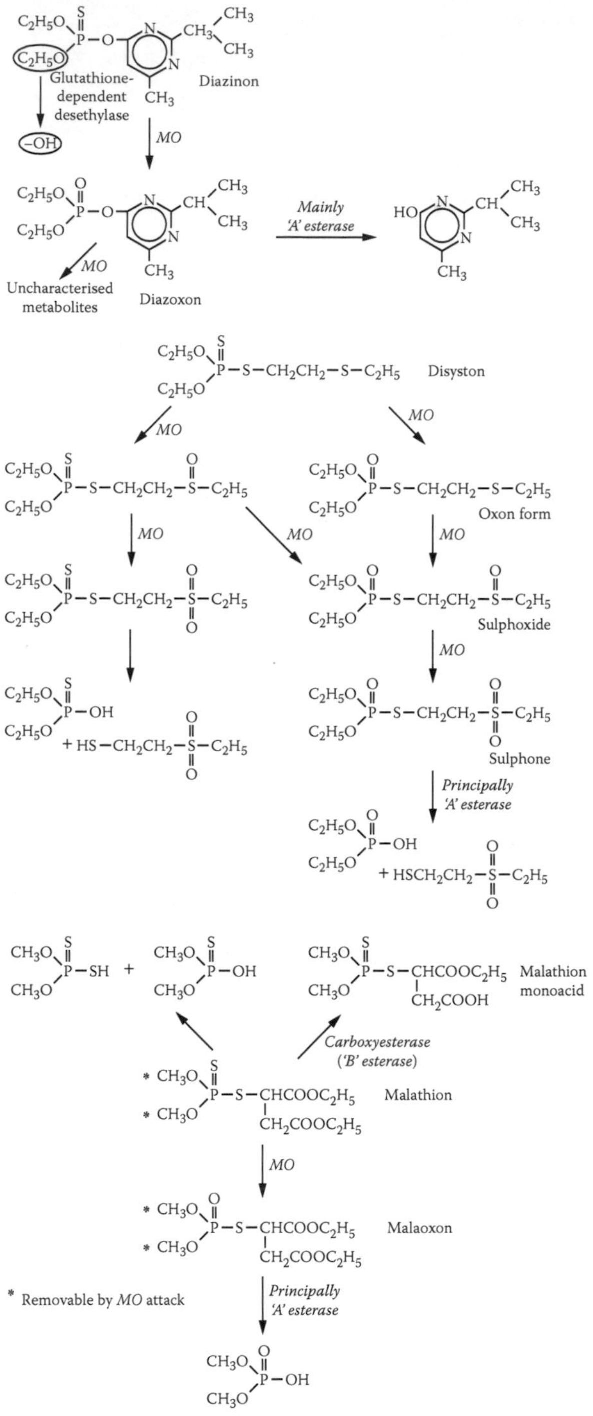

FIGURE 10.2 Metabolism of OPs.

mechanism (Devonshire and Sawicki 1979; Devonshire 1991). AChE, the site of action of OPs, is a B-esterase, which is highly sensitive to inhibition by oxons.

In addition to ester bonds with P (Section 10.2.1, Figures 10.1 and 10.2), some OPs have other ester bonds not involving P, which are readily broken by esteratic hydrolysis to bring about a loss of toxicity. Examples include the two carboxylester bonds of malathion, and the amido bond of dimethoate (Figure 10.2). The two carboxylester bonds of malathion can be cleaved by B-esterase attack, a conversion that provides the basis for the marked selectivity of this compound. Most insects lack an effective carboxylesterase, and for them malathion is highly toxic. Mammals and certain resistant insects, however, possess forms of carboxylesterase that rapidly hydrolyze these bonds, and are accordingly insensitive to malathion toxicity.

OP compounds are also susceptible to glutathione-S-transferase attack. Both R groups and X groups can be removed by transferring them to reduced glutathione to form a glutathione conjugate. As with oxidative dealkylation, an ionizable P-OH group remains after removal of the substituted group, and the result is detoxication. Diazinon, for example, can be detoxified by glutathione-dependent desethylase in mammals and resistant insects.

Looking at the overall pattern of OP metabolism, it can be seen that there is often competition between activating and detoxifying metabolic processes. Moreover, many of these processes occur relatively rapidly. There are often marked differences in the balance of these processes between species and strains, differences that may be reflected in marked selectivity. As mentioned earlier, malathion is highly selective between insects and mammals because most insects lack a carboxylesterase that can detoxify the molecule. Some strains of insects (e.g., of *Tribolium castaneum*) owe their resistance to the presence of such an esterase. Inhibition of B-esterase activity with another OP (e.g., EPN) can remove this resistance mechanism and make the resistant strain susceptible to malathion. Likewise, malathion becomes highly toxic to mammals if administered together with a B-esterase inhibitor. The inhibitor acts as a synergist. When rapid detoxication by carboxylesterase is blocked, considerable quantities of malathion are activated by monooxygenase to form malaoxon, and toxicity is enhanced.

Diazinon, and the related insecticides pirimiphos-methyl and pirimiphos-ethyl, are selectively toxic between birds and mammals (Environmental Health Criteria 198). All possess leaving groups derived from pyrimidine, and their oxon forms are excellent substrates for mammalian A-esterases. Selectivity is largely explained by the absence of significant A-esterase activity from the plasma of birds, an activity well represented in mammals (Machin et al. 1975; Brealey 1980; Brealey et al. 1980; Walker 1991; Machin et al. 1975). A-esterase activity is also low in avian liver relative to that in mammalian liver. Diazinon is activated to diazoxon in the liver, and toxicity then depends on the efficiency with which the latter can be transported by the blood to its site of action (primarily AChE in the brain). In mammals, rapid detoxication of oxons in the liver and blood gives effective protection against low doses of these OPs. Birds are not so well protected; many species lack detectable plasma A-esterase activity against oxon substrates (Mackness et al. 1987) and, on available evidence, activity in liver is relatively low (Brealey 1980; Walker 1991). Other OPs whose oxons are not good substrates for A-esterase (e.g., parathion) do not show such selectivity between birds and mammals, providing further evidence for the importance of A-esterase activity in determining the relatively low toxicity of diazinon and related insecticides to mammals. A number of cases of diazinon resistance have been reported in insects (Brooks 1972). Resistance mechanisms include detoxication by deethylation of diazinon mediated by glutathione-S-transferase, and oxidative detoxication of diazoxon mediated by monooxygenase.

10.2.3   Environmental Fate

In general, the OPs differ from the persistent OCs in their environmental fate and distribution. Because they are degraded relatively rapidly by most animals, they tend not to undergo biomagnification in the higher levels of terrestrial or aquatic food chains. However, some of them can be bioconcentrated by aquatic invertebrates from ambient water. Chlorpyrifos, for example, can be bioconcentrated by the eastern oyster (*Crassostrea virginica*) some 225-fold in comparison with ambient water (Woodburn et al. 2003). This is in keeping with the very limited metabolic capacity of mollusks (see Box 4.1). They appear to lack the effective esterases and monooxygenases, which rapidly biotransform OPs to polar metabolites in terrestrial animals. Interestingly, a lipophilic metabolite was bioconcentrated to a somewhat greater extent than the parent compound by the oysters. This metabolite, O,O,diethyl,-O-(3,5-dichloro-6-methylthio-2-pyridyl-O-phosphorothioate), was evidently formed as a result of glutathione-mediated dechlorination of the leaving group (see Chapter 2, Figure 2.15 for examples of dechlorination reactions mediated by reduced glutathione).

OPs are not very persistent in soils; hydrolysis, volatilization, and metabolism by soil microorganisms and soil animals ensure relatively rapid removal. Persistence in surface waters and sediments is also limited because of relatively rapid degradation and metabolism. Although most OPs do tend to volatilize as a consequence of their appreciable vapor pressures, they are susceptible to photodecomposition and to hydrolysis when in the atmosphere. Thus, they are not stable enough to undergo extensive long-range transport (cf. many polyhalogenated compounds). For these reasons, most harmful effects produced by OPs are likely to be limited both in time and space; limited, that is, to the general area in which they are applied, and to a relatively short period of time following their release.

The release of OPs into the environment has been very largely intentional, with the objective of controlling pests, parasites, and vectors of disease, mainly on land. Invertebrate pests of crops, forest trees, and stored products, as well as invertebrate vectors of disease, have been the principal targets. The organisms in question are mainly insects, but other types of invertebrates (e.g., *Acarina*) are sometimes controlled with OPs. Some (e.g., chlorfenvinphos) have been used to control ectoparasites of sheep and other livestock and there have been problems arising from the illegal disposal of residual sheep dips into water courses. A further limited use of OPs on land has been for the control of vertebrate pests. Birds regarded as pests (e.g., *Quelea* spp. in Africa) have been controlled by aerial spraying of roosts with parathion and fenthion (Bruggers and Elliott 1989). The use of poisoned bait containing phosdrin to control predators of game birds has become a contentious issue in Western countries. In Britain, the poisoning of protected species, such as the red kite

(*Milvus milvus*) and the golden eagle (*Aquila chrysaetos*), is illegal, and gamekeepers following this practice have been prosecuted and fined.

Although OPs have mainly been used for pest or vector control on land, there has been limited use of them in the aquatic environment, for example, to control parasites of salmon farmed in the marine environment (Grant 2002). Dichlorvos and azamethiphos have been used for this purpose, although this practice has been restricted by legislation to protect the environment in certain countries. OPs of relatively low mammalian toxicity (e.g., malathion) have sometimes been released into surface waters to control insect pests, for example, in water cress beds. Apart from the very small direct application of OPs to surface waters, there is continuing concern about unintentional contamination. Overspraying of surface waters, runoff from land, and movement of insecticides through fissures in agricultural soil and so into water courses are all potential sources of contamination with OPs, as indeed they are for agricultural pesticides more generally.

OPs are often applied as sprays. Commonly, the formulations used for spraying are emulsifiable concentrates, where the OP is dissolved in an organic liquid that acts as a carrier. OPs are also used as seed dressings and as components of dips used to protect livestock against ectoparasites. Some highly toxic OPs have been incorporated into granular formulations for application to soil or to certain crops.

Some OPs, such as chlorfenvinphos, are more persistent than most, having greater chemical stability and lower vapor pressures than is usual. Such compounds have been used where some persistence in the soil is desirable, as in the case of insecticidal seed dressings. Also, some OPs have been formulated in a way that increases their persistence. Thus, the highly toxic compounds disyston and phorate are formulated as granules for application to soil or directly to certain crops. The insecticides are incorporated within a granular matrix from which they are only slowly released, to become exposed to the usual processes of chemical and biochemical degradation. Insecticidal action may thereby be prolonged for a period of 2–3 months, much longer than would occur if they were formulated in other ways (e.g., as emulsifiable concentrates), where release into the environment is more rapid.

Notwithstanding the limited persistence of OPs generally, and the fact that they do not tend to biomagnify in the higher trophic levels, they have sometimes been implicated in the poisoning of predatory birds (for examples from the United States, United Kingdom, and Canada, see Mineau et al. 1999). Most reported cases have involved OPs of very high acute toxicity. Cases of poisoning as the result of approved use of insecticides have been explained on the grounds of a few predisposing causes. These have included direct contact of predators with spray residues and consumption of prey carrying sufficiently high pesticide burdens to poison the predators. The latter may be the consequence of prey (e.g., large insects or earthworms), immediately after OP spraying, carrying quantities of insecticide externally which are far in excess of the levels needed to poison them. If predation occurs very soon after exposure of prey to OP, tissue levels of insecticide in prey may sometimes be high enough to cause poisoning because there has been insufficient time for effective detoxication. Even though insects generally are poor vectors of insecticides because of their sensitivity to them, some strains have acquired resistance to OPs as they have insensitive forms of ChE (see Section 10.2.4) so are able to tolerate relatively high tissue levels of insecticide. Consequently, the development of this type of resistance may increase the risk of secondary poisoning of insectivores by OPs. Thus, a number of different routes of transfer need to be taken into account when considering the fate of OP insecticides applied on agricultural land.

BOX 10.1   ORGANOPHOSPHORUS "NERVE GASES"

Chemical warfare agents, such as soman and sarin, sometimes termed *nerve gases*, are powerful anticholinesterases, which bear some resemblance in structure and properties, to the OP insecticides. A major difference from most insecticides is their high volatility. These agents were possessed by the major powers during World War II, although they were never employed in warfare.

More recently, with the end of the Cold War, there has been a reduction in their stockpiles, in keeping with arms reduction treaties. At the same time, it has come to light that badly disposed canisters containing chemical weapons and originating from World War II are still around, for example, in some areas of the Baltic Sea. Thus, questions have been asked about their possible importance as environmental pollutants.

There continues to be public concern about the possibility of their being used in future. When Saddam Hussein was in power in Iraq, there was evidence that a chemical weapon of this type was used against Kurdish villagers. Subsequently, it was widely believed that these were among the weapons of mass destruction held by Saddam Hussein's regime; weapons that failed to materialize after the invasion of Iraq in 2003. Since these events, there has been concern that weapons of this type may be in the possession of "rogue" states—or individual terror groups.

There have been suspected cases of human exposure to these compounds. One issue has been the possible exposure of soldiers to them during the Gulf War of 1991. Some have suggested that this may have contributed to what has been termed the *Gulf War syndrome*, a condition reported in some NATO soldiers serving in the Gulf War. Also, during the post–Cold War era, there has been discussion about the safe disposal of the large stockpiles of chemical weapons held by the major powers (see also Chapter 1).

10.2.4   TOXICITY

The primary site of action of OPs is AChE, with which they interact as suicide substrates (see also Section 10.2.2 and Chapter 2, Figure 2.9). Similar to other B-type esterases, AChE has a reactive serine residue located at its active site, and the serine hydroxyl is phosphorylated by organophosphates. Phosphorylation causes loss of AChE activity and, at best, the phosphorylated enzyme reactivates only slowly. The rate of reactivation of the phosphorylated enzyme depends on the nature of the X groups, being relatively rapid with methoxy groups (t_{50} 1–2 h), but slower with larger alkoxyalkyl groups. Alkyl groups of phosphoryl moieties bound to AChE tend to be lost with time, leaving behind the charged group P-O⁻. The process is termed aging, and once it has occurred, reactivation virtually ceases.

In AChE isolated from *Torpedo californica,* reactive serine is one of three amino acids constituting a catalytic triad (Sussman et al. 1991, 1993; Figure 10.3). The catalytic triad is located at the bottom of a deep and narrow hydrophobic gorge lined with the rings of 14 aromatic amino acids. The catalytic triad is composed of residues of serine, histidine, and glutamic acid. Histidine is in close proximity to serine (Figure 10.3), and may therefore draw protons away from serine hydroxyl groups, thereby facilitating ionization and electrophilic attack of acetylcholine upon CO⁻. During normal hydrolysis of acetylcholine, which occurs very rapidly, the ester bond is broken, the serine residue is acetylated, and choline is released. Finally, acetate is released from the enzyme, a proton is returned to serine, and activity is quickly restored. Organophosphates are also treated as substrates by AChE, but the essential difference here is that the phosphorylated enzyme is only reactivated very slowly, if at all.

The inhibition of AChE can cause disturbances of transmission across cholinergic synapses. AChE is bound to the postsynaptic membrane (Figure 10.4), where it has an essential role in hydrolyzing acetylcholine released into the synaptic cleft from the presynaptic membrane. The rapid destruction of such acetylcholine is necessary to ensure that synaptic transmission is quickly terminated. Acetylcholine interacts with nicotinic and muscarinic receptors of the postsynaptic membrane to generate action potentials that pass along postsynaptic nerves. If stimulation of these cholinergic receptors is not quickly terminated, synaptic control is lost. If synaptic transmission is prolonged, depolarization of the postsynaptic membrane and synaptic block will follow. Synaptic block of the neuromuscular junction results in tetanus, and death due to asphyxiation follows if the diaphragm muscles of vertebrates are affected. OPs can disturb synaptic transmission in both central and peripheral nerves (Box 10.2).

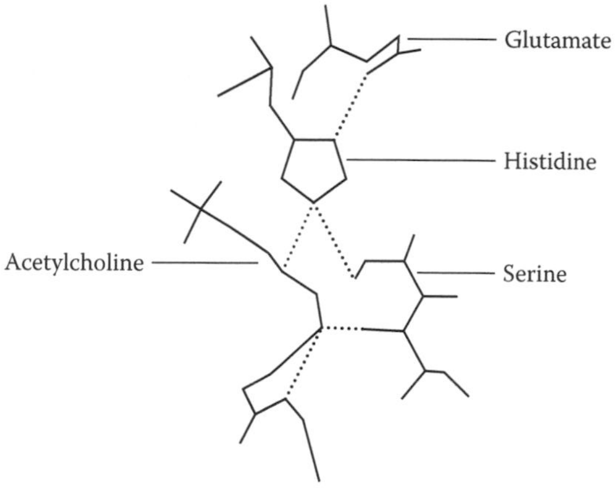

FIGURE 10.3 Acetylcholinesterase: structure of catalytic triad. The structure of the catalytic triad of the active center of the enzyme is shown (from Sussman et al. 1991).

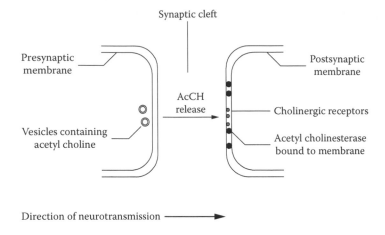

FIGURE 10.4   Diagram of cholinergic synapse.

BOX 10.2   ANTIDOTES TO CHOLINESTERASE POISONING BY ORGANOPHOSPHORUS INSECTICIDES

Because of the high human risks associated with both OP insecticides and the related nerve gases, antidotes have been developed to counteract poisoning by them. Basically, these are of two different kinds:

1. Reactivators of phosphorylated ChE. Pyridine aldoxime methiodide (PAM) and related compounds are the best known. They reactivate the phosphorylated enzyme so long as aging has not occurred. They do not, however, reactivate the aged enzyme. ChE which has been phosphorylated by certain nerve gases ages rapidly!
2. Atropine acts as an antagonist of acetylcholine at muscarinic receptors, but not at nicotinic receptors. By acting as an antagonist, it can prevent overstimulation of muscarinic receptors by the excessive quantities of acetylcholine remaining in the synaptic cleft when AChE is inhibited. The dose of atropine needs to be carefully controlled because it is toxic.

Antidotes are administered to patients after there has been exposure to OPs. They are also sometimes given as a protective measure when there is a risk of exposure, for example, to troops fighting in the Gulf War. Of the two types of antidote mentioned earlier, only atropine is effective against carbamate poisoning.

Vertebrates can tolerate a certain degree of inhibition of brain AChE before toxic effects are apparent. A typical dose–response curve for the inhibition of AChE by an OP is shown in Figure 10.5. The relationship between the degree of inhibition and the nature and severity of toxic effects is indicated in the figure. In general, effects increase in severity with increasing dose, but the quantitative relationship between percentage inhibition and effects is subject to considerable variation between compounds and between species. A typical situation in an avian species is as follows: At around 40–50% inhibition, mild physiological and behavioral disturbances are seen. Above this, more serious disturbances occur; and above 70% inhibition, deaths from anticholinesterase poisoning begin to occur (Grue et al. 1991).

There is much evidence from studies with laboratory animals that mild neurophysiological effects and associated behavioral disturbances are caused by levels of OPs well below lethal doses (see, e.g., Environmental Health Criteria 63). These include effects on EEG patterns, changes in conditioned motor reflexes, and in performance in behavioral tests (e.g., maze running by rats). Many of these observations were made after exposures too low to cause overt symptoms of intoxication. In a study with rainbow trout (*Onchorhynchus mykiss*), diazinon and malathion caused behavioral disturbances at quite low levels of brain AChE inhibition (Beauvais et al. 2000). With diazinon, the maximum level of inhibition (mean value) of brain ChE was less than 50%. There was a strong negative correlation between speed and distance of swimming, and brain AChE inhibition even down to values of about 20%. Similar results were obtained with malathion. These issues will be discussed further in Chapter 16.

The measurement of inhibition of brain AChE is a valuable biomarker assay for OPs and carbamates and is not just an index of exposure. Being an assay based on

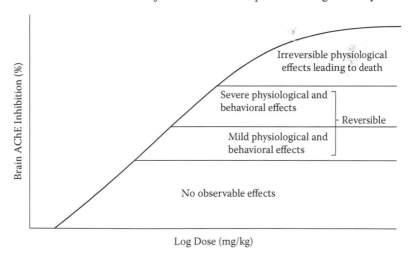

FIGURE 10.5   Stages in the progression of OP intoxication.

the principal molecular mechanism of toxicity, it has the advantage of providing an index of the different stages of the manifestation of toxicity, including early neurophysiological and behavioral effects; it can also provide an index of the potentiation of the toxicity of anticholinesterases by other environmental chemicals (see Chapter 14, Section 14.4; Walker et al. 1996). A major shortcoming is that the assay depends on destructive sampling and cannot, therefore, be used serially on individual animals. The determination of plasma butyryl cholinesterase inhibition is another biomarker assay that is nondestructive and can be used serially. However, this is only a biomarker of exposure, and there are no general rules linking inhibition of ChEs of the blood with inhibition of that of the brain, because of the complexity of the toxicokinetics, which differ markedly between species.

A few OPs have been shown to cause toxicity by interacting with a receptor other than AChE. Mipafox, leptophos, and methamidophos can all cause delayed neuropathy (Johnson 1992). The target in this case is neuropathy target esterase (NTE). This enzyme is membrane-bound, and yields a subunit of about 155 kDa when solubilized, which is high for an esterase. No symptoms are seen immediately following phosphorylation of the enzyme. Two to three weeks after exposure, long after residues of the chemicals have disappeared from the body, paralysis of muscles of distal extremities of limbs are seen together with a selective degeneration of the neurons that supply them. Distal degeneration of long axons of spinal cord and peripheral nervous system occurs. In most, but not all, examples of OP-induced delayed neuropathy (OPIDN), symptoms only start to appear with aging of the phosphorylated enzyme.

With concern about possible long-term neurological effects on sheep farmers exposed to OPs, there has been increased interest in alternative sites of action of OPs. It has been shown that there are sites in the rat brain that are more sensitive than AChE to phosphorylation by certain OPs (Richards et al. 1999). This, and the knowledge of OPIDN, has led to speculation about longer-term sublethal effects of OPs in the natural environment, due to mechanisms other than AChE inhibition. The chicken is particularly sensitive to OPIDN, which is why it is the species of choice when testing pesticides for their ability to cause this condition. Mallard ducks have also been shown to be sensitive to OPIDN. Hoffman et al. (1984) demonstrated that hen birds developed ataxia 38 days after dosing with 30 ppm of EPN and 25 days after dosing with leptophos. Does this indicate that birds, more generally, are particularly sensitive to OPIDN? Questions such as these strengthen the case for reducing the use of OPs in agriculture.

Apart from the wide range of neurotoxic and behavioral effects caused by OPs, many of which can be related to inhibition of AChE, other symptoms of toxicity have been reported. These include effects on the immune system of rodents (Galloway and Handy 2003), and effects on fish reproduction (Cook et al. 2005; Sebire et al. 2008). In these examples, the site of action of the chemicals is not identified. Indirect effects on the immune system or on reproduction following initial interaction with AChE of the nervous system cannot be ruled out. It is also possible that OPs act directly on the endocrine system or the reproductive system, and phosphorylate other targets in these locations (Galloway and Handy 2003).

Some acute toxicity data for OPs are given in Table 10.2, and a few compounds of very high OP toxicity are highlighted in Table 10.3.

TABLE 10.2
Toxicity of Organophosphorus Insecticides

Compound	Species	Type of Measurement (units)	Value
Parathion	Rat	Acute oral LD_{50} (mg/kg)	3–6
Malathion	Rat	Acute oral LD_{50} (mg/kg)	480–5600
Dimethoate	Rat	Acute oral LD_{50} (mg/kg)	150–300
Dimethoate	Birds (4)	Acute oral LD_{50} (mg/kg)	(26)
Diazinon	Rat	Acute oral LD_{50} (mg/kg)	235–1250
Diazinon	Birds (4)	Acute oral LD_{50} (mg/kg)	(4.5)
Diazinon	Fish	96 h LC_{50} (mg/L)	0.09–2.76
Demeton-S-methyl	Rat	Acute oral LD_{50} (mg/kg)	35–129
Demeton-S-methyl	Birds	Acute oral LD_{50} (mg/kg)	10–50
Demeton-S-methyl	Fish	96 h LC_{50} (mg/L)	0.6–60
Disyston	Rat	Acute oral LD_{50} (mg/kg)	12.5

TABLE 10.3
Properties of Some Carbamate Insecticides

Compound	Water Solubility (μg/mL @ 25°C)	log K_{ow}	Vapor Pressure (mm Hg @ 25°C)
Carbaryl	40	2.36	3×10^{-3}
Propoxur	1000		3×10^{-5} (30°C)
Aldicarb	6000	1.36	1×10^{-4}
Carbofuran	700		1×10^{-5}

Some OPs listed in Table 10.3 that are of exceptionally high toxicity (e.g., parathion) are no longer used in Western countries because they are considered too hazardous. In Table 10.2, the toxicity of diazinon and dimethoate to birds is given as a mean value for four different species. This is to emphasize the relatively high avian toxicity of these compounds in birds in comparison with rats and other mammals (see Environmental Health Criteria 198). The reason for this selectivity is discussed in Section 10.2.2. Similar selectivity is shown by the OPs pirimiphos-methyl and pirimiphos-ethyl, which are related structurally to diazinon. These compounds can present a serious hazard to birds when used in agriculture. Disyston (disulfoton) and the related compound phorate (thimet) are highly toxic to vertebrates generally and are normally formulated as granules, which only slowly release the insecticide. Granules limit the availability of the insecticide, and are therefore safer to use and present less risk to the environment than more readily available formulations such as emulsifiable concentrates. OPs are often very toxic to fish (see, for example, the data for demeton-S-methyl and diazinon) and aquatic invertebrates. In the case of the latter, toxic effects have been reported following exposure to levels as low as 0.01 mg/L in ambient water (Environmental Health Criteria 63).

Some of the variability found between laboratories for the rat relates to differences in the composition of the formulation used.

LD_{50}s of diazinon and dimethoate to four species of birds have been expressed as a mean to facilitate comparison with data for the rat.

Estimation of the toxicity expressed by OPs in the field is complicated by the fact that the presence of other compounds (e.g., in tank mixes or on seed dressings) may cause potentiation. As explained earlier (Chapter 2, Section 2.6), the fact that OP toxicity is regulated by relatively rapid metabolic transformation brings the risk or other compounds may cause potentiation of toxicity by inhibiting detoxication and enhancing activation. For example, it has been shown in laboratory studies that preexposure of the red-legged partridge (*Alopecurus rufus*) to prochloraz or other EBI fungicides can enhance the toxicity of malathion and dimethoate (Johnston et al. 1989, 1994a, 1994b). This effect was attributed to induction of forms of P450 by the fungicides and the consequent increased activation of the OPs. The extent to which such interactions may occur under field conditions remains a matter of speculation. It is difficult to establish the occurrence of potentiation in the field.

10.2.5 ECOLOGICAL EFFECTS

10.2.5.1 Toxic Effects in the Field

There have been many examples of birds and other vertebrates dying as a consequence of exposure to OP insecticides in the field. Worldwide, hundreds of incidents have been reported involving the poisoning of birds on agricultural land by OPs or carbamates (Hill 1992; Mineau et al. 1999). In a survey of the literature, Grue et al. (1991) concluded that birds dying of acute OP poisoning were found to have at least 50% inhibition of brain AChE activity in the great majority of cases, most of them showing 70% inhibition or more. In fact, 70% inhibition of brain AChE or more has sometimes been regarded as diagnostic of OP poisoning in birds. A similar picture emerges from laboratory studies. Identifying OPs as the cause of acute poisoning in the field depends on linking exposure to an OP (e.g., by analysis of crop or gut contents) to a level of inhibition of brain AChE high enough to suggest lethal toxicity (Grue et al. 1991; Greig-Smith et al. 1992b; Thompson and Walker 1994). Many poisoning incidents in the field reported in Western Europe and North America have been traced to the consumption by birds of material containing high levels of OP, such as seed dressed with carbophenothion, chlorfenvinphos, or fonophos, baits containing phosdrin and, occasionally, granules containing OPs. Species involved range from granivorous species poisoned by the dressed seed, to eagles, kites, and buzzards killed by poisoned baits. Spraying of OPs in the field has also been linked to lethal anticholinesterase poisoning. Thus, red-tailed hawks were found to be poisoned by OP sprays in almond orchards of California (*Buteo jamaicensis*; Hooper et al. 1989), and canopy-living birds were poisoned by fenitrothion applied to forests in New Brunswick, Canada, to control spruce bud worm (see Chapter 15 in Walker et al. 2006). Mammals have also been affected by OPs in the field. Sheffield et al. (2001) review a number of cases of such lethal and sublethal poisoning of free-living lagomorphs and rodents.

Although inhibition of brain AChE is a valuable biomarker assay for identifying lethal and other toxic effects of OPs in vertebrates, it needs to used with discretion. The degree of inhibition of brain AChE associated with lethality varies between species and compounds. Sometimes cholinergic effects appear to be more important at extra-cerebral sites, including the peripheral nervous system, than they are in the brain itself. Also, postmortem changes can occur in the field to confound analysis; loss of enzyme activity and reactivation of the inhibited can occur postmortem. Identifying OPs as the cause of lethal intoxication in the field is made easier if typical symptoms of ChE poisoning are observed prior to death.

The emphasis has thus far been on lethal effects of OPs in the field. These have been much easier to recognize than sublethal ones. The latter are much harder to detect; but the very fact that animals die from poisoning is a good indication that they will have experienced sublethal intoxication beforehand. After individuals have taken up lethal doses, they inevitably pass through a stage when effects are sublethal before they enter the final stage of lethal intoxication (Figure 10.5). Also, in the field there will be a range of exposures, from high doses that will eventually prove lethal to lower doses that will have some effects upon the animals, from which they later recover. In one study of woodpigeons (*Columba palumbus*) that had been exposed to grain dressed with chlorfenvinphos in eastern England (Cooke 1988), birds were found that behaved abnormally, were uncoordinated, and were reluctant or unable to fly. On examination of some of these birds, substantial residues were found of chlorfenvinphos in their crops and gizzards (50–170 mg/kg). Some of the birds that were affected in this way recovered and eventually flew away. Examination of brains of other birds that had displayed severe symptoms of poisoning revealed 83–88% inhibition of AChE. This and other studies bear witness to the occurrence of transitory sublethal effects when birds and mammals have been exposed to OP compounds in the field.

Inevitably, terrestrial invertebrates are susceptible to the toxicity of OPs used in the field. The honeybee is one species of particular importance, and the use of OPs and other insecticides on agricultural land has been restricted to minimize toxicity to this species. One practice has been to avoid application of hazardous chemicals to crops when there are foraging bees. The use of some compounds, for example, triazophos, has been restricted because of very high toxicity to honeybees.

As noted earlier, OPs are known to be highly toxic to aquatic invertebrates and to fish. This has been demonstrated in field studies. For example, malathion applied to watercress beds caused lethal intoxication of the freshwater shrimp *Gammarus pulex* located downstream (Crane et al. 1995). Kills of marine invertebrates have been reported following the application of OPs. Accidental release of OPs into rivers, lakes, and bays has sometimes caused large-scale fish kills (see Environmental Health Criteria 63).

10.2.5.2 Population Dynamics

Some OPs are prime examples of pollutants that are highly toxic but of low persistence, and serve as useful models for other compounds of that ilk that have been less well investigated. Because of their limited persistence, toxic effects are expected to be localized and of limited duration. As the compounds degrade quickly in tissues, residues in carcasses of animals or birds found in the field do not provide reliable evidence of the cause of death (cf. the persistent OCs). Supporting evidence, such as inhibition of brain AChE activity, is usually needed to establish causality. From an ecological point of view, such compounds appear less hazardous than compounds such as dieldrin or heptachlor epoxide, which are both highly toxic and persistent. There are, however, situations in which they may still cause ecological problems. If they are applied to an area of farmland or to an orchard several times a year over several years, effects may be seen on species of limited mobility, which are slow to recolonize treated areas after OP residues have declined. This problem may be compounded if other nonpersistent insecticides (e.g., carbamates and pyrethroids) are also used. Effects of this kind have been reported from the Boxworth Experiment—a long-term field experiment conducted by the Ministry of Agriculture, Fisheries, and Food (MAFF) in Eastern England during 1982–1990 (Greig-Smith et al. 1992a). In areas where OPs, pyrethroids, and carbamates were extensively used ("insurance areas"), some nondispersive species, such as the ground beetles *Bembidium obtusum* and *Notiophilus biguttatus,* fell drastically in numbers during the first 3 years, and remained low or totally absent until the end of the experiment. In general, there was a decline of predatory invertebrates in the area receiving the highest input of pesticide (cf. the control area).

There is concern from an ecological point of view if a high proportion of the population of a protected species is present in a particular area when a highly toxic chemical is being used. An example of this problem was the heavy mortality of wintering greylag geese (*Anser anser*) and pink-footed geese (*Anser brachyrhynchus*) in east central Scotland during 1971–1972 (Hamilton et al. 1976). Deaths were due to consumption by the geese of the OP carbophenothion, used as a seed dressing for winter wheat and barley. The geese consumed uncovered seed, and also seedlings with the contaminated seed coat still attached. It transpired that carbophenothion was particularly toxic to geese belonging to the genus *Anser*, more toxic than had been realized in the original risk assessment of the OP. *Branta* geese, such as the Canada goose (*Branta canadensis*), were found to be less susceptible. It was estimated that 60,000–65,000 wintering greylag geese, representing about two thirds of the entire British population, came to this area of Scotland during autumn in the early 1970s. Hundreds of birds died, and it was concluded that carbophenothion represented an unacceptable hazard to wintering *Anser* geese in east central Scotland. Subsequently, the use of carbophenothion as a seed dressing for winter wheat or barley was banned in the affected area.

Another example where OP spraying evidently caused ecological problems was the large-scale application of fenitrothion to forests in New Brunswick, Canada (Ernst et al. 1989; Chapter 15 in Walker et al. 2000, 2006). As described earlier (Section 10.2.4), deaths of individual birds were attributed to acute poisoning by the OP. The mortality rate due to poisoning, however, was not known, although the levels of ChE inhibition measured in surveys suggested that it may have been high. There was evidence for severe reproductive impairment in the white-throated sparrow (*Zonotrichia albicollis*) associated with a mean brain AChE inhibition of 42%. In general, many birds sampled in the area had 50% inhibition of brain AChE or more, and it was suspected that sublethal effects on birds were widespread. Apart from birds, there was clear evidence for declines in populations of honeybees and wild bees due to the application of fenitrothion.

10.2.5.3 Population Genetics

There have been many reports of insect pest species developing resistance to OP insecticides, to the extent that control of the pest has been lost. A detailed account of resistance lies outside the scope of the present book, and readers are referred to specialized texts by Georghiou and Saito (1983), Brown (1971), and Otto and Weber (1992). A few examples will now be considered that illustrate the mechanisms by which insects become resistant to OP insecticides.

In Europe, one of the most widely studied cases of resistance was that developed to OP insecticides in general by cereal aphids (Devonshire 1991). Existing OP insecticides became ineffective for aphid control in some areas, and there was a need to find suitable alternatives, for example, carbamates or pyrethroids, which were not susceptible to the same resistance mechanism. It was found that there were a number of different clones of the peach potato aphid (*Myzus persicae*) with differing levels of resistance to OPs. The level of resistance was related to the number of copies of a gene for a carboxylesterase (see earlier discussion under Section 10.2.2). In general, the larger the number of copies of the gene, the greater the activity of the carboxylesterase and the greater the level of OP resistance (in certain instances, transcriptional control was also found to be important). Resistance had evidently been acquired through gene duplication, not through the appearance of a novel esterase gene absent from susceptible aphids. Some strains of mosquitoes have also been shown to develop OP resistance by this mechanism.

A number of other examples are known in which genetically based resistance was due to enhanced detoxication of OPs. These include malathion resistance in some stored product pests owing to high carboxylesterase activity, and resistance of strains of the housefly to diazinon due to detoxication by specific forms of a glutathione-S-transferase and monooxygenase (Brooks 1972).

In some strains of insects, resistance has been related to the presence of genes that code for insensitive "aberrant" forms of the AChE. Interestingly, it has been shown that resistance may be the consequence of the change of a single amino acid residue in AChE. Sequence analysis of the AChE gene from resistant strains of *Drosophila melanogaster* and the housefly has identified six point mutations that are associated with resistance (Salgado 1999; Devonshire et al. 2000). All these mutations bring changes in amino acid residues located near the active site of AChE, according to the Torpedo enzyme model described earlier (Section 10.2.4). According to the model, all the changes would cause steric hindrance of the relatively bulky insecticides, but not to any important extent of acetylcholine itself. Thus, the insensitive enzyme could continue to function as AChE. The existence of more than one of these point mutations brings a higher level of resistance than does a single-point mutation.

Apart from the importance of OP resistance in pest control, ecotoxicologists have become interested in the development of resistance as an indication of the environmental impact of insecticides. Thus, the development of esteratic resistance mechanisms by aquatic invertebrates may provide a measure of the environmental impact of OPs in freshwater (Parker and Callaghan 1997).

10.3 CARBAMATE INSECTICIDES

The chemical and biological properties of carbamate insecticides (CBs) are described in some detail in the texts of Kuhr and Dorough (1976) and Ballantyne and Marrs (1992). An early model for their development was physostigmine, a natural product found in Calabar beans. Many CBs came into use during the 1960s, sometimes as substitutes for banned OC insecticides.

10.3.1 CHEMICAL PROPERTIES

The general structure of CBs, with some examples, are given in Figure 10.6. As can be seen, CBs are derivatives of carbamic acid, the unstable monoamide of carbonic acid. CBs have one, occasionally two, methyl groups attached to the nitrogen atom. A range of different organic groups are linked to the oxygen atom. The nature of the R group is an important determinant of the properties of CBs. Distinct from these compounds are carbamate herbicides, many of which have relatively complex R groups attached to nitrogen (see Figure 13.1 and Hassall 1990 for further details).

The properties of some CBs are given in Table 10.3.

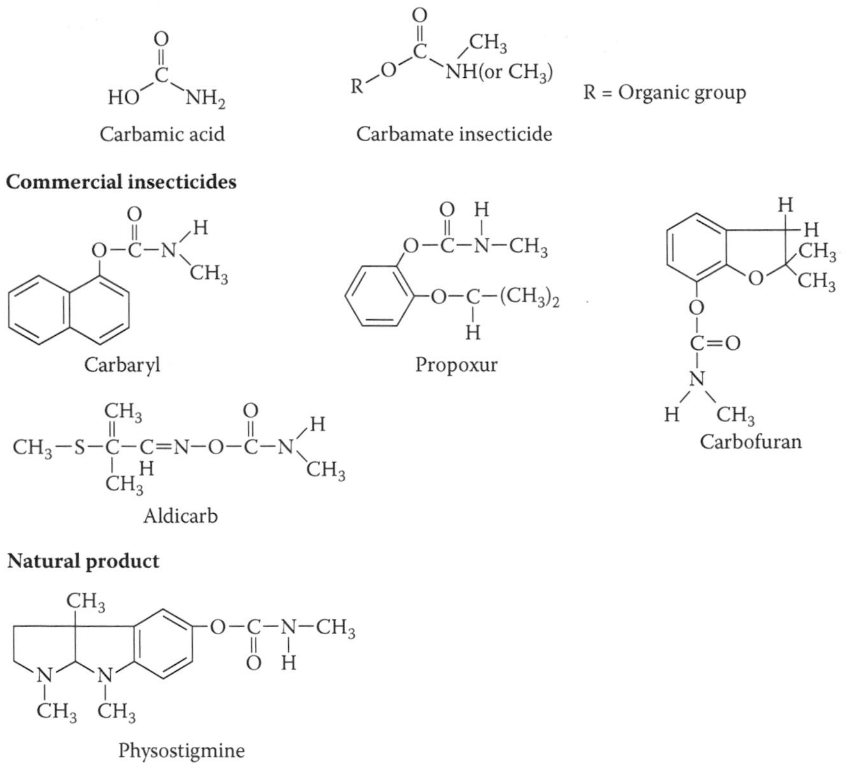

FIGURE 10.6 Some insecticidal carbamates.

Of the examples given, carbaryl is less polar and accordingly less water soluble and more volatile than the other three compounds. The reason for this is the nonpolar character of the naphthyl R group, which contrasts with the more polar groups of the other compounds. In general, CBs are more polar and water soluble than OCs, although there are marked contrasts between different members of the group. In most cases, the R group links to oxygen through carbon as with carbaryl, propoxur, and carbofuran, as shown in Figure 10.6. A few compounds such as aldicarb are linked through nitrogen, and the R group is an oxime residue. Apart from the examples given, pirimicarb, methiocarb, and methomyl are also CBs that have been widely used in pest control. Methomyl is a further example of an oxime carbamate.

Carbamates are subject to chemical hydrolysis, which takes place relatively slowly under neutral or acid conditions, but more rapidly under alkaline conditions.

10.3.2 METABOLISM

Carbamates are metabolized relatively rapidly, and metabolism tends to be complex. The present account will be restricted to major routes of biotransformation that are important in determining toxicity. The metabolism of carbaryl, aldicarb, and carbofuran is outlined in Figure 10.7. With carbaryl, primary attack in vertebrates is principally oxidation by the monooxygenase system (Hutson and Paulson 1995). This can occur both in the naphthyl ring and on the methyl group attached to nitrogen. Attack on ring positions leads to the formation of unstable epoxides that may either rearrange to form phenols, undergo hydration to diols by the action of epoxide hydrolase, or be converted to glutathione conjugates. The importance of primary oxidative attack in certain insects is illustrated by the fact that inhibitors of P450-based monooxygenases, such as piperonyl butoxide, are powerful synergists of carbaryl (synergistic ratios of more than several hundreds have been reported; Kuhr and Dorough 1976). There is uncertainty about the importance of hydrolysis as a primary mechanism of metabolic attack in vertebrates in vivo. It may be that oxidative attack tends to precede hydrolytic cleavage. For example, the hydrolysis of carbaryl by rat liver microsomes has a requirement for NADPH as a cofactor (see Environmental Health Criteria 153). Oxidative metabolites are more polar than the original carbaryl molecule and may be better substrates for esterases. All the biotransformations of carbaryl featured here cause an increase in polarity and have a detoxifying function.

With aldicarb, primary metabolic attack is again by oxidation and hydrolysis. Hydrolytic cleavage yields an oxime and represents a detoxication. Oxidation to aldicarb sulfoxide and sulfone, however, yields products that are active anticholinesterases. Carbofuran is detoxified by both hydrolytic and oxidative attack.

10.3.3 ENVIRONMENTAL FATE

CBs have been widely used in agriculture as insecticides, molluskicides, and acaricides. They have been applied as sprays and as granules or pellets. Highly toxic compounds, such as aldicarb and carbofuran, are usually only available as granules,

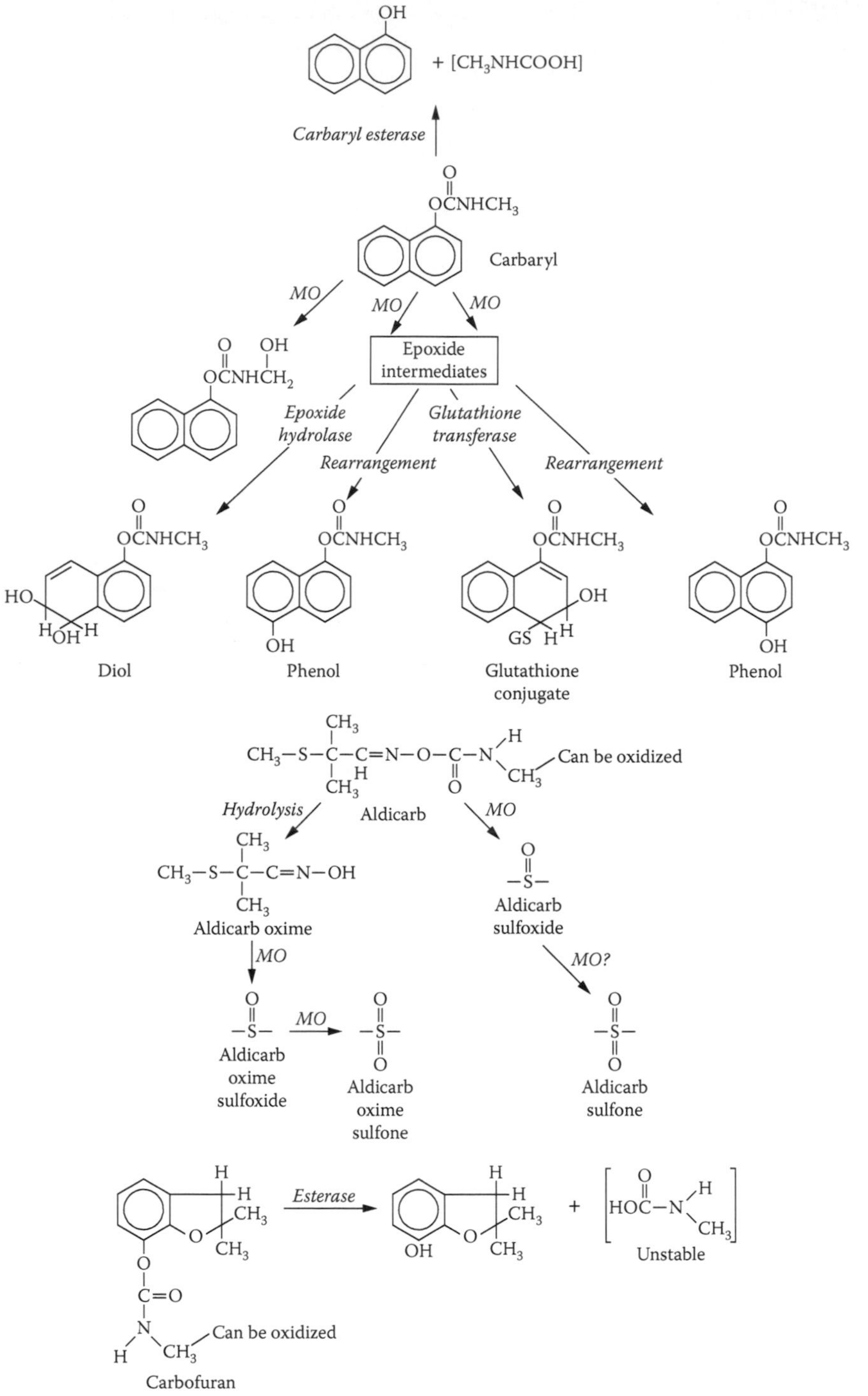

FIGURE 10.7 Metabolism of carbamates.

because other formulations are regarded as being too hazardous. Granules ensure slow release of the active ingredient under field conditions, which can have the benefits of longer-term control of pests and of reduced environmental hazards. However, there have occasionally been environmental problems with granules because (1) birds sometimes consume them; and (2) when fields are flooded, the active ingredient may be dissolved in the water of large puddles on the land surface, making it available to animals and birds (Hardy 1990).

Similar to OPs, CBs tend not to be very persistent in food chains, and do not undergo biomagnification with passage along them. Some of them (e.g., aldicarb and carbofuran) are systemic, and so may be taken up by insects feeding on plant sap. This may occur with nontarget species feeding on weeds, as well as pest species feeding on crops.

10.3.4 Toxicity

CBs, like OPs, act as inhibitors of ChE. They are treated as substrates by the enzyme and carbamylate the serine of the active site (Figure 10.8). Speaking generally, carbamylated AChE reactivates more rapidly than phosphorylated AChE. After aging has occurred, phosphorylation of the enzyme is effectively irreversible (see Section 10.2.4). Carbamylated AChE reactivates when preparations are diluted with water, a process that is accelerated in the presence of acetylcholine, which competes as a substrate. Thus, the measurement of AChE inhibition is complicated by the fact that reactivation occurs during the course of the assay. Carbamylated AChE is not reactivated by PAM and related compounds that are used as antidotes to OP poisoning (see Box 10.1).

Carbamates vary greatly in their toxicity to vertebrates. Some examples are given in Table 10.4. The most striking feature of the data is the very high acute toxicity of the two systemic carbamates aldicarb and carbofuran; carbaryl is far less toxic than either of these two compounds to mammals, birds, or fish. Propoxur is substantially less toxic to rats and birds. The toxicity to mammals and birds looks broadly similar for the four compounds represented here. However, a comparison of the toxicity of 20 CB insecticides to the red-winged blackbird (*Agelaius phoeniceus*) and starling (*Sturnus vulgaris*) with toxicity to the rat showed that in over 85% of cases the carbamate was more toxic to the bird than to the rat (Walker 1983). A possible factor here is the relatively low monooxygenase activity found in these and many other species of birds compared with the activity in the rat (see Chapter 2). Detoxication

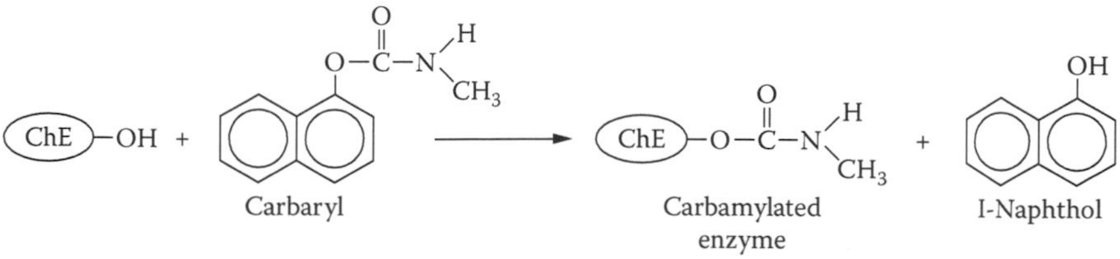

FIGURE 10.8 Carbamylation of cholinesterase.

TABLE 10.4
Toxicity of Some Carbamates

Compound	Species	Toxicity Test	Value	Units
Carbaryl	Rodents	Acute oral LD_{50}	206–963	mg/kg
	Other mammals	Acute oral LD_{50}	700–2000	mg/kg
	Birds	Acute oral LD_{50}	56–>5000	mg/kg
	Fish	96 HR LC_{50}	0.7–108	mg/L
Propoxur	Rat	Acute oral LD_{50}	95–175	mg/kg
	Birds	Acute oral LD_{50}	12–60	mg/kg
Aldicarb	Rat	Acute oral LD_{50}	0.1–7.7	mg/kg
	Birds	Acute oral LD_{50}	0.8–5.3	mg/kg
	Fish	96 HR LC_{50}	0.05–2.4	mg/L
Carbofuran	Rat	Acute oral LD_{50}	6–14	mg/kg
	Birds	Acute oral LD_{50}	0.4–4.2	mg/kg

Source: Environmental Health Criteria 64, 121, and 153; Walker (1983); and Kuhr and Dorough (1976).

by monooxygenase appears to be an important factor determining CB toxicity (see earlier discussion and Chapter 2, Section 2.6).

As noted earlier, the toxicity of CBs can be potentiated by other compounds, a consequence of their relatively rapid metabolic detoxication. Synergists, such as piperonyl butoxide and other methylene dioxyphenyl compounds, can greatly increase the toxicity of carbaryl and other CBs by inhibiting detoxication by monooxygenases up to the extent of several hundredfold (Kuhr and Dorough 1976). Although synergized CBs have not been marketed because of the environmental risks associated with the release of piperonyl butoxide and related compounds, the possibility remains that CBs may be potentiated by other pesticides under field conditions. Multiple exposures often occur in the field because of the use of pesticide mixtures (in formulations, tank mixes, and seed dressings), or due to sequential exposure as birds and other mobile species move from field to field. In one study with the red-legged partridge (*Alectoris rufa*), preexposure to the OP malathion markedly increased the toxicity of carbaryl (Johnston et al. 1994c). Dosing with malathion alone produced no inhibition of brain ChE and no symptoms of ChE poisoning. Dosing with carbaryl alone caused 56% inhibition of brain ChE, but again, no visible symptoms of toxicity. The combination of malathion with carbaryl caused 88% inhibition of brain ChE and extensive toxicity. Almost 33% of birds died, and a further 50% showed clear symptoms of ChE poisoning. Birds dosed with both compounds contained more than sevenfold higher carbaryl residues in brain than did those dosed with carbaryl alone. The potentiation of toxicity was attributed to a failure of oxidative detoxication in the liver. Malathion, like other thions, can deactivate P450 forms during the course of oxidative desulfuration (de Matteis 1974). This finding raises wider questions about potentiation of the toxicity of pesticides in the field. Many OPs are thions (see Section 10.2.1), and inhibition of vertebrate monooxygenases in the field may occur when animals or birds experience sufficiently high sublethal exposures to them. It should be emphasized that lethal exposures to OPs have been widely reported (Section 10.2.4); sublethal exposures must have been more common than these. Also, the inhibition of oxidative detoxication can bring potentiation of other readily degradable insecticides apart from CBs, such as pyrethroids and certain OPs. The problem is that such potentiation is difficult to establish in the field.

CBs, like OPs, can cause a variety of sublethal neurotoxic and behavioral effects. In one study with goldfish (*Carrasius auratus*), Bretaud et al. (2002) showed effects of carbofuran on behavioral end points after prolonged exposure to 5 µg/L of the insecticide. At higher levels of exposure (50 or 500 µg/L), biochemical effects were also recorded, including increases in the levels of norepinephrine and dopamine in the brain. The behavioral endpoints related to both swimming pattern and social interactions. Effects of CBs on the behavior of fish will be discussed further in Chapter 16, Section 16.6.1.

10.3.5 Ecological Effects

There have been a number of examples of birds and mammals being poisoned in the field by the more toxic CBs when used in the recommended way. One such example was the poisoning of about 100 black-headed gulls (*Larus ridibundus*) on agricultural land by aldicarb (Hardy 1990). The compound had been applied as a granular formulation to wet soil to control nematodes and insects in a sugar beet crop. Birds apparently died from consuming granules directly and from feeding upon contaminated earthworms. In a field study conducted at Boxworth farm, Cambs, the highly toxic CB methiocarb was shown to cause lethal poisoning of wood mice (*Apodemus sylvaticus*) when used as a molluskicide (Greig-Smith et al. 1992b). The compound had been broadcast on the soil surface as a 4% pelleted formulation. In a further example, the movement of pesticides was studied from the land surface through a soil that developed fissures, and then into neighboring water courses (Matthiessen et al. 1995). Following heavy rains, the elution of carbofuran from a granular formulation into water courses was sufficiently high to kill freshwater shrimps (*Gammarus spp.*) that had been deployed there.

Although the ability of highly toxic CBs to cause lethal poisoning in the field has been clearly demonstrated, the ecological significance of such effects remains unclear. In the case of the methiocarb poisoning of wood mice mentioned earlier, population numbers before and after the application of the molluskicide were estimated by trapping (Greig-Smith et al. 1992b). There was a rapid decline in numbers immediately following application, but there was also a rapid recovery within a week or so. In the longer run, the use of the molluskicide did not affect the size of the mouse population in the treated area. In a more detailed study, it was found that the broadcasting of molluskicide pellets altered the structure of the population; and there was a higher proportion of juveniles in the wood mouse population following broadcasting than in a control population. This change was not seen if pellets were drilled instead of being broadcast.

The repeated use of carbofuran and other carbamates has been associated with changes in the metabolic capacity of soil microorganisms (Suett 1986). Carbofuran granules were found to lose their effectiveness in some "problem" soils where the insecticide was regularly used. The soils showed enhanced capacity to degrade carbofuran, an effect attributed to either or both of the following: (1) the increase in numbers of preexisting species or strains capable of metabolizing the carbamate and, presumably, using it as a nutrient and energy source; and (2) the induction of enzyme systems capable of metabolizing the carbamate (adaptive enzymes). Such effects have also been found with herbicides, such as MCPA and 2,4-D, where the enhancement of metabolic capacity of the soil is lost after a period of time if there is no further exposure to the insecticide.

10.4   SUMMARY

The organophosphorus and carbamate insecticides are anticholinesterases, are readily biodegradable, and do not tend to bioaccumulate in food chains. They are generally seen as being more environment friendly than the persistent OC insecticides that they came to replace for many purposes during the 1960s, and have been widely used in agriculture and for certain other purposes worldwide. However, some of them are highly toxic to vertebrates and beneficial invertebrates, and have caused mortality of animals and birds in the field, albeit in limited areas and over limited time spans. There can be adverse effects on immobile invertebrates when these and other nonpersistent insecticides are repeatedly used during the course of intensive agriculture. A further cause of concern is that they are neurotoxic, and can have behavioral effects in the field, effects that are difficult to quantify but may be ecologically important. Recently, there has been growing concern about effects on sites other than AChE in the nervous system (e.g., on NTE), effects that may have longer-term consequences for affected individuals than sublethal ChE inhibition.

A few OP compounds are "chemical warfare agents."

FURTHER READING

Ballantyne, B. and Marrs, T.C. (1992). *Clinical and Experimental Toxicology of Organophosphates and Carbamates*, Butterworth/Heinemann, Oxford—Gives an indepth account of the toxicology of both groups of compounds including some information about the so-called nerve gases.

Eto, M. (1974). *Organophosphorus Insecticides: Organic and Biological Chemistry*, CRC Press, Cleveland, OH.

Fest, C. and Schmidt, K.-J. (1982). *Chemistry of the Organophosphorus Pesticides*, Springer, Berlin—This and the Eto book are valuable texts on the basic properties of the OP insecticides.

Kuhr, R.J. and Dorough, H.W. (1976). *Carbamate Insecticides*, CRC Press, Boca Raton, FL—Is a good text of similar ilk on carbamate insecticides.

11 Anticoagulant Rodenticides

11.1 BACKGROUND

Warfarin, which was introduced to the market in the late 1940s, was the first of a series of anticoagulant rodenticides (ARs) related in structure to dicoumarol (Figure 11.1; Buckle and Smith 1994; Meehan 1986). All of them are toxic because they act as anticoagulants, extending the clotting time of blood and thus causing hemorrhaging. The anticoagulant properties of naturally occurring dicoumarol were discovered in the United States early in the 20th century, when it was found to be the causal agent in cases of fatal hemorrhaging of cattle fed with spoiled clover. Subsequently, it was discovered that dicoumarol and rodenticides related to it have anticoagulant action because they act as vitamin K antagonists.

For some years, warfarin was by far the most widely used rodenticide of this type. In time, however, strains of rats resistant to warfarin began to appear, and the compound became ineffective in some areas where it had been regularly used. Resistance was overcome, at least in the short term, by a second generation of ARs sometimes called *superwarfarins*. Examples include brodifacoum, difenacoum, flocoumafen, and bromodiolone. The superwarfarins are more hydrophobic, persistent, and toxic than warfarin itself. They have usually been effective in overcoming resistance to warfarin and, consequently, have come into more widespread use with time. Although knowledge about their environmental fate and effects is at present very limited, enough is known to raise questions about the risks that might be associated with their increasing use. The combination of persistence with very high vertebrate toxicity has set the alarm bells ringing. The ensuing account will be principally concerned with the second-generation rodenticides.

11.2 CHEMICAL PROPERTIES

The formulas of some ARs are given in Figure 11.1, where it can be seen that they have some structural resemblance to both dicoumarol and vitamin K in its quinone form. All possess quinone rings linked to unsubstituted phenyl rings. The phenyl rings of the rodenticides confer hydrophobicity, especially in the relatively large and complex molecules of brodifacoum and flocoumafen. The chemical properties of some ARs are given in Table 11.1

All the compounds listed in Table 11.1 are solids. Flocoumafen and brodifacoum have particularly low vapor pressures. The hydrophobicity of brodifacoum and flocoumafen is reflected in their low water solubility. It should be remembered,

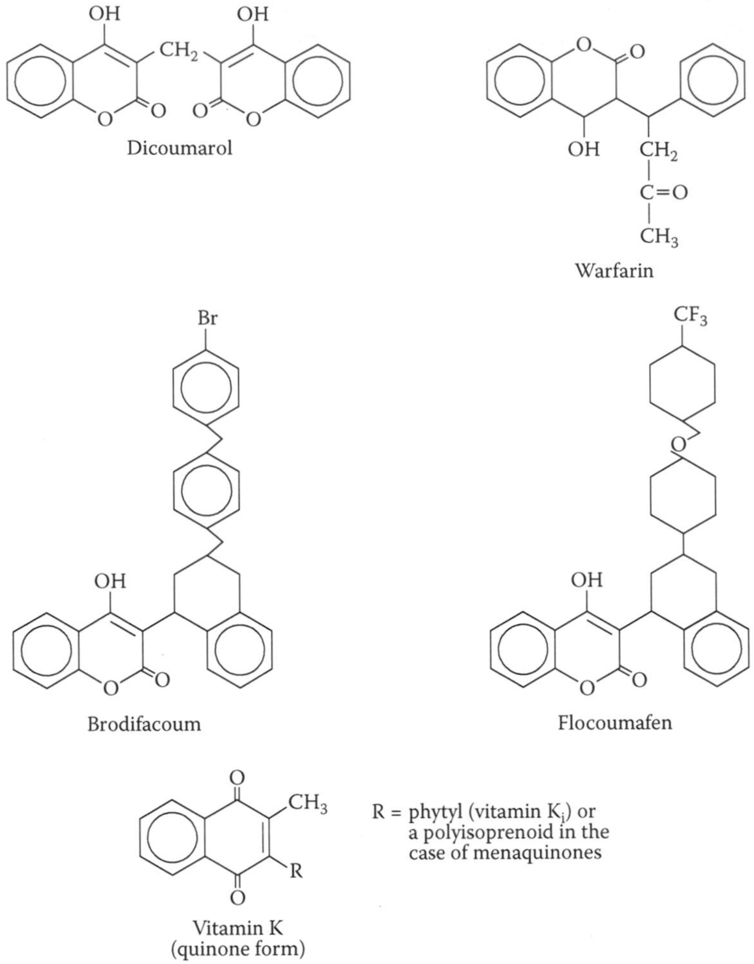

FIGURE 11.1   Anticoagulant rodenticides.

TABLE 11.1
Properties of Anticoagulant Rodenticides

Compound	Water Solubility (mg/L)	log K_{ow}	Vapor Pressure (mPa)
Warfarin	17		1.5×10^{-3}
Flocoumafen	1.1	4.7	1.3×10^{-7}
Brodifacoum	0.24 (pH 7.4)	8.5	$\ll 1 \times 10^{-3}$
Difenacoum	2.5 (pH 7.3)		

however, that because they possess ionizable hydroxyl groups, water solubility is pH dependent. With brodifacoum, for example, water solubilities are 3.8×10^{-3} and 10 mg/L at pH values of 5.2 and 9.3, respectively. An increase in pH encourages some ionization, and solubility increases accordingly.

11.3   METABOLISM OF ANTICOAGULANT RODENTICIDES

Metabolism has been studied in more detail for warfarin than for other related rodenticides. The metabolism of warfarin appears to be essentially similar to that of related compounds, and will be taken as a model for the group. Two main types of primary metabolic attack have been recognized: (1) monooxygenase attack upon diverse positions on the molecule to yield hydroxy metabolites and (2) conjugation of the hydroxyl group to yield glucuronides. The hydroxy metabolites formed by monooxygenase attack are also subject to conjugation in the case of warfarin and related ARs. More than one form of P450 is involved in warfarin metabolism. Hepatic cytochrome P450 2 C9 appears to be the most important of them, and is involved in warfarin hydroxylation in both rats and humans (Aithal et al. 1999; Wadelius and Pirmohamed 2007). Other forms, including CYP2e1, CYP2c13, CYP2A2, and CYP3a3, have been implicated in resistance mechanisms to ARs (see Section 11.6).

In a study of metabolism of 14C-flocoumafen by the Japanese quail (Huckle et al. 1989), biotransformation was extensive and rapid, with eight metabolites detected in excreta. The elimination of radioactivity from the liver of Japanese quail was biphasic (Figure 11.2). After an initial period of rapid elimination, there followed a

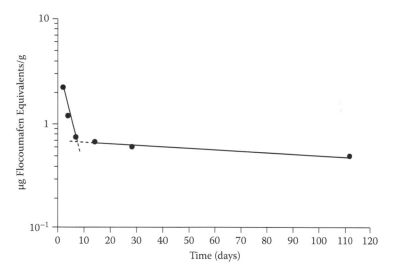

FIGURE 11.2   Loss of flocoumafen residues from quail liver. Depletion of radioactivity from Japanese quail after a single oral dose (14 mg/kg). Data are presented as microgram equivalents of *f* per gram of tissue and are mean values of two animals. Data collected at day 7 and day 12 were from four animals and three animals, respectively (from Huckle et al. 1989).

period of slow exponential decline, indicating a half-life of more than 100 days. Slow elimination in the second phase has also been demonstrated in rodents and attributed to very strong binding to one or more proteins of the hepatic endoplasmic reticulum, which greatly reduces the availability of flocoumafen to enzymes that metabolize it. Other highly toxic superwarfarins, such as brodifacoum and difenacoum, also show very slow "second phase" elimination from the liver, associated with strong protein binding. Depending on compound and species, <2μg/g (2 ppm by weight) of the residue of superwarfarins can be strongly bound in the liver. Most other tissues show less capacity for protein binding.

Some of the strong protein binding in liver is to the target site of the rodenticide, the reductase of vitamin K 2,3-epoxide (see Section 11.2.3). There is also evidence that flocoumafen can bind strongly to cyto P450 1A1. In the Japanese quail, the binding of flocoumafen to hepatic microsomes was linked to strong inhibition of EROD-ase activity, both in vivo and in vitro (Fergusson 1994). Strong binding to two sites in the liver, the vitamin K binding site and cyto P450 1A1, could explain the biphasic elimination of flocoumafen by this species (Figure 11.2). Initial relatively rapid elimination could have been mainly of rodenticide released from the site that was binding it more weakly, whereas the later much slower release could have been from the site where binding was strongest.

The marked persistence of the readily biodegradable superwarfarins in liver is something of a paradox. Ready biodegradibility does not guarantee a short half-life. Superwarfarins are much more biodegradable than refractory pollutants, such as pp-DDE, dioxin, and higher chlorinated PCBs. Yet they can still be highly persistent at low concentrations in certain tissues because tenacious binding ensures that they are virtually unavailable to the P450s that can degrade them. Such long-term storage brings the risk of long-term toxic effects in individuals and in food chains.

11.4   ENVIRONMENTAL FATE

Two contrasting uses of ARs have caused environmental problems. First, their widespread employment on farms and in areas of human habitation to control rodents; second, their use to control rats and mammalian predators, which pose a threat to other species in conserved areas (Howald et al. 1999; Eason et al. 2002). These two scenarios will now be considered in the order given.

When used on farms or in inhabited areas, warfarin and related rodenticides are usually incorporated into baits that are placed in locations where they will be found by rats, mice, and certain other vertebrate pests. There is a potential risk that baits will be eaten by pets, farm animals, and other nontarget species. To guard against this, baits are often made inaccessible to vertebrates other than the target species. For example, baits may be put in short lengths of piping, wide enough for rats and mice to enter, but narrow enough to exclude cats and dogs. In Western Europe, warfarin is used both inside and outside of farm buildings. The use of more toxic rodenticides, such as brodifacoum and flocoumafen, is often restricted to the interior of buildings, to reduce the risk to nontarget species.

When used in this controlled way, there is still concern that these compounds may be transferred via rodents to terrestrial predators and scavengers that feed upon them. Avian species at risk include members of the crow family, such as raven (*Corvus corax*), magpie (*Pica pica*), carrion crow (*Corvus corone*), and rook (*Corvus frugilegus*); owls, such as barn owl (*Tyto alba*) and tawny owl (*Strix aluca*); and large predators that are also carrion feeders, such as buzzard (*Buteo buteo*) and red kite (*Milvus milvus*). Mammalian predators, such as stoats (*Mustela erminea*), weasels (*Mustela nivalis*), and polecats (*Mustela putorius*), are also vulnerable (McDonald et al. 1998; Shore et al. 1999).

A major concern about ARs is the possibility of their being passed from rodents and other target species to other vertebrates higher in the food chain. This is particularly the case with second-generation ARs, such as brodifacoum and flocoumaphen, which have long half-lives in the liver, half-lives that can exceed 100 days during the later stages of elimination. Thus, owls and other predators may acquire these compounds from their live prey. Also, scavengers, such as corvids and kites (*Milvus* spp.), may feed on the carcasses of animals that have been poisoned by rodenticides.

Several factors contribute to the risk that nontarget predators and scavengers will acquire substantial doses of these compounds in the field. First, animals that have consumed lethal doses of ARs survive for 5 days or more before they die of hemorrhaging. Thus, they may continue feeding during this period, building up residues in the body that exceed the levels needed to lethally poison them. Second, some resistant strains of rodents can tolerate relatively high levels of rodenticide and so act as more efficient vectors of the pesticide than susceptible strains. Third, in the case of predation, rodents, etc., which are in the later stages of poisoning are likely be more vulnerable than normal healthy individuals; in other words, there may be selective predation upon the most contaminated members of the prey population.

A number of reports have established the presence of rodenticides in predators and scavengers found dead in the field (see, for example, reports of U.K. Wildlife Incident Investigation Scheme [WIIS]). Brodifacoum, difenacoum, bromodiolone, and flocoumafen have all been found, albeit at low levels in most cases (<1 ppm in liver). Sometimes, more than one type of rodenticide has been found in one individual. The toxicological significance of these residues will be discussed in Section 11.5.

One study conducted in Britain between 1983 and 1989 was of barn owls found dead in the field; 10% of the sample of 145 birds contained anticoagulant rodenticide residues in their livers, and difenacoum and brodifacoum were prominent among them (Newton et al. 1990). In another study, barn owls were fed rats that had been dosed with flocoumafen. It was found that a substantial proportion of the rodenticide ingested by owls was eliminated in pellets (Eadsforth et al. 1991). The authors suggest that exposure of owls to rodenticides in the field may be monitored by analysis of pellets dropped at roosts or regular perching places.

In contrast to the controlled use of these compounds in the neighborhood of farms and human habitation, they have sometimes been used in a less controlled way against rodents and vertebrate predators, which causes problems in conserved areas. In a number of conserved islands in New Zealand, for example, bait containing brodiphacoum has been used for rodent control, both at bait stations and by aerial distribution (Eason et al. 2002). In the latter case, poisoned bait is freely available, and herbivores and omnivores, as well as predators and scavengers are at high risk. This problem will be discussed further in Section 11.6.

11.5   TOXICITY

The mode of action of warfarin and related rodenticides is illustrated in Figure 11.3. Vitamin K in its reduced hydroquinone form is a cofactor for a carboxylase located in the rough endoplasmic reticulum of hepatocytes. It is converted into an epoxide when it participates in the conversion of glutamate residues of certain proteins to gamma-carboxy-glutamate (gla). The regeneration of the hydroquinone form of vitamin K from the 2,3-epoxide is dependent on the function of a reductase. When ARs bind to the reductase, they prevent the conversion of the epoxide to the quinone. They also inhibit the subsequent reduction of the quinone to the hydroquinone. Thus, ARs can prevent the cyclic regeneration of the hydroquinone form so that carboxylation of glutamate slows down or ceases. A gene has now been cloned that encodes for the subunit of vitamin K 2,3-epoxide reductase to which warfarin and other coumarin-related anticoagulants bind (Robertson 2004; Goodstadt and Ponting 2004). This component is termed the *vitamin K epoxide reductase subunit 1* (VKORC1). VKORC1 is now known to belong to a family of homologues that are found in vertebrates, arthropods, bacteria, and plants.

In the case of prothrombin and related clotting factors, interruption of the vitamin K cycle leads to the production of nonfunctional, undercarboxylated proteins, which are duly exported from hepatocytes into blood (Thijssen 1995). They are nonfunctional because there is a requirement for the additional carboxyl residues in the clotting process. Ionized carboxyl groups can establish links with negatively charged sites on neighboring phospholipid molecules of cell surfaces via calcium bridges,

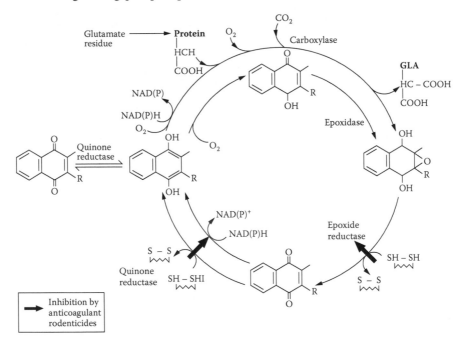

FIGURE 11.3   Action of warfarin and related rodenticides on the vitamin K cycle.

and these links are necessary for the formation of blood clots. Because clotting proteins turn over slowly (half-life of prothrombin in the rat is about 10 h), several days will elapse after inhibition of the vitamin K cycle before the level of functional clotting proteins is sufficiently low for severe hemorrhaging to occur. Typically, rats and mice begin to die from hemorrhaging 5 days or more after exposure to ARs. Owing to the strong affinity of superwarfarins for the reductase, the available binding sites may be progressively occupied by ARs over a period of weeks or even months of continuing exposure to low levels of the compounds. Evidently, all the superwarfarins bind to the same site, and it is to be expected that by so doing they will have an additive toxic effect. What is unclear, and can make interpretation of residue data difficult, is what degree of occupancy of the reductase binding sites by ARs will lead to serious hemorrhaging. Some individuals appear perfectly healthy when carrying liver residues high enough to cause hemorrhaging in others.

Because of the difficulties of interpretation of residue data, it is important to have other evidence to aid the recognition of toxic effects in the field. Hemorrhaging is usually easy to identify in a postmortem examination if carcasses are in reasonable condition. If early toxic effects are to be identified in live vertebrates, however, a biomarker assay is needed (Fergusson 1994). The detection of undercarboxylated gla proteins in blood has already been used to monitor human exposure to warfarin and other ARs (Knapen et al. 1993). The development of such an assay, for example, an ELISA assay, which could be used in predators and scavengers exposed to rodenticides in the field, has obvious attractions. It should then be possible to establish when levels of exposure in the field are high enough to begin to inhibit the vitamin K cycle, and increase the blood level of undercarboxylated clotting proteins. Toxic effects could be identified at an early stage before the occurrence of deaths due to hemorrhaging.

Because of the delay in the appearance of hemorrhaging following exposure to warfarin and related ARs, a suitable interval must elapse between exposure of experimental animals to the chemical and the assessment of mortality in toxicity testing. Typically, this period is at least 5 days. Some values of acute oral LD50 of rodenticides to vertebrates are given in Table 11.2.

Looking at the data overall, brodifacoum and flocoumafen are more toxic than warfarin to mammals. However, toxicity of the two former compounds to birds varies considerably between species. On available evidence, the galliform birds chicken (*Gallus domesticus*) and Japanese quail (*Coturnix coturnix japonica*) are much less sensitive to flocoumafen than are mammals. The chicken is less sensitive to brodifacoum than mammals. The mallard duck (*Anas platyrhynchus*), however, is just as sensitive as mammals to brodifacoum, whereas studies with the barn owl (*Tyto alba*) indicate that this species has comparable sensitivity to the rat or mouse. In toxicity tests on 13 species of birds native to New Zealand, LD50s for brodifacoum ranged from 0.95 mg/kg in the pukeko to >20 mg/kg in the Paradise shelduck (Eason et al. 2002). Thus, it appears that flocoumafen and brodifacoum are generally more toxic to mammals than to birds, although some birds are of comparable sensitivity to mammals.

Newton et al. (1990) fed mice containing brodifacoum to barn owls, and estimated that birds lethally poisoned by the rodenticide ($n = 4$) had consumed 0.150–0.182 mg/kg of the compound. The birds died within 6–17 days of receiving a single dose

TABLE 11.2
Acute Oral LD50 Values for Some Anticoagulant Rodenticides

Compound	Species	Acute Oral LD50 (mg/kg)
Warfarin	Rat	1
Warfarin	Pig	1
Brodifacoum	Rat (male)	27
Bridifacoum	Rabbit (male)	0.3
Brodifacoum	Mouse (male)	0.4
Brodifacoum	Chicken	4.5
Brodifacoum	Mallard duck	0.31
Brodifacoum	Pukeko[a]	0.95
Brodifacoum	Paradise Shelduck[b]	> 20
Flocoumafen	Rat	0.25
Flocoumafen	Mouse	0.8
Flocoumafen	Chicken	> 100
Flocoumafen	Japanese quail	> 300
Flocoumafen	Mallard duck	c 24

[a] *Porphyrio p melanotus*
[b] *Tadorna variegata*
Source: Data from Tomlin (1997) and Eason et al. (2002).

of brodifacoum and contained 0.63–1.25 mg/kg of the rodenticide in the liver. In this study, owls were also dosed with difenacoum, which was found to be less toxic than brodifacoum.

11.6 ECOLOGICAL EFFECTS

11.6.1 Poisoning Incidents in the Field

A number of studies have shown that predatory and scavenging birds can acquire liver residues of second-generation ARs when these compounds are used on farmland (Newton et al. 1990; Annual Reports of the U.K. WIIS), and that the levels are sometimes high enough to cause death by hemorrhaging (Merson et al. 1984). In a field trial with brodifacoum conducted in Virginia, United States, five screech owls (*Otus asio*) found dead 5–37 days posttreatment contained 0.4–0.8 μg/g of the rodenticide in liver (Hegdal and Colvin 1988). These residues are of similar magnitude to the levels found in poisoned barn owls in the study mentioned earlier.

When ARs have been used to control rodents and vertebrate predators in conserved areas, there have been instances of both primary and secondary poisoning. In New Zealand, such incidents have been observed on islands where bait treated with brodifacoum has been used. Casualties have included native raptors, such as the Australasian harrier (*Circus approximans*) and morepork (*Ninox novaeseelaniae*), as well as other species, such as the pukeko, Western weka (*Galliralus australis*), and southern black-backed gull (*Larus dominicanus*) (Eason and Spurr 1995; Eason et al. 2002). Three indigenous species of birds were found to be severely reduced in numbers immediately after use of brodifacoum on a number of islands. These were the Western weka, pukeko, and Stewart Island weka (*Galliralis a. scotti*). In one case, the entire island population of Western wekas was wiped out by primary and secondary poisoning following the application of brodifacoum (Eason and Spurr 1995). In another study, on Langara Island, British Columbia, Canada, ravens died of brodifacoum poisoning during a rat control program (Howald et al. 1999). They had acquired the rodenticide both directly from bait and indirectly by predating or scavenging poisoned rats. Postmortem examination established that the birds had died of severe hemorrhaging, and contained 0.98–2.52 mg/kg brodifacoum in the liver, mean value 1.35 mg/kg ($n = 13$).

As discussed earlier, a problem with these field incidents is that the low levels of rodenticides found in many of the poisoned birds are of similar magnitude to those in birds surviving exposure. A low residue level may signify everything or nothing. Additional evidence is needed to establish that the concentrations of rodenticide present in the livers of birds or mammals found in the field are sufficient to have caused death, for example, the presence of hemorrhaging in the carcasses.

The demonstration that owls as well as other predators and scavengers can acquire lethal doses of superwarfarins in the field following normal patterns of use has raised questions about the possibility of their causing widespread population declines. In one case reported in Malaysia, a population decline of owls was related to the use of superwarfarins (see Newton et al. 1990). In Britain, however, a widespread decline of the barn owl population during the 1980s could not be explained in terms of rodenticide use. Only a small proportion (2%) of the barn owls found dead during the period 1983–1989 contained residues of brodifacoum+difenacoum of 0.1 ppm or above in the liver. Thus, no more than 2% of the dead owls contained residues of rodenticides high enough to suggest poisoning. However, it should also be pointed out that the use of superwarfarins was restricted at the time of the survey. The recommended use of brodifacoum, for example, was (and still is) restricted to the interior of buildings. Superwarfarins have been used mainly in areas where resistance has developed to warfarin. The critical question is whether increasing use of superwarfarins would bring a significant risk to owls and other species exposed to them. During the 1990s, cases have been reported of red kites and other birds of prey dying as a consequence of poisoning by superwarfarins (see reports of WIIS). This is a controversial issue because red kites have been reintroduced into England in recent years. The population is still small but growing. It has been suggested that mortality due to rodenticide poisoning may prevent the reestablishment of the species in certain areas of the country where these chemicals are coming to be more widely used.

In summary, there is as yet no clear evidence that superwarfarins have caused any widespread declines of predators and scavengers that feed on rodents. However, the persistence and very high cumulative toxicity of these compounds suggest that they could pose a serious hazard to such species if they were to be more widely used. The situation needs to be kept under close review.

11.6.2 POPULATION GENETICS

Resistance to warfarin has developed in populations of rats after repeated exposure to the rodenticide (Thijssen 1995). One resistant strain discovered in Wales was found to have a much reduced capacity to bind warfarin to liver microsomes, in comparison to susceptible rats. Resistance was due to a gene that encoded for a form of vitamin K epoxide reductase that was far less sensitive to warfarin inhibition than the form in susceptible rats. Another resistant strain, which arose in the area of Glasgow, was found to differ from the resistant Welsh strain. Resistance was again due to an altered form of vitamin K epoxide reductase. However, the strain from the Glasgow area contained a form of the enzyme that bound warfarin just as strongly as that from susceptible rats. The difference was that the binding was readily reversible (Thijssen 1995). Recent work has established that a number of mutations of the VKORC1 of vitamin K epoxide reductase are associated with resistance in rodents (Pelz et al. 2005). Different mutations of VKOCR1 could explain the contrasting warfarin-binding properties in the two resistant strains mentioned earlier.

Much of the known resistance to warfarin and related ARs in rodents has been attributed to mutant forms of vitamin K epoxide reductase, but a few strains show evidence of a metabolic mechanism. For example, a resistant strain of rats from the area of Andover, United Kingdom, appears to owe its resistance to enhanced detoxication by a P450-based monooxygenase. It is known that CYP2C9, the principal cytochrome P450 form concerned with the hydroxylation of warfarin, exists in different forms with differing catalytic properties. Indeed, there is evidence that it is differences in these forms that underlie corresponding differences between individual humans in their sensitivity to this anticoagulant (Aithal et al. 1999). Thus, it seems clear that some warfarin resistance is due to relatively rapid hydroxylation by monooxygenase.

There is also evidence from a Danish study suggesting that cytochrome-P450-mediated detoxication may contribute to bromodiolone resistance in one strain of rats. In comparison with susceptible rats, resistant ones showed overexpression of the genes CYP2e1, CYP2c13, CYP3a2, and CYP3a3 (Markussen et al. 2008). Although these results strongly suggest the involvement of cytochrome P450 forms in conferring resistance, the role of these forms in metabolism of bromodiolone has not been clarified.

11.7 SUMMARY

Warfarin and the second-generation superwarfarins are ARs that have a structural resemblance to dicoumarol and vitamin K. They act as vitamin K antagonists, thereby retarding or stopping the carboxylation of clotting proteins in the hepatic endoplasmic reticulum. The buildup of nonfunctional, undercarboxylated clotting proteins in the blood leads eventually to death by hemorrhaging.

Brodifacoum, difenacoum, flocoumafen, and other superwarfarins bind strongly to proteins of the hepatic endoplasmic reticulum and consequently have long half-lives in vertebrates, often exceeding 100 days. Thus, they present a hazard to predators and scavengers that feed on rodents which have been exposed to superwarfarins. A number of species of predatory and scavenging birds have died as a consequence of secondary poisoning by superwarfarins in field incidents, and questions are asked about the long-term risks associated with expanding use of these compounds.

When used to control rodents and predatory mammals in conserved areas, they have caused both primary and secondary poisoning of nontarget species, sometimes associated with population declines.

FURTHER READING

There is a shortage of appropriate texts on the ARs. Buckle and Smith (1994) and Mechin (1986) describe the use of ARs in rodent control. Thijssen (1995) gives a concise account of mode of action and resistance mechanisms. For effects on nontarget species, reference should be made to the individual citations given in the foregoing text.

12 Pyrethroid Insecticides

12.1 BACKGROUND

The insecticidal properties of pyrethrum, a product prepared from the dried and powdered heads of flowers belonging to the genus Chrysanthemum, have long been recognized. First introduced into Europe in the middle of the 19th century, early sources were the region of the Caucasus and the Adriatic coast. Subsequently, the major source of commercial pyrethrum was the species *Chrysanthemum cinerari-aefolium* grown in East Africa. In the course of time the insecticidal ingredients of pyrethrum, the pyrethrins, were chemically characterized. Six pyrethrins were identified, all of them lipophilic esters (Figure 12.1). They are formed from two acids—chrysanthemic acid and pyrethric acid—in combination with three bases: pyrethrolone, cinerolone, and jasmolone.

A serious limitation of pyrethrins as commercial insecticides is their instability. On the one hand they are photolabile and have only limited life when applied to surfaces, for example, plant leaves, exposed to direct sunlight. On the other, they are readily biodegradable and often have only a short "knockdown" effect on target insects unless they are synergized with compounds such as piperonyl butoxide that will repress their oxidative metabolism. The important point is that they have served as models for the development of the synthetic pyrethroids, one of the most widely used types of insecticide at the present time. The first synthetic pyrethroids, compounds such as allethrin and bioallethrin, were not sufficiently photostable to have great commercial potential (Leahey 1985). Subsequently, a series of compounds were discovered that had greater stability, which came to achieve great commercial success. Included among these are permethrin, cypermethrin, deltamethrin, fenvalerate, cyfluthrin, cyhalothrin, and others. Their widespread introduction during the 1970s came on the heels of the environmental problems associated with the persistent organochlorine insecticides. Although the synthetic pyrethroids have sufficient metabolic stability to be effective insecticides, they are, nevertheless, readily biodegradable by vertebrates and do not tend to be biomagnified in food chains. At the time of their introduction, they were seen to be environmentally friendly insecticides, which, for some purposes, were effective alternatives to organochlorine insecticides.

12.2 CHEMICAL PROPERTIES

The structures of some pyrethroid insecticides are shown in Figure 12.1. They are all lipophilic esters showing some structural resemblance to the natural pyrethrins. They can all exist in a number of different enantiomeric forms. Permethrin, cypermethrin, and deltamethrin, for example, all have three asymmetric carbon atoms

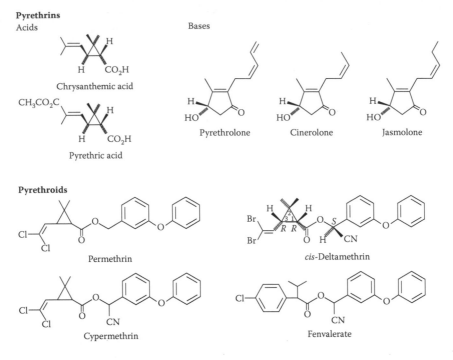

FIGURE 12.1   Structure of pyrethrins and pyrethroids.

and, consequently, eight possible enantiomers (Leahey 1985; Environmental Health Criteria 94; Environmental Health Criteria 95; Environmental Health Criteria 97). Their enantiomers fall into two categories, cis or trans, depending on the stereochemistry of the 1 relative to the 3 position of the three-membered ring of the acid moiety. Thus, there are four possible cis enantiomers and four possible trans enantiomers for each of these three compounds. Commercial products are usually racemic mixtures of different enantiomers. A notable exception is deltamethrin, which is marketed as a single cis-isomer (Environmental Health Criteria 97). Fenvalerate differs from the other pyrethroids featured in Figure 12.1 on account of the structure of its acid moiety. Nevertheless, it has similar biological properties to the other pyrethroids. The properties of some pyrethroids are given in Table 12.1.

Taking the pyrethroids, apart from fenvalerate, they are solids with low water solubility, marked lipophilicity, and low vapor pressure. Fenvalerate is a viscous liquid with an appreciable vapor pressure. Being esters, the pyrethroids are subject to hydrolysis at high pH. They are sufficiently stable to heat and light to be effective insecticides in the field.

12.3   METABOLISM OF PYRETHROIDS

The metabolism of permethrin will be taken more generally as an example of the metabolism of pyrethroids (Figure 12.2). The two types of primary metabolic attack are by microsomal monooxygenases and esterases. Monooxygenase attack involves

TABLE 12.1
Properties of Some Pyrethroid Insecticides

Compound	Water Solubility µg/mL @ 20 or 25°C	log K_{ow}	Vapor Pressure #Pa @ 20 or 25°C
Permethrin (racemate)	0.2	6.5	1.3×10^{-6}
Cypermethrin (racemate)	0.009	6.3	1.9×10^{-7}
alpha Cypermethrin (2 cis isomers)	0.005–0.01	5.16	1.7×10^{-7}
Deltamethrin	<0.002	5.43	2×10^{-6}
Fenvalerate	0.002	6.2	3.7×10^{-5}

Note: #1 Pascal (Pa) = 0.0075 torr (i.e., mms of Hg).

Source: Data from Environmental Health Criteria 82, Environmental Health Criteria 94, Environmental Health Criteria 95, Environmental Health Criteria 97, and Environmental Health Criteria 142.

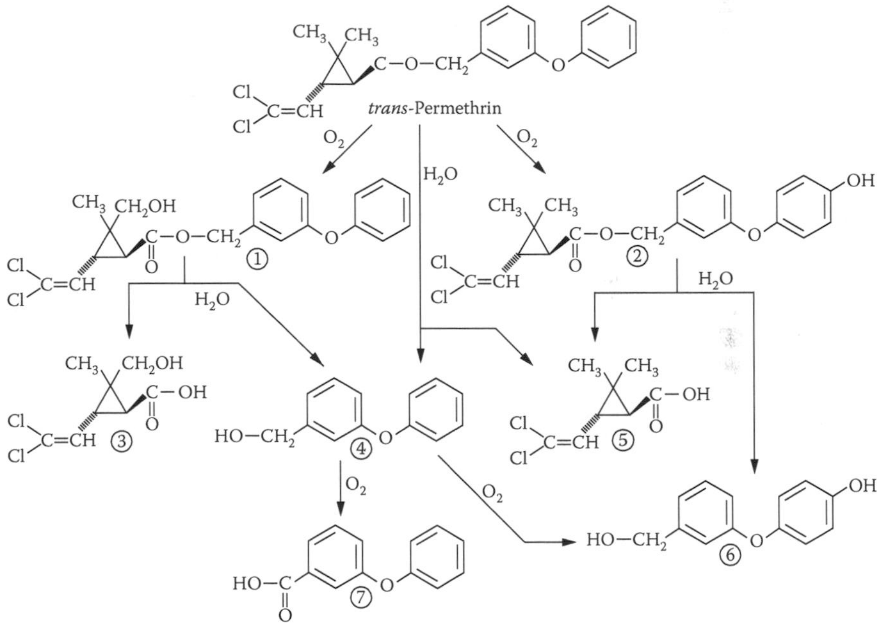

FIGURE 12.2 The metabolism of *trans*-permethrin.

different forms of cytochrome P450 and yields metabolites with hydroxyl groups substituted in both the acidic and basic moieties. The principal metabolites formed by primary oxidation are compounds 1 and 2 in the figure. Hydroxylation occurs on a methyl group of the acid moiety and on a free para ring position in the basic moiety. Esteratic hydrolysis of permethrin yields metabolites 4 and 5. Metabolite 4 is a base, and metabolite 5 is an acid. The oxidative metabolites 1 and 2 are also subject to esteratic hydrolysis. Hydrolysis of oxidative metabolite 1 yields again the base, metabolite 4, whereas hydrolysis of oxidative metabolite 2 yields again the acid, metabolite 5. In addition to these, oxidative metabolite 1 yields the hydroxy base, metabolite 6, whereas oxidative metabolite 2 yields the hydroxy acid, metabolite 3. Thus, taken together, the esteratic hydrolysis of metabolites 1 and 2 yields, on the one hand, the same two metabolites that arise from the hydrolysis of permethrin itself and, additionally, two further metabolites (3 and 6) that contain hydroxyl groups that were introduced by oxidative attack upon the parent compound. In summary, metabolites 4 and 5 are the products of esteratic hydrolysis of permethrin; metabolite 4 is also generated by the hydrolysis of metabolite 1 and metabolite 6 by the hydrolysis of metabolite 2. Metabolites 3 and 6 contain additional hydroxy groups introduced by oxidative attack. The hydroxyl groups are then available for conjugation with glucuronide, sulfate, peptide, etc., depending on species. In both insects and vertebrates the excreted products are mainly conjugates.

There has been some controversy over the relative importance of oxidation and esteratic hydrolysis in primary metabolic attack. The strong potentiation of toxicity of certain pyrethroids to insects by piperonyl butoxide (PBO) and other P450 inhibitors (see Chapter 2, Section 2.5) suggests the dominance of oxidation over hydrolysis as a detoxication mechanism. However, the interpretation of metabolic studies has sometimes been complicated by the shortage, even the apparent absence, of primary oxidative metabolites such as those shown in Figure 12.2. One problem has been identification and quantification of conjugates that can be rapidly formed from the various metabolites containing hydroxy groups, in both in vivo and in vitro studies on insects.

When trying to elucidate the metabolic regulation of toxicity, a difficulty had been establishing the metabolic pathways by which hydroxylated metabolites such as compounds 3 and 6 were formed. Did hydroxylation occur before or after hydrolytic cleavage of the ester bond? In most cases, available evidence strongly suggests that oxidation predominates over hydrolysis as a primary mode of metabolic attack. In insects, the marked synergistic action of P450 inhibitors such as PBO and ergosterol biosynthesis inhibitors (EBIs) (see Chapter 2, Section 2.6) is not consistent with esterase attack, the dominant mechanism of primary metabolism of pyrethroids. Further, the products of esteratic cleavage are strongly polar in character and are hardly ideal substrates for the hydrophobic active centers of cytochrome P450s. It should also be mentioned that the primary products of oxidative attack are more polar than the original insecticides, and are likely, on that account, to be better substrates for esterase attack (cf. OP hydrolysis, Chapter 10). Such a mechanism can explain an observation made by several workers studying microsomal metabolism of pyrethroids—that switching on P450 oxidation by addition of NADPH can increase the rate of hydrolysis (Lee et al. 1989).

12.4   ENVIRONMENTAL FATE OF PYRETHROIDS

Pyrethroids are extensively used in agriculture, so agricultural land is often contaminated by them. They can also reach field margins and hedgerows through spray drift. Because of their high toxicity to aquatic organisms, precautions are taken to prevent their entering surface waters, which can be a consequence of spray drift or soil run off. In normal agricultural use, it is important that they are not applied too close to surface waters including ditches and water courses. Their use in sheep dips, to control ectoparasites, has raised concern over the safe disposal of unused dipping liquids; unused dips should not be discharged into adjacent water courses.

Two major factors determining the environmental fate of pyrethroids are marked lipophilicity and rapid biodegradation by many animals. Even fish can degrade them reasonably rapidly. When pyrethroids reach soils or aquatic systems, they become strongly adsorbed to the colloidal fraction—mineral particles and associated organic matter. Consequently, if they do reach surface waters, their initial concentrations in water fall rapidly because of adsorption to this colloidal material. In most aquatic organisms, they are metabolized rapidly enough to limit the degree of bioconcentration that occurs. Bioconcentration studies with fish have shown bioconcentration factors (BCFs) in the steady state ranging from 50 to several thousand, depending on species, age, etc. These are, of course, concentrations in tissue relative to concentrations in water, and they are, in the main, considerably below the values predicted from the high K_{ow} of the insecticides (Environmental Health Criteria 97; Environmental Health Criteria 142). The lower values are considerably below the BCFs reported for persistent organochlorines (OCs) such as dieldrin and p,p'-DDE, which usually exceed 1000 (see Chapter 5, Section 5.3.3). The major factors responsible for this are believed to be (1) reasonably rapid metabolism by many fish (Chapter 4, Table 4.1) and (2) strong adsorption to colloids in some test systems that contain suspended material such as sediment. In some studies, the pyrethroid concentrations measured in water by chemical analysis included considerable amounts of insecticide in the adsorbed state, which was not readily available to the fish. Here, the concentration of pyrethroid determined by chemical analysis considerably overestimated the levels that fish were effectively exposed to, and consequently underestimated the BCFs that were achieved (Leahey 1985).

In a laboratory study of the persistence of five pyrethroids in soil, the rates of loss followed the order fenpropathrin>permethrin>cypermethrin>fenvalerate>delta methrin (Chapman and Harris 1981). Microbial degradation was an important factor determining their rate of disappearance. Half-lives of deltamethrin determined in two German soils were found to be 35 days in a sandy soil but 60 days in a sandy loam. Hill and Schaalje (1985) showed that deltamethrin applied in the field underwent a biphasic pattern of loss—an initial rapid loss being succeeded by a slower first-order degradation. This is essentially similar to the pattern of loss of another group of hydrophobic insecticides—the OC compounds—except that the latter are eliminated much more slowly, especially in the later stages of the process (see Chapter 4, Section 4.2, and Chapter 5). Both types of insecticide have pK_{ow} in the range 5–7, but OCs are metabolized much more slowly than pyrethroids by soil microorganisms.

Pyrethroids can also persist in sediments. In one study, alpha-cypermethrin was applied to a pond as an emulsifiable concentrate (Environmental Health Criteria 142). After 16 days of application, 5% of the applied dose was still present in sediment, falling to 3% after a further 17 days. This suggests a half-life of the order of 20–25 days—comparable in magnitude to half-lives measured in temperate soils.

The general picture, then, is that pyrethroids are reasonably persistent in soils and sediments but not to the same degree as OC compounds such as dieldrin and p,p'-DDE. They do undergo bioconcentration from water by fish and other aquatic organisms. However, because of their ready biodegradability, they are not biomagnified with movement through the upper trophic levels of food chains in the way that persistent OCs are. There is, however, concern that residues in sediments may continue to be available to certain bottom feeders long after initial contamination, and that some aquatic invertebrates of lower trophic levels, which are deficient in detoxifying enzymes, may bioconcentrate/bioaccumulate them to a marked degree.

12.5 TOXICITY OF PYRETHROIDS

Pyrethroids, such as p,p'-DDT, are toxic because they interact with Na^+ channels of the axonal membrane, thereby disturbing the transmission of nerve action potential (Eldefrawi and Eldefrawi 1990, and Chapter 5, Section 5.2.4 of this book). In both cases, marked hydrophobicity leads to bioconcentration of the insecticides in the axonal membrane and reversible association with the Na^+ channel. Consequently, both DDT and pyrethroids show negative temperature coefficients in arthropods; increasing temperature brings decreasing toxicity because it favors desorption of insecticide from the site of action.

Pyrethroids show very marked selective toxicity (Table 12.2). They are highly toxic to terrestrial and aquatic arthropods and to fish, but only moderately toxic to rodents, and less toxic still to birds. The selectivity ratio between bees and rodents is 10,000- to 100,000-fold with topical application of the insecticides. They therefore appear to be environmentally safe so far as terrestrial vertebrates are concerned. There are, inevitably, concerns about their possible side effects in aquatic systems, especially on invertebrates.

A field problem that has emerged is the synergistic action of certain EBI fungicides upon pyrethroids. Some combinations of EBIs with pyrethroids are highly toxic to bees, with synergistic ratios of the order 10–20 (Pilling 1993; Colin and Belzunces 1992; Meled et al. 1998). There have been reports from France and Germany of kills of bees in the field attributable to synergistic effects of this kind, following the use of tank mixes by spray operators. The enhancement of the toxicity of lambda cyhalothrin

TABLE 12.2
Toxicity of Some Pyrethroids

Compound	LD_{50} Rat (mg/kg)	LD_{50} Birds (mg/kg)	96 h LC_{50} Fish (µg/L)	LC_{50} Aquatic Invertebrates (µg/L)
Permethrin	500	>13,000 [4]	0.6–314	0.018–1.2
Cypermethrin	250	>10,000 [1]	0.4–2.8	0.01–5
Fenvalerate	451	>4,000 [3]	0.3–200	0.008–1
Deltamethrin	129	4,000 [1]	0.4–2.0	5 (Daphnia)

Note: Mean values given for birds; the number of species tested is given in brackets.
Source: Data from Environmental Health Criteria 82, Environmental Health Criteria 94, Environmental Health Criteria 95, and Environmental Health Criteria 97.

to bees by the EBI fungicide prochloraz has been demonstrated in a semi field trial (Bromley-Challenor 1992). The synergistic action of EBIs has been attributed, largely or entirely, to inhibition of detoxication by cytochrome P450 (see, for example, Pilling et al. 1995). Questions are now being asked about possible hazards to wild bees and other pollinators posed by pyrethroid/EBI mixtures.

12.6 ECOLOGICAL EFFECTS OF PYRETHROIDS

12.6.1 POPULATION DYNAMICS

Because of the high toxicity of pyrethroids to aquatic invertebrates, these organisms are likely to be adversely affected by contamination of surface waters. Such contamination might be expected to have effects at the population level and above, at least in the short term. In one study of a farm pond, cypermethrin was applied aerially, adjacent to the water body (Kedwards et al. 1999a). Changes were observed in the composition of the macroinvertebrate community of the pond that were related to levels of the pyrethroid in the hydrosoil. *Diptera* were most affected, showing a decline in abundance with increasing cypermethrin concentration. Chironimid larvae first declined and later recovered.

Harmful effects on macroinvertebrate communities have also been demonstrated in mesocosm studies, and will be discussed briefly here for comparison with field studies. In one study, cypermethrin and lambda cyhalothrin were individually applied to experimental ponds at the rates of 0.7 and 1.7 g a.i./ha and the results subjected to multivariate analysis (Kedwards et al. 1999b). Treatment with pyrethroid caused a decrease in abundance of *gammaridae* and *asellidae* but a concomitant increase in *planorbidae, chironimidae, hirudinae,* and *lymnaeidae. Gammaridae* were found to be more sensitive to the chemicals than *asellidae*, their numbers remaining depressed until the termination of the experiment (15 weeks) with both treatments. This may have been because they inhabit the sediment surface where there would have been relatively high levels of recently adsorbed pyrethroid, whereas the *asellidae* are epibenthic, burrowing into the hydrosoil, where lower levels of insecticide should have existed, at least in the short term. In a wide-ranging study of the impact of pyrethroids used to control pests of cotton, three different pond systems in Great Britain and the United States were employed in seven separate experiments (Giddings et al. 2001). Results from mesocosm studies were compared with those from related field studies that also utilized toxicity data for the insecticides (Solomon et al. 2001). The different taxa showed the following range of sensitivity to cypermethrin and esfenvalerate, measured in terms of abundance: amphipods, isopods, midges, mayflies, copepods, and cladocerans (most sensitive) ranging to fish, snails, oligochaetes, and rotifers (least sensitive). Values for lowest-observed effect concentrations were derived from this investigation.

Considering evidence from both field and mesocosm studies, it may be concluded that certain groups of aquatic macroinvertebrates are sensitive to pyrethroids and that there can be changes, in the short term, at the population level and above with exposure to environmentally realistic concentrations of them. It should be possible to pick up effects of this kind in natural waters using ecological profiling, for example, the River Invertebrate Prediction and Classification System (RIVPACS). There is a need here for combining ecological profiling with chemical analysis, to facilitate the detection of chemicals that cause changes in community structure in natural waters.

12.6.2 POPULATION GENETICS

The continuing use of pyrethroids in agriculture has led to the emergence of resistant strains of pests. One of the best-studied examples is the tobacco budworm (*Heliothis virescens*), a very serious pest of cotton in the southern United States (McCaffery 1998). Indeed, the resistance problem has sometimes been severe enough to threaten a loss of control over the pest. A study of a number of resistant strains from the field has revealed two major types of resistance mechanism. Some individuals possess aberrant forms of the target site, the Na$^+$ channel. At least two forms are known that confer either "kdr" (<100-fold) or "super kdr" (>100-fold) resistance, which is the consequence of the presence of insensitive forms of the Na$^+$ channel protein (McCaffery 1998, and Chapter 4, Section 4.4 of this book).

This type of resistance has been found in a number of species of insects, including *Musca domestica, Heliothis virescens, Plutella xylostella, Blatella germanica, Anopheles gambiae,* and *Myzus persicae*. Kdr has been attributed to three different changes of single amino acids of the voltage-dependent sodium channel, and super kdr to changes in pairs of amino acids, also located in the sodium channel (Salgado 1999). Interestingly, it appears that earlier selective pressure by dichlorodiphenyl trichloroethene (DDT) raised the frequency of kdr genes in the population before pyrethroids came to be used. Thus, some "pyrethroid resistance" already existed before these insecticides were applied in the field.

The other major mechanism of pyrethroid resistance found in some field strains of *Heliothis virescens* was enhanced detoxication due to a high rate of oxidative detoxication, mediated by a form of cytochrome P450 (McCaffery 1998). Some strains, such as PEG 87, which was subjected to a high level of field and laboratory selection, possessed both mechanisms. Other example of pyrethroid resistance due to enhanced detoxication may be found in the literature on pesticides.

12.7 SUMMARY

Pyrethroid insecticides were modeled upon naturally occurring pyrethrins, which were once quite widely used as insecticides but had the disadvantages of being photochemically unstable and susceptible to rapid metabolic detoxication. Pyrethroids are more stable than pyrethrins and, like DDT, act upon the voltage-dependent sodium channel of the nerve axon. They are lipophilic but are readily biodegradable by most organisms of higher trophic levels. Although they can undergo bioconcentration in the lower trophic levels of aquatic food chains, unlike OC insecticides, they are not prone to biomagnification in the upper trophic levels of either aquatic or terrestrial food chains. They are, however, strongly adsorbed in soils and sediments where they can be persistent.

Pyrethroids are much more toxic to invertebrates than to most vertebrates. They can have serious effects upon aquatic invertebrates, at least in the short term. They can be synergized by inhibitors of cytochrome P450, such as EBI fungicides, and so there are potential hazards associated with the use of mixtures of these two types of pesticides. Resistance to pyrethroids has developed in a number of pest species due to both insensitive forms of the target site (sodium channel) and/or enhanced metabolic detoxication.

FURTHER READING

Environmental Health Criteria 82 [Cypermethrin], 94 [Permethrin], 95 [Fenvalerate], 97 [Deltamethrin], and 142 [Alphacypermethrin] are all valuable sources of information on the environmental toxicology of pyrethroids.

Leahey, J.P. (Ed.) (1985). *The Pyrethroid Insecticides*—A multiauthor work that covers many aspects of the toxicology and ecotoxicology of the earlier pyrethroids.

Part 3

Further Issues
and Future Prospects

The first part of this text dealt with basic principles determining the distribution and effects of organic pollutants in the living environment. The second focused on major groups of organic pollutants, describing their chemical and biological properties and showing how these properties were related to their environmental fate and ecological effects. Attention was given to case histories, especially to long-term studies conducted in reasonable depth and detail, which illustrate how some of these principles work out in practice in the complex and diverse natural environment. The importance of these case studies should be strongly emphasized because, despite the shortcomings of many of them and the often only limited conclusions that can be drawn from them, they do provide insights into what happens in the "real world" as opposed to the theoretical one represented by the model systems that are necessarily employed during the course of risk assessment. Consideration of these "case histories" can give valuable guidelines with regard to the development of improved new ecotoxicity tests and testing systems (e.g., microcosms and mesocosms).

Since the recognition in the 1960s of the environmental problems presented by some persistent organochlorine insecticides, there have been many restrictions and bans placed upon these and other types of organic pollutants in western countries. These restrictions have been in response to perceived environmental problems posed by an individual compound or classes of compounds. As we have seen, some restrictions/bans have been followed, in the shorter or longer term, by the recovery of populations that were evidently adversely effected by them. Such was the case with various species of predatory vertebrates following restrictions on organochlorine insecticides, or on dog whelks and other aquatic mollusks following restrictions on the use of organotin compounds. Thus, in more recent times, some relatively clear-cut cases of pollution problems associated with particular compounds or classes of compounds appear to have been resolved—at least in more developed countries of the world where there have been strong initiatives to control environmental pollution.

Consequently, in developed countries, there has been much less evidence for the existence of such relatively clear-cut pollution problems in recent years. On the other hand, concern has grown that there may be more insidious long-term effects that have thus far escaped notice. Interest has grown in the possible effects of mixtures of relatively low levels of contrasting types of pollutants, to which many free-living organisms are exposed. In the extreme case, it has been suggested that ecotoxicology might be regarded as just one type of stress, alongside others such as temperature, disease, nutrients, etc. (Van Straalen 2003).

This third part of the book will be devoted mainly to the problem of addressing complex pollution problems and how they can be studied employing new biomarker assays that exploit new technologies of biomedical science. Chapter 13 will give a broad overview of this question. The following three chapters, "The Ecotoxicological Effects of Herbicides," "Endocrine Disruptors," and "Neurotoxicity and Behavioral Effects," will all provide examples of the study of complex pollution problems.

The concluding chapter will attempt to look into the future. What changes are we likely to see in pollution caused by organic compounds and in the regulatory systems designed to control such pollution? What improvements may there be in testing procedures having regard for ethical questions raised by animal welfare organizations? Can ecotoxicity testing become more ecologically relevant? Can more information be gained by making greater use of field studies?

13 Dealing with Complex Pollution Problems

13.1 INTRODUCTION

In the second part of the text, attention was focused on particular pollutants or groups of pollutants. Their chemical and biochemical properties were related to their known ecotoxicological effects. Sometimes, with the aid of biomarker assays, it has been possible to relate the responses of individuals to consequent effects at the level of population and above. Biomarker assays provided the essential evidence that adverse effects on populations, communities, and ecosystems were being caused by environmental levels of particular chemicals. The examples given included population declines of raptors due to eggshell thinning caused by p,p'-DDE, and decline or extinction of dog whelk populations due to imposex caused by tributyl tin (TBTs). These were relatively straightforward situations where much of the adverse change was attributable to a single chemical. In other cases, as with the decline of raptors in the U.K., effects were related to one group of chemicals, in this case the cyclodienes. Since these events, there have been extensive bans on certain chemicals, and there is less evidence of harmful effects due to just one chemical or group of related chemicals. Interest has moved toward the possible adverse effects of complex mixtures of chemicals, sometimes of contrasting modes of action, often at low levels.

Establishing the effects of combinations of chemicals in the field is no simple matter. There are many cases where adverse effects at the level of population or above have been shown to correlate with levels of either individual pollutants or combinations of pollutants, but the difficulty comes in establishing causality, in establishing that particular chemicals at the levels measured in environmental samples are actually responsible for observed adverse effects. This question has already been encountered in the studies of pollution of the Great Lakes of North America by polychlorinated aromatic compounds and other pollutants (this book, Chapters 5, 6, and 7). A major difficulty is that many factors other than the pollutants actually determined by chemical analysis may contribute to population declines, including shortage of food, habitat change, disease, and climatic change—and other pollutants that have not been analyzed. Such factors may very well correlate with the measured pollutant levels, especially where comparison is simply being made between the populations in one or two polluted areas and a "reference" population in a "clean" area. In badly polluted areas, there may be elevated levels of other pollutants in addition to those determined by chemical analysis, and these may have direct effects upon the species being studied, or indirect ones by causing changes in food supply.

13.2   MEASURING THE TOXICITY OF MIXTURES

As explained earlier, toxicity testing of pesticides and industrial chemicals for the purposes of statutory environmental risk assessment is very largely done on single compounds (Chapter 2, Section 2.5). For reasons of practicality and cost, only a minute proportion of the combinations of pollutants that occur in the natural environment can be tested for their toxicity. This dilemma will be discussed further in Section 13.3. A different and more complex situation exists, however, in the real world; mixtures of pollutants are found in contaminated ecosystems, in effluents discharged into surface waters, for example, sewage and industrial effluents, and in waste waters from pulp mills. The tests or bioassays employed here usually measure the toxicity expressed by mixtures, and investigators are presented with the problem of identifying the contributions of individual components of a mixture to this toxicity. Simple toxicity tests/bioassays often establish the presence of toxic chemicals without identifying the mechanisms by which toxicity is expressed. The issue is further complicated by the possibility that naturally occurring xenobiotics, such as phytoestrogens taken up by fish, may contribute significantly to the toxicity that is measured.

In the simplest situation, chemicals in a mixture will show additive toxicity. If environmental samples are submitted for both toxicity testing and chemical analysis, the toxicity of the mixture may be estimated from the chemical data, to be compared with the actual measured toxicity. As explained earlier for the estimation of dioxin equivalents (Chapter 7, Section 7.2.4), the toxicity of each component of a mixture may be expressed relative to that of the most toxic component (toxic equivalency factor or TEF). Using TEFs as conversion factors, the concentration of each component can then be converted into toxicity units (toxic equivalents or TEQs) the summation of which gives the predicted toxicity for the whole mixture. Often, the estimated toxicity of mixtures of chemicals in environmental samples falls short of the measured toxicity. Two major factors contribute to this underestimation of toxicity: first, failure to detect certain toxic molecules (including natural xenobiotics), and second, the determination by analysis of chemicals that are of only limited availability to free-living organisms, as when there is strong adsorption in soils or sediments. In the latter case, analysis overestimates the quantity of a chemical that is actually available to an organism. Potentiation (synergism) between pollutants can also contribute to the underestimation of toxicity when making calculations based on chemical analysis (see Doi, Chapter 12 in Volume 2 of Calow 1994). Sometimes, during the course of analysis, mixtures of pollutants present in environmental samples are subjected to a fractionation procedure in an attempt to identify the main toxic components. By a process of elimination, toxicity can then be tracked down to particular fractions and compounds.

The advantages of combining toxicity testing with chemical analysis when dealing with complex mixtures of environmental chemicals are clearly evident. More useful information can be obtained than would be possible if one or the other were to be used alone. However, chemical analysis can be very expensive, which places a limitation on the extent to which it can be used. There has been a growing interest in the development of new, cost-effective biomarker assays for assessing the toxicity of mixtures. Of particular interest are bioassays that incorporate mechanistic biomarker responses and are inexpensive, rapid, and simple to use (see Section 13.5). These can be used alone or in combination with standard toxicity tests, and some of them identify the mechanisms responsible for toxic effects, thus indicating the types of compounds involved.

13.3 SHARED MECHANISM OF ACTION—AN INTEGRATED BIOMARKER APPROACH TO MEASURING THE TOXICITY OF MIXTURES

A very large number of toxic organic pollutants, both manmade and naturally occurring, exist in the living environment. However, they express their toxicity through a much smaller number of mechanisms. Some of the more important sites of action of pollutants were described earlier (Chapter 2, Section 2.3). Thus, a logical approach to *measuring* the toxicity of mixtures of pollutants is to use appropriate mechanistic biomarker assays for monitoring the operation of a limited number of mechanisms of toxic action and to relate the responses that are measured to the levels of individual chemicals in the mixtures to which organisms are exposed (Peakall 1992, Peakall and Shugart 1993). Such an approach can provide an index of additive toxicity of mixtures, which takes into account any potentiation of toxicity at the toxicokinetic level (Walker 1998c). Mechanistic biomarkers can be both qualitative and quantitative; they identify a mechanism of toxic action and measure the degree to which it operates. Thus, they can provide an integrated measure of the overall effect of a group of compounds that operate through the same mechanism of action. Where the mechanism of action is specific to a particular class of chemicals, it can be related to the concentrations of components of a mixture which belong to that class.

Four examples will now be given of such mechanistic biomarker assays that can give integrative measures of toxic action by pollutants, all of which have been described earlier in the text. Where the members of a group of pollutants share a common mode of action and their effects are additive, TEQs can, in principle, be estimated from their concentrations and then summated to estimate the toxicity of the mixture. In these examples, toxicity is thought to be simply related to the proportion of the total number sites of action occupied by the pollutants and the toxic effect additive where two or more compounds of the same type are attached to the binding site.

1. **The inhibition of brain cholinesterase** is a biomarker assay for organophosphorous (OP) and carbamate insecticides (Chapter 10, Section 10.2.4). OPs inhibit the enzyme by forming covalent bonds with a serine residue at the active center. Inhibition is, at best, slowly reversible. The degree of toxic effect depends upon the extent of cholinesterase inhibition caused by one or more OP and/or carbamate insecticides. In the case of OPs administered to vertebrates, a typical scenario is as follows: sublethal symptoms begin to appear at 40–50% inhibition of cholinesterase, lethal toxicity above 70% inhibition.
2. The **anticoagulant rodenticides** warfarin and superwarfarins are toxic because they have high affinity for a vitamin K binding site of hepatic microsomes (Chapter 11, Section 11.2.4). In theory, an ideal biomarker would

measure the percent of vitamin K binding sites occupied by rodenticides. However, the technology is not currently available to do that. On the other hand, the measurement of increases in plasma levels of undercarboxylated clotting proteins some time after exposure to rodenticide provides a good biomarker for this toxic mechanism.

3. Some hydroxy metabolites of coplanar PCBs, such as 4-OH and 3,3′4,5′-tetrachlorobiphenyl, act as **antagonists of thyroxin** (Chapter 6, Section 6.2.4). They have high affinity for the thyroxin-binding site on transthyretin (TTR) in plasma. Toxic effects include vitamin A deficiency. Biomarker assays for this toxic mechanism include percentage of thyroxin-binding sites to which rodenticide is bound, plasma levels of thyroxin, and plasma levels of vitamin A.

4. Coplanar PCBs, PCDDs, and PCDFs express **Ah-receptor-mediated toxicity** (Chapter 6, Section 6.2.4). Binding to the receptor leads to induction of cytochrome P450 I and a number of associated toxic effects. Again, toxic effects are related to the extent of binding to this receptor and appear to be additive, even with complex mixtures of planar polychlorinated compounds. Induction of P4501A1/2 has been widely used as the basis of a biomarker assay. Residue data can be used to estimate TEQs for dioxin (see Chapter 7, Section 7.2.4).

In addition to the foregoing, three further examples in this list (numbers 5–7) deserve consideration. These are (5) interaction of endocrine disruptors with the estrogen receptor, (6) the action of uncouplers of oxidative phosphorylation, and (7) mechanisms of oxidative stress. Until now only the first is well represented by biomarker assays that have been employed in ecotoxicology.

5. **Interaction with the estrogen receptor (ER)** has been important in the development of biomarker assays for endocrine disrupting chemicals (EDCs), and will be discussed in Chapter 15. The considerable range of biomarker assays (including bioassays) already developed is reviewed by Janssen, Faber, and Bosveld (1998). A surprisingly diverse range of chemicals can act as agonists or antagonists for the estrogen receptor, producing "feminizing" or "masculinizing" effects. These include *o,p′*-DDE, certain PAHs, PCBs, PCDDS, PCDFs, alkylphenols, and naturally occurring phyto- and myco-estrogens. However, it should be borne in mind that some EDCs (e.g., *o,p′*-DDE, PCBs) probably act through their hydroxymetabolites, which bear a closer resemblance to natural estrogens than the parent compounds and, second, that others (e.g., alkylphenols) are only very weak estrogens.

A number of biomarker assays have been developed for fish. Apart from a variety of nonspecific endpoints such as organ weight and histochemical change, vitellogenin synthesis has provided a specific and sensitive endpoint, which has been very useful for detecting the presence of environmental estrogens at low concentrations. A number of different cell lines have been developed for use in bioassays for rapid screening of environmental samples. These include fish and bird hepatocytes, mouse hepatocytes, human mammary tumor cells, and yeast cells (Janssen, Faber, and Bosveld 1998). The endpoints include vitellogenin production, competitive binding to ER, the activation of galactosidase, the generation of light through the intermediacy of reporter genes, and the elevation of mRNA levels. The diversity of the available bioassays reflects the high profile that endocrine disruptors have been given in recent years. Some of these assays are described in more detail in Section 13.5.

6. **Uncouplers of oxidative phosphorylation.** Oxidative phosphorylation of ADP to generate ATP is a function of the mitochondrial inner membrane of animals and plants. Compounds that uncouple the process are general biocides, showing toxicity to animals and plants alike. For oxidative phosphorylation to proceed, a proton gradient must be built up across the inner mitochondrial membrane. The maintenance of a proton gradient depends on the inner mitochondrial membrane remaining impermeable to protons. Most uncouplers of oxidative phosphorylation are weak acids that are lipophilic when in the undissociated state. Examples include the herbicides dinitro ortho cresol (DNOC) and dinitro secondary butyl phenol (dinoseb), and the fungicide pentachlorophenol (PCP). The proton gradients across inner mitochondrial membranes are built up by active transport, utilizing energy from the electron transport chain that operates within the membrane. The direction of the gradient falls from the outside of the membrane to the inside (Figure 13.1). The dissociated forms (conjugate bases) of the weak acids combine with protons on the outside of the membrane to form undissociated lipophilic acids, which then dissolve in the membrane and diffuse across to the inside. Here, where the H^+ concentration is lower than on the outside of the membrane, they dissociate to release protons, and so act as proton translocators. They run down proton gradients, and hence "uncouple"

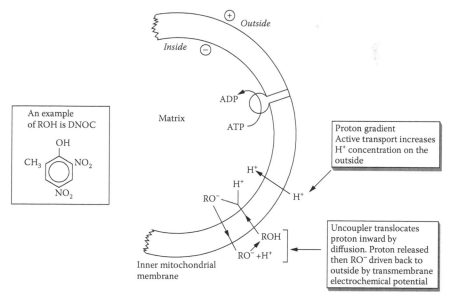

FIGURE 13.1   Uncouplers of oxidative phosphorylation.

oxidative phosphorylation, dissipating the energy that would otherwise have driven ATP synthesis. The action of uncouplers can be measured in isolated mitochondria with an oxygen electrode that follows the rate of oxygen consumption in relation to the rate of NADH consumption (see Nicholls 1982). Thus, the combined toxic action of mixtures of "uncouplers" can be studied in isolated mitochondria. Such studies can be used to investigate the significance of tissue levels of mixtures of, for example, substituted phenols, found in tissues of animals after exposure to them in vivo.

7. The participation of some OPs in redox cycling with consequent **oxidative stress**. It has become increasingly apparent that the toxicity of certain compounds is due to their ability to facilitate the generation in tissues of highly unstable oxyradicals, such as the superoxide anion $O_2.^-$ and the hydroxyl radical OH, as well as hydrogen peroxide, H_2O_2. These reactive species can cause cellular damage including lipid peroxidation and DNA damage, and have been implicated in certain disease states such as atherosclerosis and some forms of cancer (Halliwell and Gutteridge 1986). Because they are so unstable, they are difficult or impossible to detect. Proof of their existence depends upon indirect evidence. The appearance of characteristic products of oxyradical attack (e.g., oxidized lipids, malonaldehyde from lipid peroxidation, and oxidative adducts of DNA), and the induction of enzymes involved in their destruction (e.g., superoxide dismutase, catalase, and peroxidase) can all provide evidence for the presence of oxyradicals and give some indication of their cellular concentrations.

These highly reactive species can be generated as a consequence of the presence of certain organic pollutants, such as bipridyl herbicides and aromatic nitro compounds (Figure 13.2). Taking as examples the herbicide paraquat (Hathway 1984), and nitropyrene (Hetherington et al. 1996), both can receive single electrons from reductive sources in the cell to form unstable free radicals. These radicals can then pass the electrons on to molecular oxygen to form superoxide anion, with regeneration of the original molecule. Thus, a cyclic process is established, the net effect being to transfer electrons from a reductive source to oxygen with generation of an oxyradical. Once formed, superoxide can undergo further reactions to form hydrogen peroxide and the highly reactive hydroxy radical. The toxicity of paraquat to plants and animals is believed to be due, largely or entirely, to cellular damage caused by oxyradicals. In the case of plants, these radicals attack the photosynthetic system (see Hassall 1990). In animals, toxic action is mainly against Type 1 and Type II alveolar cells, which take up the herbicide by a selective active transport system (see Timbrell 1999).

There is evidence that mechanisms other than the production of free radicals of nitrogen-containing aromatic compounds are important in the case of pollutants. Refractory substrates for cytochrome P450, such as higher chlorinated PCBs, may facilitate the release into the cell of active forms of oxygen (e.g., the superoxide ion) by, in effect, blocking binding sites for substrates to be oxidized and thereby deflecting activated oxygen produced by the heme nucleus. The unused activated oxygen may then escape from the domain of the cytochrome P450 in the form of superoxide to cause oxidative damage elsewhere in the cell.

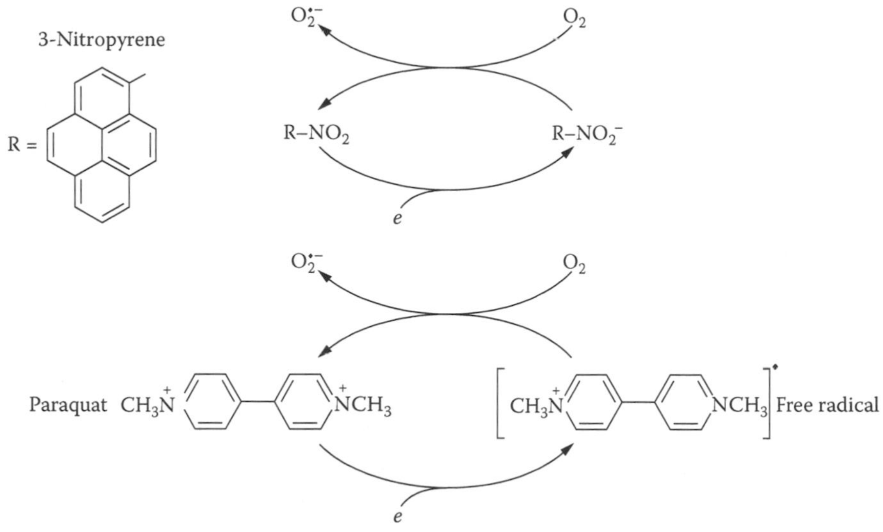

FIGURE 13.2 Superoxide generation by 3-nitropyrene and paraquat.

At the time of writing, the toxicity of oxyradicals generated by the action of pollutants is highly topical because of the relevance to human diseases. It is not an easy subject to investigate because of the instability of the radicals and the different mechanisms by which they may be generated. Hopefully rapid progress will be made so that monitoring the effects of oxyradicals will make an important contribution to the growing armory of mechanistic biomarkers for the study of environmental effects of organic pollutants.

Viewing the foregoing examples overall, the first five all involve interaction between organic pollutants and discrete sites on proteins, one of them the active site of an enzyme, the others being "receptors" to which chemicals bind to produce toxicological effects. Knowledge of the structures and properties of such receptors facilitates the development of QSAR models for pollutants, where toxicity can be predicted from chemical parameters (Box 17.1). Indeed, new pesticides are sometimes designed on the basis of models of this kind. For example, some ergosterol synthesis inhibitor fungicides (EBIs) that can lock into the catalytic site of P450s have been discovered by following this approach. Interactions such as these are essentially similar to the interactions of agonists and/or antagonists with receptors in pharmacology.

The last two examples do not belong in the same category, there being no discrete single binding site on a protein. Uncouplers of oxidative phosphorylation operate across the inner mitochondrial membranes, their critical properties being the ability to reversibly interact with protons and their existence in the uncharged lipophilic state after protons are bound. Oxyradicals can, in principle, be generated by a variety of redox systems in differing locations, which are able to transfer single electrons to oxygen under cellular conditions. The systems that carry out one electron reduction of nitroaromatic compounds and aromatic amines have yet to be properly elucidated.

Neither of these mechanisms of toxic action is susceptible to the kind of QSAR analysis referred to earlier, the employment of which depends on knowledge of the structure of particular binding sites.

13.4   TOXIC RESPONSES THAT SHARE COMMON PATHWAYS OF EXPRESSION

When chemicals have toxic effects, the initial molecular interaction between the chemical and its site of action (receptor, membrane, redox system, etc.), is followed by a sequence of changes at the cellular and whole-organism levels that eventually lead to the appearance of overt symptoms of intoxication. Until now, discussion has been focused upon mechanisms of toxicity, that is, on the primary interaction of toxic chemicals with their sites of action. As we have seen, biomarker assays such as the measurement of acetylcholinesterase inhibition can monitor this initial interaction in a causal chain that leads to the overt expression of toxicity. Such mechanistic biomarkers are specific for particular types of chemicals acting at particular sites. By contrast, other biomarkers that measure consequent changes at higher levels of organization, for example, the release of stress proteins, damage to cellular organelles, and disturbances to the nervous system or endocrine system are less specific, and can, in principle, provide integrated measures of the effects of diverse chemicals in a mixture operating through contrasting mechanisms of action. It is possible, therefore, to measure the combined effects of chemicals working through different modes of action if these effects are expressed through a common pathway (e.g., the nervous system or the endocrine system) that can be monitored by a higher-level biomarker assay. For example, two chemicals may act on different receptors in the nervous system, but they may both produce similar disturbances such as tremors, hyperexcitability, and even certain changes in the EEG pattern, all of which can be measured by higher-level biomarker assays.

When moving from the primary toxic lesion to the knock-on effects at higher levels of organization, the higher one goes, the harder it becomes to relate measured effects to particular mechanisms of toxic action. Thus, it is advantageous to use combinations of biomarkers operating at different organizational levels rather than single biomarker assays when investigating toxic effects of mixtures of dissimilar compounds; it becomes possible to relate initial responses to higher-level responses in the causal chain of toxicity. Although they often do not give clear evidence of the mechanism of action, higher-level biomarker assays (e.g., scope for growth in mollusks, or behavioral effects in vertebrates) have the advantage that they can give an integrated measure of the toxic effects caused by a mixture of chemicals.

Taken together, combinations of biomarker assays working at different organizational levels can give an "in-depth" picture of the sequence of adverse changes that follows exposure to toxic mixtures, when compounds in the mixture with different modes of action cause higher-level changes through a common pathway of expression. Two prime examples are (1) chemicals that cause endocrine disruption, and (2) neurotoxic compounds. To illustrate these issues further in more depth and detail, later chapters are devoted to endocrine disruptors (Chapter 15) and neurotoxicity and behavioral effects (Chapter 16).

Thus far, the discussion has dealt primarily with biomarker responses in living organisms. In the next section, consideration will be given to the exploitation of this principle in the development of bioassay systems that can be used in environmental monitoring and environmental risk assessment.

13.5 BIOASSAYS FOR TOXICITY OF MIXTURES

Both cellular systems and genetically manipulated microorganisms have been used to measure the toxicity of individual compounds and mixtures present in environmental samples such as water, soil, and sediment. Such bioassays can have the advantages of being simple, rapid, and inexpensive to use. They can provide evidence for the existence in environmental samples of chemicals with toxic properties, acting either singly or in combination. Some of them provide measures of the operation of certain modes of action, thus giving evidence of the types of compounds responsible for toxic effects; simple bioassays that use broad indications of toxicity such as lethality or reduction of growth rate as end points do not do this.

A shortcoming of bioassay systems is the difficulty of relating the toxic responses that they measure to the toxic effects that would be experienced by free-living organisms if exposed to the same concentrations of chemicals in the field. These simple systems do not reproduce the complex toxicokinetics of living vertebrates and invertebrates. As explained earlier in Chapter 2, toxicokinetic factors are determinants of toxicity, and there are often large metabolic differences between species that cause correspondingly large differences in toxicity. With persistent pollutants, this problem may be partially overcome by conducting bioassays upon tissue extracts, but even here there are complications. How closely does the use of an extract reproduce the actual cellular concentrations at the site of action in the living animal? How similar are the toxicodynamic processes of a test system to those operating in the living animal? The site of action may very well differ when, as is usually the case, the species represented in the test system differs from the species under investigation. This may also be the case when comparing a resistant with a susceptible strain of the same species. It is clear from many examples of resistance to pesticides that a difference of just one amino acid residue of a target protein can profoundly change the affinity for the pesticide, and consequently the toxicity (see Chapter 2, Section 2.4, and various examples in Chapters 5–14). Thus, the use of material from a susceptible strain in a test system raises problems when dealing with resistant strains from the field.

These things said, bioassay systems have considerable potential for biomonitoring and environmental risk assessment. By giving a rapid indication of where toxicity exists, they can identify "hot spots" and pave the way for the use of more sophisticated methods of establishing cause and effect, including chemical analysis and biomarker assays on living organisms. In the context of biomonitoring, they are useful for checking the quality of surface waters and effluents, and giving early warning of pollution problems. In these respects they have considerable advantages over chemical analysis. They can be very much cheaper and, because chemical analysis is not comprehensive, they can measure the toxicity of compounds that escape detection in the chemical laboratory.

A number of bioassays utilize microorganisms. Some, such as the Microtox test system, give a nonspecific measure of toxicity. This system utilizes a bioluminescent marine organism *Vibrio fischeri*, which emits light due to the action of the enzyme luciferase (see Calow 1993). Toxicity is measured by the degree of inhibition of light. A more specific type of test is the bacterial mutagenicity assay, the best-known example of which is the Ames test (Maron and Ames 1983). This type of test has been widely used by the chemical industry to screen pesticides and drugs for mutagenic properties. In the Ames test, histidine dependent strains of the bacterium *Salmonella typhimurum* are exposed to individual chemicals or mixtures. Mutation is shown by a loss of histidine-dependence, and mutation rates are related to dose. An important feature of the Ames test is that it incorporates a metabolic activation system, usually a preparation of mammalian hepatic microsomes with high monooxygenase activity. Thus, a distinction can be made between pollutants that are themselves mutagenic and others that require metabolic activation by the P450 system.

A number of mammalian and fish cell lines have been used to test for toxicity, some of them measuring particular mechanisms. Bioassay systems have been developed that test for Ah-receptor-mediated toxicity (Chapter 7, Section 7.2.4). Some cell hepatoma lines, such as one from mice, contain the Ah receptor, and cells of this type have been transfected with reporter genes (Garrison et al. 1996). An example is the CALUX system, where interaction of coplanar PCBs, dioxins, etc., with the Ah receptor of hepatoma cells triggers the synthesis of luciferase and consequent light emission. The degree of occupancy of the Ah receptor by these compounds determines the quantity of light that is emitted. Thus, the CALUX system can give an integrated measure of the effects of mixtures of polyhalogenated compounds on the Ah receptor, and an indication, therefore, of the potential of such mixtures to cause Ah-receptor-mediated toxicity.

In another example, fish hepatocyte lines have been used to detect the presence of environmental estrogens. Primary cultures of rainbow trout hepatocytes containing the estrogen receptor can show elevated levels of vitellogenin when exposed to environmental estrogens (Sumpter and Jobling 1995). Assays with this system, together with assays for vitellogenin production in caged male fish, have demonstrated the presence of estrogenic activity at sewage outfalls. Subsequent investigation established that much of the estrogenic activity was due to natural estrogens in sewage, but there was also evidence that nonyl phenols derived from detergents had an estrogenic effect in a highly polluted stretch of river. The estrogen receptor is responsive to a number of environmental compounds, including organochlorine compounds such as dicofol and *o,p'*-DDT, nonyl phenols (rather weak), and naturally occurring phytoestrogens (IEH Assessment 1995 and Chapter 15). Once again, an assay system that is mechanistically based can give an integrated measure of the adverse effects of mixtures of environmental chemicals.

Fish hepatocyte lines have also been developed, which can show cytochrome P450 1A1 induction due to PAHs and planar polychlorinated aromatic compounds binding to the Ah receptor (Vaillant et al. 1989, Pesonen et al. 1992).

Apart from the scientific advantages offered by this new technology, it has also been welcomed by organizations seeking a reduction in the number of animals used in toxicity testing (see Chapter 15, Section 15.6, and Walker 1998b).

13.6 POTENTIATION OF TOXICITY IN MIXTURES

The problem of potentiation was discussed earlier (Chapter 2, Section 2.5). Potentiation is often the consequence of interactions at the toxicokinetic level, especially inhibition of detoxication or increased activation. The consequences of such potentiation may be evident not only at the whole animal level but also in enhanced responses of biomarker assays that measure toxicity (Figure 13.3). By contrast, biomarkers of exposure alone are unlikely to give any indication of potentiation at the toxicokinetic level.

The real problem about potentiation is anticipating where it may occur when the only available toxicity data is for the individual compounds that will constitute the mixture. This is a frequent issue in the regulation of pesticides. When should the use of new mixtures of old pesticides be approved? When considering mixtures, it tends to be assumed that toxicity will be additive unless there is clear evidence to the contrary, an approach that has worked out reasonably well in practice. However, there are important exceptions (Chapter 2, Section 2.5, and Chapters 10 and 12). Full consideration should be given to known mechanisms of potentiation when questions are raised about the possible toxicity of mixtures. Where, on sound mechanistic grounds, there appears to be a clear risk of potentiation, appropriate tests should be carried out to establish the toxicity of the mixture in question. In this way, the very limited resources available for testing mixtures would be targeted on the most important cases. With the very rapid growth of understanding of the mechanistic basis of toxicity, it should become increasingly possible to anticipate where substantial potentiation of toxicity will occur. In this field there is no substitute for expert knowledge. The resources do not exist for any general statutory requirement for the toxicity testing

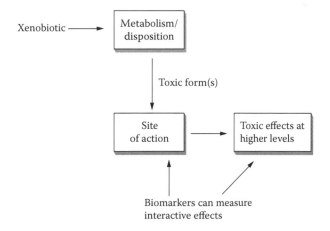

FIGURE 13.3 Biomarkers of toxic effect.

of mixtures of industrial chemicals that may be released into the environment. Even if such resources did exist, such an exercise would be very largely a waste of time because substantial potentiation of toxicity in mixtures is a rare event.

As understanding grows of biochemical mechanisms that lead to strong potentiation of toxicity, more attention should be given to the release of compounds that are not toxic in themselves but that may increase the toxicity of pollutants already present in the environment. Recognition of the synergistic action of the P450 inhibitor piperonyl butoxide has made it unlikely that insecticide formulations containing it would be approved for use on food crops. Another example is the EBI fungicides that can potentiate the toxicity of pyrethroid and phosphorothionate insecticides (2.6). There may be situations in which their use in agriculture increases the hazards presented by certain commonly applied insecticides to wild life.

13.7 SUMMARY

Most statutory toxicity testing is done on individual compounds. In the natural environment, however, organisms are exposed to complex mixtures of pollutants. Toxicity testing procedures are described for environmental samples that contain mixtures of different chemicals.

Particular attention is given to the development of new mechanistic biomarker assays and bioassays that can be used as indices of the toxicity of mixtures. These biomarker assays are typically based on toxic mechanisms such as brain acetylcholinesterase inhibition, vitamin K antagonism, thyroxin antagonism, Ah-receptor-mediated toxicity, and interaction with the estrogenic receptor. They can give integrative measures of the toxicity of mixtures of compounds where the components of the mixture share the same mode of action. They can also give evidence of potentiation as well as additive toxicity.

In more complex scenarios, the components of a mixture that work by different mechanisms may still express toxicity through the same pathway. Here, biomarker assays operating at higher levels of biological organization can give integrative measures of toxic effect. Endocrine disruptors and neurotoxic compounds provide examples of groups of pollutants whose effects can be studied in this way. A case is made for using carefully selected combinations of biomarkers operating at different organizational levels when investigating complex pollution problems. Such an approach can give an in-depth picture of toxic effects.

Bioassays can be used for cost-effective biomonitoring and rapid screening of environmental samples to detect the presence of mixtures of toxic chemicals and to identify hot spots.

FURTHER READING

Fossi, M.C. and Leonzio, C. (1994). *Nondestructive Biomarkers*—A collection of detailed reviews of nondestructive biomarkers.

Janssen, P.A.H., Faber, J.H., and Bosveld, A.T.C. (1998). *[Fe]male*—Reviews biomarker assays and bioassays for ERs in reasonable detail.

Peakall, D.B. (1992). *Animal Biomarkers as Pollution Indicators*—An authoritative text on biomarkers.

Persoone, G., Janssen, C., and De Coen, W. (Eds.) (2000). *New Microbiotests for Routine Toxicity Testing and Screening*—An extensive review of bioassay systems and toxicity tests that also contains some valuable discussion of the principles that underlie them.

Walker C.H. (1998b). *Alternative Approaches and Tests in Ecotoxicology*—Gives a broad review of biomarker assays in ecotoxicology.

14 The Ecotoxicological Effects of Herbicides

14.1 INTRODUCTION

Herbicides constitute a large and diverse class of pesticides that, with a few exceptions, have very low mammalian toxicity and have received relatively little attention as environmental pollutants. Much of the work in the field of ecotoxicology and much environmental risk assessment has focused on animals, especially vertebrate animals. There has perhaps been a tendency to overlook the importance of plants in the natural world. Most plants belong to the lowest trophic levels of ecosystems, and animals in higher trophic levels are absolutely dependent on them for their survival.

The direct environmental effects caused by herbicides appear to have been very largely upon plants. Mostly, these have been on target weed species, but sometimes upon nontarget species as well, due to spray drift. A few have had toxic effects upon animals. Paraquat and other bipyridyl herbicides have appreciable mammalian toxicity and have been implicated in poisoning incidents involving lagomorphs on agricultural land (see De Lavaur et al. 1973, Sheffield et al. 2001). The mode of action of these compounds was discussed in Chapter 13. Dinoseb and related dinitrophenols, which act as uncouplers of oxidative phosphorylation in mitochondria, are general biocides that are little used today because of their hazardous nature. The mode of action of these compounds was also discussed in Chapter 13. Also, the herbicide 2,4,5,T, a constituent of Agent Orange, has sometimes been contaminated with significant amounts of the highly toxic compound tetrachlorodibenzodioxin (TCDD) (see Chapter 7, Section 7.2.1).

There have been some reports of herbicides disturbing the metabolism of microorganisms in soil. For example, dichlobenil was shown to increase the rate of CO_2 formation from glucose in soil (Somerville and Greaves 1987).

Herbicides are, in general, readily biodegradable by vertebrates and are not known to undergo substantial biomagnification in food chains. Their principal use has been for weed control in agriculture and horticulture, although they have also been employed as defoliants in forests (e.g., in the Vietnam War), controlling weeds in gardens, on roadside verges, and in water courses, and as management tools on estates and nature reserves. This chapter will be mainly concerned with their impact on the agricultural environment where they have been extensively used. Brief mention will be made of their wider dispersal in the aquatic environment.

14.2 SOME MAJOR GROUPS OF HERBICIDES AND THEIR PROPERTIES

The following brief account identifies only major groups of herbicides not mentioned elsewhere in the text, and is far from comprehensive. Their mode of action is only dealt with in a superficial way. From an ecotoxicological point of view, there has not been as much concern about their sublethal effects upon plants as there has been in the case of mammals, and there has not been a strong interest in the development of biomarker assays to establish their effects. The major concern has been whether weeds, or nontarget plants, have been removed following herbicide application—a rather easy matter to establish as plants are fairly sedentary. For a more detailed account of herbicide chemistry and biochemistry, see Hassall (1990).

The phenoxyalkane carboxylic acids are among the most successful and widely used herbicides. They act as plant growth regulators and produce distorted growth patterns in treated plants. Compounds such as 2,4-D, MCPA, and mecoprop (Figure 14.1) are used as selective herbicides to control dicotyledonous weeds in monocotyledonous crops such as cereals and grass. They are formulated as water-soluble potassium or sodium salts or as lipophilic esters, and are frequently sprayed in combination with other types of herbicides having different modes of action and patterns of weed control. They are applied to foliage, and are not soil acting.

Ureides (e.g., diuron, linuron) and triazines (e.g., atrazine, simazine, ametryne) all act as inhibitors of photosynthesis and are applied to soil (see Figure 14.1 for structures). They are toxic to seedling weeds, which they can absorb from the soil. Some of them (e.g., simazine) have very low water solubility and, consequently, are persistent and relatively immobile in soil (see Chapter 4, Section 4.3, which also mentions the question of depth selection when these soil-acting herbicides are used for selective weed control).

Sulphonylurea herbicides such as chlorsulfuron and sulfometuron are also soil acting, have effects upon cell division, and are highly phytotoxic. Indeed, they can be toxic to plants when present in soil at levels low enough to make chemical analysis difficult. Carbamate herbicides constitute a relatively diverse group. Some, such as barban (Figure 14.1), are applied to foliage, whereas others (e.g., chlorpropham) are soil acting. The latter type have effects upon cell division. Other important herbicides or groups of herbicides include glyphosate, aminotriazole, chlorinated benzoic acids (e.g., dicamba), and halogenated phenolic nitriles (e.g., ioxynil, bromoxynil).

14.3 IMPACT OF HERBICIDES ON AGRICULTURAL ECOSYSTEMS

Since World War II, herbicides have come to be widely used in agriculture and horticulture in the developed world. Frequently, they have been used in "cocktails" containing several ingredients of contrasting modes of action, thus giving control over a wide range of weed species. The effectiveness of the application of herbicides together with cultivation of the land is evident in many agricultural areas of Western Europe and North America, where few weeds are seen. It is easier to control plants, which are stationary, than to control mobile insect or vertebrate pests. By the same token, it is also easier to judge the population effects of control measures (e.g., use of

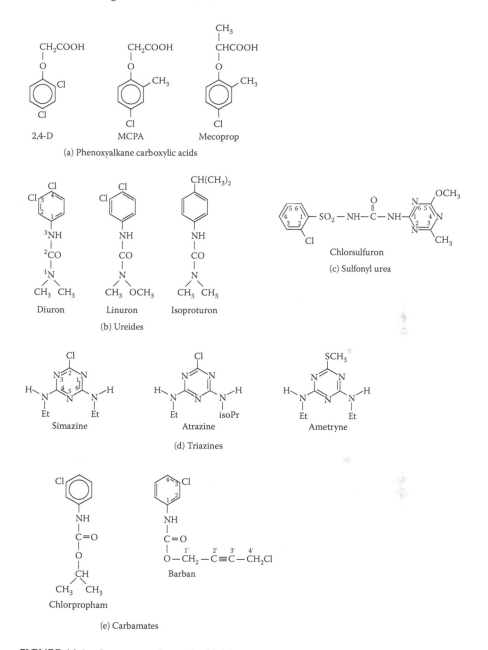

FIGURE 14.1 Structures of some herbicides.

pesticides) than in the case of mobile animals. Weed species have been very effectively controlled over large areas of agricultural land. In Britain, concern has been expressed about the near extinction of certain once-common farmland species that are of botanical interest, for example, corn cockle (*Agrostemma githago*) and pheasants eye (*Adonis annua*).

Ecologically, such a large reduction of weed species represents a major change in farmland ecosystems and may be expected to have knock-on effects on other species. Certain problems have come to light with the investigation of the status of birds on farmland. In one study, the Game Conservancy investigated the reasons for a severe and continuing decline of the grey partridge (*Perdix perdix*) on farmland in Britain. The study commenced in the late 1960s and established that the decline was closely related to increased chick mortality (Potts 1986, Potts 2000; also Chapter 12 of Walker et al. 2006). The high chick mortality was largely explained by a short-age of their insect food (e.g., sawflies) due, in turn, to the absence of the weeds upon which the insects themselves feed. An effect at the bottom of the food chain led to a population decline further up. It is worth reflecting that such an effect by herbicides could not have been forecasted by normal risk assessment (see Chapters 14 and 15). The herbicides responsible are in general of very low avian toxicity, and ordinary risk assessment would have declared them perfectly safe to use so far as partridges and other birds are concerned. Subsequent work has shown that partridge populations can continue to survive on agricultural land if headlands are left unsprayed, thereby allowing weeds to survive, weeds that will support the insects on which young partridges feed.

This study helped to ring the alarm bells about possible other indirect effects of the wide use of herbicides in agriculture. More recently, further evidence has been gained of the reduction in populations of insects and other arthropods on farmland that may relate, at least in part, to the removal of weeds by the use of herbicides. A study of farmland birds in Britain established the marked decline of several species in addition to the grey partridge, which may be the consequence of the indirect effects of herbicides and other pesticides (Crick et al. 1998, and Chapter 12 of Walker et al. 2006). Species affected include tree sparrow (*Passer montanus*), turtle dove (*Streptopelia purpur*), spotted flycatcher (*Musciapa striata*), and skylark (*Alauda arvensis*). A study is currently in progress to attempt to establish the cause of these declines.

Recently, controversy about the possible side effects of herbicides used on agricultural land has intensified with the development of genetically modified (GM) crops. Some GM crops are relatively insensitive to the action of herbicides, thus permitting the application to them of unusually high levels of certain herbicides. The advantage of increasing the dose, from the agricultural point of view, is the control of certain difficult weeds. From an ecotoxicological point of view, though, increasing dose rates of herbicides above currently approved levels raises the possibility that this may cause undesirable ecological side effects. It is very important that any such change in practice is rigorously tested in field trials as part of environmental risk assessment before approval for marketing is given by regulatory authorities. Such new technology, based on GM crops, should only be introduced if it is shown to be environmentally safe.

One problem that has arisen with the use of herbicides in agriculture is spray or vapor drift. When fine spray droplets are released, especially if applied aerially, they may be deposited beyond the target area due to air movements to cause damage there. In the first place, this is a question of application technique. Herbicides, like other pesticides, should not be applied as sprays under windy conditions. In most situations, herbicides are not applied aerially because of the danger of drift. Where herbicides have appreciable vapor pressure, there may be problems with vapor drift. Under hot conditions, volatile herbicides may go into the vapor state, and the vapor may drift farther than the spray droplets. This happened with early volatile ester formulations of phenoxyalkanecarboxylic acids (Hassall 1990). Nowadays, formulations are of less volatile esters, or as aqueous concentrates of Na or K salts, which are of low volatility. Spray drift of herbicides can result in damage to crops and wild plants outside the spray area. The cause of such damage can be hard to establish with highly active herbicides (e.g., sulfonylureas) where their phytotoxic concentrations are low enough to make chemical detection difficult.

14.4 MOVEMENT OF HERBICIDES INTO SURFACE WATERS AND DRINKING WATER

As discussed earlier (Chapter 4, Section 4.2), pesticides have a very limited tendency to move through soil profiles into drainage water because of the combined effects of adsorption by soil colloids (important for herbicides such as simazine, which have relatively high K_{ow}), metabolism (important for water-soluble and readily biodegradable herbicides such as 2,4-D and MCPA), and in some cases volatilization. In reality, however, there are complications. In the first place there may be runoff from agricultural land into neighboring water courses after heavy rainfall. Soil colloids with adsorbed herbicides can be washed into drainage ditches and streams. There is an additional problem with certain soils high in clay minerals (Williams et al. 1996, and Chapter 4, Section 4.2). During dry periods these soils shrink and develop deep cracks. If heavy rains follow, free herbicides located near the soil surface and colloids contaminated with adsorbed herbicides can be quickly washed down into the drainage system without passing through the soil profile. In the Rosemaund experiment, the herbicides atrazine, simazine, isoproturon, trifluralin, and MCPA were all detected in drainage water following heavy rain. The respective maximum concentrations in µg/L (ppb) were 81, 68, 16, 14, and 47 (Williams et al. 1996). These levels were reached following normal approved rates of application of the herbicides and raise questions about possible effects on aquatic plants growing in receiving waters. As mentioned elsewhere (Chapter 10, Section 10.3.4), the level of carbofuran found during the same study was sometimes high enough to kill freshwater shrimp (*Gammarus pulex*) used as a monitoring organism (Matthiessen et al. 1995).

Since this study was undertaken, surveys have been carried out that provide more information on the levels of herbicides in British rivers. In one study, a number of different herbicides were detected in the Humber rivers (House et al. 1997). Several triazines were found in the rivers Aire, Calder, Trent, Don, and Ouse, the most abundant of them being simazine and atrazine. The results for simazine showed peaks in spring and again in early autumn of 1994 for the Trent and Aire, the autumn peak coinciding with the first major storm of the year (Figure 14.2). The maximum level recorded for atrazine was > 8 µg/L in the river Calder in spring 1994. This was high enough to be toxic to phytoplankton and algae but was not sustained. It was not regarded as high enough to be toxic to aquatic invertebrates or fish. Phenyl ureas and

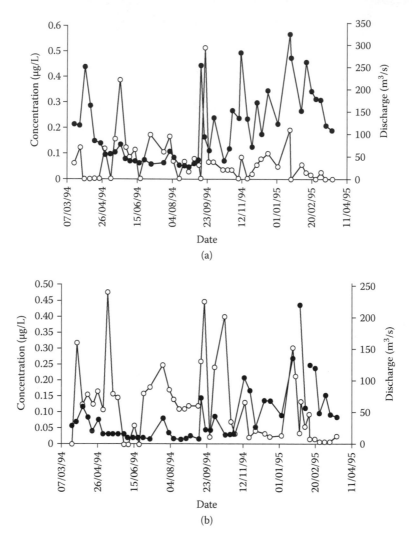

FIGURE 14.2 Atrazine levels in the Humber River area. Comparison of the concentration of simazine and river discharge over one annual cycle for (a) River Trent at Cromwell Lock, and (b) River Aire at Beale. ○, Simazine concentration; ●, river discharge. (From House et al. [1977]. With permission.)

phenoxyalkanoic acids were also detected. Concentrations were generally low, but levels of the following herbicides were detected up to maximum values (μg/liter), which are given in parentheses.

Diuron (< 8.7)
Chlortoluron (< 0.67)
Mecoprop (< 8.2)

These high levels were sporadic and transitory. However, some of them were high enough to have caused phytotoxicity, and more work needs to be done to establish whether herbicides are having adverse effects upon populations of aquatic plants in areas highlighted in this study. It should also be borne in mind that there may have been additive or synergistic effects caused by the combinations of herbicides found in these samples. For example, urea herbicides such as diuron and chlortoluron act upon photosynthesis by a common mechanism, so it seems likely that any effects upon aquatic plants will be additive. Similarly, simazine and atrazine share a common mechanism of action.

With the acceptable concentrations of herbicides in drinking water being taken to very low levels by some regulatory authorities (e.g., the EC), there has been interest in very low levels of atrazine present in some samples of groundwater and in drinking water. This finding illustrates the point that mobility of pesticides becomes increasingly evident as sensitivity of analysis improves.

14.5 SUMMARY

Contamination of agricultural land by herbicides is an example of the complexity of pollution in the real world. A wide variety of compounds of diverse structure, chemical properties, and mechanism of action are used as herbicides. Important groups of herbicides include phenoxyalkane carboxylic acids, ureides, triazines, and carbamates. They are often applied as mixtures of compounds with contrasting properties and modes of action. Very few of them have appreciable toxicity to animals, so initial toxic effects are mainly restricted to plants. Being generally biodegradable, they do not usually undergo significant biomagnification with movement along food chains.

The successful use of herbicides and associated cultivation procedures has greatly reduced the populations of weed species in many agricultural areas, sometimes bringing species of botanical interest to near extinction. Intensive weed control in cereal farming has been shown to cause the reduction of certain insect populations, and consequently also of the grey partridge, whose chicks are dependent on insect food. The reported decline of some other insectivorous birds on agricultural land in Britain may have a similar cause. The introduction of GM crops with high tolerance to herbicides may lead to increases in dose rates of herbicides on agricultural land with attendant ecotoxicological risks.

Significant levels of herbicides have also been detected in rivers, although these are usually transitory. Heavy rainfall can move herbicides from agricultural land to nearby ditches and streams due to runoff, and in soils that are high in clay, percolation of water occurs through deep fissures with consequent movement into neighboring water courses. Such events under extreme weather conditions are likely to have contributed to the pulses of herbicide contamination observed in some rivers. Questions have been asked about possible effects of such episodic pollution on populations of aquatic plants.

FURTHER READING

Ashton, F.M. and Crafts, A.S. (1973). *Mode of Action of Herbicides*—Describes the mode of action of major types of herbicides.

Hassall, K.A. (1990). *The Biochemistry and Use of Pesticides*—Includes a readable account of the biochemistry of herbicides.

Potts, G.R. (1990). *The Partridge*—An authoritative account of the factors responsible for the decline of the grey partridge on agricultural land.

Williams, R.J. et al. (1996). *Report of the Rosemaund Study*—Describes the movement of herbicides through cracks and fissures in heavy soil into neighboring water courses following heavy rainfall.

15 Endocrine-Disrupting Chemicals and Their Environmental Impacts

R. M. Goodhead and C. R. Tyler

15.1 INTRODUCTION

There is substantial and increasing evidence that endocrine disruption—defined here as a hormonal imbalance initiated by exposure to a pollutant and leading to alterations in development, growth, and/or reproduction in an organism or its progeny—is impacting wildlife adversely on a global scale (Tyler et al. 1998; Taylor and Harrison 1999; Vos et al. 2000). The causative chemicals of endocrine disruption in wildlife populations are wide ranging and include natural and synthetic steroids, pesticides, and a plethora of industrial chemicals. The effects induced range from subtle changes in biochemical pathways to major disruptions in reproductive performance. In the worst-case scenarios, endocrine disruption has led to population crashes and even the localized extinctions of some wildlife species.

This chapter aims to provide the reader with an insight into the phenomenon of endocrine disruption, detailing its emergence as a major research theme. We present some of the better-known examples of endocrine disruption in wildlife populations, identifying the causative chemicals and explaining (when known) their mechanisms of action. Increasingly, it is being realized that some endocrine disrupting chemicals (EDCs) have multiple mechanisms of action, and we discuss how genomics is starting to unravel the complexity of their biological effect pathways. The identification of various EDCs has principally arisen from observations of adverse effects in wildlife populations, but more recently, chemicals have been screened systematically for endocrine-disrupting activity and this approach has added, very considerably, to the list of EDCs of potential concern for wildlife and human health. This chapter also considers the interactive effects of EDCs—most wildlife species are exposed to complex mixtures of EDCs and their effects in combination can differ compared with exposure to single chemicals—and highlights differences in both species and life stage sensitivity to the effects of EDCs. Although the focus for endocrine disruption has been on disturbances in the physiology of animals, studies have also shown that some EDCs can alter behavior, including reproductive behavior, and we discuss some of the potential impacts of these effects on breeding dynamics and population genetic structure. Finally, we provide an analysis on the lessons learned from endocrine disruption in the context of ecotoxicology more broadly.

15.2 THE EMERGENCE OF ENDOCRINE DISRUPTION AS A RESEARCH THEME

Endocrine disruption as a research theme emerged at the Wingspread Conference in 1991 and through the publications that resulted from this meeting (Colborn and Clement 1992; Colborn et al. 1993). Knowledge that chemicals can modify hormonal systems, however, was known for many years prior to this, and as early as in the 1930s, Cook and associates noted that injection of certain "estrus producing compounds" initiated a sex change in the plumage of Brown Leghorn chickens (Cook et al. 1933). Dodds and associates, in a series of papers, also in the 1930s (Dodds 1937a, 1937b; Dodds et al. 1937, 1938) similarly identified various synthetic compounds that had estrogenic activity. Furthermore, natural estrogens in plants (so-called phytoestrogens) were suspected of causing reproductive disturbances in sheep feeding on clover-rich pastures 25 years before the Wingspread Conference (Coop and Clark 1966).

15.3 MODES OF ACTION OF ENDOCRINE-DISRUPTING CHEMICALS

To date, most EDCs that have been identified work by mimicking endogenous hormones. These chemicals can act as agonists or antagonists of hormone receptors to either generate or block hormone-mediated responses. Other mechanisms identified include inhibiting or inducing enzymes associated with hormone synthesis, metabolism, or excretion. Less well-characterized effect pathways include reacting directly or indirectly with endogenous hormones or altering hormone receptor numbers or affinities.

The most commonly reported EDCs in the environment are estrogenic in nature (McLachlan and Arnold 1996), and feminization in exposed males has been reported in a wide range of wildlife species. The most comprehensively researched case on the feminization of wildlife is for the intersex (the simultaneous presence of both males and female sex cells within a single gonad) condition in fish living in U.K. rivers, described later in this chapter. There is a wide body of literature on the subject of environmental estrogens, including whole journal issues and special reports dedicated to the subject and to which we would refer the reader for in-depth analyses (e.g., *Pure and Applied Chemistry*, 1998 volume 70 [9]; *Pure and Applied Chemistry*, 2003 volume 75 [11–12]; *Ecotoxicology*, 2007 volume 16 [1]; EPA Special Report on Environmental Endocrine Disruption 1997; *Molecular and Cellular, Endocrinology*, 2005 volume 244 [1–2]; *Water Quality Research Journal Canada*, 2001 volume 36 [2]). The list of known estrogenic chemicals spans pharmaceuticals, various classes of pesticides, plasticizers, resins, and many more, and this list has increased considerably with the systematic screening of chemicals for this activity (see the following text).

Chemicals with antiestrogenic chemicals have been known to exist for 50 years (Lerner et al. 1958, in Wakeling 2000). These chemicals exert their effects by blocking the activation of the estrogen receptor or by binding the aryl hydro-carbon (Ah) receptor, in turn leading to induction of Ah-responsive genes that can have a spectrum of antiestrogenic effects (Lerner et al. 1958, in Wakeling 2000). Antiestrogens create an androgenic environment, producing symptoms similar to those of androgenic exposure. Antiestrogenic chemicals known to enter the environ-ment include pharmaceuticals, such as tamoxifen and fulvestrant, used to treat breast cancer; raloxifene, which is used in the prevention of osteoporosis; and some of the polyaromatic hydrocarbons (PAH) such as anthracene (Tran et al. 1996).

Chemicals with antiandrogenic activity include pharmaceuticals developed as anticancer agents (e.g., flutamide, Neri and Monahan 1972; Neri et al. 1972, in Lutsky et al. 1975) and 179-methyltestosterone used to treat testosterone deficiency (Katsiadaki et al. 2006). Other antiandrogens include various pesticides such as the *p,p'*-DDE metabolite of DDT, the herbicides linuron and diuron, and metabolites of the fungicide vinclozolin (Gray et al. 1994). Antiandrogens create a similar over-all effect to estrogens (Kelce et al. 1995), and it been hypothesized that some of the feminized effects seen in wildlife populations may result from chemicals block-ing the androgen receptor rather than as a consequence of exposure to (or possibly in addition to) environmental estrogens (Sohoni and Sumpter 1998; Jobling et al., submitted). An extensive study on wastewater treatment works (WWTW) effluents in the United Kingdom has found very widespread antiandrogenic activity in these discharges (Johnson et al. 2004; see case example for the feminization of fish later in chapter). There has also been increasing evidence to support links between increases in the group of disorders referred to as testicular dysgenesis syndrome (TDS) in humans, which originate during fetal life, and exposure to environmental chemicals with antiandrogenic activity (Fisch and Golden 2003; Sharpe and Skakkebaek 2003; Sharpe and Irvine 2004; Giwercman et al. 2007).

Few environmental androgens have been identified, but one of the best examples of hormonal disruption in wildlife is an androgenic effect, namely, the induction of imposex in marine gastropods exposed to the antifouling agent tributyl tin (TBT, discussed in detail in Section 15.4). Androgenic responses in vertebrate wildlife are also known to occur, and reported examples include the masculinization of female mosquito fish, *Gambusia affinis holbrooki*, living downstream of a paper mill effluent (Howell et al. 1980), and the masculinization of fathead minnow, *Pimephales promelas*, living in waters receiving effluent from cattle feedlots in the United States (Jegou et al. 2001). In the latter case, the causative chemical was identified as 17β-trenbolone (TB), a metabolite of trenbolone acetate, an anabolic steroid used as a growth promoter in beef production (Wilson et al. 2002; Jensen et al. 2006).

Several groups of chemicals are known that can disrupt thyroid function. Some of these chemicals have a high degree of structural similarity to thyroid hormones and act via binding interference with endogenous thyroid hormone receptors. Thyroid hormones are fundamental in normal development and function of the brain and sex organs, as well as in metamorphosis in amphibians, and in growth and regulation of metabolic processes (Brouwer et al. 1998) and, thus, chemicals that interfere with their functioning can potentially disrupt a very wide range of biological processes. Developmental effects in wildlife populations indicative of disruptions in the thyroid system are widely reported, and they include malformation of limbs due to excessive or insufficient retinoic acid (structurally similar to thyroid hormones) in birds and mammals, the production of small eggs and chicks in birds, and impaired metamorphosis in amphibians (reviewed in Rolland 2000). Known thyroid-disrupting chemicals include many members of the polyhalogenated aromatic hydrocarbons (PHAHs) such as PCBs (polychlorinated biphenyls; see Chapter 6, Section 6.2.4), dioxins, PAHs, polybrominated dimethylethers (PBDEs, flame retardants), and phthalates (Brouwer et al. 1998; Zhou et al. 1999; Rolland 2000; Boas et al. 2006).

Other modes of hormonal disruption identified, but for which there is considerably less data, include those acting via the progesterone or Ah receptors, corticosteroid axis, and the enzyme systems involved with steroid biosynthesis. Chemicals interacting with the progesterone receptors can impact both reproductive and behavioral responses, notably in fish in which progesterones can function as pheromones (Zheng et al. 1997; Hong et al. 2006). Various progesterones are used in contraceptive pharmaceuticals such as norethisterone, levonorgestrel, desogestrel, and gestodene, and find their way into the aquatic environment via WWTW discharges. The fungicide vinclozolin and the pyrethroid insecticides fenvalerate and permethrin have also been shown to interfere with progesterone function (Kim et al. 2005; Buckley et al. 2006; Qu et al. 2008).

It has long been recognized that the Ah receptor (AhR) is a ligand-activated transcription factor that plays a central role in the induction of drug-metabolizing enzymes and hence in xenobiotic activation and detoxification (Marlowe and Puga 2005; Okey 2007; see Chapter 6, Section 6.2.4). Much of our understanding of AhR function derives from analyses of the mechanisms by which its prototypical ligand 2,3,7,8 tetrachlorodibenzo-p-diosin (TCDD) induces the transcription of CYP1A1 (Pocar et al. 2005), which encodes for the microsomal enzyme cytochrome P4501A1 that oxygenates various xenobiotics as part of their step-by-step detoxification (Conney 1982; see Chapter 6, Section 6.2.4.). Most effects on the endocrine systems of organisms exposed to halogenated and polycyclic aromatic hydrocarbons such as benzopyrene, polybrominated dimethylethers (PBDEs), and various PCBs are mediated by the Ah receptor (Pocar et al. 2005).

Interference with corticosteroid function and the stress response has been shown for a variety of chemicals, including the pharmaceutical salicylate (Gravel and Vijayan 2006) and the PAH, phenanthrene (Monteiro et al. 2000a, 2000b). Other classes of chemicals shown to have significant effects on cortisol levels include PCBs and PAHs (Hontela et al. 1992, 1997). The precise mechanisms for these effects are poorly understood, but for PCBs, are believed to be via their actions through the Ah receptor (Aluru and Vijayan 2006).

Studies on the endocrine-disrupting effects of chemicals via enzyme biosynthesis pathways have focused on cytochrome P450 aromatase, encoded by the CYP19 gene, and involved with the production of estrogens from androgens (Cheshenko et al. 2008). Modulation of aromatase CYP19 expression and function can dramatically alter the rate of estrogen production, disturbing the local and systemic levels of estrogens that play a critical role in vertebrate developmental sex differentiation and reproductive cycles (Simpson et al. 1994). Natural and synthetic chemicals, including certain xenoestrogens, phytoestrogens, pesticides, and organotin compounds, are able to inhibit aromatase activity, both in mammals and fish (reviewed in Kazeto et al. 2004 and Cheshenko et al. 2008). Another enzyme in the sex steroid biosynthesis pathway that can be disrupted by EDC exposure effects is cytochrome P450 17 alpha-hydroxylase/C17-20-lyase (P450c17), which catalyzes the biosynthesis of dehydroepiandrosterone (DHEA) and androstenedione in the adrenals (Canton et al. 2006) and testosterone in the Leydig cells within the testis (Majdic et al. 1996). Maternal treatment with diethylstilbestrol (DES) or the environmental estrogen, 4-octylphenol (OP), has been shown to reduce expression of P450c17 in fetal Leydig cells (Majdic et al. 1996), which can have subsequent adverse affects on fetal steroid synthesis and the masculinization process. PAHs and Di (n-butyl) phthalate (DBP) also cause dose-dependent reductions in P450c17 expression in fetal testis of rats (Lehmann et al. 2004).

Some EDCs have been shown to have multiple hormonal activities (Sohoni and Sumpter 1998). Examples of this include bisphenol A, o,p'-DDT, and butyl benzyl phthalate, which possess both estrogenic and antiandrogenic activity, acting both as an agonist at the estrogen and antagonist at the androgen receptor. Other examples include the PCBs that can alter the estrogenic pathway, interfere with thyroid function, and disrupt corticosteroid function via the Ah receptor pathway. Some estrogens are even agonists in one tissue yet antagonists in another (Cooper and Kavlock 1997). Adding further to this complexity, disruptions to the endocrine system can affect the functioning of the nervous and immune systems and the processes they control (and vice versa). Examples of this include increases in autoimmune diseases in women that result from exposure to the clinical estrogen DES, and suppression in the expression of a gene associated with immune function (Williams et al. 2007), modifications in phagocyte cells to the point of suppressing phagocytosis (Watanuki et al. 2002), and decreases in IgM antibody concentrations (Hou et al. 1999) in fish exposed to the steroid estrogen 17β-oestradiol (E$_2$).

In an attempt to unravel the pathways of effect of some EDCs and the biological systems affected, toxicogenomics, most notably transcriptomics, are being increasingly explored. Different mechanisms of toxicity can generate specific patterns of gene expression indicative of the mode of action (and the biological processes affected; Tyler et al. 2008). Expanded PCR-based methodologies have been used to highlight the complex nature of the estrogenic effect of the pesticides p,p'-DDE and dieldrin in fish (Garcia-Reyero et al. 2006a, 2006b; Garcia-Reyero and Denslow 2006; Barber et al. 2007). In the Garcia-Reyero et al. 2006a study, three different modes of action were identified, namely, direct interactions with sex steroid receptors, alteration of sex steroid biosynthesis, and alterations in sex steroid metabolism. Expanded PCR-based methodologies have similarly been applied to illustrate the multiple mechanisms of action of environmental steroidal estrogens (E$_2$ and the pharmaceutical estrogen ethinyloestradiol, EE$_2$) and the antiandrogen flutamide in fish (Filby et al. 2006; Filby et al. 2007b). In that work E$_2$ was shown to trigger a cascade of genes regulating growth, development, thyroid, and interrenal function. Responses were noted across six different tissues, with implications of more wide-ranging effects of these chemicals beyond their well-documented effects on reproduction. Santos et al. (2007), employing an oligonucleotide gene array (with 16,400 identified gene targets), recently discovered alterations in the expression of cascades of genes associated with cell cycle control, energy metabolism, and protection against oxidative stress in zebrafish exposed to environmentally relevant concentrations of EE_2.

As our knowledge of the pathways of effects for EDCs has evolved, the number of chemicals classified as EDCs has increased, and a more extensive list of these chemicals is detailed later in this chapter. The terminology used to describe chemicals that affect the endocrine system has also changed over time. Some now refer to EDCs as *endocrine-active* or *endocrine-modulating*, rather than *endocrine-disrupting* chemicals, as they do not necessarily always have deleterious effects.

15.4   CASE STUDIES OF ENDOCRINE DISRUPTION IN WILDLIFE

Most examples of endocrine disruption have been reported in wildlife living in, or closely associated with, the aquatic environment. This is perhaps not surprising given that our freshwater and marine systems act as a sink for most chemicals we discharge into the environment. This section describes some of the better-known examples of endocrine disruption in wildlife populations, and assesses the strength of the associations with specific chemicals. Few studies have been able to provide an unequivocal link between a specific EDC and a population-level impact, in part because of the complexity of the chemical environment to which wildlife is exposed. The exceptions to this are for DDT and its metabolites, responsible for the decline of raptor populations and for TBT in the localized extinctions of some marine mollusks. There are, however, other examples from wildlife studies in which very strong associations have been established between specific chemicals, or groups of chemicals, and endocrine-disrupting effects, in some cases at levels likely to impact populations. Examples include exposure to PCBs and developmental abnormalities in fish-eating birds in and around the Great Lakes (considered elsewhere in Chapter 6), exposure to DDT and its metabolites and altered sexual endocrinology in alligators living in lakes in Florida, exposure to environmental estrogens, including steroidal estrogens, and the feminization of fish living in U.K. rivers. A further case study that we consider here, in part to highlight some of the complexities and controversies surrounding the issues of endocrine disruption in wildlife populations, is the link proposed between exposure to the herbicide atrazine and adverse effects in frog populations in the United States.

15.4.1   DDT (AND ITS METABOLITES) AND DEVELOPMENTAL ABNORMALITIES IN BIRDS AND ALLIGATORS

Effects of chemicals on the endocrine systems leading to reproductive disturbances were documented in wildlife populations in the 1960s. These original studies included work on osprey populations, where it was hypothesized that detectable levels of DDT and its metabolites could be responsible for hatching failure (Ames and Mersereau 1964). Although DDT was known to be fatal to birds in areas of high contamination (Wurster et al. 1965), it was not until Ratcliffes' landmark paper (1967) that the role of DDT in eggshell thinning was discovered, and its significance in the decline of some species of predatory birds was elucidated. In this work, it was established that exposure to DDT, when high enough, could cause eggshell thinning of 18% or more, so that egg shells were simply crushed during incubation (see Chapter 5, Section 5.2.5.1; Peakall 1993). Population-level impacts of DDT, particularly on raptors and shore birds, led to the intense study of its toxicology, which continues today. Even now there is still controversy about the mechanism through which the active metabolite of DDT, p,p'-DDE, causes eggshell thinning. Lundholm (1997) has suggested that it may involve disturbance of prostaglandin metabolism, putting it squarely into the arena of endocrine disruption (see Chapter 5, Section 5.2.4). Other studies on fish-eating birds have further shown that DDT and it metabolites can also disrupt sexual development in birds through their action as environmental estrogens (Welch et al. 1969; Fry and Toone 1981; Gilbertson et al. 1991; Fry 1995).

The case for DDT-induced modifications to the endocrine systems of the American alligator (*Alligator mississippiensis*) emerged from field observations on a heavily polluted lake in Florida, Lake Apopka. The study lake was originally subjected to a pesticide spill in 1980 and additionally received extensive agricultural, nutrient, and pesticide runoff. In the 5 years following the spill, juvenile recruitment plummeted owing to decreased clutch viability and increased juvenile mortality (Woodward et al. 1993, in Guillette et al. 2000). In the early 1990s, there was a population recovery but a number of sublethal problems were then reported, including alterations in plasma E_2, testosterone, and thyroxine concentrations, as well as morphological changes in the gonads (Guillette et al. 1994). Guillette and colleagues (1996) subsequently reported an altered sexual endocrinology in exposed alligators, and a correlation was established between reduced phallus (penis) size and low plasma testosterone levels. A study by Heinz and coworkers (1991) found elevated levels of p,p'-DDE, a metabolite of DDT that is known to have antiandrogenic effects (Kelce et al. 1995), in alligator eggs collected during 1984–1985 when compared to two other reference lakes.

Alligators express environmental sex determination, where temperature has a significant influence on the sex of the offspring (Ferguson and Joanen 1982) and this can be overridden by exposure of the embryo to environmental estrogens (or antiandrogens), resulting in sex reversal (Bull et al. 1988). This feature makes sexual development in alligators especially sensitive to EDCs and has been used to screen for chemicals that can that interfere with sex steroid and/or thyroid hormone function to influence sex (Guillette et al. 1995). In laboratory-based studies, topologically applying ("painting") p,p'-DDE and 2,3,7,8-tetrachlorodibenzo-p-diosin (TCDD) on to the shells of alligator eggs has been shown to alter the subsequent sex ratio and sexual endocrinology of the resulting embryos and juveniles (Matter et al. 1998). Together, the combined field and laboratory studies have provided a persuasive argument that the metabolites of DDT contributed to the altered sexual endocrinology and development in the alligators of Lake Apopka. Further work on Lake Apopka, however, found elevated levels of other organochlorine pesticides and PCBs in the serum of juvenile alligators, and egg-painting studies have similarly found that some of these chemicals too can alter sexual development (Crain et al. 1997). Thus, in the case of the alligators in Lake Apopka, although the metabolites of DDT have likely contributed to the alterations in sexual development seen, they are likely not the sole contributing factor. As a final note in the alligator story, it is very difficult to control the dose to the embryo for applications of chemicals to the outside of the egg, and as no dose verifications were provided in the previously mentioned studies, some have questioned the robustness of the cause–effect relationship drawn between DDT/PCBs and sexual disruption in alligators in Lake Apopka (Muller et al. 2007).

15.4.2   TBT and Imposex in Mollusks

In 1981, Smith reported the occurrence of imposex, the expression of a penis and/or a vas deferens in females of the marine gastropod *Nassarius obstoletus,* and hypothesized that the antifouling agent tributyl tin (TBT) was responsible. This was subsequently proved by Gibbs and Bryan for imposex in the dog whelk (*Nucella lapillus*), which resulted in reproductive failure and population level declines in this species (Bryan et al. 1986; Gibbs and Bryan 1986). The imposex condition has now been reported over extensive geographical regions and in over 150 species of marine mollusks (Matthiessen and Gibbs 1998). Extensive laboratory-based exposures have shown that imposex is induced by TBT in adults at concentrations as low as 5 ng TBT/L in the water (Gibbs et al. 1988) and in juvenile or larval dog whelks at exposure concentrations of only 1 ng TBT/L (Mensink et al. 1996). Concentrations of TBT in some harbors and in busy shipping lanes exceeded 30 ng/L (Langston et al. 1987). Environmental concentrations of TBT therefore were, and in some areas still are, sufficient to induce imposex in some marine mollusks. The unequivocal evidence that TBT has caused population-level declines, and even localized population extinctions, in marine mollusks, led to its ban from use on ships less than 25 m in the United Kingdom in 1987 and from 1982 in France. The International Maritime Organisation has phased out TBT, and a complete ban of TBT on all European vessels was imposed in January 2008. In areas where TBT is no longer used, there has been recovery in the populations of marine mollusks (Waite et al. 1991; Rees et al. 2001).

The mechanisms through which TBT masculinizes female gastropods are still uncertain. One hypothesis is that TBT acts as an inhibitor of aromatase, restricting the conversion of androgen to estrogen and/or that it inhibits the degradation of androgen, both of which would cause higher levels of circulating androgen (Bettin et al. 1996). Oberdörster and McCelland-Green (2002) argued that TBT acts as a neurotoxin to cause abnormal release of the peptide hormone Penis Morphogenic Factor. A more recent study (Horiguchi et al. 2007) provides persuasive evidence that TBT acts through the retinoid X receptor (RXR); the suggestion is that RXR plays important roles in the differentiation of certain cells required for the development of imposex symptoms in the penis-forming area of females. Interestingly, despite the known harmful effects of TBT, it is still used widely as an antibacterial agent in clothes, nappies, and sanitary towels, providing further routes of entry into the environment (via landfills, etc.). Triphenyl-tin (TPT), which is widely used as a fungicide on potatoes and to control algae in rice fields (Strmac and Braunbeck 1999), has been shown to induce sexual disruption in various species of gastropods at environmentally relevant concentrations (Schulte-Oehlmann et al. 2000; Horiguchi et al. 2002; Santos et al. 2006) as well as causing a delay in hatching and producing histological alterations in the gonads of zebrafish (Strmac and Braunbeck 1999). We may therefore not as yet have realized the wider endocrine-disrupting impacts of the organotins.

15.4.3 ESTROGENS AND FEMINIZATION OF FISH

The story of the feminization of fish in the United Kingdom originated from observations of intersex in roach (*Rutilus rutilus*) living in WWTW settlement lagoons. Roach are normally single-sexed, and intersex is an unusual occurrence. Another finding independently established that effluents from WWTW were estrogenic to fish, inducing the production of a female yolk protein (vitellogenin, VTG) in male fish (Purdom et al. 1994). Some of the effluents surveyed were extremely estrogenic, inducing up to 10^6-fold induction of VTG in caged fish for a 3-week exposure (Purdom et al. 1994). The estrogenic activity at some of these sites was shown to persist for kilometers downstream of the effluent discharge into the river (Harries et al. 1996). Major surveys have since established a widespread occurrence of intersex in roach populations living in U.K. rivers (86% of the 51 study sites; Jobling et al. 2006) and gonadal effects range from single oocytes, or small nests of oogonia, interspersed throughout an otherwise normal testis (Nolan et al. 2001), to the most extreme cases where half the testis comprised ovarian tissue. In some individuals, the sperm duct that enables the sperm to be released is absent and replaced by an ovarian cavity (Jobling et al. 1998; van Aerle et al. 2001; Nolan et al. 2001). Other biological effects recorded in the wild roach and other fish species attributed to WWTW effluent exposure include abnormal concentrations of blood sex steroid hormones, altered spawning times and fecundity in females, as well as reduced testicular development in males (Nolan et al. 2001; Jobling et al. 2002b; Tyler et al. 2005; Jobling et al. 2006). Definitive evidence that effluents from WWTWs induce sexual disruption has been established through a series of controlled exposures, where all of the feminine characters seen in wild roach have been experimentally induced (Rodgers-Gray et al. 2000, 2001; Liney et al. 2005, 2006; Tyler et al. 2005; Gibson et al. 2005; Lange et al. 2008).

In theory, intersex wild roach could arise as a consequence of the exposure of males to estrogens or the exposure of females to androgens, as sex can be altered through either of these exposure scenarios. The evidence supporting the hypothesis that intersex roach in English rivers arise from the feminization of genetic males, however, is substantive and is based on the following facts: (1) the number of roach with normal testes in the wild populations studied is inversely proportional to the number of intersex roach (Jobling et al. 1998, 2006); (2) WWTW effluent discharges into U.K. rivers are estrogenic (Purdom et al. 1994; Harries et al. 1997, 1999; Rodgers-Gray et al. 2000, 2001; Jobling et al. 2003) and/or antiandrogenic, which would further enhance any feminization of males, but rarely androgenic (Johnson et al. 2004); (3) wild male and intersex roach contain VTG in the plasma (Jobling et al. 1998, 2002a, 2002b); and (4) wild intersex roach generally have plasma levels of 11-ketotestosterone, the main male sex hormone in fish, and E_2 levels more similar to those in normal males than in normal females.

Through fractionating WWTW effluents and screening those fractions with cell-based bioassays responsive to estrogens, the natural steroidal estrogens E_2 and estrone (E$_1$; Desbrow et al. 1998; Rodgers-Gray et al. 2000, 2001), together with EE$_2$ (a component of the contraceptive pill) have been identified as the major contributing agents in the feminization of wild roach. These natural and synthetic steroidal estrogens, derived from the human population, are predominantly excreted as inactive glucuronide conjugates (Maggs et al. 1983), but they are biotransformed back into the biologically active parent compounds by bacteria in WWTWs (van den Berg et al. 2003; Panter et al. 2006). Horse estrogens used in hormone replacement therapy (Gibson et al. 2005) and alkylphenolic chemicals, derived from the breakdown of industrial surfactants (see later text), have also been shown to contribute to the estrogenic activity of some WWTW effluents and are biologically active in fish at environmentally relevant concentrations (Gibson et al. 2005). Alkylphenolic chemicals have been shown to be especially prevalent in WWTW receiving significant inputs from the wool scouring industries (Jones and Westmoreland 1998; Sun and Baird 1998).

Laboratory exposures of roach and other fish species to steroidal estrogens and alkylphenolic chemicals have induced VTG synthesis, gonad duct disruption, and oocytes in the testis, albeit for the latter effect, at concentrations generally higher than that found in effluents and receiving rivers (Blackburn and Waldock 1995; Tyler and Routledge 1998; Metcalf et al. 2000; Yokota et al. 2001; van Aerle et al. 2002; Hill and Janz 2003). EE$_2$ is present at considerably lower concentrations in the aquatic environment than for the natural steroidal estrogens, but it is exquisitely potent in fish, inducing VTG induction at only 0.1 ng/L EE$_2$ in rainbow trout (*Oncorhynchus mykiss*) (Purdom et al. 1994) for a 3-week exposure, inducing intersex in zebrafish at 3 ng EE$_2$/L, and causing reproductive failure in zebrafish for a lifelong exposure to 5 ng EE$_2$/L (Nash et al. 2004). In a recent study in which a whole experimental lake was contaminated with 5–6 ng EE$_2$/L, over a 7-year period there was a complete population collapse of fathead minnow (*Pimephales promelas*) fishery (Kidd et al. 2007). Adding further to the hypothesis that steroidal estrogens play a major role in causing intersex in wild roach in U.K. rivers, the incidence and severity of intersex in roach from a study on 45 sites (39 rivers) found they both were significantly correlated with the predicted concentrations of the E$_1$, E$_2$, and EE$_2$ present in the rivers at those sites (Jobling et al. 2006). Adding further complexity to the story of the feminization of roach in U.K. rivers, and as mentioned earlier, U.K. WWTW effluents are also antiandrogenic (Johnson et al. 2004), and this activity is likely to contribute to the feminization phenomenon (Jobling et al., submitted; see Section 15.7).

Importantly, the intersex condition in roach has been shown to affect their ability to produce gametes, which is dependent on the degree of disruption in the reproductive ducts and/or altered germ cell development (Jobling et al. 2002a, 2002b). Small numbers of wild roach occur in affected wild populations that cannot produce any gametes owing to the presence of severely disrupted gonadal ducts. In the majority of intersex roach found, male gametes were produced that, although viable, were of poorer quality than those from normal males obtained from aquatic environments that do not receive WWTW effluent (Jobling et al. 2002b). Fertilization and hatchability studies showed that intersex roach even with a low level of gonadal disruption were compromised in their reproductive capacity and produced less offspring than roach from uncontaminated sites under laboratory conditions (Jobling et al. 2002b).

In that study there was an inverse correlation between reproductive performance and severity of gonadal intersex (Jobling et al. 2002b).

The phenomenon of estrogens in WWTW effluents is not unique to the United Kingdom and occurs more widely in Europe [Germany (Hecker et al. 2002), Sweden (Larsson et al. 1999), Denmark (Bjerregaard et al. 2006), Portugal (Diniz et al. 2005), Switzerland (Vermeirssen et al. 2005), and the Netherlands (Vethaak et al. 2005)] and in the United States (Folmar et al. 1996), Japan (Higashitani et al. 2003), and China (Ma et al. 2005). The level of estrogenic impact seen in fish in U.K. rivers, however, appears to be greater than for elsewhere in Europe, and globally. Why this is the case is not known, but it may relate to the fact that often a considerable proportion of the flow of rivers in the United Kingdom is made up of treated WWTW effluent; 10% WWTW effluent is a common level of contamination, and for some rivers it is more normally 50% of the flow. In extreme cases in the United Kingdom, and generally in the summer months during periods of low rainfall, treated wastewater effluent can make up the entire flow of the river.

15.4.4 ATRAZINE AND ABNORMALITIES IN FROGS

Globally, many amphibian populations are suffering drastic declines (Wake 1991; Houlahan et al. 2000). Causation ascribed to these declines include effects on environmental conditions induced by climate change, introduction of alien predators, overharvesting, habitat destruction, the increase of various diseases, and effects of UV light on embryo development (Alford et al. 2007; Gallant et al. 2007; Skerratt et al. 2007; van Uitregt et al. 2007). In some cases chemical exposures have been implicated, but not proved. As an example, in some parts of the United States, deformities seen in frog populations, in which individuals either lack or have additional limbs, have been associated with exposure to PCBs that can interfere with normal thyroid hormone and/or retinoic acid signaling and function. Controlled laboratory studies have shown that some PCB congeners can induce some of the limb abnormalities seen in the wild (Gutleb et al. 2000; Qin et al. 2005); however, these effects are only induced at exposure concentrations exceeding those normally seen in the wild. In fact, the case for limb deformities in frogs is becoming an increasingly complex story and causative agents ascribed are now wide-ranging and include the chytrid fungus, parasitic trematodes, UV radiation and a wide range of chemical contaminants (Meteyer et al. 2000; Loeffler et al. 2001; Johnson et al. 2002; Ankley et al. 2002; 2004; Davidson et al. 2007).

More controversially, endocrine disruption as a consequence of exposure to the herbicide atrazine (2-chloro-4-ethylamine-6-isopropylamine-s-triazine), one of the most widely used herbicides in the world, has also been hypothesized to explain various adverse biological effects in frog populations in the United States. Exposure to atrazine in the laboratory at high concentrations, far exceeding those found in the natural environment, has been reported to induce external deformities in the anuran species *Rana pipiens, Rana sylvatica,* and *Bufo americanus* (Allran and Karasov 2001). Studies by Hayes et al. have suggested that atrazine can induce hermaphroditism in amphibians at environmentally relevant concentrations (Hayes et al. 2002; Hayes et al. 2003). Laboratory studies with atrazine also indicated the herbicide could affect the development of the larynges in exposed males and thus affect the capability for vocalization in male frogs (Hayes et al. 2002). From their combined field and laboratory studies, these authors developed the hypothesis that exposure to atrazine might account for population level declines in leopard frogs (*Rana pipiens*) (Hayes et al. 2003). In their field studies, it was shown that the maximal seasonal concentration of atrazine coincided with the breeding period of these frogs. However, the laboratory studies on the affects of atrazine in frogs by Hayes and colleagues have not been replicated, and in more recent studies, no effects of atrazine were found on gonad development, growth, or metamorphosis for exposures to ecologically relevant concentrations (Coady et al. 2004; Jooste et al. 2005b). There are also data suggesting that alterations in gonadal development and the proposed population-level impacts in amphibians do not correlate with areas receiving atrazine application (Du Preez et al. 2005). The inability to deduce the mechanism of action of atrazine for the biological effects reported creates a significant level of uncertainty regarding the association between atrazine and disruptions in sexual development in frogs, and is a topic of considerable scientific debate (see Hayes 2005 and Jooste et al. 2005a for more detailed analyses).

15.4.5   EDCs AND HEALTH EFFECTS IN HUMANS

Vertebrates share many functional similarities in their endocrine systems, including their regulatory control and the nature of the hormones and their receptors (Munkittrick et al. 1998). The reproductive abnormalities observed in wildlife populations may therefore potentially be extrapolated to effects in the reproductive health of human populations, if similar exposures to EDCs occur.

A paper from Elizabeth Carlsen and colleagues heightened an awareness of the potential impact of EDCs in humans in 1992, when evidence was provided for a decreasing quality of semen in humans spanning a 50-year period. The interest became more intense when, in the following year, there was speculation that increased estrogen exposure, possibly from environmental estrogens in utero, could be a contributory factor. One of the strongest associations shown in this regard is between reduced sperm counts in men and exposure to estrogenic pesticides (Swan et al. 2003a, 2003b). In addition, and as mentioned earlier, there has also been increasing evidence to support links between increased TDS in humans and exposure to environmental antiandrogens (Fisch and Golden 2003; Sharpe and Irvine 2004; Giwercman et al. 2007). The contributing role of EDCs to these reproductive disorders in the general human population, however, is still largely unknown and is complicated by the social, dietary, and behavioral changes that have occurred over the period during which sperm counts have declined and the incidence of TDS has increased.

15.5   SCREENING AND TESTING FOR EDCs

The findings of harmful effects of chemicals acting via the endocrine system in wildlife populations (and the potential for inducing harm to human health) has highlighted the inadequacies of the screening and testing procedures to protect wildlife (and possibly humans) against the endocrine-disrupting effects of chemicals. As a consequence, in 1996, the U.S. EPA established the Endocrine Disruptor Screening and Testing Advisory Committee (EDSTAC), which recommended the use of a battery of tests applied in a tiered approach to identify chemicals affecting the estrogen, androgen, or thyroid hormone systems. Similar, although less intensive, activities were activated in Europe and in Japan. The list of recommended screens and tests included in silico, in vitro, and in vivo approaches. In silico approaches included the use of quantitative structure–activity relationships (QSAR), in which the specific chemical structures are modeled to establish if chemicals have a high probability of binding to and activating a specific receptor, or posses functional groups that exist in other known EDCs. In an analysis of the QSAR approach as applied to EDCs, it was shown to have a high predictive capability for chemicals interacting with a specific hormone receptor. There were exceptions to this, however, and they include kepone, where the structure of the chemical is not especially similar to endogenous steroid estrogen, yet it binds effectively to the estrogen receptors and triggers an estrogenic response. Furthermore, for chemicals active via pathways other than receptors, for example, via affecting hormone biosynthesis, chemical structure, and thus the QSAR approach, is far less predictive.

The list of in vitro assays for EDCs includes competitive ligand-binding assays, which investigate binding interactions of chemicals with specific hormone receptors, and hormone-dependent cell proliferation or gene expression assays. The cell-based assays include primary cultures, for example, fish hepatocytes that express VTG mRNA/protein when exposed to estrogen, and immortalized cell lines such as human breast cancer MCF-7 cells (Balaguer et al. 2000), yeasts (*Saccharomyces cerevisiae;* Metzger et al. 1995; Routledge and Sumpter 1996) transfected with plasmids carrying the estrogen receptor (E-SCREEN) or androgenic receptor (A-SCREEN) and a reporter gene incorporating a DNA response element responsive to estrogens or androgens, respectively. Other cell systems have been developed that are responsive to chemicals that interact with progesterone receptors (Soto et al. 1995) and responsive to thyroid hormone mimics (see Zoeller et al. 2007 for a critical review on these).

The yeast reporter gene assays not only assess for the interaction of the chemical with the hormone receptor, but also the ability of that receptor–chemical ligand interaction to activate the hormone DNA response element. It should be realized, however, that most of these systems have been developed with human and mammalian hormone receptors and differences in ligand potencies can occur between different animal species. A comprehensive review of in vitro assays for measuring estrogenic activity, and some of the issues of comparability, is provided by Zacharewski (1997).

A major limitation of in vitro screening systems is that endocrine modifications can be complex, and they are not necessarily limited to a specific organ, molecular mechanism, or exposure route. As an example, an estrogenic effect could potentially come about owing to an increase in gonadal estrogen production, a decrease in gonadal androgen production, an increase in the production of gonadotrophin from the anterior pituitary, a decrease in hepatic enzymatic degradation of estrogen, an increase in the concentration of serum sex-hormone-binding proteins limiting free hormone in the serum, a decrease in cytostolic binding proteins that potentially limit free estrogen in the cell and/or agonistic binding of the compound to an estrogen receptor (Guillette et al. 2000). In vivo tests in mammals for quantifying the effects of EDCs include the Hershberger assay, where the principle relies on castration of male rats to remove the source of endogenous androgens; thus, any androgenic response is due to the test chemical, the uterotrophic assay, in which uterus growth is measured as a response to estrogens (Yamasaki et al. 2003; Clode 2006) and various reproductive performance tests. In fish, in vivo tests for EDCs include short-term exposures assays that measure VTG induction and effects on the development of secondary sex features (that are sex hormone dependent), various tests to measure effects on reproductive performance, and full fish life-cycle tests. In amphibians, larval development tests are being devised to test for chemicals with thyroid activity (Gutleb et al. 2007). For invertebrates, in vivo tests for EDCs are generally focused on development and reproductive endpoints and measured over at least one generation (Gourmelon and Ahtiainen 2007). These tests, however, are often not specific for EDCs and this, together with a general lack of knowledge regarding the hormone systems of most invertebrates, has in many cases made interpretations on the mechanism of the biological effects difficult.

Advantages of in vivo test systems for EDCs are that they allow for metabolism and bioconcentration of the compound of interest. The importance of this is illustrated by the fact that we now know that there are a variety of EDCs for which it is their products of metabolism rather than the parent compound that are endocrine-active (e.g., for the products of metabolism of the pesticides DDT, vinclozolin, and methoxychlor), and many are lipophilic and bioconcentrate/bioaccumulate. The disadvantages of in vivo approaches compared with in silico and in vitro approaches are associated with their inherent higher costs and the desire to reduce the number of animals used in chemical testing. Furthermore, endpoints measured in some of the in vivo tests for EDCs are not necessarily specific for a single mode of action (e.g., for reproduction). Thus, ideally included in in vivo tests when assessing for effects of EDCs on growth, development, and reproduction are biomarkers that inform on the mode of action. For more detailed assessments on the various assays for testing and screening EDCs, we would refer the reader to the following articles: Zacharewski (1997), Gray (1998), O'Connor et al. (2002), and Clode (2006).

What is important to emphasize is that, given the range of known EDCs, their potential to act with more than one mechanism of action, and ability of some chemicals to mediate effects via multiple tissues, no single effective test exists for an "endocrine disrupter"; rather, a suite of approaches is required to capture the spectrum of possible effects.

15.6   A LENGTHENING LIST OF EDCs

The list of chemicals with endocrine-disrupting activity has increased considerably with the systematic screening of chemicals employing some of the methods described in the previous section. Here we expand on the list of known EDCs to illustrate the diversity of chemicals of concern, but the list is by no means exhaustive.

15.6.1 Natural and Pharmaceutical Estrogens

As illustrated in the case for feminization of fish in U.K. rivers, both natural and synthetic steroidal estrogens present in the environment are impacting wildlife adversely. The source of natural estrogens in surface waters is predominantly via the human population and via discharges through WWTW. Other sources, however, include both diffuse and point sources from livestock practices, including from poultry and cattle farms (Shore et al. 1998; Lange et al. 2002). Concentrations of individual steroidal estrogens at some point source discharges are sufficient to inhibit sexual development and function. For a review on steroidal estrogens and their effects in fish, we would refer the reader to Tyler and Routledge (1998).

15.6.2 Pesticides

DDT and its metabolites have been shown to have various endocrine-disrupting effects, including acting as an estrogen (*o,p'-DDT* and *p,p'*-DDT) and an antiandrogen (*p,p'*-DDE). Concentrations in most environments, however, are generally now at, or below, the no-observed-effect levels. Nevertheless, in developing countries where DDT-based pesticide use still occurs, concentrations in water sources have been recorded as high as 10 µg/L, and at these levels they would induce endocrine disturbances in exposed animals (Begum et al. 1992). Many other pesticides have now been reported with endocrine activity (Short and Colborn 1999, reviewed in Bretveld et al. 2006), and they include other organochlorine pesticides, such as methoxychlor (its hydroxy metabolites; Thorpe et al. 2001), and lindane and kepone, which are structurally similar to DDT (Eroschenko 1981). These chemicals can still be found in surface waters at biologically effective concentrations, and can induce gonadal developmental aberrations, VTG induction, behavioral changes, such as exploring and learning responses, and disrupt ionic regulation in fish (Davy et al. 1973; Weisbart and Feiner 1974; McNicholl and Mackay 1975; Begum et al. 1992; Donohoe and Curtis 1996; Metcalfe et al. 2000). The photosynthesis-inhibiting herbicides linuron and diuron, and metabolites of the fungicide vinclozolin are also all endocrine-active, acting as antiandrogens (Kelce et al. 1994; Thorpe et al. 2001). Other pesticides reported to have endocrine-disrupting activity, with (anti)estrogenic and/or (anti)androgenic activity, include some of the pyrethroids, including permethrin, fenvalerate, and cypermethrin, albeit weakly (Tyler et al. 2000; McCarthy et al. 2006; Jaensson et al. 2007; Sun et al. 2007).

15.6.3 PCBs

The estrogenic activity of PCBs was brought to light in 1970, and it was quickly established that some are also toxic and bioaccumulative to wildlife (Hansen et al. 1971; Hansen et al. 1974) and could impair the reproductive performance capabilities in a variety of animals (Platonow and Reinhart 1973; Nebeker et al. 1974). The estrogenic nature of some PCBs has been shown to reverse gonadal sex in turtles (Crews et al. 1995), and affect uterine development in rats (Gellert 1978). Some of the 209 PCB congeners also have thyroid-disrupting activity (Darnerud et al. 1996), and are especially bioaccumulative (Guiney et al. 1979; Tyler et al. 1998). PCBs were used widely in industry as coolants and insulating fluids for transformers and capacitors, but production was greatly reduced in the 1970s, partly because they were found to be highly mobile via atmospheric transport and were detected in remote (e.g., arctic) wildlife populations (Muir et al. 1992). PCB residues in the environment have declined since the 1980s (Fensterheim 1993), and rarely do PCB concentrations in the aquatic environment now exceed 1 µg/L. Nevertheless, it is estimated that 70% of the world's production of PCBs is still in use or in stock, and therefore has the potential to enter the environment. Chapter 6 provides more detailed information on the chemistry and ecotoxicology of PCBs.

15.6.4 Dioxins

Dioxins are important environmental pollutants. Of the 75 congeners, 2,3,7,8-tetra-chloro-dibenzo-p-dioxin (TCDD) is considered the most reproductively toxic. Uses included as a herbicide, and exposure of organisms to extremely low concentrations of TCDD (0.1–1 µg/kg/d) can lead to alterations in the reproductive systems of the subsequent offspring that are consistent with demasculinization (Gray et al. 1995). Gray and colleagues (1995) showed that a dose of 1 µg TCDD/kg on a single day during gestation delayed puberty and testicular descent in rats and hamsters and caused a 58% reduction in the ejaculated sperm count, all consistent with demasculinizing modes of action. Polychlorinated dibenzodioxins (PCDDs) are lipophillic and in the aquatic environment are generally found at low concentrations, although some ecosystems are highly contaminated, for example, paper mill effluents, where concentrations may rise to 40 µg/L (Merriman et al. 1991). In nonaquatic environments, PCDDs reach their highest concentrations in landfill sites (U.S. EPA 2006), and leaching from these landfill sites will inevitably have adverse effects for the local wildlife. Chapter 7 deals with the general chemistry and wider ecotoxicology of PCDDs.

15.6.5 Polybrominated Diphenyl Ethers

Polybrominated diphenyl ethers (PBDEs) appear to disrupt thyroid function (Carlsson et al. 2007) via their interactions with thyroid hormone receptors (Marsh et al. 1998). In rats, exposure to low doses of 2,2′,4,4′,5-pentabromodiphenyl ether (PBDE-99) has been shown to reduce the concentration of circulating thyroid hormones (Kuriyama et al. 2007). Some PBDE congeners have also been tested for carcinogenicity and shown significant dose-related increases in liver tumors (Hooper and McDonald 2000). PBDEs are similar in structure to PCBs. They are used extensively as flame retardants in a wide range of products, including electrical equipment, textiles, plastics, and building materials, and they leach into the environment from these products (de Wit 2002; McDonald 2002). First detected in the environment in 1979, PBDEs have now been found very widely in the environment (Allchin et al. 1999), and levels are generally higher in aquatic species than in terrestrial species (Sellström et al. 1993; Pijnenburg et al. 1995). PBDEs are highly resistant to degradation and bioaccumulate in animal tissue (Allchin et al. 1999; Gustafsson et al. 1999). Concentrations of PBDEs in invertebrates derived from industrial areas have been recorded up to 480 ng/g lipid (Yunker and Cretney 1996), and in fish at concentrations of up to 27,000 ng/g lipid in muscle tissue and 110,000 ng/g lipid in liver (Andersson and Blomkvisit 1981). Tetrabromodiphenyl (TeBDE) is consistently the predominant congener in body tissues (Jansson et al. 1993; Sellström et al. 1993; Luross et al. 2002). As a measure of their ubiquity in the environment, 90% of freshwater fish in the study by Hale et al. (2001) had detectable levels of TeBDE exceeding those of PCB-153, typically the most abundant PCB congener.

Public concern about PBDE levels in the environment was heightened when it was shown that a sharp increase in the concentration of certain PBDEs had occurred in human breast milk over only a 10-year period (Meironyté et al. 1999; Norén and Meironyté 2000), and the levels of exposure in some infants and toddlers were similar to those shown to cause developmental neurotoxicity in animal experiments (Costa and Giordano 2007). As a result of these concerns, the majority of commercial PBDE mixtures have been banned from manufacture, sale, and use within the European Union.

15.6.6 BISPHENOLS

Bisphenol A (BPA) was first discovered as an estrogen back in the 1930s, when it was developed as an estrogen for clinical use (Dodds and Lawson 1938). The discovery of DES, however, a far more effective synthetic estrogen, meant that the use of bisphenol A as a clinical estrogen was quickly superseded. Then, in the 1950s, BPA was reacted with phosgene by a Bayer chemist, Hermann Schnell, to produce polycarbonate plastic and, subsequently, to synthesize epoxy resins, which are now used widely, including as lacquer preservatives in the lining of food cans, in automotive parts, and in compact discs (reviewed in Oehlmann et al. 2008). The estrogenic activity of bisphenol A was "rediscovered" in 1993 when it leached out of polycarbonate flasks during autoclaving and subsequently had a stimulatory effect in an estrogen-dependent cell culture system (Krishnan et al. 1993). Through in vitro screenings, a wide range of other bisphenols have been shown to be estrogenic (Brotons et al. 1995; Fernandez et al. 2001). Bisphenols are only weakly active as estrogens in in vitro assays, with a potency of approximately 1:5000 compared to that of E_2 (Krishnan et al. 1993). In vivo effects in rats occur at doses of tens of μg BPA/d. In the United States, concentrations of BPA in surface waters have been recorded up to 8 μg/L (Staples 1998). In vivo studies in fish have shown that concentrations of 16 μg/L in the water can affect the progression of spermatogenesis, and inhibition of gonadal growth occurred in both males and females at concentrations of 640 and 1280 μg/L (Sohoni et al. 2001). BPA has also been shown to invoke an antiandrogenic response in the A-SCREEN (Sohoni and Sumpter 1998).

15.6.7 ALKYLPHENOLS

Alkylphenol ethoxylates (APEs) are nonionic surfactants that are used in the manufacturing of plastics, agricultural chemicals, cosmetics, herbicides, and industrial detergent formulations. Alkylphenols such as nonylphenol (NP) are the products of microbial breakdown of APEs and have been known to be estrogenic since 1938 (Dodds and Lawson 1938). As with bisphenol A, NP was rediscovered more recently as an estrogenic xenobiotic by Soto et al. (1991) in an estrogen-dependent cell prolif- eration assay. Alkylphenols are relatively weak estrogens, with an affinity for the ER 2,000–100,000-fold less than E_2 (reviewed in Nimrod and Benson 1996). However, alkyphenols can also interact with the androgen receptor to induce antiandrogenic effects (Gray et al. 1996). NP induces VTG synthesis in fish at concentrations as low as 6.1 µg/L for a 14-day exposure and 650 ng/L over for a 3-week exposure (Harries et al. 2000; Thorpe et al. 2001). NP has also been shown to affect pituitary function and the release of gonadotrophins (which control the whole reproductive cascade in vertebrates) in fish at concentrations of only 0.7 µg/L for an 18-week exposure (Harris et al. 2001). As for many other EDCs, longevity of exposure affects both the threshold and magnitude of the response; alkylphenols such as NP have been shown to bioconcentrate in fish up to 34,000-fold (Smith and Hill 2004). In the aquatic environment, rivers and the sea receive substantial amounts of APEs from WWTWs and industrial effluent discharges. It is estimated that 60% of the world's production ends up in the aquatic environment (Uguz et al. 2003). Domestic effluent can contain up to hundreds of µg APEs/L (Naylor 1995), whereas industrial effluent, especially that from pulp and textile industries, can contain mg/L concentrations. In some riv- ers in the United Kingdom that have historically received high-level discharges from the textile industry, alkylphenolic chemicals were shown to be some of the major contaminants inducing feminized responses in exposed fish (Harries et al. 1996; Sheahan et al. 2002).

15.6.8 PHTHALATES

Phthalates are the most abundant synthetic chemicals in the environment (Peakall 1974). Used in lubricating oils, insect repellents, and cosmetics and predominantly to impart flexibility to plastics, phthalates have been measured in rivers (Sheldon and Hites 1978; Fatoki and Vernon 1990), drinking waters (Suffet et al. 1980), and marine environments (Jobling et al. 2002b). Many thousands of tons of plastics are also disposed of annually in landfill sites, resulting in phthalate esters leach- ing into and contaminating groundwaters. The estrogenic activity of two phthalate esters, di-n-butylphthlate (DBP) and butylbenzyl phthalate (BBP), was discovered by Jobling et al. (1995). Further studies in fish have shown that BBP and diethyl phthalate (DEP) both induce VTG at an exposure concentration around 100 µg/L via the water (Harries et al. 2000; Barse et al. 2007). Various in vitro screens and tests have shown that phthalates mediate their effects via binding to the estrogen receptor (Jobling et al. 1995; Harries et al. 1997). However, antiestrogenic effects of phtha- lates also occur in vivo (inhibition of VTG expression). In addition to these estrogenic effects, some phthalates are also known to be toxic to aquatic organisms (Mayer and Sanders 1973, in Giam et al. 1978) and mammals (Lee et al. 2007). Recent studies have shown that DBP, monoethyl phthalate (MEP, a metabolite of DEP), and mono- (2-ethylhexyl) phthalate (MEHP) can induce DNA damage in human sperm and/ or male rats (Wellejus et al. 2002; Hauser et al. 2007). Environmental concentra- tions of dimethyl phthalate (DMP), DEP, DBP are reported to be between 0.3 to

30 μg/L (Sheldon and Hites 1978; Fatoki and Ogunfowokan 1993) in the Western world, which is below the no-effect level for wildlife species studied, but concentrations in developing countries are often significantly higher. In Nigeria, for example, DBP concentrations have been reported to be as high as to 1472 mg/L in river water (Fatoki and Ogunfowokan 1993).

15.6.9 NATURAL EDCs

All the foregoing examples, with the exception of some of the steroidal estrogens, are synthetic chemicals. Naturally occurring estrogens produced by fungi (myco-estrogens) and plants (phytoestrogens) can also have endocrine-disrupting effects. We described one example of this earlier, where exposure to high level of phytoestrogens affects the reproductive biology of sheep (Adams 1998). Phyto- and mycoestrogens enter the aquatic environment, and although detailed studies are lacking, concentrations recorded range between 4 to 157 ng/L (Stumpf et al. 1996; Erbs et al. 2007). Their effects in wildlife are largely unstudied when compared to that for other EDCs, but controlled exposures to phytoestrogens have been shown to induce estrogenic responses, including induction of VTG synthesis in fish (Pelissero et al. 1991; Pelissero et al. 1993) and infertility and liver disease in captive cheetahs (Setchell et al. 1987).

Although the likelihood for biologically harm has not been assessed fully, for most EDCs the exposure concentrations in ambient environments (away from hotspots of chemical discharges) would suggest that they are insufficient to do so. Exceptions to this include the case studies detailed in the previous section. It should, however, also be emphasized that most studies on the effects of EDCs under controlled laboratory conditions have not considered long-term chronic exposures encompassing full life cycles, and some wildlife species are exposed lifelong to some of the EDCs described earlier.

15.7 EFFECTS OF MIXTURES

Wildlife, especially organisms living in and/or closely associated with the aquatic environment, are often exposed to highly complex mixtures of EDCs, and this complicates identification of the causality of the physiological disruptions seen. As an example, although steroid estrogens and alkylphenolic chemicals have been identified as major contributors to the feminization of wild fish, many other estrogenic (and antiandrogenic) chemicals occur in WWTW effluents, including plasticizers such as phthalates, bisphenols, and various pesticides and herbicides, which could potentially contribute to the feminized effects. Individually, these chemicals are unlikely to play a significant role in the disruption of sex in wild fish, given their relatively lower estrogenic potency compared with steroidal estrogens, but as part of a mixture they may contribute to a more significant effect. Exposure studies that replicate environmentally relevant mixtures of EDCs are lacking, but simple environmental mixtures of akylphenolic chemicals, pesticides, and plasticizers have been shown to be additive in their estrogenic effects, both in vitro (Silva et al. 2002) and in vivo in fish (Thorpe et al. 2001, 2003, 2006; Brian et al. 2005). Indeed, it has also been shown that concentrations of individual EDCs insufficient to induce a biological response on their own can do so when added together as part of a mixture (Silva et al. 2002). These studies have used simple endpoints, such as VTG induction, but similar deleterious interactive (additive) effects of EDCs have also been shown for fitness and fecundity endpoints in fathead minnows exposed via the water (Brian et al. 2007). A further complication in the analysis of EDC mixtures is that the biological responses induced by a single EDC can be modified when part of an environmental mixture. As an example, Filby et al. (2007a) recently found that EE_2 impacted health-related endpoints differently in fish (fathead minnow) when exposed as part of an environmentally relevant mixture compared to exposure to the EE_2 alone.

Toxicogenomics has the potential to advance our understanding of the mixture effects of EDCs, as it does for identifying modes of action of individual EDCs. A prerequisite for this, however, is experimental data describing precisely "omics" responses to single EDCs and for simple mixtures. These studies are needed to provide robust reference data sets before graduating to studies of more complex and environmentally relevant mixtures. It is also the case that effects resulting from mixture exposure may result in the activation of pathways different from that observed for the individual compounds (Tyler et al. 2008) and producing a transcriptomic signature very different for the mixture compared to that for the individual chemicals (Finne et al. 2007).

The sheer complexity of environmental mixtures of EDCs, possible interactive effects, and capacity of some EDCs to bioaccumulate (e.g., in fish, steroidal estrogens and alkylphenolic chemicals have been shown to be concentrated up to 40,000-fold in the bile [Larsson et al. 1999; Gibson et al. 2005]) raises questions about the adequacy of the risk assessment process and safety margins established for EDCs. There is little question that considerable further work is needed to generate a realistic picture of the mixture effects and exposure threats of EDCs to wildlife populations than has been derived from studies on individual EDCs. Further discussion of the toxicity of mixtures will be found in Chapter 2, Section 2.6.

15.8   WINDOWS OF LIFE WITH ENHANCED SENSITIVITY

Some life stages of organisms are more susceptible to the effects of EDCs than others. Sensitive periods include embryogenesis, when up to 90% of the genome is transcribed, and early life, when fundamental features of developmental programming, including sex assignment and various behaviors, are defined. In humans the adverse effects on reproductive development induced by DES resulted from exposures in utero (Herbst et al. 1971). Many of the reproductive abnormalities comprising TDS in humans too are believed to manifest during embryogenesis and early fetal development (Fisch and Golden 2003; Sharpe and Irvine 2004; Giwercman et al. 2007). Studies in rats have also shown that extremely small differences in the concentrations of endogenous sex hormones surrounding developing embryos can have profound effects on subsequent sex-related behaviors, further illustrating the susceptibility of early life stages to EDCs that affect sex steroid hormone concentrations at this time. There is even evidence that some EDCs can induce epigenic effects, through altering DNA methylation (Anway et al. 2005).

In wildlife organisms, too, early life appears to be one of the more sensitive periods for EDC effects. The developmental abnormalities seen in fish-eating birds living in and around the Great Lakes, for example, resulted from exposure of the embryos to thyroid-disrupting chemicals deposited into the eggs during oogenesis (Spear et al. 1990 and Murk et al. 1996, in Rolland 2000). Similarly, the effects on limb formation in birds and mammals and impaired metamorphosis in amphibians reported earlier (Rolland 2000) all resulted from exposure to thyroid-disrupting chemicals during early life. Alteration to the development of the gonad and voice box in frogs exposed to atrazine, and steroidal estrogens, also resulted from exposures of embryos and early life stages (Hayes et al. 2002). In the alligator case study, persistent differences in development, posthatch survivorship, and gene expression in animals from a contaminated environment are hypothesized to be embryonic in their initiation (Milnes et al. 2008)

The heightened sensitivity of early life stages in fish to sex steroid hormones is well established and has been exploited in the aquaculture industry in the production of monosex populations (Piferrer 2001). In many fish, even gonochorists (single-sexed fish), complete sex reversal can be induced by hormonal treatments during early life (Pandian and Sheela 1995). This greater plasticity of the sexual phenotype in fish makes them especially susceptible to the effects of environmental estrogens and other sex hormone mimics (Devlin and Nagahama 2002). Fish are responsive to estrogens as embryos, well before the sex differentiation process, and estrogen receptors are expressed in zebrafish from 1 day postfertilization (dpf) (Legler et al. 2000). Controlled laboratory studies exposing fish early life stages to a wide range of estrogenic EDCs, including at environmentally relevant concentrations, have disrupted gonadal development (e.g., alkylphenols; Gray and Metcalfe 1997, 1999, in van Aerle et al. 2002), PCBs (Matta et al. 1997; Billsson et al. 1998) pesticides (Nimrod and Benson 1998), and steroidal estrogens (van Aerle et al. 2002; Nash et al. 2004). Early-life-stage fish are exquisitely sensitive to EE_2. As an example, exposure of zebrafish to only 1 ng EE_2/L in the water between 20 and 60 days posthatch (dph) was shown to disrupt sexual development (Orn et al. 2003), and an exposure concentration of 2 ng EE_2/L resulted in an all-female population. These concentrations are environmentally relevant, emphasizing the concern for exposure of early-life-stage fish to this EDC.

The exposure effects of EDCs induced during early life are especially noteworthy as they can be long lasting and, in some cases, irreversible. As an example, exposure of male fish during early life to environmental estrogens can induce an ovarian cavity in the testis that remains with them throughout their lives (Rodgers-Gray et al. 2001). Deep-penetrating effects of exposure to estrogen during early life are not restricted to effects on male fish. Rodgers-Gray et al. (2001), for example, showed that exposure of roach to an estrogenic WWTW effluent for between 50 and 100 days posthatch subsequently affected the timing of sexual differentiation and accelerated the development of female sex characteristics. In laboratory studies with roach, we have recently found that short-term exposure to exogenous steroidal estrogen during early life can sensitize the female to estrogen effects in later life (more than one year after the original exposure [Lange et al., submitted]).

Other life stages potentially susceptible to the effects of exposure to EDCs include puberty, for chemicals that interfere with the sex steroid hormone pathways; and final maturation, for chemicals that have progestagen activity. Altered timing of puberty and/or timing of gamete production could affect seasonal breeding animals; timing of reproduction is critical for ensuring maximal survival of their offspring (i.e., where there is maximal food availability). Another life stage potentially susceptible to the effects of EDCs is the smolting event in some salmonid fish when they transfer from freshwater to saltwater. This process is associated with changes in a suite of hormones, including corticosteroids and prolactin, and it has been shown to be disrupted on exposure to both alkylphenols and E_2 (McCormick et al. 2005; Bangsgaard et al. 2006). None of these potentially sensitive biological processes and life stages have been well studied.

In summary, this section highlights that timing of exposure to EDCs can be critical in the nature and magnitude of the effects produced.

15.9   SPECIES SUSCEPTIBILITY

Most vertebrates are responsive to steroidal hormones and their mimics, and the hormonal systems signaling these responses, and their controlling factors, are highly conserved (Sumida et al. 2003). It is perhaps not surprising, therefore, that studies in vitro have shown that an EDC interacting with a specific receptor in one species, will also do so with another, and sometimes with very similar affinity, even across divergent organisms. As an example, White et al. (1994) demonstrated that a number of alkylphenolic compounds were estrogenic to bird, fish, and mammalian cell lines at equivalent potencies. From such studies, it would be easy to assume that an EDC effective in one vertebrate organism will be equally effective in another. It is the case that rodent models are thought of as being sufficiently similar to humans to make them suitable for informing on the effects of EDCs for the protection of human health.

Care should be taken in making general statements on EDCs effects, however, as despite the many similarities in hormones and their receptors in vertebrates, there are some clear distinctions too. For example, sexual development is estrogen dependent in birds but not mammals (Tyler et al. 1998), and thus, bird reproductive development may be more sensitive to estrogen mimics compared to that for mammals (Ottinger et al. 2002). In mammals, too, the developing fetus is protected from abnormal hormonal exposure by high maternal levels of α-fetoprotein and sex-steroid-binding globulin (Vannier and Raynaud 1975, in Westerlund et al. 2000), but for oviparous organisms there is no equivalent system; eggs deposited into the aquatic environment can be exposed constantly to EDCs and other chemicals, although the chorion offers some degree of protection against chemical uptake (Finn 2007).

Examples of differences in the responses of wildlife organisms to EDCs include the differences in sensitivity to phthalates and bisphenols among mollusks, crustaceans, and amphibians compared to fish. In invertebrates, biological effects are observed at exposures in the ng/L to low μg/L range, compared to high μg/L for most effects in fish (reviewed in Oehlmann et al. 2008). In addition, aquatic mollusks tend to bioconcentrate and bioaccumulate pollutants to a greater level than fish, possibly owing to poorer capabilities for metabolic detoxification (see Chapter 4, Section 4.3).

Even among more closely related species, differences in responses to EDCs are sometimes apparent, possibly owing to differences in metabolic capability. As an example, rainbow trout show a greater sensitivity to steroid estrogens compared to roach and carp, *Cyprinus carpio* (Desbrow et al. 1998; Jobling et al. 2003). As a further example in fish, studies of two species of sucker exposed to effluent from kraft mills gave contrasting effects on the secondary sex characteristic (tubercle formation) (Kloeppersams et al. 1994), but the mechanistic basis for these differences was not established. It is worth emphasizing that comparing responses and relative sensitivities of different species to EDCs in field studies is notoriously difficult, however, and it is critical that seasonal and temperature-dependent fluctuations be taken into account, as well as the stage in the reproductive cycle. Such factors can alter the number and affinity of receptors and, as such, alter sensitivity of the organism studied (Campbell et al. 1994).

The effects of some EDCs appear to be specific to a particular organism or group of organisms. As an example, TBT is associated with imposex in over 150 species of prosobranchs, yet none or very little data exists on TBT inducing imposex on any other groups of invertebrates (Sumpter and Johnson 2005). Furthermore, studies in the laboratory have not provided conclusive evidence for an effect of TBT on sexual development in vertebrates. This might suggest that TBT is more likely to be mediating its effect through the RXR receptor in mollusks rather than via signaling pathways common to both vertebrates and invertebrates (e.g., via effects on aromatase activity; see earlier discussion on TBT).

Thus, both the nature and the severity of the biological effects of EDC can differ greatly among species, making it difficult to make any generalizations and highlighting the need for a battery of tests and range of different organisms in the hazard identification process. Without this approach, unforeseen populational consequences can result. A classic example of this is the catastrophic decline in the populations of three vulture species in South Asia, where accidental exposure to the veterinary anti-inflammatory drug diclofenac through their scavenging on livestock carcasses caused death from kidney failure and severe visceral gout within days of exposure (Oaks et al. 2004; Taggart et al. 2007, see also Chapter 17, Section 17.1), effects that were not seen in mammals exposed in the laboratory.

Factors other than differences in animal physiology that can affect species susceptibility to EDCs, include differences in their habitat, and ecological niche. Considering the aquatic environment, organisms living in or closely associated with the sediments are more likely to be exposed to high concentrations of EDCs compared to pelagic animals, as most EDCs are hydrophobic in nature and build up in the sediments. Animals at higher trophic levels are also likely to be at greater risk from EDCs compared to animals at lower levels, as EDCs can bioaccumulate to very high levels. This is not always the case, however, and in studies on pike (*Esox lucius*) derived from the same rivers in the United Kingdom where there was a high incidence of sexual disruption in roach and on which pike prey, only a very low incidence of sexual disruption was observed (Vine et al. 2005). The reason for this lower-level effect in pike was not established, but it again serves to further illustrate that care should be taken in generalizing on species sensitivity to the effects of EDCs in wildlife populations.

15.10   EFFECTS OF EDCs ON BEHAVIOR

Most studies on EDCs and their effects have focused on disruptions to growth, development, and reproduction. Increasingly, however, it is being realized that EDCs can have significant, and sometimes profound, effects on behavior, and some of these effects on individuals could have population-level consequences. EDCs have been reported to affect a wide range of behaviors, including those associated with sex, dominance, and aggression, and in rodent/human models, with motivation, general communication, and with learning abilities (Sideroff and Santolucito 1972). For the most part, the effects reported for EDCs on behaviors are suppressive. An exception to this is for exposure to the environmental estrogen bisphenol A in rats, where an increased boldness has been reported (Farabollini et al. 1999). For a comprehensive overview on the effects of EDCs on behavior, we would refer the reader to the review by Zala and Penn (2004). Here we present a few key examples of EDCs on behavior and discuss the possible implications for wildlife populations.

The knowledge that exposure to chemicals that mimic hormones, such as DDT and other pesticides, affected behavior in birds was reported over 50 years ago when abnormal changes in nesting and courtship behaviors were recorded in populations of bald eagles in Florida. White and coworkers (1983) also found altered breeding behaviors in a population of laughing gulls exposed to PCBs and organochlorine pesticides. The consequences of both these perturbations was significant, with decreases in the size of the populations. Further behavioral changes recorded in birds exposed to EDCs and other chemicals include a decreased sexual arousal, altered coordination, memory effects, alterations in response to maternal calls and fright stimuli and effects in males on courtship behavior, including singing and displaying (reviewed in Zala and Penn 2004). Recently, it was reported that exposure of the American robin (*Turdus migratorius*) to DDT resulted in a reduction in the size of song nuclei within the brain and a drastic reduction in the *nucleus intercollicularis*, a structure that is critical for normal sexual behavior (Iwaniuk et al. 2006). In contrast, in another study it was shown that the feeding of male starlings (*Sturnus vulgaris*) with invertebrates dosed with natural and synthetic estrogen mimics resulted in an increase in song length and complexity (Markman et al. 2008; although these males also showed reduced immune function). In frogs, too, exposure to EDCs, including steroidal estrogens and atrazine, has been shown to affect the vocalization capabilities of males through disruptions in the development of their larynges, as detailed earlier. The implications of these effects of EDCs have not been addressed for either birds or amphibians, but they could be very significant given the importance of song and vocalization on the breeding dynamics of these animals.

The effects of EDCs on behavior in fish have been more extensively studied than in birds. Examples of the effects of EDCs seen in fish include "profound alterations" in courtship behavior in male guppies (*Poecilia reticulate*) exposed to vinclozolin and DDE, including at environmentally relevant concentrations (Baatrup and Junge 2001) and altered courtship behavior in three-spined stickleback exposed to environmentally relevant concentrations of EE_2 (Bell 2001). In the stickleback studies, exposed males became less aggressive and had a reduced nesting activity, and this was linked with reduced concentrations of the male sex androgen 11-ketotestosterone. Recently, we found that fenitrothion (FN), a widely used organophosphorous pesticide, with structural similarities with the model antiandrogen flutamide, impaired the breeding biology of sticklebacks, including effects on breeding behavior in males, similar to that seen for exposures to EE_2 (Sebire et al. 2008). Environmentally relevant concentrations of FN severely impacted the male, impairing nest-building activity and reducing both the intensity of the zig-zag dance and aggressiveness toward the females in the courtship activity (Sebire et al. 2008). FN is an acetylcholinesterase (AChE) inhibitor (Fleming and Grue 1981; Morgan et al. 1990) and has this activity in fish (Sancho et al. 1997; Morgan et al. 1990), and in theory therefore, the effects reported could result from a general disruption of neural function. The behavioral effects, however, were specific to reproductive behaviors, and feeding, swimming, etc., were not affected. Furthermore, there were concentration-related inhibitions of FN on spiggin (a glue produced in the kidney and used to make the nest, and under androgenic control), supporting the hypothesis that effects of FN on male reproductive behavior were due to an antiandrogenic mode of action.

Changes in reproductive behaviors have the potential to impact genetics of populations more widely. As an example, recently we found that exposure to EE_2 disrupted normal reproductive hierarchies in group-spawning zebrafish, altering the normal pattern of genetic diversity among offspring, even in the absence of impacts on survival or egg production. Using microsatellite-based paternity analyses, we found that a short-term exposure (17 days) to 10 ng EE_2/L resulted in a suppression of the reproductive dominance normally enjoyed by behaviorally dominant male fish in the breeding colony. This reduction in male dominance was again linked to a reduction in plasma-concentrations of the hormone 11-ketotestosterone (Coe et al., in press). Why there should be a differential effect of the EDC on the dominant versus subordinate males allowing for the increased paternity in otherwise subordinate males is not known, but it might relate to differential impacts of EE_2 on the reproductive physiology of dominants and subordinates, or indirectly via condition-dependent mate selection by females, or both. Whatever the mechanism, the outcome is important to the population genetics of group-spawning fish exposed to EDCs. The suppression of reproductive success in dominant males may act to increase and maintain genetic diversity, by reducing the normal skew in reproductive success between dominants and subordinates. As natural populations may be exposed for long periods of time, including lifelong exposure, these effects are likely to be even more pronounced in the wild.

The foregoing example is one example of how altered behavior as a consequence of exposure to an EDC could impact the population breeding dynamics. Another example is where there are in fact no changes in the reproductive behavior in animals that are sexually deficient physically, due to EDC exposure (e.g., males are unable to produce viable sperm). In such a scenario in group-breeding fish, the sexually compromised individuals could affect the ability of "healthy" males to breed successfully, by preventing them gaining access to females during the spawning act, or at the very least reducing their fertilization success rates. Indeed, this has been shown for sexually compromised fish in colonies of breeding zebrafish (Nash et al. 2004).

Here we illustrate that the population-level impacts of exposure to EDCs cannot easily be extrapolated from bioassays on individuals. Even when total fecundity is not affected, changes in dominance hierarchies could threaten the biological integrity of normal populations and disrupt patterns of genetic diversity. Considering these potential effects on reproductive behaviors, the potential hazard of EDCs in the environment is not captured fully through present test guidelines. Although behaviors, and alterations thereof, can provide an integrative measure for assessments on organism health (Clotfelter et al. 2004), their effective use for quantifying EDCs effects requires detailed knowledge of the behavioral traits of interest and, ideally, a mechanism to link these effects with the chemicals modes of action. There are very few examples where this level of detail is available for wildlife species. An exception might be the breeding behaviors associated with reproduction in sticklebacks, which have been studied for over 60 years (since Tinbergen and Vaniersel 1948).

15.11   LESSONS LEARNED FROM ENDOCRINE DISRUPTION AND THEIR WIDER SIGNIFICANCE IN ECOTOXICOLOGY

Rachel Carson in her book *Silent Spring* raised public consciousness on the wider implications to wildlife populations of exposure to synthetic chemicals, most notably for pesticides. The issue of endocrine disruption has served to heighten this awareness and has illustrated that even without overt toxicity, some chemicals can nevertheless induce harm to individuals, or indeed whole populations, through more subtle mechanisms that modify hormone systems. Endocrine disruption research has identified some weaknesses in the existing environmental risk assessments of chemicals. An illustration of this is the case of imposex in mollusks induced by TBT, whose mode of action was not predicted using the traditional risk assessment process; however, its use in the environment subsequently has had population-level consequences. Traditional markers of toxicology, including overt toxicity, cellular damage, DNA damage, aberrant features of development, etc., principally derived from acute studies, do not necessarily capture the potential for significant harm of a chemical. This shortfall adds weight to the need to develop and apply a wider battery of biomarkers to inform on biochemical effect pathways. These biomarkers in turn need to be linked with health impacts on individuals and population-level effect measures. In this context, there is a strong argument for the need to embrace modern molecular approaches, including genome-wide approaches to enhance our capability to screen and test for chemical effect mechanisms much more widely.

Endocrine disruption has emphasized that in making our assessments on the potential hazards of EDCs, we need to be especially mindful of potentially sensitive life stages, latency of effects, and differences in species sensitivity that relate to both differences in their biology and ecological niche, and these will apply equally to many other chemicals. It is important to emphasize also that different conclusions on chemical potency can potentially be reached depending on whether the assessments have been made in vitro versus in vivo. As an example, for studies on environmental estrogens, interactions with estrogen receptors in vitro have shown strong similarities between divergent species, but in vivo studies have shown that responses can differ considerably (in sensitivity) even in closely related species. There are at least three estrogen receptor subtypes in many vertebrates that differ in their tissue distributions and strength of interactions with different environmental estrogens. This illustrates the need for care when extrapolating for effects of chemicals between in vitro and in vivo studies and across species. The dangers in cross-species extrapolations for the effects of chemicals is especially well illustrated in the case of diclofenac, where adverse effects are bringing about potential extinction of wild vulture populations in Asia owing to their high susceptibility to the drug, which was not predicted from laboratory studies on mammals.

Research into the biological effects of EDCs has clearly shown they can have interactive effects, and this is a priority for further work if we are to develop an accurate picture of their potential effects on wildlife populations. Studies on EDCs have also served to emphasize that chemicals that affect behaviors can potentially impact populations through altering breeding patterns and changing the normal genetic structure of populations. It is the author's opinion that testing strategies for EDCs should take these facts into account and incorporate a more intelligent approach to the hazard identification process, rather than necessarily simply applying present standardized tests. Modifications of the present OECD and USEPA testing programs for chemicals are presently under way to incorporate endpoints more suitable for detecting EDCs, but they do not include some important systems affected, including behavior.

It is unlikely that any chemical-testing strategy developed will predict all adverse outcomes, especially those that arise as a consequence of chronic exposures to low levels of chemicals. The American Chemical Society lists over 246,000 inventoried or regulated chemicals worldwide (http://www.cas.org/cgi-bin/cas/regreport.pl), and this does not include their products of degradation, conjugates, or metabolites. The challenges for testing and screening of chemicals for endocrine-disrupting activity are therefore considerable. Endocrine disruption as an environmental issue arose through observations of adverse effects in wildlife populations, not through systematic screening and testing of chemicals, and it is the opinion of these authors that long-term environmental monitoring programs of wildlife populations need to be included as part of any effective and well-integrated hazard identification and risk management program for EDCs. We accept that the challenges associated with disentangling adverse effects in wildlife populations with causative agents are considerable, but such programs can nevertheless provide warning signals regarding which pragmatic chemical identification programs can be mounted. Environmental monitoring programs ideally need also to include analytical chemistry to measure for exposure concentrations. This, however, also involves considerable associated challenges. For example, EDCs such as EE_2 can induce biological effects at concentrations difficult to quantify accurately in environmental samples (at parts of a ng/L in water), even using some of the most sophisticated analytical techniques. Further support for the need for wildlife monitoring programs comes from the argument that individual-level consequences of toxicant exposure can be weak predictors of population-level consequences (Forbes and Calow 1999). Modeling approaches are gaining increasing favor to support investigations into the potential for population-level effects of EDCs and other chemicals. These approaches, however, require high-quality empirical data for both their development and validation, again drawing on the need for data from monitoring programmes of wildlife populations.

Finally, and to end this chapter on a positive note, in a few instances data derived from case studies on endocrine disruption in wildlife have led to the adoption of the precautionary principle, where even in the absence of population-level effects data, sufficient likelihood for harm to wildlife has been shown to support programs to remediate for the adverse effects seen. Examples of this in the United Kingdom include chemicals shown to induce feminized responses in fish, and include the phasing out of some alkylphenolic chemicals from use in the textile industry and a program to investigate and implement the most effective treatment process for the removal of steroidal estrogens from WWTW effluents before their discharge into rivers, both activities of which have been driven by the U.K. Environment Agency.

15.12   SUMMARY

Endocrine disruption is a chemically induced hormonal imbalance resulting in alterations in development, growth, and/or reproduction. Evidence for endocrine disruption in some wildlife populations is substantive, and in a few cases the identity of the chemicals inducing the effect have been proved. Systematic screening of chemicals for endocrine-disrupting activities using a suite of bioassays has identified an extensive (and increasing) list of chemicals with a diverse range of effects. Early life stages and developmental processes controlled by thyroid hormones and sex steroids appear to be especially sensitive to endocrine disruption. A further, and one of the important, findings from endocrine disruption research is that chemicals without overt toxicity can induce more subtle alterations in an animals' physiology that can have far-reaching biological effects, including population-level consequences for wildlife.

FURTHER READING

Carson, R. (1963). *Silent Spring*. London, Hamish Hamilton.

Oehlmann, J. and Schulte-Oehlmann, U. (2003). Endocrine disruption in invertebrates. *Pure and Applied Chemistry* **75**, 2207–2218.

Skakkebaek, N.E., Jorgensen, N., Main, K.M., Rajpert-De Meyts, E., Leffers, H., Andersson, A.M., Juul, A., Carlsen, E., Mortensen, G.K., Jensen, T.K., and Toppari, J. (2006). Is human fecundity declining? *International Journal of Andrology* **29**, 2–11.

Tyler, C.R., Jobling, S., and Sumpter, J.P. (1998). Endocrine disruption in wildlife: A critical review of the evidence. *Critical Reviews in Toxicology* **28**, 319–361.

Zala, S.M. and Penn, D.J. (2004). Abnormal behaviours induced by chemical pollution: a review of the evidence and new challenges. *Animal Behaviour* **68**, 649–664.

16 Neurotoxicity and Behavioral Effects of Environmental Chemicals

16.1 INTRODUCTION

In the previous chapters there have been many examples of environmental chemicals, both natural and human-made, that have harmful effects on the nervous system of animals. Many of these compounds are toxic both to vertebrates and invertebrates. Interestingly, five major groups of insecticides, organochlorine insecticides (OCs), organophosphorous insecticides (OPs), carbamate insecticides, pyrethroids, and neonicotinoids, all owe their insecticidal toxicity largely or entirely to their action on sites in the nervous system. A few of these compounds have also been used to control vertebrate pests (e.g., the cyclodiene endrin has been used for vole control, and the OP insecticides fenthion and parathion for controlling birds). Separate chapters have been devoted to the OCs (Chapter 5), OPs and carbamates (Chapter 10), and the pyrethroids (Chapter 12). Other human-made pollutants also have harmful effects on the nervous system of animals, although they are not used with the intention of doing so. Examples include the organomercury fungicides and tetraethyl lead, which has been used as an antiknock in petrol (both in Chapter 8). It would appear, therefore, that the nervous system represents an "Achilles heel" within both vertebrates and invertebrates when it comes to the toxic action of chemicals. When pesticide manufacturers have screened for insecticidal activity across a wide diversity of organic chemicals, many of the substances that have proved successful in subsequent commercial development have been neurotoxic.

This line of argument can be extended to natural toxins as well (Chapter 1). Thus, many plant toxins such as the pyrethrins, physostigmine, strychnine, veratridine, aconitine, etc., all act upon the nervous system. As discussed earlier, the presence of such compounds in plants is taken as evidence for a coevolutionary arms race between higher plants and the animals that graze upon them. The production of these compounds may protect the plants against grazing by vertebrates and invertebrates. Apart from plants, animals and microorganisms also produce neurotoxins that have deadly effects upon vertebrates or invertebrates or both in the living environment. For example, snakes, spiders, and scorpions all produce neurotoxins, which they inject into their prey to immobilize them (see Chapter 1, Section 1.3.1). Also, tetrado-toxin is stored within the puffer fish, and ergot alkaloids are produced by the fungus

Claviceps purpurea. Indeed, these natural toxins have given many useful leads in the design of new pesticides, biocides, or drugs.

In earlier chapters, many examples were given of lethal effects and associated neurotoxic or behavioral effects or both caused by pesticides in the field. These included effects of organomercury fungicides upon birds (Chapter 8, Section 8.2.5), organochlorine insecticides on birds, and both organophosphorous and carbamate insecticides upon birds (Chapter 10, Section 10.2.4). Also, a retrospective analysis of field data on dieldrin residues in predatory birds in the U.K. suggested that sublethal neurotoxic effects were once widespread and may have contributed to population declines observed at that time (Chapter 5, Section 5.3.5.1). Lethal and sublethal effects of neurotoxic insecticides upon bees is a long-standing problem (see Chapter 10, Section 10.2.5). Speaking generally, it has been difficult to clearly identify and quantify neurotoxic and behavioral effects caused by pesticides to wild populations, especially where the compounds in question have been nonpersistent (e.g., OP, carbamate, or pyrethroid insecticides), and where any sublethal effects would have been only transitory.

It is very clear, therefore, that there have been many examples of neurotoxic effects, both lethal and sublethal, caused by pesticides in the field over a long period of time. Far less clear, despite certain well-documented cases, is to what extent these effects, especially sublethal ones, have had consequent effects at the population level and above. Interest in this question remains because neurotoxic pesticides such as pyrethroids, neonicotinoids, OPs, and carbamates continue to be used, and questions continue to be asked about their side effects, for example, on fish (Sandahl et al. 2005), and on bees and other beneficial insects (see, for example, Barnett et al. 2007).

The present account will consider, in a structured way, how neurotoxic compounds may have effects upon animals, and how these effects can progress through different organizational levels, culminating in behavioral and other effects at the "whole animal" level. Emphasis will be placed upon the identification and quantification of these effects using biomarker assays, and upon attempts to relate these biomarker responses to consequent effects at the population level and above, referring to appropriate examples. The concluding discussion will focus on the use of this approach to identify and quantify existing pollution problems and on its potential in environmental risk assessment.

In the first place, there are a number of different sites of action for toxic chemicals within the central and peripheral nervous system of both vertebrates and invertebrates. When studying the effects of neurotoxic compounds, it is desirable to monitor the different stages in response to them using appropriate biomarker assays, beginning with initial interaction at the target site (site of action), progressing through consequent disturbances in neurotransmission, and culminating in effects at the level of the whole organism, including effects upon behavior. Thus, in concept, a suite of biomarker assays can be used to measure the time-dependent sequence of changes that follows initial exposure to a neurotoxic compound—changes that constitute the process of toxicity. From integrated studies of this kind should come principles and techniques that can be employed to develop and validate new approaches and assays for the purpose of environmental monitoring and environmental risk assessment. In reality, however, only a very limited range of biomarker assays are available at the time of writing, and much work still needs to be done to realize this objective.

An overview will first be given of the interaction of neurotoxic compounds with target sites within the nervous system before moving on to discuss disturbances caused in neurons and, finally, effects at the whole-organism level; prominent among the latter will be behavioral effects. Throughout, consideration will be given to biomarker assays that may be used to monitor the toxic process. Examples will be given of the successful use of biomarker assays, where, by judicious use of such assays, effects observed in the field have been attributed to neurotoxic chemicals. In conclusion, there will be a discussion of attempts to relate biomarker responses to consequent effects upon populations and above.

16.2   NEUROTOXICITY AND BEHAVIORAL EFFECTS

Animal behavior has been defined by Odum (1971) as "the overt action an organism takes to adjust to its environment so as to ensure its survival." A simpler definition is "the dynamic interaction of an animal with its environment" (D'Mello 1992). Another, more elaborate, one is, "the outward expression of the net interaction between the sensory, motor arousal, and integrative components of the central and peripheral nervous systems" (Norton 1977). The last definition spells out the important point that behavior represents the integrated function of the nervous system. Accordingly, disruption of the nervous system by neurotoxic chemicals may be expected to cause changes in behavior (see Klaasen 1996, pp. 466–467).

Throughout the present text, toxicity is described as a sequence of changes initiated by the interaction of a chemical with its site (or sites) of action, progressing through consequent localized effects and culminating in adverse changes seen at the level of the whole organism. Thus, in what follows, the description of the biochemical mode of action of neurotoxic compounds will be followed by an account of localized effects before concluding with effects seen at the level of the whole animal, particularly behavioral effects.

By approaching neurotoxicity in this way, it should be possible, in the longer term, to develop biomarker assays that can monitor the different stages in toxicity and to produce combinations of biomarker assays that will give a quantitative in-depth picture of the sequence of changes that occurs when an organism is exposed to a neurotoxic compound *or a mixture of neurotoxic compounds*. In following this progression, one moves from biochemical interactions, which are particular for a certain type of compound, to behavioral effects that are far less specific. However, by following this integrated approach, it should be possible to distinguish the contribution of individual members of a mixture to a common effect at a higher level of biological organization, for example, an alteration in the conduction of nervous impulse or a change in behavior. Later in this account, examples will be given describing experiments that have successfully linked mechanistic biomarker assays to behavioral changes despite the complexity of the nervous system.

Following from the above, behavioral assays, which can be relatively simple and cost-effective, can be very useful as primary screens when testing chemicals for their neurotoxicity in the context of medical toxicology (see Dewar 1983, Atterwill et al. 1991, and Tilson 1993). Where disturbances of behavior are identified, subsequent more specific tests, including in vitro assays, may then be performed to establish where and how damage is being caused to the nervous system. It should be added that behavioral effects of chemicals may be very important in ecotoxicology. They may be critical in determining adverse changes at the population level (Walker 2003, Thompson 2003).

Some authors have drawn attention to evidence for the greater sensitivity of early developmental stages of mammals to neurotoxins in comparison to adults (Colborn et al. 1998, Eriksson and Talts 2000). It has been claimed that neurotoxic and endocrine-disrupting chemicals are most damaging if there is exposure during embryonic, fetal, or postnatal life stages. This is a point to be borne in mind when investigating the long-term effects of neurotoxins using biomarker strategies.

16.3   THE MECHANISMS OF ACTION OF NEUROTOXIC COMPOUNDS

The principal, known mechanisms of action of some neurotoxic environmental chemicals are summarized in Table 16.1. In considering these, it needs to be borne in mind that the interactions between chemicals and the nervous system in vivo can be very complex, and there is a danger of oversimplification when arguing from mechanisms of action shown to occur in vitro. It is very important to relate results obtained in vitro to interactions that occur in vivo, taking into account toxicokinetic factors. The distribution of chemicals over the entire nervous system and the concentrations reached at different sites within it are critical in determining the consequent interactions and toxic responses. Further, any given neurotoxic compound may interact not just with one well-defined target but with contrasting target sites in different parts of the nervous system. Thus, one chemical may interact with two or more quite different receptor sites (e.g., Na^+ channel and GABA receptor) at the same time, albeit in different parts of the nerve network. Also, there may be different forms of the same type of active site—with contrasting affinities for neurotoxic compounds. That said, this account will attempt to focus on the principal modes of action that particular chemicals have shown to particular species of animals in vivo.

Taking first the **voltage-sensitive Na^+ channels** (Chapter 5, Figure 5.4) that are found in the plasma membranes of nerve and muscle cells of both vertebrates and invertebrates, it is seen that these are regulated by two separate processes: (1) activation, which controls the rate and voltage-dependence of the opening of this hydrophobic channel, and (2) inactivation, which controls the rate and voltage-dependence of the closure of the channel. These channels are known to exist in many different forms despite the fact that they all have the same common function, that is, the regulation of sodium currents across the plasma membrane. Three different types are recognized in rat brain, and strongly contrasting forms are recognized in different strains of the same species. Resistant strains of houseflies and other insects have different forms from susceptible strains of the same species. For example, kdr and super kdr strains have forms of the proteins constituting Na^+ channels which are different from those found in susceptible strains (see Chapter 5, Section 5.2.5.2), and the forms present in these resistant strains are insensitive to both DDT and pyrethroid insecticides; that is, they provide the basis for resistance to the insecticides.

TABLE 16.1

Neurotoxic Action of Some Environmental Chemicals

Sites of Action	Human-Made Chemicals	Notes	Natural Toxins	Notes
Na⁺ Channels	DDT Pyrethroids	Both can prolong the passage of Na⁺ current	Pyrethrins Veratridine	Veratridine appears to act at a different part of pore channel from DDT or pyrethroids
Nicotinic acetylcholine receptors	Neonicotinoids	Similar action to Nicotine	Nicotine	Act as agonists causing desensitization of receptor
Gamma aminobutyric acid (GABA) receptors	Dieldrin, endrin, gamma HCH (BHC), toxaphene	Inhibitors of receptor, reducing chloride influx	Picrotoxinin	Inhibitor of GABA receptors
Acetylcholinesterase	OP and carbamate insecticides	Inhibitors of enzyme causing buildup of acetylcholine in synapses	Physostigmine	Inhibitor of acetylcholinesterase
Neuropathy target esterase	Certain OP compounds including DFP, mipafox, and leptophos	Aging of inhibited enzyme leads to degeneration of peripheral nerves		
Cause damage to CNS of vertebrates	Organomercury and organolead compounds	Toxicity may be connected with ability to combine with SH groups	Methyl mercury	Occurs naturally as well as being human made

Sources: Eldefrawi and Eldefrawi (1990), Johnson (1992), Ballantyne and Marrs (1992), and Salgado (1999).

The Na$^+$ channel is the target for certain naturally occurring toxins (see Chapter 5, Figure 5.4). The lipid-soluble alkaloid veratridine can activate the channel by binding to it and stabilizing it in a permanently open conformation (Eldefrawi and Eldefrawi 1990). This causes a prolongation of the sodium current and disruption of the action potential—typically, repetitive firing of the action potential. The marine toxins tetrodotoxin and saxitoxin have the opposite effect. They are organic ions bearing a positive charge that can bind to the channel near its extracellular opening and thereby block the movement of sodium ions. Of the insecticides, the principal mode of action of both DDT and the pyrethroid insecticides is thought to be upon Na$^+$ channels. Rather like veratridine, they bind to the channel causing a prolongation of the Na$^+$ current, although they appear to bind to a different part of the protein than does this alkaloid (Chapter 5, Figure 5.4). Nerves poisoned by DDT typically produce multiple rather than single action potentials when they are electrically stimulated (Figure 16.1).

A. Action Potential Passed Along Nerve following Single Voltage Stimulus

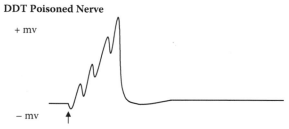

B. Current Generated on Postsynaptic Membrane of Inhibitory Synapse following Stimulation with Gab

FIGURE 16.1   Generation of action potentials.

The **nicotinic receptor for acetylcholine** is located on postsynaptic membranes of nerve and muscle cells. It is found in both the central and peripheral nervous system of vertebrates, but only in the central nervous system of insects (Eldefrawi and Eldefrawi 1990). A hydrophobic cationic channel is an integral part of this transmembrane protein. With normal synaptic transmission, acetylcholine released from nerve endings interacts with its binding site on the receptor protein, and this leads to an opening of the pore channel and an influx of cations. The consequent depolarization of the membrane triggers the generation of an action potential by neighboring sodium channels, and so the message is passed on. The natural insecticide nicotine acts as an agonist for acetylcholine and can cause desensitization of the receptor. Neonicotinoid insecticides such as imidacloprid act in a similar way to nicotine. They are more lipophilic than the natural compound and are more effective as insecticides.

Gamma aminobutyric acid (GABA) receptors are located on the postsynaptic membranes of inhibitory synapses of both vertebrates and insects and contain within their membrane-spanning structure a chloride ion channel. They are found in both vertebrate brains and invertebrate cerebral ganglia (sometimes referred to as brains) as well as in insect muscles. Particular attention has been given to one form of this receptor—the GABA-A receptor—as a target for novel insecticides (Eldefrawi and Eldefrawi 1990). It is found both in insect muscle and vertebrate brain. The remainder of this description will be restricted to this form.

GABA-A possesses a variety of binding sites (Chapter 5, Figure 5.4). One of them is for the natural transmitter GABA, an interaction that leads to the opening of the pore channel and the influx of chloride ions (Figure 16.1). Another, close to or in the chloride ion channel, binds the naturally occurring convulsant picrotoxinin, the cyclodiene insecticides (e.g., dieldrin, endrin), gamma HCH (lindane), and toxaphene. Convulsions accompany severe poisoning by these insecticides. The GABA-A receptor of mammalian brain is believed to be the primary target for cyclodiene insecticides in that organ. Binding of picrotoxinin and cyclodiene insecticides to the receptor retards the influx of chloride ions through the pore channel following stimulation with GABA; that is, they inhibit the normal functioning of the receptor.

Acetylcholinesterase is a component of the postsynaptic membrane of cholinergic synapses of the nervous system in both vertebrates and invertebrates. Its structure and function has been described in Chapter 10, Section 10.2.4. Its essential role in the postsynaptic membrane is hydrolysis of the neurotransmitter acetylcholine in order to terminate the stimulation of nicotinic and muscarinic receptors (Figure 16.2). Thus, inhibitors of the enzyme cause a buildup of acetylcholine in the synaptic cleft and consequent overstimulation of the receptors, leading to depolarization of the postsynaptic membrane and synaptic block.

The carbamate and OP insecticides and the organophosphorous "nerve gases" soman, sarin, and tabun all act as anticholinesterases, and most of their toxicity is attributed to this property. The naturally occurring carbamate physostigmine, which has been used in medicine, is also an anticholinesterase. Some OP compounds can cause relatively long-lasting inhibition of the enzyme because of the phenomenon of

---

BOX 16.1   TECHNIQUES FOR MEASURING THE INTERACTION OF NEUROTOXIC CHEMICALS WITH THEIR SITES OF ACTION

A central theme of this text is the development of biomarker assays to measure the extent of toxic effects caused by chemicals both in the field studies and for the purposes of environmental risk assessment.

Considering the examples given in Table 16.1, a number of possibilities present themselves. In the first place, competitive binding studies may reveal the extent to which a toxic compound is attached to a critical binding site. For example, the convulsant TBPS binds to the same site on GABA-A receptors of rat brain as do cyclodiene insecticides such as dieldrin. In samples preexposed to dieldrin, the binding of radiolabeled TBPS will be less than in controls not exposed to the cyclodiene (Abalis et al. 1985). The difference in binding of the radioactive ligand to the treated sample in comparison to binding to the control sample provides a measure of the extent of binding of dieldrin to this target. Similarly, the competitive binding of tetrodotoxin and saxitoxin to the Na^+ channel may be exploited to develop an assay procedure.

In cases where the mode of action is the strong or irreversible inhibition of an enzyme system, the assay may measure the extent of inhibition of this enzyme. This may be accomplished by first measuring the activity of the inhibited enzyme and then making comparison with the uninhibited enzyme. This practice is followed when studying acetylcholinesterase inhibition by organophosphates (OP). Acetylcholinesterase activity is measured in a sample of tissue of brain from an animal that has been exposed to an OP. Activity is measured in the same way in tissue samples from untreated controls of the same species, sex, age, etc. Comparison is then made between the two activity measurements, and the percentage inhibition is estimated.

"aging"; the inhibited enzyme undergoes chemical modification, and inhibition then becomes effectively irreversible.

A few OP compounds cause delayed neuropathy in vertebrates because they inhibit another esterase located in the nervous system, which has been termed **neuropathy target esterase (NTE).** This enzyme is described in Chapter 10, Section 10.2.4. OPs that cause delayed neuropathy include diisopropyl phosphofluoridate (DFP), mipafox, leptophos, methamidophos, and triorthocresol phosphate. The delay in the appearance of neurotoxic symptoms following exposure is associated with the aging process. In most cases, nerve degeneration is not seen with initial inhibition of the esterase but appears some 2–3 weeks after commencement of exposure, as the inhibited enzyme undergoes aging (see Section 16.4.1). The condition is described as OP-induced delayed neuropathy (OPIDN).

Organometallic compounds such as alkylmercury fungicides, and tetraethyl lead, used as an antiknock in petrol, are neurotoxic, especially to the central nervous system of vertebrates (Wolfe et al. 1998, Environmental Health Criteria 101, and Chapter 8,

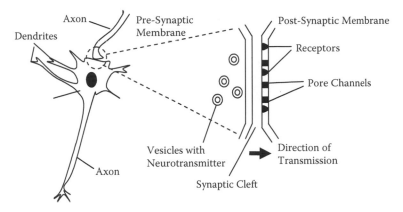

Cholinergic (Nicotinic) Synapse

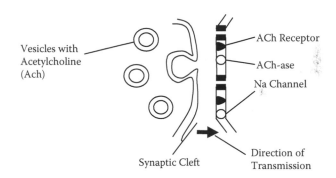

FIGURE 16.2   Schematic diagram of synapse.

Section 8.2.4 and Section 8.2.5 in this book). Neurotoxic effects in adult mammals include ataxia, difficulty in locomotion, neurasthenia, tremor, impairment of vision and, finally, loss of consciousness and death. Necrosis, lysis, and phagocytosis of neurons are effects coinciding with these symptoms of toxicity. As described earlier, sublethal neurotoxic effects on humans and wild vertebrates have occurred and still occur as the result of environmental contamination by methylmercury. The mechanism of neurotoxic action is complex and is not well understood. There is strong evidence that methylmercury compounds can have adverse effects upon a number of proteins, including enzymes and membrane-spanning proteins involved in ion transport (ETAC 101). It seems probable that the strong tendency of these compounds to bind with—and thereby render ineffective—functional –SH groups of the proteins is the main reason for this (see, for example, Jacobs et al. 1977, who studied the inhibition of protein synthesis by methylmercury compounds). There is also evidence that exposure to sublethal levels of methyl mercury can cause changes in the concentration of neurochemical receptors in the brains of mammals and birds (Basu et al. 2006, Scheuhammer et al. 2008). Thus, an increase in concentration of brain muscarinic receptors for acetylcholine and a decrease in the concentration of brain receptors for glutamate was observed following exposure to environmentally realistic levels of methylmercury. This observation was made both in mink and common loons.

In summary, the toxic effects of methylmercury on vertebrates are complex and wide ranging, and with the present state of knowledge it is not possible to ascribe this neurotoxicity to one clearly defined mode of action.

16.4 EFFECTS ON THE FUNCTIONING OF THE NERVOUS SYSTEM

Following combination with their sites of action, the main consequent effects of the neurotoxic compounds described here are upon synaptic transmission or propagation of action potential. In some cases (e.g., methylmercury and some OPs) there are signs of physical damage such as demyelination, phagocytosis of neurons, etc. The following account will be mainly concerned with effects of the first kind—that is, electrophysiological effects—which may provide the basis for assays that can monitor the progression of toxicity from an early stage and thus provide a measure of sublethal effects caused by differing levels of exposure. Effects on the peripheral nervous system and the central nervous system will now be considered separately.

16.4.1 Effects on the Peripheral Nervous System

Electrical impulses are passed along nerves as a consequence of the rapid progression of a depolarization of the axonal membrane. In the resting state, a transmembrane potential is maintained on account of the impermeability of the nerve to ions such as Na^+ and K^+. Were the membrane freely permeable, these ionic gradients could not be sustained. Active transport processes maintain ionic gradients in excess of those that could be achieved purely by passive diffusion. However, when Na^+ channels open in the axonal membrane, a very brief inwardly flowing Na^+ current causes a transient depolarization. This is rapidly corrected by a subsequent outward flow of K^+ ions. The Na^+ current is terminated when the pore channel closes, and the succeeding K^+ current flows briefly until the transmembrane potential returns to its resting state (Figure 16.1).

The passage of action potentials along a nerve can be recorded by inserting microelectrodes across the neuronal membrane and using them to record changes in the transmembrane potential in relation to time. This has been done in a variety of ways. Microelectrodes can be inserted into nerves of living animals, or into isolated nerves, or cellular preparations of nerve cells (see Box 16.2). An important refinement of the technique involves "voltage clamping." This permits the "fixing" of the transmembrane potential, which restricts the movement of ions across the membrane. Thus, it is possible to measure just the Na^+ current or the K^+ current in control and in "poisoned" nerves, thereby producing a clearer picture of the mechanism of action of neurotoxic compounds that affect the conduction of action potentials along nerves. Measurements of this kind may be just of spontaneous action potentials or of potentials that are elicited by electrical or chemical stimulation. Chemical stimulation may be accomplished using natural neurotransmitters such as acetylcholine.

The effects of neurotoxic chemicals upon nerve action potential have been measured both in vertebrates and insects. Of particular interest has been the comparison

---

BOX 16.2   IN VITRO ASSAYS FOR NEUROTOXICITY

There has long been an interest in the development of in vitro assays for detecting neurotoxic effects of chemicals from the point of view of both human risk assessment and environmental risk assessment. The effects of neurotoxic chemicals on laboratory animals is a major concern of animal welfare organizations. An outstanding problem is that, because of the complexity of the nervous system, some neurotoxic effects can only be detected in vivo—in whole animal systems (Dewar 1983, Atterwill et al. 1991). Thus, it is difficult to foresee the total banning of in vivo tests. However, in vitro assays can still make an important contribution to testing protocols for chemicals. These protocols can include a combination of in vivo and in vitro tests, with a consequent reduction in the use of animals for testing procedures (Atterwill et al. 1991).

Atterwill et al. (1991) list six categories of nervous system culture that have been used in in vivo testing procedures. These are dispersed cell cultures, explant cultures, whole organ cultures, reaggregate cultures, whole embryo models, and cell lines. It is possible in cultures such as these to measure the cellular response to neurotoxic chemicals. Electrophysiological measurements can be made even on single cells, revealing effects of chemicals upon ion currents and transmembrane potential. Also, there is the possibility of following effects on the release of chemical messengers such as cyclic AMP from postsynaptic membranes, when neurotransmitters interact with their receptors.

In one example (Lawrence and Casida 1984, Abalis et al. 1985) rat brain microsacs were used to test the action of cyclodiene insecticides such as dieldrin and endrin on the GABA receptors contained therein. The influx of radiolabeled Cl^- into the microsacs via the pore channel of the receptor was inhibited by these chemicals. A similar assay was developed using microsacs from cockroach nerve. Assays with this preparation showed again the inhibitory effect of a cyclodiene (this time heptachlor epoxide) on Cl^- influx. Also, that microsacs from cyclodiene resistant cockroaches were insensitive to the inhibitory effect of picrotoxinin, which binds to the same site on the GABA receptor (Kadous et al. 1983).

---

of the responses of different species and strains of insects to insecticides. Returning to the examples given in Table 16.1, both DDT and pyrethroid insecticides interact with the Na^+ channel of the axonal membrane of insects. With repeated use of DDT, insects such as houseflies came to develop kdr and super kdr resistance against the insecticide. Both types of resistance are due to the appearance of forms of the Na^+ channel that are insensitive to the insecticide (see Chapter 4, Section 4.5, and Chapter 12, Section 12.6). The fact that these strains also show marked cross-resistance to pyrethroids is compelling evidence that this pore channel represents the principal site of action for both types of insecticide in insects.

The effects of DDT on nerve action potential are illustrated in Figure 16.1. In nerves poisoned by the insecticide, there is a prolongation of the sodium current and a consequent delay in returning to the resting potential. This can result in the generation of further spontaneous action potentials, that is, there can be repetitive action potentials following a single stimulus.

As described earlier, the **chloride channels,** which are associated with GABA receptors, are affected by the action of cyclodienes and certain other chlorinated insecticides. These chemicals can inhibit the action of the neurotransmitter GABA by binding to a site in or near the pore channel, with consequent reduction in the inward flow of Cl⁻ (see Figure 16.1). Electrophysiological studies have been carried out that involve the stimulation of GABA receptors of insect muscle (e.g., of the locust). Treatment with GABA causes hyperpolarization of the membrane, an effect that is retarded when the receptors are preexposed to cyclodienes, or to the natural product, picrotoxinin. The action of picrotoxinin on GABA receptors of the locust *Calliphora erythrocephala* and the resulting neurophysiological effects are described by Von Keyserlingk and Willis (1992). So, again, the interaction of a neurotoxic compound with a receptor can be related to consequent electrophysiological effects (see also Box 16.2).

The neurophysiological effects of anticholinesterases have been studied in the peripheral nervous system of experimental animals and humans. In some cases of human poisoning, effects on motor conduction were measured using electromyography (EMG), which involves the insertion of a needle-recording electrode into muscle (Misra 1992). In cases of OP poisoning, there was evidence of several types of neurophysiological effects, including repetitive activity. Poisoning in vertebrates leads to a buildup of the neurotransmitter on cholinergic junctions, which, if severe enough, will cause a depolarization of the synaptic membrane and loss of synaptic transmission. Thus, the later stages of poisoning should be evident from measurement of the postsynaptic signal by EMG. Effects of anticholinesterases on the sensory system of the mammalian PNS have also been monitored using electrophysiological methods.

The neurophysiological effects of nicotine have been widely reported in the pharmacological literature, and the neonicotinoid insecticides are known to act in a similar way. Initially, these compounds act as agonists of nicotinic receptors of acetylcholine, but this interaction leads to desensitization of the receptor, resulting in a loss of synaptic transmission. Thus, their effects can be monitored by recording the signals from cholinergic synapses such as the neuromuscular junction of vertebrates and testing responsiveness to acetylcholine stimulation by EMG measurements. This can be done, for example, with denervated muscle of the rat.

The delayed neuropathy caused by certain OPs that inhibit neuropathy target esterase is characterized by a number of pathological changes in the peripheral nervous system of vertebrates (Johnson 1992, Veronesi 1992). Electrophysiological measurements on the sciatic nerve of hens have shown a significant increase in excitability 24 hours after dosing with one of these compounds. The hen is used as a test organism on account of its high susceptibility to this type of poisoning. In the longer term (2–3 weeks), degenerative changes appear in peripheral nerves that are characteristic of this type of poisoning, changes that affect the distal extremities and are associated with a sensory–motor deficit. These later effects have been observed in mammals, including humans.

16.4.2   Effects on the Central Nervous System

The spontaneous electrical activity of the brain can be measured by electroencephalography (EEG), a technique that has been widely employed to study neurotoxic effects of chemicals both in humans and in experimental animals. EEG waves represent summated synaptic potentials generated by the pyramidal cells of the cerebral cortex (Misra 1992). These potentials are the responses of cortical cells to rhythmical changes arising from thalamic nuclei. The signals recorded can be separated into frequency bands—faster waves exceeding 13 Hz, and slower ones below 4 Hz.

Changes in EEG patterns have been observed when humans and experimental animals are exposed to neurotoxic compounds. Thus, humans occupationally exposed to aldrin or dieldrin showed characteristic changes in EEG patterns (Jaeger 1970). These changes were sometimes accompanied by symptoms of intoxication such as muscle twitching and convulsions. Many studies have shown changes in EEG patterns following exposure of experimental animals to OP insecticides. Rats exposed to parathion showed a damping of all EEG frequencies and reduction of amplitude, changes that were dose related (Vajda et al. 1974). Experiments with primates showed that acute exposure to OPs can cause desynchronization of EEG patterns, including increased higher frequency activity and decreased lower frequency activity. Increased exposure led to slowing of the EEG, followed by spike–wave discharges that accompany convulsions (Burchfield et al. 1976).

Neurotoxicity has often been associated with lesions in the central nervous system. Methyl mercury, for example, has been shown to cause progressive destruction of cortical structures and cerebral edema in mammals (see Wolfe et al. 1998). O'Connor and Nielsen (1981) reported necrosis, astrogliosis, and demyelination in otters dosed with methylmercury. Also, organophosphate-induced delayed neuropathy (OPIDN) caused by certain OPs can lead to degenerative changes in fibers of the spinal cord of rats in addition to the peripheral effects mentioned earlier (Veronesi 1992). However, these effects appear at a relatively late stage in the progression of toxicity. Thus, they do not have the same potential as biochemical or electrophysiological effects when it comes to developing biomarker assays. The latter can provide early sensitive indications of toxic disturbances before there is physical evidence of damage.

Recently, there has been a growth of interest in the development of in vitro methods for measuring toxic effects of chemicals on the central nervous system. One approach has been to conduct electrophysiological measurements on slices of the hippocampus and other brain tissues (Noraberg 2004, Kohling et al. 2005). An example of this approach is the extracellular recording of evoked potentials from neocortical slices of rodents and humans (Kohling et al. 2005). This method, which employs a three-dimensional microelectrode array, can demonstrate a loss of evoked potential after treatment of brain tissue with the neurotoxin trimethyltin. Apart from the potential of in vitro methods such as this as biomarkers, there is considerable interest in the use of them as alternative methods in the risk assessment of chemicals, a point that will be returned to in Section 16.8.

16.5   EFFECTS AT THE LEVEL OF THE WHOLE ORGANISM

In the first place, severe neurotoxicity can cause gross neurophysiological disturbances at the whole organism level, such as convulsions, paralysis, and inability to walk (or, in the case of birds, to fly). In vertebrates, convulsions are symptomatic of poisoning by dieldrin and related insecticides that act upon GABA receptors of the central nervous system. As we have seen, inhibition of GABA receptors can cause disruption of transmission across inhibitory synapses that are mediated by gamma amino butyric acid, and a consequence of this can be coordinated muscular disturbances, including convulsions. Damage to the nervous system caused by organomercury compounds or OPIDN, caused by compounds such as DFP and Mipafox, can lead to paralysis and locomotor failure. These severe effects, which are associated with the later stages of poisoning, can cause disruption of patterns of behavior utilized in testing procedures. However, more subtle changes of behavior, which give early indications of toxic action, are of particular interest in the present context and will be the subject of the remainder of this section.

Many tests have been devised to provide quantitative measures of behavioral disturbances caused by neurotoxic chemicals. Tests have been devised that assess the effects of chemicals on four behavioral functions (D'Mello 1992). These are sensory, cognitive, motor, and affective functions. However, because the entire nervous system tends to work in an integrated way, these functions are not easily separable from one another. For example, the outcome of tests focused on sensory perception by rats may be influenced by effects of the test chemical on motor function.

Speaking generally, many laboratory studies have shown behavioral effects in vertebrates or invertebrates or both exposed to organochlorine, carbamate, OP, pyrethroid, and neonicotinoid insecticides. However, the critical questions are: (1) to what extent have these effects been demonstrated at normal levels of exposure in the field? and (2), if such effects have occurred in the field, have there been knock-on effects at the population level? These issues will be returned to in Section 16.7.

Fish have proved to be sensitive test organisms for the detection of behavioral effects. In an early paper, Warner et al. (1966) studied the effects of some pesticides on the behavior of goldfish (see also Chapter 5, Section 5.3.4). They measured several behavioral responses, including spontaneous activity and response to stimuli such as light and shock. Toxaphene, a chlorinated insecticide that acts upon GABA receptors, caused behavioral changes down to a concentration of 0.4 µg/L, which is far below the median lethal concentration. They also reported behavioral effects caused by the OP insecticide tetraethylpyrophosphate (TEPP) at concentrations that produced no overt signs of intoxication. Subsequently, other workers, including Beauvais et al. (2000) and Sandahl et al. (2005) have also demonstrated quantifiable sublethal behavioral effects of OPs on fish at low levels of exposure. Scholtz, Truelove, and French et al. (2000) studied the sublethal effects of diazinon on Chinook salmon (*Onchorhynchus tshawytscha*) and found disturbances in antipredator and homing behavior. Speaking more generally, anticholinesterases have been shown to cause a variety of sublethal disturbances in fish, including on swimming performance, swimming stamina, prey capture, predator detection, predator avoidance, migration, learning, and conspecific social interactions (Sandahl et al. 2005). Some of these studies related behavioral effects to levels of inhibition of acetylcholinesterase, and will be discussed further in the next section.

Behavioral effects of OP insecticides have also been shown in birds (see review by Grue et al. 1991). Behavioral effects of OCs, OPs, and methylmercury on birds have been reviewed by Peakall (1985, 1996). A remarkably wide range of behavioral tests were used in these studies. Tests employed included the following:

Adaptive behavior	Introduction of chicks to hens preadapted to brooding cages.
Approach behavior	Reaction to taped maternal call.
Avoidance behavior	Distance run after fright stimulus.
Detour learning	Food-deprived chicks learning to detour away from sight of food. Through tunnel to obtain food.
Dominance-subordinate pattern	Placing bird on either side of divided area. Raising wall and finding dominance.
Nest attentiveness	Use of telemetered eggs to record core temperature.
Nest defense	Classed as "aggressive," "moderate," or "weak."
Open field behavior	Movement of chicks monitored by sensors.
Operant behavior	Conditioning to response to lighted key to obtain food.
Predatory behavior	Attack on moving prey model with hidden meat reward.

Many of these tests gave evidence for changes in behavior following exposure to neurotoxic pesticides. The author concludes that significant behavioral effects were often recorded down to one order of magnitude below the LC_{50} in question. Some tests, such as operant tests, were relatively simple and gave reproducible results, but it was difficult to evaluate the relevance of these to survival in the wild. Other tests, such as breeding behavior and prey capture, were more complex and less reproducible, but more relevant to the natural world.

A wide range of sublethal effects of pyrethroids, carbamates, OPs, and neonicotinoids have been demonstrated in bees (Thompson 2003). With honeybees (*Apis mellifera*), effects have been shown on division of labor, conditioned responses, foraging, colony development, larval behavior, repellency, and nest mate recognition. Many effects occurred at or below levels of exposure anticipated in the field. OP, carbamate, and neonicotinoid insecticides had effects on the "wagtail" dance by which bees communicate the direction of a source of food to other bees. There has been considerable interest in developing tests for behavioral effects of pesticides upon bees. However, there have been reservations about including them in regulatory testing protocols (Thompson and Maus 2007). It is argued that any behavioral effects that are ecologically important will be picked up in field or semi-field trials.

In medical toxicology, there have been many reports of humans showing behavioral disturbances following exposure to sublethal levels of neurotoxic compounds. With cases of OP poisoning in humans, symptoms have included anxiety, emotional lability, giddiness, and insomnia (Lotti 1992). Early symptoms of cyclodiene poisoning in occupationally exposed workers have included dizziness, drowsiness, hyperirritability, and anorexia (Jaeger 1970).

16.6   THE CAUSAL CHAIN: RELATING NEUROTOXIC EFFECTS AT DIFFERENT ORGANIZATIONAL LEVELS

It is clear from the last section that the action of many neurotoxic compounds finds ultimate expression at the level of the whole organism, and there are many instances of effects on behavior. Linking responses at different organizational levels culminating in effects upon behavior is of considerable interest and importance. The value of adopting this approach when studying the effects of pollutants upon complex fish behavior has been reviewed by Scott and Sloman (2004). These matters said, there is seldom a clear picture of the sequence of changes that leads to toxic manifestations. Consideration will now be given to examples where there is some evidence of links between responses at different levels of biological organization and the possibilities of using biomarker assays to monitor them.

16.6.1   CHEMICALS SHARING THE SAME PRINCIPAL MODE OF ACTION

Some of the best evidence of links between effects at different organizational levels comes from studies with OPs, where levels of AChE inhibition have been compared with associated neurophysiological and behavioral effects. In adopting this approach, however, the picture is complicated by mounting evidence for these compounds acting on target sites other than AChE, as discussed in Section 16.3. Thus, behavioral disturbances caused by an OP may be the outcome of interaction with both AChE and one or more other sites of action. The following account, however, will be concerned with situations where effects of OPs are closely related to levels of AChE inhibition. More complex scenarios will be discussed in the next section.

Reviewing the effects of OPs on humans and experimental animals, Lotti (1992) states that neurotoxic and behavioral disturbances are found when there is 50–80% inhibition of acetylcholinesterase of the nervous system, 85–90% inhibition of brain cholinesterase is associated with severe toxicity, and over 90% inhibition with respiratory failure and death. Both chronic and acute exposure can produce a range of symptoms of neurotoxicity, including behavioral disturbances. In one study with experimental animals, prolonged exposure to OPs caused typical patterns of behavioral and physiological change related to AChE inhibition, which were followed by recovery (Banks and Russell 1967). Behavioral effects have sometimes been observed at very low levels of exposure, raising again the question whether there are sites of action for OPs in the CNS other than AChE.

In one study with common marmosets, the animals were dosed with diazinon (10, 90, or 130 mg/kg i.m.) and measurements of erythrocyte cholinesterase inhibition recorded, together with effects on EEG pattern and cognitive performance over a 12-month period (Muggleton et al. 2005). Initial inhibition of AChE was <82%, but quickly returned to normal. Short-term changes in steep pattern were seen, but there were no long-term changes in any of the measures made. It should be noted, however, that inhibition of erythrocyte acetylcholinesterase is likely to be much higher than inhibition of brain cholinesterase when animals are dosed in this way. The authors note that there have been reports of long-term effects of low doses of OPs on central nervous system (CNS) function, including steep, cognitive performance, and EEG changes.

Turning now to effects upon fish, Beauvais et al. (2000) showed behavioral effects following exposures of rainbow trout (*Oncorhynchus mykiss*) to diazinon and malathion. In the case of the malathion treatment, no fish died. Sandahl et al. (2005) obtained similar results when studying the response of juvenile coho salmon (*Oncorhynchus kisutch*) to chlorpyrifos. The exposures were all sublethal with no deaths of fish even at the highest exposure (2.5 µg/L). Significant correlations were observed between percentage inhibition of brain cholinesterase and spontaneous feeding and swimming behaviors. At the lowest level of exposure (0.6 µg/L), there was a significant reduction in AChE activity and, associated with that, significant alterations in swimming and feeding behaviors. Brain ACh-E was inhibited by 23 (±1%), whereas spontaneous swimming rate was reduced by 27 (±5%) in the same treatment group (standard errors in parentheses). Regarding feeding behavior, both the latency to strike and the striking rate were also significantly affected at this low dose.

In a further study on effects of anticholinesterases on the behavior of fish, cutthroat trout (*Oncorhynchus clarki clarki*) were exposed to sublethal levels of the carbamate insecticide carbaryl (Labenia et al. 2007). In this case, however, significant effects upon behavior were only demonstrated at high levels of brain cholinesterase inhibition (above 70%). At this high level, effects were reported on both swimming performance and avoidance of predation by lingcod (*Ophiodon elongates*). It is worth mentioning that inhibition of the enzyme by carbamates is more readily reversible than inhibition by OPs (Chapter 10, Section 10.3.4).

In a wide-ranging review, Grue et al. (1991) give many examples of studies that have attempted to relate inhibition of cholinesterases by pesticides to physiological and behavioral effects in mammals and birds. Behavioral effects measured in birds included changes in walking, singing, and resting. Despite some examples from well-designed studies in which a relationship was shown, generalizations proved difficult, and there was much evidence of intraspecific and interspecific variation in responses to anticholinesterases. It was not possible to define critical levels of brain cholinesterase activity across species with regard to sublethal effects. However, the conclusion was that there were examples of the impairment of physiological function and behavior once inhibition exceeds about 40%. Thus, these studies did not show such high sensitivity to cholinesterase inhibition as was demonstrated in some of the behavioral tests upon fish OPs discussed earlier, where there were clear indications of effects at below 25% inhibition.

Hart (1993) reports a study of behavioral effects of the OP insecticide chlorfenvinphos on captive starlings (*Sturnus vulgaris*). Birds were dosed with 3–9 mg/kg of the insecticide presented orally in the form of capsules. Behavioral effects were related to brain cholinesterase levels. The most sensitive parameter was posture, which was found to change when brain cholinesterase activity fell below 88% of the control value. Reductions in flying and singing, and increased resting were associated with inhibition to below 61% of the normal level. Within 5 hours, behavior returned to normal, reflecting the relatively rapid metabolic detoxication of this insecticide, as with most other OP insecticides.

Linkages between cholinesterase inhibition and behavior have also been studied in terrestrial arthropods that are exposed to OP and carbamate insecticides on agricultural land (Engenheiro et al. 2005). In a study with the isopod *Porcellio dilatatus* exposed to soil contaminated with dimethoate, measurements were made of locomotor activity. A relationship was found between several locomotor parameters and the degree of cholinesterase inhibition. Locomotor behavior is crucial in this species for burrowing, avoiding predation, seeking food, migration, and reproduction.

16.6.2 Effects of Combinations of Chemicals
with Differing Modes of Action

As has already been discussed, in heavily polluted areas, disturbances of the nervous system of free-living animals may be caused by chemicals with contrasting modes of action interacting with more than one site of action at the same time. For examples of sites of action, see Table 16.1. It should be emphasized that some of these sites of action can be responsive to naturally occurring as well as human-made neurotoxins. Thus, when measuring responses of animals at the whole-organism level to complex mixtures of chemicals (e.g., in fish deployed into polluted waters), the effects of chemicals acting through different pathways are difficult if not impossible to distinguish using assay systems that operate at higher organizational levels. For example, in a study of the effects of neurotoxic compounds on primates, dieldrin and the OP sarin produced similar effects on EEG patterns even though one chemical was acting through the GABA receptor, whereas the other was causing cholinesterase inhibition (Burchfield et al. 1976).

This complication aside, assays at the whole-organism level do have the advantage of presenting an integrated measurement of the effects of one or more compounds. It should be added that a better in-depth picture can be obtained by using such assays in combination with others that operate at lower organizational levels. In the aforementioned example given, inclusion in the study of assays for brain acetylcholinesterase inhibition and binding to critical sites on the GABA receptor should give a more complete picture of the toxic effects caused by the chemicals, thereby allowing some distinction to be made between the respective contributions of sarin and dieldrin to disturbances of the EEG pattern.

Because of their wide-ranging and "holistic" character, assays of behavioral effects have been used as screening procedures when testing for neurotoxicity (see, for example, Iversen 1991, Tilson 1993). They can provide sensitive indications of neurotoxic disturbances, which can then be traced back to their ultimate cause by using mechanistic biomarker assays.

16.7 RELATING NEUROTOXICITY AND BEHAVIORAL EFFECTS TO ADVERSE EFFECTS UPON POPULATIONS

Broadly speaking, the direct behavioral effects of neurotoxic pollutants on wild animals may be on feeding, breeding, or avoidance of predation (Beitinger 1990), or any combination of these. Any of these changes may have adverse effects on populations. Additionally, in the natural world, populations may be affected indirectly because of neurotoxic and behavioral effects on other species. Thus, a population decline of one species due to a behavioral effect of a pollutant may lead to a consequent decline of its parasites or predators, even though they are not themselves directly affected by the chemical. Direct effects will now be discussed before considering indirect ones.

As explained earlier, a number of examples of population declines have been related to neurotoxic pesticides. These have included the decline of predatory birds in Britain caused by cyclodiene insecticides, local declines of buzzards in the Netherlands related to dieldrin (Koeman 1972), the decline of Western grebes on Clear Lake, United States, caused by DDD (Hunt and Bischoff 1960), and local declines of migrating geese caused by carbophenothion in Northern England and Scotland (Hamilton et al. 1976). This latter incident was investigated in some detail (see Chapter 10, Section 10.2.5), and provides a good example of the use of a biomarker assay (acetylcholinesterase inhibition) to confirm the cause of toxic effects in the field and a consequent local reduction in population (Stanley and Bunyan 1979). There is also some evidence for adverse population effects of methylmercury fungicides on predatory birds in Sweden during the 1960s, although this conclusion did not have the support of biomarker assays (Borg et al. 1969). In all of these examples, there was clear evidence of lethal toxicity; there was much evidence, too, of sublethal effects in these different scenarios, but it was not clear at the time to what extent they contributed to mortality.

Neurotoxicity and behavioral disturbances can adversely affect feeding in different ways. In the case of predators, feeding behavior includes the components—searching for, encountering, choosing, capturing, and handling of prey (Atchison et al. 1996). All of these functions may be adversely affected by neurotoxic effects. Effects of pollutants upon the capture and handling of prey by certain aquatic species have been reported (Atchison et al. 1996). Predators that rely on highly developed hunting skills to catch mobile prey may die of starvation because of an inability to catch prey. Evidence that this may have been an important factor in the decline of raptors caused by cyclodiene insecticides was presented earlier (Chapter 5, Section 5.3.5.1). Similarly, predatory birds unable to fly after exposure to methylmercury could not have caught mobile prey (Chapter 8, Section 8.2.4).

At a more subtle level, behavioral disturbances may make it more difficult for animals to find food. Pyrethroids, carbamates, OPs, and neonicotinoids can disturb the foraging activity of bees (Thompson 2003). Interestingly, effects have been shown upon the wagtail dance of bees, and this disrupts communication between individuals as to the location of nectar-bearing plants. Also, the neonicotinoid imidacloprid has been shown to adversely affect conditioned responses such as proboscis extension of honeybees (Guez et al. 2001). Nicotinoids can disturb the functioning of cholinergic synapses, which are involved in the operation of the proboscis reflex as well as in learning and memory in the honeybee. Again, effects of this kind can have a detrimental impact on foraging.

Behavioral effects of pollutants may also disrupt reproduction. In principle, it seems reasonable to suppose that behavioral effects upon birds may lead to disturbances of pairing or mating, nest desertion, incubation of eggs, or failure to protect nest and young, although there is a shortage of solid evidence for this happening in the natural world. In one study with four different species of ducks (Brewer et al. 1988), application of methyl parathion led to reduced survival of ducklings, and this was attributed to brood abandonment. Exposure to sublethal levels of methylmercury has sometimes been associated with behavioral effects and reduced reproductive success in birds (see Chapter 8 of this book). In a study of common loons in North America, there was evidence of aberrant breeding behavior (e.g., reduced nest occupancy) that was related to levels of exposure to this pollutant. There was also evidence of reduced reproductive success related to methylmercury exposure in the same population (Evers et al. 2008).

Another adverse behavioral effect of neurotoxic compounds can be reduced ability to avoid predation. In a study of predation of newts on tadpoles, Cooke (1971) demonstrated that tadpoles that had been exposed to DDT were less able to avoid predation than controls. Further, because of the persistence of DDT and its metabolites in the tadpoles, the predator was itself selecting a diet high in persistent neurotoxic compounds—an act of self-destruction. It has been argued that such selective predation on prey highly contaminated by persistent neurotoxic pollutants may have been quite widespread when these compounds were in regular use. Raptorial birds such as the peregrine, for example, are attracted to prey that behaves abnormally; for example, an individual bird fluttering on the ground can attract the attention of a predator. Thus, when one considers the marked biomagnification of such compounds that has occurred in food chains (see Chapter 2, Figure 2.8), selective predation may have accentuated the problem of bioaccumulation.

Turning now to indirect effects of neurotoxic pollutants, the status of predators and parasites can be affected by reductions in numbers of the species that they feed upon. Thus, the reduction in numbers of a prey species due to a behavioral effect can, if severe enough, cause a reduction in numbers of a predator. Also, as mentioned earlier, behavioral effects upon a prey species may lead to selective bioaccumulation of persistent neurotoxic pollutants such as DDT and dieldrin by predators; thus, a behavioral effect may be hazardous for predator and prey alike!

When neurotoxic pollutants interact with their sites of action, consequent effects on the functioning of the nervous system may be manifest in a variety of disturbances in behavior. Many of the latter have the potential to cause knock-on effects at the level of population because of disruption of such activities as feeding, breeding, and avoidance of predation. The question remains: to what extent were such effects important in cases where population declines were attributed to neurotoxic pollutants? In many instances, there is inadequate evidence to answer this question retrospectively. Looking ahead, however, the development of biomarker strategies and new biomarker assays could provide the technology for tackling future ecotoxicological problems of this kind.

Considering population effects of neurotoxic pollutants more generally, persistence is clearly an important factor. With pollutants of short biological half-life, effects will tend to be transitory, whereas persistent pollutants are likely to produce longer-lasting behavioral disturbances. Thus, the environmental risks presented by recalcitrant OCs such as dieldrin and DDT would appear to be greater than those presented by readily biodegradable OPs, carbamates, or pyrethroids, from the point of view of neurotoxic and behavioral effects. The use of persistent OCs has now been largely discontinued, their global sales being estimated at only 2.1% of all insecticides in 2003 (Nauen 2006). Thus, interest in them is mainly retrospective. However, the use of nonpersistent neurotoxic pesticides is still widespread. OPs, carbamates, pyrethroids, and neonicotinoids accounted for 24.7%, 10.5%, 19.5%, and 15.7% of global sales of insecticides, respectively, in 2003 (Nauen 2006). Taken collectively, this represents some 70% of all insecticide sales during that year. So, questions remain because little is known about the importance or otherwise of sublethal neurotoxic and behavioral effects or consequent population effects that these compounds may be having in the natural environment.

16.8 CONCLUDING REMARKS

There is much evidence that neurotoxic pollutants, mainly pesticides, have had both lethal and sublethal effects upon free-living vertebrates in the natural environment. Lethal effects have, for obvious reasons, been much easier to recognize than sublethal ones. At the same time, the mere fact that neurotoxic compounds have caused mortality is, in itself, clear evidence that there must have been sublethal effects as well, although the latter were seldom recognized at the time. As has been shown in many well-designed studies, there are a variety of readily measurable neurotoxic and behavioral effects in the early stages of poisoning by OCs, OPs, carbamates, pyrethroids, and neonicotinoids before the onset of symptoms of severe poisoning and death. Animals dying from poisoning in the field would have shown these symptoms in the early stages of intoxication, as in the case of birds and mammals showing convulsions before succumbing to dieldrin poisoning in field incidents during the late 1950s and early 1960s. By contrast, there would have been many cases of individuals experiencing lower exposures and showing early symptoms of poisoning but not receiving high enough doses to kill them outright. Such individuals may have recovered completely and gone on to lead "normal" lives, or the sublethal effects may have had harmful consequences in the shorter or longer term by reducing ability to feed, breed, or avoid predation.

With recent advances in biochemical toxicology, incorporating new techniques of molecular biology, it is now possible to develop better mechanistic biomarker assays that will facilitate the identification and quantification of the different changes in the sequence of events that underlie neurotoxicity. In this respect, medical toxicology is much further advanced than ecotoxicology. However, techniques developed for the former should be applicable to the latter. Microarray assays to monitor changes at the level of the gene can run alongside assays to show changes at the cellular level (e.g., interaction with sites of action, electrophysiological responses). Appropriate combinations of assays can give an in-depth picture of the operation of this causal chain, which can then be related to behavioral and other whole-organism responses to neurotoxic pollutants.

In the first place, this approach can be adopted in field studies of polluted areas where neurotoxic effects are suspected on the basis of circumstantial evidence, ecological profiling, or the results of bioassays, or any combination of these (see Chapter 13, Section 13.4). Once a polluted area has been identified, "clean" indicator organisms may be deployed from the laboratory into this area. For comparison, the same indicator organisms can also be deployed to a reference area that is relatively unpolluted and can act as a control. Biomarker responses such as acetylcholinesterase inhibition or changes in the electrophysiological properties of nerves can then be measured in the deployed individuals. Thus, evidence may be sought for the operation of neurotoxic mechanisms—as explained in the foregoing text—and those pollutants responsible for the toxic effects identified and quantified by chemical analysis. Apart from investigations of this kind, this approach is also useful in field trials of pesticides and other chemicals. Fish and other aquatic species have been studied in this way (see Chapter 15 for examples).

Arguments are bound to be raised about the cost of such an approach, but the important point is that much may be learned about the ecotoxicology of neurotoxic pollutants from a few well-designed long-term investigations that can act as case studies to give guidance when dealing with pollution problems with neurotoxic compounds more generally. Knowledge gained in this way will be valuable—and should be cost effective—in the longer term. A lot of money is spent on limited short-term tests and short-term projects in ecotoxicology that contribute little or nothing to a more fundamental understanding of the harmful effects of chemicals upon natural ecosystems in the longer term.

In a similar way, an integrated biomarker approach has a role when carrying out experiments in mesocosms. Under these controlled conditions, behavioral effects of neurotoxic pollutants, acting singly or in combination, can be monitored and compared with data on predator–prey relationships and effects at the population level. The employment of mechanistic biomarker assays can facilitate comparisons between results obtained in mesocosms and other data obtained in the field or in laboratory tests. Here is one way of attempting to answer the difficult question— "how comparable are mesocosms to the real world"?

There is a continuing interest in the development of biomarker assays for use in environmental risk assessment. As discussed elsewhere (Section 16.6), there are both scientific and ethical reasons for seeking to introduce in vitro assays into protocols for the regulatory testing of chemicals. Animal welfare organizations would like to see the replacement of toxicity tests by more animal-friendly alternatives for all types of risk assessment—whether for environmental risks or for human health.

Considering risk assessment generally, Dewar (1983) and Atterwill et al. (1991) have reviewed the subject of alternative procedures for testing neurotoxic compounds. Atterwill et al. (1991) give details of a number of in vitro tests that might be developed for this purpose and propose a stepwise scheme for neurotoxicity testing that incorporates some of them. However, they and other authorities on the subject stress the difficulty of devising a testing protocol based on in vitro assays alone because of the complexity of the nervous system. More recently, in a report by the

European Centre for the Validation of Alternative Methods (ECVAM), six in vitro systems for chronic neurotoxicity testing are recommended for further consideration (Worth and Balls 2002). These are described as in vitro models that may be suitable for long-term toxicity testing. The systems are

1. Primary neuronal cells (rat) and their reaggregates
2. Permanent neuronal cells
3. Astrocytes
4. Oligodendrocytes
5. Microglia
6. Brain slices from hippocampus

In the summary of the aforementioned report, the authors recommend, as did earlier reviewers of this subject, the development and evaluation of a tiered testing strategy for neurotoxicity. The further development of in vitro models for establishing mechanisms of neurotoxicity should be part of this strategy. Full consideration should also be given to advances in the "omics" and other technological fields.

Iversen (1991) stresses the need for some in vivo testing for neurotoxicity and emphasizes the value of sensitive behavioral tests. Behavioral tests are described for mice and rats, which provide measures of mood, posture, CNS excitation, motor coordination, sedation, exploration, responsiveness, learning, and memory function. Such assays can function as primary screens for neurotoxicity before adopting a "stepwise" scheme of in vitro tests to discover more about the initial site of action of neurotoxic compounds. It is argued that the requirement for animal testing can be drastically reduced by adopting structured in vitro protocols such as these.

The foregoing proposals were made particularly with the requirements of human risk assessment in mind. There are differences when considering tests for environmental risk assessment. The ultimate concern here is about the risk of causing adverse effects at the population level rather than about effects upon individuals. As we have seen, some population declines of birds, have been explained, at least in part, by the lethal toxicity of neurotoxic compounds (e.g., effects of dieldrin upon certain raptors). In the case of the lethal poisoning of geese in the U.K. by carbophenothion, local population declines were related to the insecticide by the use of a biomarker assay, acetylcholinesterase inhibition, thus ruling out other possible causes of mortality (Stanley and Bunyan 1979). However, it is now clear that some population declines caused by pollutants in the natural environment have been due to sublethal effects rather than lethal ones (e.g., organotin compounds causing imposex in the dog whelk, p,p'-DDE causing eggshell thinning in some predatory birds). Thus, measuring biomarker responses and relating them to population effects should be of greater value than simply using lethality as an end point in ecotoxicity testing; lethality is only one of the factors that can cause a population to decline.

The relationship between biomarker responses and effects at the population level can be tested in both field experiments and more controlled experiments in mesocosms. It may be possible to define thresholds for biomarker assays performed on indicator species, above which population effects have been shown to occur. Indicator species may be either free living or deployed. The advantage of the latter is greater experimental control. Laboratory-reared fish, for example, can be deployed into polluted and clean waters and comparisons made between biomarker responses in the two cases. The importance of behavioral factors in population ecology needs to be emphasized. So, too, does the sensitivity of certain behavioral functions to the effects of neurotoxic compounds, a sensitivity that is evident from some of the assays (e.g., on fish) that were described earlier.

From a scientific point of view, behavioral assays can provide a sensitive indication of neurotoxic effects and, moreover, one that presents an integrated measure of a response of the whole organism to a pollutant or mixture of pollutants. As such, it seems reasonable to suppose that behavioral effects might well relate to consequent changes at the population level. Also, behavioral assays should be useful as primary screens for detecting neurotoxicity before testing with other, mechanistic biomarker assays to identify mode of action. From the ethical point of view, however, questions remain about whether a greater use of behavioral assays would lead to a reduction in use of animals in toxicity testing procedures and animal suffering.

16.9　SUMMARY

The nervous systems of vertebrates and invertebrates are susceptible to the toxic action of many chemicals, both human-made and naturally occurring. Five major classes of insecticides—the organochlorine, carbamate, organophosphorous, pyrethroid, and nicotinic insecticides—all act in this way. So, too, do a variety of naturally occurring neurotoxins, including pyrethrins, nicotine, physostigmine, picrotoxinin, and many others. Acetylcholinesterase, sodium channels, GABA receptors, and nicotinic receptors are all examples of sites of action in the nervous system for both human-made and naturally occurring neurotoxic compounds.

Toxic effects and sometimes associated population declines in wild vertebrates and invertebrates have been attributed to the action of neurotoxic pesticides, and a number of examples have been discussed here. In these field studies, it has been much easier to recognize lethal toxic effects than sublethal ones, although it is clear that the former could not have occurred without the latter. Thus, interest has grown in the use of biomarker assays that have the potential to measure the sequence of changes that occur in animals exposed to neurotoxic compounds—changes that may lead to neurophysiological and behavioral disturbances and finally death. Such an approach can give a better understanding of the phenomenon of neurotoxicity in the earlier stages of intoxication and an ability to recognize it and quantify it in the laboratory and in the field. The employment of combinations of biomarker assays operating at different levels of biological organization can be used to assess the overall effect of mixtures of neurotoxic chemicals acting through contrasting mechanisms.

Apart from the use of this approach to study the ecotoxicology of neurotoxic pollutants in the field, it also has potential for use during the course of environmental risk assessment. An understanding of the relationship between biomarker responses to neurotoxic compounds and effects at the population level can be gained from both field studies and the use of mesocosms and other model systems. From these it may be possible to define critical thresholds in biomarker responses of indicator species above which population effects begin to appear. In the longer term, this approach should yield new assays and strategies for environmental risk assessment that will be better scientifically and more acceptable ethically than many of the practices followed at the present time.

FURTHER READING

Atterwill, C.K. et al. (1991). *Alternative Methods and Their Application in Neurotoxicity Testing*—Describes a range of in vitro tests for neurotoxicity and proposes a "stepwise scheme" for neurotoxicity testing.

Ballantyne, B.C. and Marrs, T.C. (Eds.) (1992). *Clinical and Experimental Toxicology of Organophosphates and Carbamates*—A wide-ranging collection of chapters giving a broad coverage of the toxicology of these important neurotoxic compounds.

Eldefrawi, M.E. and Eldefrawi, A.T. (1991). *Nervous-System-Based Insecticides*—Describes the mechanisms of action of a wide range of neurotoxic compounds, both human-made and naturally occurring.

17 Organic Pollutants
Future Prospects

17.1 INTRODUCTION

During the second half of the 20th century, it was discovered that a number of organic pollutants (OPs) were having harmful side effects in natural ecosystems, prominent among which were chemicals that combined high toxicity (lethal or sublethal) with marked biological persistence. Examples such as dieldrin, DDT, TBT, and methyl mercury are represented in Part 2 of the present book. Following these discoveries, restrictions and bans on the release of these chemicals into the environment were introduced in many countries. Persistent organochlorine (OC) insecticides, for example, were withdrawn from many uses and replaced by less persistent OP and carbamate insecticides. More stringent legislation was brought in to control the production and marketing of new chemicals, with clearer guidelines for environmental risk assessment. Particularly strict rules were applied to new pesticides—something of a double-edged weapon. This tightening of regulations has reduced the risk of new pesticides creating new problems, but it may also have impeded the discovery and registration of newer, more environmentally friendly compounds by making research and development too expensive. In spite of the immediate advantages tighter regulations bring, they can, in the long run, be counterproductive.

Since the introduction of these restrictions and bans on persistent pesticides, it has been discovered that other less persistent compounds can also cause environmental problems. Some highly toxic insecticides, including the carbamate aldicarb used as a granular formulation, and the organophosphorous compounds carbophenothion and chlorfenvinphos used as seed dressings, have been responsible for poisoning incidents on agricultural land (Hardy 1990). Also, tributyl tin, an antifouling agent used in marine paints, has been shown to have serious effects upon aquatic mollusks, including oysters and dog whelks (Chapter 8, Section 8.3 of this book). So, with the tightening of the rules, other compounds of lesser persistence have been subject to restrictions and bans. At the same time, some new pesticides have come on to the market that are regarded as being more environmentally friendly. Among the insecticides, newer pyrethroids and neonicotinoids fall into this category.

With these developments, it would appear that many of the more obvious environmental problems relating to particular compounds or groups of compounds have now disappeared. At least, this seems to be so in the developed world where there are now strict controls of environmental pollution that are reasonably well enforced. However, this does not necessarily apply to third-world countries where there is not such strict control. One consequence of this trend in developed countries has been for ecotoxicological research and the development of ecotoxicity testing procedures to move toward strategies for testing the toxicity of complex mixtures, the components of which are usually at low concentration. More attention, for example, has been paid to environmental contamination by low levels of pharmaceuticals. Where these have been found to be present in complex mixtures, questions have been asked about the possibility of potentiation of toxicity.

Among pharmaceuticals, EE2 has been the subject of particular recent attention because of its ability to cause endocrine disruption in fish, as has been described in Chapter 15. Low levels of mixtures of beta blockers, such as propranolol, metoprol, and nadolol have been detected in surface waters, and there have been investigations of their possible effects on aquatic invertebrates (Huggett et al. 2002). Veterinary medicines, too, have come under scrutiny: for example, the dramatic effects of diclofenac on vultures, which will be discussed shortly. Many questions remain to be answered about the possible ecological effects of complex mixtures of pharmaceuticals and veterinary medicines.

This trend toward studying the effects of low levels of chemicals existing in complex mixtures has been evident in much of the third part of the present text. Some researchers have gone so far as to suggest that ecotoxicology might now be seen as an aspect of stress ecology: that the toxic action of environmental chemicals is just one of a number of stress factors to which free-living organisms are exposed, and that stress factors should be considered as a whole when studying populations (Van Straalen 2003).

This approach has scientific merit, and the desirable situation may exist in some areas of the world where the effect of pollutants is no more important than that of other stress factors. However, there are other areas where this is not the case, and serious problems of pollution still exist—areas where, in other words, chemical stress substantially outweighs other stress factors and is the driving force behind changes in population numbers or the genetic composition of populations. Even in North America, there are areas, some of them Superfund sites, where serious pollution still exists, and certain organic pollutants are having effects on natural populations. In the present text, examples were given of ongoing problems in certain areas with high levels of organomercury (Chapter 8, Section 8.2) and of PCBs and dioxins (Chapters 6 and 7). More dramatically, there have been other instances, sometimes with chemicals not previously considered as important pollutants, in countries less well developed than the United States or Canada. For example, a few years ago, a pollution problem came to light in India and Pakistan that was as disturbing for the Zoroastrian community as it was for conservationists. Three species of vulture declined rapidly during the decade 1997–2006 (Green, Taggart, and Das 2006). The species affected were *Gyps bengalensis, Gyps indicus*, and *Gyps tenuirostris*. Further investigation has strongly implicated the nonsteroidal antiinflammatory drug diclofenac in the death of these birds over a wide area (Schultz et al. 2004 and Green, Taggart, and Das 2006). Many of the birds found dead over this large area contained residues of diclofenac. Many also had visceral gout, which is a condition strongly associated with diclofenac toxicity. They evidently obtained residues of this compound by feeding on dead cattle that had recently been dosed with it. A toxicological investigation into the uptake of diclofenac by vultures concluded that more than 10% of the birds could acquire a lethal dose in a single meal if they were feeding on cattle that had died within two days of receiving of the drug. So, once again, a chlorinated compound that shows a fair degree of persistence in vertebrates has been implicated in lethal poisoning and population decline in the field. Speaking more widely, there continue to be cases of gross misuse of pesticides and other organic pollutants in some parts of the world, including on humans who use them. In India, for example, there are still reports of organophosphorous poisoning of spray operators working in crops such as cotton. We know little of the ecological consequences of the misuse of such pesticides.

The following sections will attempt to look ahead to likely future problems with organic pollution, to probable changes in the ways in which it is studied and monitored, and in the tests and strategies used for environmental risk assessment of organic chemicals.

17.2   THE ADOPTION OF MORE ECOLOGICALLY RELEVANT PRACTICES IN ECOTOXICITY TESTING

Currently, the environmental risk assessment of chemicals for registration purposes depends on the comparison of two things: (1) An estimate (sometimes a measure) of the environmental concentration of a chemical, and (2) an estimate of the environmental toxicity of this chemical. Environmental concentration is difficult to estimate, especially for mobile species in terrestrial ecosystems. In the present example, the estimation of environmental toxicity is expressed as a concentration, which may be a No Observable Effect Concentration (NOEC) or an LC_{50} for the most sensitive organism found in a series of ecotoxicity tests. Very seldom is the species used in the toxicity test one of those most at risk in the natural environment. Nearly always a surrogate is used, and there is the immediate question of species differences in susceptibility, which are largely unknown. In the case of birds, for example, two or three species are used in ecotoxicity testing, whereas some 8700 species exist in nature. For further discussion of these issues, see Chapter 6 in Walker et al. (2000), Calow (1993), and Walker (1998b). Because of the high levels of uncertainty involved, the estimate of environmental toxicity is divided by a large safety factor, commonly 1000. Following these computations, there is perceived to be a risk if the estimate of environmental concentration (1) exceeds the estimate of environmental toxicity (2).

The limitations of this approach are not difficult to appreciate (see, for example, Kapustka, Williams, and Fairbrother 1996). It is based on the approach to risk assessment used in human toxicology and has been regarded as the best that can be done with existing resources. It is concerned with estimating the likelihood that there will be a toxic effect upon a sensitive species following the release of a chemical into the environment. With the very large safety factors that are used, it may well seriously overestimate the risks presented by some chemicals. More fundamentally, it does not address the basic issue of effects upon populations, communities, or ecosystems. Small toxic effects may be of no significance when it comes to possible harmful effects at these higher levels of biological organization, where population numbers are often controlled by density-dependent factors (Chapter 4). Also, it does not deal with the question of indirect effects. As mentioned earlier (Chapter 14), standard environmental risk assessment of herbicides would have given no indication that they could be the indirect cause of a decline of the grey partridge on agricultural land.

There has been growing pressure from biologists for the development of more ecologically relevant end points when carrying out toxicity testing for the purposes of environmental risk assessment (see Walker et al. 1998b, and Chapter 12 in Walker et al. 2000). In concept, populations will decline when pollutants, directly or indirectly, have a sufficiently large effect on rates of mortality and/or rates of recruitment to reduce population growth rate (Chapter 4, Section 4.4). Thus, sublethal effects, such as on reproduction or behavior, can be more important than lethal ones. If pollutant effects can be quantified in this way, for example, through the use of biomarker assays for toxic effect (see Chapter 15, Section 15.4), then better risk assessment is made possible by including them in appropriate population models. In practice, this approach is still at an early stage of development; it is a research strategy that can only be used in a few cases, and cannot yet deal with the large numbers of compounds submitted for risk assessment. Nevertheless, looking at the problem from this more fundamental point of view does suggest certain improvements that could be made in the protocols for environmental risk assessment.

A large proportion of the resources currently being spent on the determination of LD_{50} levels for birds or LC_{50} levels of fish could be diverted to more relevant testing procedures. At best these values give only a rough indication of lethal toxicity in a small number of species. A ranking of compounds with respect to toxicity, for example, low toxicity, moderate toxicity, etc., is good enough for such a crude and empirical approach; knowing particular values a little more precisely does practically nothing to improve the quality of this type of environmental risk assessment where the uncertainties are so big. In the first place, greater consideration of ecological aspects before embarking on testing should lead to the selection of more appropriate species, life stages, and end points in the testing protocol. It might be useful, for example, to include tests on behavioral effects if testing a neurotoxic pesticide, or of reproductive effects if testing a compound that can disturb steroid metabolism. These are mechanisms the operation of which might be expected to have adverse ecological effects. In a number of instances, population declines have been the consequence of reproductive failure (e.g., effects of p,p'-DDE on shell thickness of raptors, effects of PCBs and other polychlorinated compounds on reproduction of fish-eating birds in the Great Lakes, and the effects of TBT on the dog whelk). Effects on behavior may affect breeding and feeding and lead to population decline.

In some species there may be good reasons for looking at the toxicity of certain types of compounds in early developmental stages (e.g., avian embryos, larval stages of amphibians) rather than adults. In short, testing protocols should be more flexible so that there can be a greater opportunity for expert judgment rather than following a rigid set of rules. Knowledge of the metabolism and the mechanism of action of a new chemical may suggest the most appropriate end points in toxicity testing. Indeed, mechanistic biomarkers can provide better and more informative end points than lethality; they can be used to monitor progression through sublethal (including subclinical) effects before lethal tissue concentrations are reached.

An approach that has gained attention recently is the use of model ecosystems: microcosms, mesocosms, and macrocosms for testing chemicals (Chapter 4,

Section 4.5). Of these, mesocosms have stimulated the greatest interest. In these, replicated and controlled tests can be carried out to establish the effects of chemicals upon the structure and function of the (artificial) communities they contain. The major problem is relating effects produced in mesocosms to events in the real world (see Crossland 1994). Nevertheless, it can be argued that mesocosms do incorporate certain relationships (e.g., predator/prey) and processes (e.g., carbon cycle) that are found in the outside world, and they test the effects of chemicals on these. Once again, the judicious use of biomarker assays during the course of mesocosm studies may help to relate effects of chemicals measured by them with similar effects in the natural environment.

During the period leading up to the implementation of the REACH (Registration, Evaluation, Authorization, and Restriction of Chemicals) proposals by the European Commission in 2006, there was considerable public debate about the procedures that are laid down in it for ecotoxicity testing of industrial chemicals (Walker 2006). One strongly debated issue relevant to the present discussion was the operation of the tiered testing system recommended in it. The recommended testing protocols were determined by the level of production or importation of particular chemicals. Although this may be a convenient system to operate from a bureaucratic point of view, it inevitably has encountered a good deal of criticism from scientists, including the U.K. Royal Commission on Environmental Pollution. The biggest objection is that the scale of production or exportation of a chemical bears a very uncertain relationship to the actual exposure of free-living organisms to it, and it is the latter issue that is important in the context of environmental risk assessment. From an ecological point of view, testing protocols should be decided on the basis of estimated environmental exposure. Despite these objections, a tiered system linked to annual production has now been accepted by the European Commission. The improvement of testing procedures is a slow business.

17.3 THE DEVELOPMENT OF MORE SOPHISTICATED METHODS OF TOXICITY TESTING: MECHANISTIC BIOMARKERS

The judicious use of mechanistic biomarkers can, in theory, overcome many of the basic problems associated with establishing causality. In the field, they can be used to measure the extent to which pollutants act upon wild species through defined toxic mechanisms, thus giving more insight into the sublethal as well as lethal effects of chemicals. Most important, they can provide measures of the integrated effects of mixtures of compounds operating through the same mechanism, measures that take into account potentiation at the toxicokinetic level (Chapter 13, Section 13.4). With the advent of new biomarker technology such as that of the "omics" (See Chapter 4, Box 4.2), they can be used to study the effect of complex mixtures containing chemicals working through contrasting mechanisms of action (see Chapters 15 and 16). In theory, biomarkers can provide the vital link between known levels of exposure and changes in mortality rates or recruitment rates; estimates of mortality rates and recruitment rates so obtained can then be incorporated into population models (Chapter 4, Section 4.3). More explicitly, graphs can be generated that link a biomarker response to a population parameter, as has already been achieved with eggshell thinning in the sparrowhawk induced by p,p'-DDE and imposex in the dog whelk caused by TBT. Currently, because of the shortage of appropriate biomarker assays, this approach lies largely in the realm of research and cannot be applied to most problems with environmental chemicals.

At the practical level, an ideal mechanistic biomarker should be simple to use, sensitive, relatively specific, stable, and usable on material that can be obtained by nondestructive sampling (e.g., blood or skin). A tall order, no doubt, and no biomarker yet developed has all of these attributes. However, the judicious use of combinations of biomarkers can overcome the shortcomings of individual assays. The main point to emphasize is that the resources so far invested in the development of biomarker technology for environmental risk assessment has been very small (cf. the investment in biomarkers for use in medicine). Knowledge of toxic mechanisms of organic pollutants is already substantial (especially of pesticides), and it grows apace. The scientific basis is already there for technological advance; it all comes down to a question of investment.

As mentioned earlier, the development of bioassay techniques is one important aspect of biomarker technology. Cell lines have been developed for species of interest in ecotoxicology, for example, birds and fish (Pesonen et al. 2000), and have sometimes been genetically manipulated (e.g., with incorporation of receptors and reporter genes) to facilitate their employment as biomarker assays (Walker 1998b). In principle, it should be possible to conserve the activities of enzymes concerned with detoxication and activation in these cellular systems so that the toxicokinetics of the in vitro assay are comparable to those of the living animal. Bioassays with such cellular systems could be developed for species of ecotoxicological interest that are not available for ordinary toxicity testing, which would go some ways in overcoming the fundamental problem of interspecies differences in toxicity. One difficulty encountered with cell lines has been that of gene expression. Enzymes concerned with detoxication or activation have sometimes not been expressed in cell systems. However, work with genetically manipulated cell lines has begun to overcome this problem (Glatt et al. 1997).

Interest has been expressed in the possibility of using biomarker assays as a part of risk assessment for regulatory purposes, and some workers have suggested tiered testing procedures that follow this approach (see, for example, Handy et al. 2003). It is to be hoped that regulatory schemes, such as that of REACH (see European Union 2003), will be sufficiently flexible to incorporate new assays and testing strategies as the science advances.

17.4   THE DESIGN OF NEW PESTICIDES

It is not surprising that many of the organic pollutants that have caused environmental problems have been pesticides. Pesticides are designed with a view to causing damage to pests, and selectivity between pests and other organisms can only be achieved to a limited degree. In designing new pesticides, manufacturers seek to produce compounds of greater efficacy, cost effectiveness, and environmental safety than are offered by existing products (for an account of the issues involved in the development of new, safer insecticides, see Hodgson and Kuhr 1990). Sometimes the driving force behind pesticide innovation is to overcome a developing resistance problem when existing products become ineffective against major pests. It may also be to provide a product that is more environmentally safe than those currently on the market. Innovation, however, is to some extent hampered by escalating costs—not least, the costs associated with ecotoxicity testing and environmental risk assessment.

With the rapid growth of knowledge in the field of biochemical toxicology, it is becoming increasingly possible to design new pesticides based on structural models of the site of action—the QSAR (Quantitative Structure–Activity Relationship) approach (see Box 17.1). Sophisticated computer graphic systems make life easier for the molecular modeler. The discovery and development of EBI fungicides as inhibitors of certain forms of P450 provide an example of the successful application of this approach.

There has also been rapid growth in understanding of the enzyme systems that metabolize pesticides and other xenobiotics (see Chapter 2 of this book, and Hutson and Roberts 1999). As more is discovered about the mechanisms of catalysis by P450-based monooxygenases, esterases, glutathione-S-transferases, etc., it becomes easier to predict the routes and rates of metabolism of pesticides. In principle, it has become easier to design readily biodegradable pesticides that have better selectivity than existing products (there are large species differences in metabolism that can be exploited). It should also be possible to design pesticides that not only overcome resistance but are also selectively toxic toward resistant strains.

**BOX 17.1   QUANTITATIVE STRUCTURE–ACTIVITY
RELATIONSHIPS (QSARS)**

There has long been an interest in mathematical relationships between chemical structure and toxicity, and the development of models from them that can be used to predict the toxicity of chemicals (see Donkin, Chapter 14 in Vol. 2 of Calow 1993). If considering groups of compounds that share the same mode of action, much of the variation in toxicity between different molecules is related to differences in cellular concentration when the same dose is given. In other words, toxicokinetic differences are of primary importance in determining selective toxicity (see Chapter 2, Section 2.2). The simplest situation is represented by nonspecific narcotics, which include general anesthetics. Toxicity here is related to the relatively high concentrations that the compounds reach in biological membranes, and is not due to any specific interaction with cellular receptors (see Chapter 2, Section 2.3). Simple models can relate chemical properties to both cellular concentration and toxicity. Good QSARs have been found for narcotics when using descriptors for lipophilicity such as log K_{ow}. For example, the following equation relates the hydrophobicity of members of a group of aliphatic, aromatic, and alicyclic narcotics to their toxicity to fish.

$$\log 1/LC_{50} = 0.871 \log K_{ow} - 4.87 \qquad \text{(Konemann 1981)}$$

Other much more toxic compounds operating through specific biochemical mechanisms (e.g., OP anticholinesterases) cannot be modeled in this way. If toxicity were to be plotted against log K_{ow}, such compounds would be represented as "outliers" in relation to the straight line provided by the data for the narcotics (Lipnick 1991). Their toxicity would be much greater than predicted by the simple hydrophobicity model for the narcotics. For such compounds, more sophisticated QSAR equations are required that bring in descriptors for chemical properties relating, for example, to their ability to interact with a site of action. An example: of such an equation relates the properties of OPs to their toxicity to bees (Vighi et al. 1991):

$$log1/LD_{50} = 1.14 \log K_{ow} - 0.28 [\log K_{ow}]^2 + 0.28^2 X$$
$$- 0.76^2 Xox - 1.09Y3 + 0.096[Y3]^2 + 12.29$$

where X and Y are chemical descriptors for reactivity with the active site of cholinesterase. The K_{ow} is for the active "oxon" form of an OP.

In general, it is easier to use models such as these to predict the distribution of chemicals (i.e., relationship between exposure and tissue concentration) than it is to predict their toxic action. The relationship between tissue concentration and toxicity is not straightforward for a diverse group of compounds, and depends on their mode of action. Even with distribution models, however, the picture can be complicated by species differences in metabolism, as in the case of models for bioconcentration and bioaccumulation (see Chapter 4). Rapid metabolism can lead to lower tissue concentrations than would be predicted from a simple model based on K_{ow} values. Thus, such models need to be used with caution when dealing with different species.

Also of continuing interest is the identification of naturally occurring compounds that have biocidal activity (Hodgson and Kuhr 1990, Copping and Menn 2000, Copping and Duke 2007). These may be useful as pesticides in their own right, or they may serve as models for the design of new products. Examples of natural products that have already served this function include pyrethrins, nicotine, rotenone, plant growth regulators, insect juvenile hormones, precocene, avermectin, ryanodine, and extracts of the seed of the neem tree (*Azadirachta indica*) (see Copping and Duke 2007, Otto and Weber 1992, and Hodgson and Kuhr 1990). It is likely that natural products will continue to be a rich source of new pesticides or models for new commercial products in the years ahead. A vast array of natural chemical weapons have been produced during the evolutionary history of the planet, and many are still awaiting discovery (Chapter 1).

17.5   FIELD STUDIES

Ecotoxicology is primarily concerned with effects of chemicals on populations, communities, and ecosystems, but the trouble is that field studies are expensive and difficult to perform and can only be employed to a limited extent. In the main, environmental risk assessment of pesticides and certain other chemicals has to be accomplished by other means. With the registration of pesticides, large-scale field studies are occasionally carried out to resolve questions that turn up in normal risk assessment (Somerville and Walker 1990) but are far too expensive and time consuming to be used with any regularity. Lack of control of variables and the difficulty of achieving adequate replication are fundamental problems. However, the development of new strategies, and the development of new biomarker assays could pave the way for more informative and cost-effective investigations of the effects of pollutants in the field. Small-scale field studies and semi-field studies are used for risk assessment of pesticides to bees (Thompson and Maus 2007).

That said, long-term case studies of pollution by chemicals can give important insights into problems with other similar chemicals that may arise later on. Cases in point include long-term studies that have been carried out on persistent lipophilic compounds such as OC insecticides, PCBs, organomercury compounds, and organotin compounds, which have been described in this second section of this book. With the advance of science, results from well-conducted field studies can be looked at retrospectively to gain new insights—with the benefit of hindsight. In the final analysis, the natural environment is too complex to just make simple predictions with laboratory-based models, and there is no adequate substitute for hard data from the real world. It is important that long term in-depth studies of pollution of the natural environment continue.

The use of biotic indices in environmental monitoring is one way of identifying existing/developing pollution problems in the field (see Chapter 11 in Walker et al. 2000). Such ecological profiling can flag up structural changes in communities that may be the consequence of pollution. For example, the RIVPACS system can identify changes in the macroinvertebrate communities of freshwater systems (Wright 1995). It is important that adverse changes found during biomonitoring are followed up by the use of biomarker assays (indicator organisms or bioassays or both) and chemical analysis to identify the cause. As noted earlier, improvements in biomarker technology should make this task easier and cheaper to perform.

Biomarker assays can be used to establish the relationship between the levels of chemicals present and consequent biological effects both in controlled field studies (e.g., field trials with pesticides) and in the investigation of the biological consequences of existing or developing pollution problems in the field. In the latter case, clean organisms can be deployed to both clean and polluted sites in the field, and biomarker responses measured in them. Organisms can be deployed along pollution gradients so that dose-response curves can be obtained for the field for comparison with those obtained in the laboratory. An example of this approach was the deployment of *Mytilus edulis* along PAH gradients in the marine environment and the measurement of scope for growth (Chapter 9, Section 9.6). The challenge here is to take the further step and relate biomarker responses to population parameters so that predictions of population effects can be made using mathematical models. The predictions from the models can then be compared with the actual state of the populations in the field. The validation of such an approach should lead to its wider employment in the general field of environmental risk assessment.

17.6 ETHICAL QUESTIONS

There has been growing opposition to the use of vertebrate animals for toxicity testing. This has ranged from the extremism of some animal rights organizations to the reasoned approach of the Fund for the Replacement of Animals in Medical Experiments (FRAME), and the European Centre for the Validation of Alternative Methods (ECVAM) (see Balls, Bridges, and Southee 1991, issues of the journal *ATLA*, and publications of ECVAM at the Joint Research Centre, Ispra, Italy). FRAME, ECVAM, and related organizations advocate the adoption of the principles of the three Rs, namely, the reduction, refinement, and replacement of testing procedures that cause suffering to animals.

Regarding ecotoxicity testing, these proposals gain some strength from the criticisms raised earlier to existing practices in environmental risk assessment. There is a case for making testing procedures more ecologically relevant, and this goes in hand with attaching less importance to crude measures of lethal toxicity in a few species of birds and fish (Walker 1998b). The savings made by a substantial reduction in the numbers of vertebrates used for "lethal" toxicity testing could be used for the development and subsequent use of testing procedures that do not cause suffering to animals and are more ecologically relevant. Examples include sublethal tests (e.g., on behavior or reproduction), tests involving the use of nondestructive biomarkers, the use of eggs for testing certain chemicals, and the refinement of tests with mesocosms. Rigid adherence to fixed rules would prolong the use of unscientific and outdated practices and slow down much-needed improvements in techniques and strategies for ecotoxicity testing. Better science should, for the most part, further the aims of the three Rs.

17.7 SUMMARY

With the restrictions and bans placed in developed countries on a considerable number of environmental chemicals—especially on persistent and/or highly toxic pesticides—many serious pollution problems have been resolved and more attention has come to be focused on the effects of mixtures of organic pollutants, often at quite low concentrations. It has been argued by some that chemical pollution should be seen as part of stress ecology, that chemicals should be considered together with other stress factors to which free-living organisms are exposed. While this trend has been marked in developed countries, it has not necessarily been true of other countries where there is less control of environmental pollution by chemicals, and there are still some serious problems with certain organic pollutants.

With improvements in scientific knowledge and related technology, there is an expectation that more environmentally friendly pesticides will continue to be introduced, and that ecotoxicity testing procedures will become more sophisticated. There is much interest in the introduction of better testing procedures that work to more ecologically relevant end points than the lethal toxicity tests that are still widely used. Such a development should be consistent with the aims of organizations such as FRAME and ECVAM, which seek to reduce toxicity testing with animals. Mechanistic biomarker assays have the potential to be an important part of this approach, especially as they incorporate new technologies such as the "omics." They have potential for employment in field studies, providing the vital link between exposure to chemicals and consequent toxicological and ecotoxicological effects.

FURTHER READING

New developments are best followed by reading current issues of the leading journals in the field, which include *Environmental Toxicology and Chemistry, Ecotoxicology, Environmental Pollution, Environmental Health Perspectives, Bulletin of Environmental Contamination and Toxicology, Archives of Environmental Contamination and Toxicology, Functional Ecology, Applied Ecology,* and *Biomarkers.*

Glossary

AChE (acetylcholinesterase): An enzyme that hydrolyzes acetylcholine.

Adducts: In the context of toxicology; products of stable linkages between xenobiotics and endogenous molecules (e.g., between PAH metabolites and DNA).

Ah receptor (Aryl hydrocarbon receptor): A receptor located on a cytoplasmic protein to which planar compounds such as PAHs, coplanar PCBs, and PCDDs bind. Binding initiates the induction of cytochrome P450IAI/2.

Ah-receptor-mediated toxicity: Toxic effects associated with the binding of polychlorinated aromatic compounds such as coplanar PCBs and PCDDs to the Ah receptor.

Alkaloids: A diverse group of nitrogen-containing organic compounds synthesized by plants, many of which show biological activity.

Antagonism: With reference to toxicity: where the toxicity of a mixture is less than the sum of toxicities of its components.

Anthropogenic: Generated by the activities of humans.

Anticoagulant rodenticides (ARs): Rodenticides that cause hemorrhaging, usually through disturbing the synthesis of clotting proteins (e.g., warfarin, brodifacoum, difenacoum).

Aryl: Aromatic moiety.

ATPases: Adenosine triphosphatases.

Bioaccumulation factor (BAF): Concentration of a chemical in an animal, or concentration of the same chemical in its food.

Bioconcentration factor (BCF): Concentration of a chemical in an organism or concentration of same chemical in the ambient medium.

Biomagnification: Increase in concentration of a chemical in living organisms with passage along a food chain.

Biomarker: A biological response to a chemical at the individual level or below, demonstrating a departure from normal status. Usually restricted to responses at the level of the whole organism or below.

Biotransformation: Conversion of a chemical into one or more products by a biological mechanism (predominantly by enzyme action).

Carbanion: Chemical moiety bearing a negative charge on a carbon atom.

Carbene: Free radical with two unpaired electrons on a carbon atom.

Carbonium ion: Chemical moiety bearing a positive charge on a carbon atom.

Carboxylesterases: Esterases that hydrolyze organic compounds with carboxylester bonds. Carboxylesterases that are inhibited by organophosphates (OPs) belong to the category EC 3.1.1.1 in the IUB classification of enzymes.

Carcinogen: A substance that can cause cancer.

Cholinesterase (ChE): A general term for esterases that hydrolyze cholinesters.

Cholinergic: Associated with the neurotransmitter acetylcholine.

Congener: A member of a group of structurally related compounds.

Conjugate: In biochemical toxicology, a structure (often an anion) formed by the combination of a xenobiotic (usually a phase I metabolite) with an endogenous component (e.g., glucuronate sulfate or glutathione).

Coordination: In chemistry, the donation of electrons by one atom to another in bond formation.

Cyclodienes: A group of organochlorine (OC) insecticides, some of which are highly toxic and persistent (e.g., aldrin, dieldrin, and heptachlor).

Cytochrome P450: A hemeprotein that catalyzes many biological oxidations (see also microsomal monooxygenases).

p,p'-DDE: p,p'-Dichlorodiphenyldichloroethylene (stable metabolite of p,p'-DDT)

p,p'-DDT: p,p'-Dichlorodiphenyltrichloroethane (main insecticidal component of the insecticide DDT).

EBI fungicides: Ergosterol biosynthesis inhibitors used as fungicides.

$EC[D]_{50}$: Concentration (dose) that has an effect on 50% of a population. Also known as *median effect concentration* (dose).

Ecotoxicogenomics: Application of genomics to ecotoxicology.

Electrophile: An electron-seeking atom or group.

Endocrine disruptors: Chemicals that cause disturbances of the endocrine system (e.g., by acting as agonists or antagonists of the estrogen receptor).

Endoplasmic reticulum: Membranous network of cells that contains many enzymes that metabolize xenobiotics. Hepatic microsomes consist mainly of vesicles derived from the endoplasmic reticulum of liver.

Epoxide hydrolase: A type of enzyme that converts epoxides to diols by the addition of water.

Ester: An organic salt that yields an acid and a base when hydrolyzed.

Esterases: Enzymes that hydrolyze esters.

Eukaryotes: Organisms that contain DNA within their nuclei.

Free radical: A molecule or atom possessing an unpaired electron.

Fugacity: Tendency of a chemical to escape from the phase in which it is located to another phase (e.g., from liquid to gas).

GABA: Gamma aminobutyric acid, a neurotransmitter. Acts upon GABA receptors located especially in the nervous system.

Genome: The complete set of genes for a particular species.

Genomics: The study of how the genome translates into biological function.

Genotoxic: Toxic by acting on genetic material, especially DNA.

Gla proteins: Proteins containing residues of gamma carboxy glutamate, including clotting proteins of the blood. They are able to bind calcium ions.

Glucuronyl transferases: A group of enzymes that catalyze the formation of conjugates between glucuronide and a xenobiotic (usually a phase I metabolite).

Glutathione-S-transferases: A group of enzymes that catalyze the formation of conjugates between reduced glutathione and xenobiotics.

Hydrophilic (water loving): Polar organic compounds tend to be hydrophilic.

Hydrophobic (water hating): Nonpolar organic compounds are hydrophobic.

Immunotoxicity: Toxicity to the immune system.

Imposex: The imposition of male characteristics on females in prosobranch mollusks, especially dog whelk (*Nucellus lapillus*).

Induction: With reference to enzymes: an increase in activity due to an increase in their cellular concentrations. This may be a response to a xenobiotic, and often involves an increased rate of synthesis of the enzyme.

Isoenzymes (isozymes): Enzymes that are very similar to one another in size and structure but with differences in catalytic ability.

K_{ow}: Octanol–water partition coefficient.

Ligand: A compound with specific binding properties.

Ligandin: A form of glutathione-S-transferase with a marked capacity for binding certain lipophilic xenobiotics.

Lipophilic (lipid loving): Such organic compounds tend to be of low polarity and are hydrophobic.

Lipoproteins: Macromolecules that are associations of lipids with proteins. Involved in the transport of both lipids and lipophilic xenobiotics in the blood.

Mechanistic biomarker: A biomarker that provides a measure of a toxic effect (some biomarkers only measure exposure). In the simplest case, this involves the direct measurement of the operation of a mechanism of toxicity (e.g., of acetylcholinesterase inhibition).

Metabolomics: The study of metabolites.

Microcosm, mesocosm, and macrocosm: Small, medium, or large multispecies system in which effects of chemicals can be studied.

Microsomes: Vesicles obtained from homogenized tissues by ultracentrifugation. They are derived mainly from the endoplasmic reticulum in the case of the liver (hepatic microsomes).

Mitochondrion: A subcellular organelle in which oxidative phosphorylation occurs, leading to the generation of ATP.

Monooxygenases (MOs): Enzyme systems of the endoplasmic reticulum of many cell types, which can catalyze the oxidation of a great diversity of lipophilic xenobiotics, are particularly well developed in hepatocytes. Forms of cytochrome P450 constitute the catalytic centers of monooxygenases.

Neuropathy target esterase (NTE): An esterase of the nervous system whose inhibition by certain OPs (e.g., mipafox, leptophos) can lead to the development of delayed neuropathy.

Neurotransmitter: Endogenous substance involved in the transmission of nerve impulses.

Nucleophile: An atom or group of atoms that seeks a positive charge.

NOE[C] D: No observed effect concentration or dose.

Oxyradical: An unstable form of oxygen, possessing an unpaired electron (e.g., superoxide anion).

PAH: Polycyclic aromatic hydrocarbon.

PCB: Polychlorinated biphenyl.

PCDD: Polychlorinated dibenzodioxin.

PCDF: Polychlorinated dibenzofuran.

Phosphorothionates: OP compounds containing thion groups (cf. organophosphates that contain oxon groups).

Phytotoxic: Toxic to plants.

Poikilotherms: Organisms that are unable to regulate their body temperatures.

Polarity: Possessing electrical charge.

Population growth rate (r): Per capita rate of increase of population.

Potentiation: With reference to toxicity: the situation where the toxicity of a combination of compounds is greater than the summation of the toxicities of its individual components.

Proteomics: Study of all the proteins in the cell.

Pyrethrins: Naturally occurring lipophilic esters that are toxic to many insects.

Pyrethroids: Synthetic insecticides having a strong resemblance to pyrethrins.

QSARs (quantitative structure–activity relationships): Relationships between structural parameters of chemicals and their toxicity.

Recalcitrant: See "Refractory."

Reductase: An enzyme catalyzing reductions.

Refractory: With reference to environmental chemicals: those that are unreactive ("difficult to manage").

Resistance: Reduced susceptibility to a chemical that is genetically determined.

RIVPACS: River Invertebrate Prediction and Classification.

Rotenone: A complex flavonoid produced by the plant *Derris ellyptica*. It has insecticidal activity due to its ability to inhibit electron transport in the mitochondrion.

Selective toxicity (selectivity): Difference in toxicity of a chemical toward different species, strains, sexes, age groups, etc.

Superoxide anion: Oxyradical implicated in oxidative stress.

Superwarfarins: Second-generation anticoagulant rodenticides related to warfarin.

Synergism: Similar to potentiation [q.v.], but some authors use the term in a more restricted way (e.g., where one component of a mixture, the synergist, would not cause toxicity if applied alone at the dose in question).

TBT: Tributyl tin.

TCDD: Tetrachlorodibenzodioxin.

Toxic equivalent (TEQ): A value that expresses the toxicity of a chemical relative to that of a reference compound.

Toxicodynamics: Relating to the toxic action of chemicals on living organisms.

Toxicogenomics: Application of genomics to toxicology.

Toxicokinetics: Relating to the fate of toxic chemicals within living organisms—that is, questions of uptake, distribution, metabolism, storage, and excretion; factors that determine how much of a toxic form reaches the site of action.

Transcriptomics: Simultaneous measurement of the expression of all genes (or a substantial part thereof) in the genome of an organism.

Transthyretin (TTR): A protein complex found in blood that binds both retinol (vitamin A) and thyroxine.

Uncouplers of oxidative phosphorylation: Compounds that uncouple oxidative phosphorylation from electron transport in the inner mitochondrial membrane. Most are weak lipophilic acids that can run down the proton gradient across this membrane.

Vitamin K: A cofactor for the carboxylase of the hepatic endoplasmic reticulum, which is responsible for completing the synthesis of blood-clotting proteins.

Vitellogenin: A protein that forms part of the yolk of egg-laying vertebrates.

Xenobiotic: A "foreign compound" that has no role in the normal biochemistry of a living organism. A "normal" endogenous compound to one species can be a xenobiotic to another species.

References

Aarts, J.M.M.J.G., Denison, M.S., and Haan, L.H.G. et al. (1993). Antagonistic effects of di orth PCBs on Ah-receptormediated induction of luciferase activity by 3,4,3',4'-TCB in mouse hepatocyte 1c1c7 cells. *Organohalogen Compounds* **14**, 69–72.

Abalis, I.M., Eldefrawi, M.E., and Eldefrawi, A.T. (1985). High affinity stereospecific binding of cyclodiene insecticides and gamma HCH to GABA receptors in rat brain. *Pesticide Biochemistry and Physiology,* **24**, 95–102.

Adams, N.R. (1998). Clover phyto-oestrogens in sheep in Western Australia. *Pure and Applied Chemistry* **70**, 1855–1862.

Agosta, W. (1996). *Bombardier Beetles and Fever Trees: A Close-up Look at Chemical Warfare and Signals in Animals and Plants.* Reading, MA: Addison Wesley.

Aguilar, A. and Borrell, A. (1994). An assessment of organochlorine pollutants in cetaceans by means of skin and hypodermic biopsies. In *Non-Destructive Biomarkers in Vertebrates.* M.J.C. Fossi and C. Leonzio (Eds.), Lewis, Boca Raton, FL. 245–267.

Aguilar, A., Borrell, A., and Pastor, T. (1996). Organochlorine compound levels in striped dolphins from the Western Mediterranean 1987–1993. *European Research on Cetaceans* **10**, 281–285.

Ahlborg, U.G., Becking, G.C., and Birnbaum, L.S. et al. (1994). Toxic equivalency factors for dioxin-like PCBs. *Chemosphere* **28**, 1049–1067.

Aithal, G.P., Day, C.P., and Kesteven, P.J. et al. (1999). Association of polymorphisms in the cytochrome P450 CYP2C9 with warfarin dose requirement and risk of bleeding complications. *Lancet* **353**, 717–719.

Aldridge, W.N. (1953). Serum esterases I. *Biochemical Journal* **53**, 110–117.

Aldridge, W.N. and Street, B.W. (1964). Oxidative phosphorylation; biochemical effects and properties of trialkyl tin. *Biochemical Journal* **91**, 287–297.

Alford, R.A., Bradfield, K.S., and Richards, S.J. (2007). Ecology—Global warming and amphibian losses. *Nature* **447**, E3–E4.

Allchin, C.R., Law, R.J., and Morris, S. (1999). Polybrominated diphenylethers in sediments and biota downstream of potential sources in the UK. *Environmental Pollution* **105**, 197–207.

Allran, J.W. and Karasov, W.H. (2001). Effects of atrazine on embryos, larvae, and adults of anuran amphibians. *Environmental Toxicology and Chemistry* **20**, 769–775.

Aluru, N. and Vijayan, M.M. (2006). Aryl Hydrocarbon receptor activation impairs cortisol response to stress in Rainbow Trout by disrupting the rate-limiting steps in steroidogenesis. *Endocrinology* **147**, 1895–1903.

Alzieu, C., Heral, M., Thibaud, Y. et al. (1982). Influence des peintures antisalissures a base d'organostanniques sur la calcification de la coquille de l'huitre *Crassostrea gigas. Rev. Trav. Inst. Peches Marit.* **45**, 101–116.

Ames, P.L. and Mersereau, G.S. (1964). Some factors in the decline of the osprey in Connecticut. *The Auk* **81**, 173–185.

Andersson, Ö. and Blomkvisit, G. (1981). Polybrominated aromatic pollutants found in fish in Sweden. *Chemosphere* **10**, 1051–1060.

Ankley, G.T., Diamond, S.A., and Tietge, J.E. et al. (2002). Assessment of the risk of solar ultraviolet radiation to amphibians. I. Dose-dependent induction of hindlimb malformations in the Northern leopard frog (*Rana pipiens*). *Environmental Science and Technology* **36**, 2853–2858.

Ankley, G.T., Degitz, S.J., and Diamond, S.A. et al. (2004). Assessment of environmental stressors potentially responsible for malformations in North American anuran amphibians. *Ecotoxicology and Environmental Safety* **58**, 7–16.

Anway, M.D., Cupp, A.S., Uzumcu, M., and Skinner, M.K. (2005). Epigenetic transgenerational actions of endocrine disruptors and mate fertility. *Science* **308**, 1466–1469.

Arenal, C.A., Halbrook, R.S., and Woodruff, J.J. (2004). European starling: avian model and monitor of PCB contamination at a superfund site. *Environmental Toxicology and Chemistry* **23**, 93–104.

Ashton, F.M. and Crafts, A.S. (1973). *Mode of Action of Herbicides.* John Wiley, New York.

Atchison, G.J., Sandheinrich, M.B., and Bryan, M.D. (1996). Effects of environmental stressors on interspecific interactions of aquatic animals. In M.C. Newman and C.H. Jagoe (Eds.), *Quantitative Ecotoxicology: A Hierarchical Approach.* Lewis, Chelsea, MI.

Atterwill, C.K., Simpson, M.G., and Evans, R.J. et al. (1991). Alternative methods and their application in neurotoxicity testing. In M. Balls, J. Bridges, and J. Southee (Eds.), *Animals and Alternatives in Toxicology.* Basingstoke, U.K, Macmillan 121–176.

Baatrup, E. and Junge, M. (2001). Antiandrogenic pesticides disrupt sexual characteristics in the adult male guppy (*Poecilia reticulata*). *Environmental Health Perspectives* **109**, 1063–1070.

Bacci, E. (1994). *Ecotoxicology of Organic Pollutants.* Lewis, Boca Raton, FL.

Bailey, S., Bunyan, P.J., Jennings, D.M., Norris, J.D, Stanley, P.I., and Williams, J.H. (1974). Hazards to wildlife from the use of DDT in orchards: II. A further study. *Agro-Ecosystems* **1**, 323–338.

Bakke, J.E., Feil, V.J., and Bergman, A. (1983). Metabolites of 2,4′,5-trichlorobiphenyl in rats. *Xenobiotica* **13**, 555–564.

Bakke, J.E., Bergman, A., and Larsen, G.L. (1982). Metabolism of 2,4′,5-trichlorobiphenyl by the mercapturic acid pathway *Science* **217**, 645–647.

Balaguer, P., Fenet, H., and Georget, V. et al. (2000). Reporter cell lines to monitor steroid and antisteroid potential of environmental samples. *Ecotoxicology* **9**, 105–114.

Ballantyne, B. and Marrs, T.C. (1992). *Clinical and Experimental Toxicology of Organophosphates and Carbamates.* Oxford, UK: Butterworth/Heinemann.

Balls, M. and Worth, A.P. (Eds.) (2002). *Alternative Methods for Chemicals Testing: Current Status and Future Prospects. ATLA* **30**, Supplement 1, 71–80.

Balls, M., Bridges, J., and Southee, J. (Eds.) (1991). *Animals and Alternatives in Toxicology.* Basingstoke, U.K. Macmillan.

Bangsgaard, K., Madsen, S.S., and Korsgaard, B. (2006). Effect of waterborne exposure to 4-tert-octylphenol and 17(beta)-estradiol on smoltification and downstream migration in Atlantic salmon, *Salmo salar. Aquatic Toxicology* **80**, 23–32.

Banks, A. and Russell, R.W. (1967). Effects of chronic reductions in acetylcholinesterase activity on serial problem-solving. *Journal of Comparative Physiology and Psychology* **64**, 262–267.

Barber, D.S., McNally, A.J., Denslow, N.D., and Garcia-Reyero, N. et al. (2007). Exposure to *p,p′*-DDE or dieldrin during the reproductive season alters hepatic CYP expression in largemouth bass (*Micropterus salmoides*). *Aquatic Toxicology* **81**, 27–35.

Barnett, E.A., Charlton, A.J., and Fletcher, M.R. (2007). Incidents of bee poisoning with pesticides in the United Kingdom, 1994–2003. *Pest Management Science* **63**, 1051–1057.

Barse, A.V., Chakrabarti, T., Ghosh, T.K. et al. (2007). Endocrine disruption and metabolic changes following exposure of *Cyprinus carpio* to diethyl phthalate. *Pesticide Biochemistry and Physiology* **88**, 36–42.

Basu, N., Klevanic, K., and Gamberg, M. et al. (2005). Effects of mercury on neurochemical receptor binding characteristics in wild mink. *Environmental Toxicology and Chemistry* **24**, 1444–1450.

Basu, N., Scheuhammer, A.M., and Rouvinen-Watt, K. et al. (2006). Methyl mercury impairs components of the cholinergic system in captive mink. *Toxicology Science* **91**, 202–209.

Batten, P.L. and Hutson, D.H. (1995). Species differences and other factors affecting metabolism and extrapolation to man. In *The Metabolism of Agrochemicals,* Vol. 8 of *Progress in Pesticide Biochemistry and Toxicology.* D.H. Hutson and G.D. Paulson (Eds.). Chichester, UK: John Wiley, 267–308.

Beauvais, S.L., Jones, S.B., Brewer, S.K., and Little, E.E. (2000). Physiological measures of neurotoxicity of diazinon and malathion to larval rainbow trout and their correlation with behavioural measures. *Environmental Toxicology and Chemistry* **19**, 1875–1880.

Begum, S., Begum, Z., and Alam, M.S. (1992). Organochlorine pesticide contamination of rainwater, domestic tap-water and well-water of Karachi-City. *Journal of the Chemical Society of Pakistan* **14**, 8–11.

Beitinger, T.L. (1990). Behavioural reactions for the assessment of stress in fish. *Journal of Great Lakes Research* **16**, 495–528.

Bell, A.M. (2001). Effects of an endocrine disrupter on courtship and aggressive behaviour of male three-spined stickleback, *Gasterosteus aculeatus. Animal Behaviour* **62**, 775–780.

Bello, S.M., Franks, D.G., and Stegemann, J.J. et al. (2001). Acquired resistance to Ah receptor agonists in a population of Atlantic killifish (*Fundulus heteroclitus*) inhabiting a Superfund site: in vivo and in vitro studies on the inducibility of drug metabolising enzymes *Toxicological Sciences* **60**, 77–91.

Bernard, R.F. (1966). DDT residues in avian tissues. *Journal of Applied Ecology* **3**, (Suppl.) 193–198.

Bettin, C., Oehlmann, J., and Stroben, E. (1996). TBT-induced imposex in marine neogastropods is mediated by an increasing androgen level. *Helgolander Meeresuntersuchungen* **50**, 299–317.

Billsson, K., Westerlund, L., Tysklind, M., and Olsson, P.-E. (1998). Developmental disturbances caused by polychlorinated biphenyls in zebrafish (*Brachydanio rerio*). *Marine Environmental Research* **46**, 461–464.

Bitman, J., Cecil, H.C., Feil, V.J., Harris, S.J. et al. (1978). Estrogenic activity of *o,p'*-DDt metabolites and related compounds. *Journal of Agriculture and Food Chemistry* **26**, 149–151.

Bjerregaard, L.B., Korsgaard, B., and Bjerregaard, P. (2006). Intersex in wild roach (*Rutilus rutilus*) from Danish sewage effluent-receiving streams. *Ecotoxicology and Environmental Safety* **64**, 321–328.

Blackburn, M.A. and Waldock, M.J. (1995). Concentrations of alkylphenols in rivers and estuaries in England and Wales. *Water Research* **29**, 1623–1629.

Bloom, N.S. (1992). On the chemical form of mercury in edible fish and marine invertebrate tissue. *Canadian Journal of Fisheries and Aquatic Science* **49**, 1010–1017.

Boas, M., Feldt-Rasmussen, U., and Skakkebaek, N.E. et al. (2006). Environmental chemicals and thyroid function. *European Journal of Endocrinology* **154**, 599–611.

Boon, J.P., Van Arnhem, E., and Jansen, S. et al. (1992). The toxicokinetics of PCBs in marine mammals with special reference to possible interactions of individual congeners with cytochrome P450 dependent monooxygenase systems: an overview. In C.H Walker and D. Livingstone (1992). *Persistent Pollutants in Marine Ecosystems* 119–160.

Borg, K., Erne, K., Hanko, E., and Wanntorp, H. (1970). Experimental secondary mercury poisoning in the goshawk. *Environmental Pollution* **1**, 91–104.

Borg, K., Wanntorp, H., Erne, K., and Hanko E. (1969). Alkyl mercury poisoning in terrestrial Swedish wildlife. *Viltrevy* **6**, 301–379.

Borga, K., Gabrielsen, G.W., and Skaare, J.U. (2001). Biomagnification of organochlorines along a Barents sea food chain. *Environmental Pollution* **113**, 187–198.

Borga, K., Hop, H., and Skaare, J.U. et al. (2007). Selective bioaccumulation of chlorinated pesticides and metabolites in Arctic seabirds. *Environmental Pollution* **145**, 545–553.

Borlakoglu, J.T., Wilkins, J.P.G., and Walker, C.H. (1988). Polychlorinated biphenyls in sea birds—molecular features and metabolic interpretations. *Marine Environmental Research* **24**, 15–19.

Bosveld, A.T.C. (2000). Biochemical and developmental effects of dietary exposure to PCBs 126 and 153 in common tern chicks. *Environmental Toxicology and Chemistry* **19**, 719–730.

Bowerman, D.A., Best, T.G., and Grubb, G.M. et al. (1998). Trends of contaminants and effects in bald eagles of the Great Lakes Basin. In M. Gilbertson et al. (Eds.) *Trends in Levels and Effects of Persistent Toxic Substances in the Great La*kes, 197–212.

Bowerman, W.M., Best, D.A., and Giesy, J.P. et al. (2003). Associations between regional differences in PCBs and DDE in blood of nestling bald eagles and reproductive productivity. *Environmental Toxicology and Chemistry* **22**, 371–376.

Boyd, I.L., Myhill, D.G., and Mitchell-Jones, A.J. (1988). Uptake of lindane by pipistrelle bats and its effect on survival. *Environmental Pollution* **51**, 95–111.

Brealey, C.J. (1980). *Comparative Metabolism of Pirimiphos-methyl in Rat and Japanese Quail.* Ph.D. Thesis, University of Reading, UK.

Brealey, C.J., Walker, C.H., and Baldwin, B.C. (1980). "A" Esterase activities in relation to differential toxicity of pirimiphos-methyl. *Pesticide Science* **11**, 546–554.

Bretaud, S., Saglio, P., and Saligaut, C. et al. (2002). Chemical and behavioural effects of carbofuran in goldfish. *Environmental Toxicology and Chemistry* **21**, 175–181.

Bretveld, R.W., Thomas, C.M.G., and Scheepers, P.T.J. et al. (2006). Pesticide exposure: the hormonal function of the female reproductive system disrupted? *Reproductive Biology and Endocrinology* **4**.

Brewer, L.W., Driver, C.J., and Kendall, R.J. et al. (1988). The effects of methyl parathion on northern bobwhite survival. *Environmental Toxicology and Chemistry* **7**, 375–379.

Brian, J.V., Harris, C.A., and Scholze, M. et al. (2005). Accurate prediction of the response of freshwater fish to a mixture of estrogenic chemicals. *Environmental Health Perspectives* **113**, 721–728.

Brian, J.V., Harris, C.A., and Scholze, M. et al. (2007). Evidence of estrogenic mixture effects on the reproductive performance of fish. *Environmental Science and Technology* **41**, 337–344.

Broley, C.L. (1958). Plight of the American Bald Eagle. *Audubon Magazine* **60**, 162–171.

Bromley-Challenor, K.C.A. (1992). *Synergistic Mechanisms of Synthetic Pyrethroids and Fungicides in Apis Mellifera.* M.Sc. Thesis, University of Reading, UK.

Brooks, G.T. (1974). *The Chlorinated Insecticides.* Vols. 1 and 2, Cleveland, OH: CRC Press.

Brooks, G.T. (1992). Progress in structure-activity studies on cage convulsants and related GABA receptor chloride ionophore antagonists. In D. Otto and B. Weber (Eds.) *Insecticides: Mechanism of Action and Resistance.* Newcastle upon Tyne, UK: Intercept Press, 237–242.

Brooks, G.T. (1972). Pathways of enzymatic degradation of pesticides. *Environmental Quality and Safety* **1**, 106–163.

Brooks, G.T., Pratt, G.E., and Jennings, R.C. (1979). The action of precocenes in milkweed bugs and locusts. *Nature* **281**, 570–572.

Brotons, J.A., Oleaserrano, M.F., and Villalobos, M. et al. (1995). Xenoestrogens released from lacquer coatings in food cans. *Environmental Health Perspectives* **103**, 608–612.

Brouwer, A., Morse, D.C., and Lans, M.C. et al. (1998). Interactions of persistent environmental organohalogens with the thyroid hormone system: Mechanisms and possible consequences for animal and human health. *Toxicology and Industrial Health* **14**, 59–84.

Brouwer, A. (1991). Role of biotransformation in PCB-induced alterations in vitamin A and thyroid hormone metabolism in laboratory and wildlife species. *Biochemical Society Transactions* **19**, 731–737.

Brouwer, A., Klasson-Wehler, E., and Bokdam, M. et al. (1990). Competitive inhibition of thyroxine binding to transthyretin by monohydroxy metabolites of 3,4,3′,4′-TCB. *Chemosphere* **20**, 1257–1262.

Brouwer, A. (1996). Biomarkers for exposure and effect assessment of dioxins and PCBs. In IEH Report on *The use of Biomarkers in Environmental Exposure Assessment.* Institute Environmental Health, Leicester, U.K.: 51–58.

Brown, A.W.A. (1971). Pest resistance to pesticides In *Pesticides in the Environment,* R. White-Stevens (Ed.) Dekker: New York, 437–551.

Bruggers, R.L. and Elliott, C.C.H. (1989). *Quelea Quelea: Africa's Bird Pest.* Oxford, U.K.: Oxford University Press.

Bryan, G.W., Gibbs, P.E., Hummerstone, L.G., and Burt, G.R. (1986). The decline of the gastropod *Nucella lapillus* around Southwest England—Evidence for the effect of tributyltin from antifouling paints. *Journal of the Marine Biological Association of the United Kingdom* **66**, 611–640.

Buckle, A.P. and Smith, R.H. (Eds.) (1994). *Rodent Pests and Their Control.* CAN International.

Buckley, J., Willingham, E., Agras, K., and Baskin, L. (2006). Embryonic exposure to the fungicide vinclozolin causes virilization of females and alteration of progesterone receptor expression in vivo: an experimental study in mice. *Environmental Health: A Global Access Science Source* **5**, 4.

Bull, J.J., Gutzke, W.H.N., and Crews, D. (1988). Sex reversal by estradiol in three reptilian orders. *General and Comparative Endocrinology* **70**, 425–428.

Bull, K.R., Every, W.J., and Freestone, P. et al. (1983). Alkyl lead pollution and bird mortalities on the Mersey Estuary, UK, 1979–1981. *Environmental Pollution (Series A)* **31**, 239–259.

Burchfield, J.L, Duffy, F.H., and Sim, V.N. (1976). Persistent effect of sarin and dieldrin upon the primate electroencephalogram. *Toxicology and Applied Pharmacology* **35**, 365–379.

Burczynski, M. (Ed.) (2003). *An Introduction to Toxicogenomics.* Boca Raton, FL: CRC Press.

Burgess, N.M. and Meyer, M.W. (2008). Methyl mercury exposure associated with reduced productivity in common loons. *Ecotoxicology* **17**, 83–92.

Burton, G. and Allen, Jr. (Ed.) (1992). *Sediment Toxicity Assessment.* Boca Raton, FL: Lewis.

Calabrese, E.J. (2001). Androgens: Biphasic dose responses. *Critical Reviews in Toxicology* **31**, 517–522.

Calow, P. (Ed.) (1993). *Handbook of Ecotoxicology,* Vols. I and 2 Oxford, U.K.: Blackwell.

Campbell, P.M., Pottinger, T.G., and Sumpter, J.P. (1994). Changes in the affinity of estrogen and androgen receptors accompany changes in receptor abundance in Brown and Rainbow Trout. *General and Comparative Endocrinology* **94**, 329–340.

Canton, R.F., Sanderson, J.T., and Nijmeijer, S. et al. (2006). In vitro effects of brominated flame retardants and metabolites on CYP17 catalytic activity: A novel mechanism of action? *Toxicology and Applied Pharmacology* **216**, 274–281.

Caquet, T., Lagadic, L., and Sheffield, S.R. (2000). Mesocosms in ecotoxicology (1) Outdoor aquatic systems. *Reviews in Environmental Contamination and Toxicology* **165**, 1–38.

Carlsen, E., Giwercman, A., and Keiding, N. et al. (1992). Evidence for decreasing quality of semen during past 50 Years. *British Medical Journal* **305**, 609–613.

Carlsson, G., Kulkarni, P., and Larsson, P. et al. (2007). Distribution of BDE-99 and effects on metamorphosis of BDE-99 and-47 after oral exposure in *Xenopus tropicalis. Aquatic Toxicology* **84**, 71–79.

Carson, R. (1963). *Silent Spring.* London: Hamish Hamilton.

Chanin, P.R.F. and Jefferies, D.J. (1978). The decline of the otter in Britain: Analysis of hunting records and discussion of causes. *Biological Journal of the Linnaean Society* **10**, 305–328.

Chapman, R.A. and Harris, C.R. (1981). Persistence of pyrethroid insecticides in a mineral and an organic soil. *Journal of Environmental Science and Health B* **16**, (5) 605–615.

Cheng, Z. and Jensen, A. (1989). Accumulation of organic and inorganic tin in the blue mussel, *Mytilus edulis*, under natural conditions. *Marine Pollution Bulletin* **20**, 281–286.

Cheshenko, K., Pakdel, F., and Segner, H. et al. (2008). Interference of endocrine disrupting chemicals with aromatase CYP19 expression or activity, and consequences for reproduction of teleost fish. *General and Comparative Endocrinology* **155**, 31–62.

Chipman, J.K. and Walker, C.H. (1979). The metabolism of dieldrin and two of its analogues: the relationship between rates of microsomal metabolism and rates of excretion of metabolites in the male rat. *Biochemal Pharmacology* **28**, 1337–1345.

Clark, R.B. (1992). *Marine Pollution,* 3rd edition. Oxford, U.K.: Oxford Scientific.

Clarkson, T.W. (1987). Metal toxicity in the central nervous system. *Environmental Health Perspectives* **75**, 59–64.

Clode, S.A. (2006). Assessment of in vivo assays for endocrine disruption: Best practice and research. *Clinical Endocrinology and Metabolism* **20**, 35–43.

Clotfelter, E.D., Bell, A.M., and Levering, K.R. (2004). The role of animal behaviour in the study of endocrine-disrupting chemicals. *Animal Behaviour* **68**, 665–676.

Coady, K.K., Murphy, M.B., and Villeneuve, D.L. et al. (2004). Effects of atrazine on metamorphosis, growth, and gonadal development in the green frog (*Rana clamitans*). *Journal of Toxicology and Environmental Health, Part A: Current Issues* **67**, 941–957.

Coe, T., Hamilton, P.B., and Hodgson, D.H. et al. An environmental estrogen alters dominance hierarchies and disrupts sexual selection in group spawning fish. *Environmental Science and Technology* (in press).

Colborn, T. and Clement, C. (1992). *Chemically-Induced Alterations in Sexual and Functional Development*: *The Wildlife/Human Connection*. Princeton, NJ: Princeton Scientific.

Colborn, T., Saal, F.S.V., and Soto, A.M. (1993). Developmental effects of endocrine-disrupting chemicals in wildlife and humans. *Environmental Health Perspectives* **101**, 378–384.

Colborn, T., Smolen, M.J., and Rolland, R. (1998). Environmental neurotoxic effects: the search for new protocols in functional teratology. *Toxicology and Industrial Health* **14**, 9–23.

Colin, M.E. and Belzunces, L.P. (1992). Evidence of synergy between prochloraz and deltamethrin: a convenient biological approach. *Pesticide Science* **36**, 115–119.

Connell, D.W. (1994). The octanol–water partition coefficient, Chapter 13 of Vol. 2, in Calow (Ed.). *Handbook of Ecotoxicology*. Oxford, U.K.: Blackwell.

Conney, A.H. (1982). Induction of Microsomal-Enzymes by Foreign Chemicals and Carcinogenesis by Polycyclic Aromatic-Hydrocarbons—Clowes, G.H.A. Memorial Lecture. *Cancer Research* **42**, 4875–4917.

Connor, M.S. (1983). Fish/sediment concentration ratios for organic compounds *Environmental Science and Technology* **18**, 31–35.

Cook, J.W., Dodds, E.C., and Hewett, C.L. (1933). A synthetic oestrus-exciting compound. *Nature* **131**, 56–57.

Cooke, A. (1988). Poisoning of woodpigeons on Woodwalton fen. M.P. Greaves, B.D. Smith, and P.W. Greig-Smith (Eds.) *Field Studies for the Study of the Environmental Effects of Pesticides*. BCPC Monograph No. 40, Thornton Heath, U.K.: British Crop Protection Council, 297–301.

Cooke, A. (1971). Selective predation by newts of tadpoles treated with DDT. *Nature* **229**, 275–276.

Cooke, B.K. and Stringer, A. (1982). Distribution and Breakdown of DDT in Orchard Soil. *Pesticide Science* **13**, 545–551.

Coop, I.E. and Clark, V.R. (1966). Influence of live-weight on wool production and reproduction in high country Flocks. *New Zealand Journal of Agricultural Research* **9**, 165–168.

Cooper, R.L. and Kavlock, R.J. (1997). Endocrine disruptors and reproductive development: A weight-of-evidence overview. *Journal of Endocrinology* **152**, 159–166.

Copping, L.G. and Duke, S.O. (2007). Natural products that have been used commercially as crop protection agents. *Pest Management Science* **63**, 524–554.

Copping, L.G. and Menn, J.J. (2000). Biopesticides: a review of their action, applications and efficacy. *Pest Management Science* **56**, 651–676.

Costa, L.G. and Giordano, G. (2007). Developmental neurotoxicity of polybrominated diphenyl ether (PBDE) flame retardants. *Neurotoxicology* **28**, 1047–1067.

Craig, P.J. (Ed.) (1986). *Organometallic Compounds in the Environment: Principles and Reactions,* Longmans.

Crain, D.A., Guillette, L.J., and Rooney, A.A. et al. (1997). Alterations in steroidogenesis in alligators (*Alligator mississippiensis*) exposed naturally and experimentally to environmental contaminants. *Environmental Health Perspectives* **105**, 528–533.

Crane, M., Delaney, P., and Watson, S. et al. (1995). The effect of malathion 60 on *Gammarus pulex* below water cress beds. *Environmental Toxicology and Chemistry* **14**, 1181–1188.

Crews, D., Bergeron, J.M., and McLachlan, J.A. (1995). The role of estrogen in turtle sex determination and the effect of PCBs. *Environmental Health Perspectives* **103**, 73–77.

Crick, H.Q.P., Baillie, S.R., Balmer, D.E., and Bashford, R.I. et al. (1998). *Breeding Birds in the Wider Countryside; Their Conservation Status*. Thetford, U.K.: Research Report 198, British Trust for Ornithology.

Crosby, D.G. (1998). *Environmental Toxicology and Chemistry*. New York: Oxford University Press.

Crossland, N.O. (1994). Extrapolation from mesocosms to the real world. *Toxicology and Ecotoxicology News* **1**, 15–22.

Custer, C.M, Custer, T.W., and Rosiu, C.J. et al. (2005). Exposure and effects of 2,3,7,8-TCDD in tree swallows nesting along the Woonasquatucket River, Rhode Island, USA. *Environmental Toxicology and Chemistry* **24**, 93–109.

D'Mello, J.P.F., Duffus, C.M., and Duffus, J.H. (Eds.) (1992). *Toxic Substances in Crop Plants*. London: Royal Society of Chemistry.

Darnerud, P.O., Morse, D.C., Klasson-Wehler, E. et al. (1996). Binding of 3,3', 4,4' TCB metabolite to fetal transthyretin and effects on fetal thyroid hormone levels in mice. *Toxicology* **106**, 105–114.

Davidson, C., Benard, M.F., and Shaffer, H.B. et al. (2007). Effects of chytrid and carbaryl exposure on survival, growth and skin peptide defenses in foothill yellow-legged frogs. *Environmental Science and Technology* **41**, 1771–1776.

Davila, D.R., Mounho, B.J., and Burchiel, S.W. (1997). Toxicity of PAH to the human immune system: models and mechanisms. *Toxicology and Ecotoxicology News (TEN)* **4**, 5–9.

Davis, D. and Safe, S.H. (1990). Immunosuppressive activities of PCBs in C57BL/ 6N mice: structure–activity relationships as Ah receptor agonists and partial agonists. *Toxicology* **63**, 97–111.

Davy, F.B., Kleereko. H., and Matis, J.H. (1973). Effects of exposure to sublethal DDT on exploratory behavior of goldfish (*Carassius auratus*). *Water Resources Research* **9**, 900–905.

Daxenberger, A. (2002). Pollutants with androgen-disrupting potency. *European Journal of Lipid Science and Technology* **104**, 124–130.

De Lavaur, E., Grolleau, G., and Siouy, G. (1973). Experimental poisoning of hares with paraquat-treated alfalfa. *Annales de Zoologie Ecologie Animale* **5**, 609–622.

De Matteis, F. (1974). Covalent binding of sulphur to microsomes and loss of cytochrome P450 during oxidative desulphuration of several chemicals. *Molecular Pharmacology* **10**, 849.

De Voogt, P. (1996). Ecotoxicology of chlorinated aromatic hydrocarbons. In *Chlorinated Organic Micropollutants*, No 6 in series *Issues in Environmental Science and Technology*. R.E. Hester and R.M. Harrison (Eds.) Royal Society of Chemistry 89–112.

de Wit, C.A. (2002). An overview of brominated flame retardants in the environment. *Chemosphere* **46**, 583–624.

Desbrow, C., Routledge, E.J., and Brighty, G.C. et al. (1998). Identification of estrogenic chemicals in STW effluent. 1. Chemical fractionation and in vitro biological screening. *Environmental Science and Technology* **32**, 1549–1558.

Devlin, R.H. and Nagahama, Y. (2002). Sex determination and sex differentiation in fish: an overview of genetic, physiological, and environmental influences. *Aquaculture* **208**, 191–364.

Devonshire, A.L. (1991). Role of esterases in resistance of insects to insecticides. In *Biochemical Society Transactions* **19**, 755–759.

Devonshire, A.L. and Sawicki, R.M. (1979). Insecticide resistant *Myzus persicae* as an example of evolution by gene duplication London. *Nature* **280**, 140–141.

Devonshire, A.L., Byrne, G.D., and Moores, G.D. et al. (1998). Biochemical and molecular characterisation of insecticide sensitive acetylcholinesterase in resistant insects. In *Structure and Function of Cholinesterases and Related Proteins*, Doctor, B.P, Quinn, D.M., Rotundo, R.L. and Taylor, P. (Eds.) New York: Plenum Press, 491–496.

Dewar, A.J. (1983). Neurotoxicity. In M. Balls, R.J. Riddell, and A.N. Worden (1983). *Animal and Alternatives in Toxicity Testing.* London: Academic Press, 229–283.

Diniz, M.S., Peres, I., and Pihan, J.C. (2005). Comparative study of the estrogenic responses of mirror carp (*Cyprinus carpio*) exposed to treated municipal sewage effluent (Lisbon) during two periods in different seasons. *Science of the Total Environment* **349**, 129–139.

Dodds, E.C. (1937a). The chemistry of oestrogenic compounds and methods of assay. *British Medical Journal* **1937**, 398–399.

Dodds, E.C. (1937b). Observations on the structure of substances, natural and synthetic, and their reactions on the body. *Lancet* **2**, 1–5.

Dodds, E.C. and Lawson, W. (1938). Molecular structure in relation to oestrogenic activity. Compounds without a phenanthrene nucleus. *Proceedings of the Royal Society of London, Series B: Biological Sciences* **125**, 222–232.

Dodds, E.C., Fitzgerald, M.E.H., and Lawson, W. (1937). Oestrogenic activity of some hydrocarbon derivatives of ethylene. *Nature* **140**, 772–772.

Dodds, E.C., Goldberg, L., and Lawson, W. et al. (1938). Oestrogenic activity of certain synthetic compounds. *Nature* **141**, 247–248.

Donohoe, R.M. and Curtis, L.R. (1996). Estrogenic activity of chlordecone, *o,p'*-DDT and *o,p'*-DDE in juvenile rainbow trout: Induction of vitellogenesis and interaction with hepatic estrogen binding sites. *Aquatic Toxicology* **36**, 31–52.

Drabek, J. and Neumann, R. (1985). Proinsecticides. In *Progress in Pesticide Biochemistry and Toxicology,* Vol. 5: *Insecticides*. D.H. Hutson and T.R. Roberts (Eds.) Chichester, U.K.: John Wiley, 35–86.

Drouillard, K.G., Fernie, K.L., and Smits, K.E. et al. (2001). Bioaccumulation and toxicokinetics of 42 PCB congeners in American kestrels. *Environmental Toxicology and Chemistry* **20**, 2514–2522.

Du Preez, L.H., Solomon, K.R., and Carr, J.A. et al. (2005). Population structure of the African Clawed Frog (*Xenopus laevis*) in maize-growing areas with atrazine application versus non-maize-growing areas in South Africa. *African Journal of Herpetology* **54**, 61–68.

Du, W., Awalola, T.S., Howell, P., and Koekemoer, L.L. et al. (2005). *Insect Molecular Biology* **14** (2), 179–183.

Eadsforth, C.V., Dutton, A.J., and Harrison, A.J. et al. (1991). A barn owl feeding study with (14C) flocoumafen-dosed mice. *Pesticide Science* **32**, 105–119.

Eason, C.T., Murphy, E.C., and Wright, G.R.G. et al. (2002). Assessment of risks of brodifacoum to non-target birds and mammals in New Zealand. *Ecotoxicology* **11**, 35–48.

Eason, C.T. and Spurr, E.B. (1995). Review of toxicity and impacts of brodifacoum on non target wildlife in New Zealand *New Zealand Journal of Zoology* **22**, 371–379.

Edson, E.F., Sanderson, D.M., and Noakes, D.N. (1966). Acute toxicity data for pesticides. *World Review of Pest Control* **5** (3) Autumn, 143–151.

Edwards, C.A. (1973). *Persistent Pesticides in the Environment,* 2nd edition. Cleveland, OH: CRC Press.

Ehrlich, P.R. and Raven, P.H. (1964). Butterflies and plants: a study in co-evolution. *Evolution* **18**, 586–608.

Eldefrawi, M.E. and Eldefrawi, A.T. (1990). Nervous-system-based insecticides. In *Safer Insecticides—Development and Use.* E. Hodgson and R.J. Kuhr (Eds.) New York: Marcel Dekker.

Elliott, J.E., Norstrom, R.J., and Keith, J.A. (1988). Organochlorines and eggshell thinning in Northern gannets (*Sula bassanus*) from Keith, Eastern Canada. *Environmental Pollution* **52**, 81–102.

Elliott, J.E., Wilson, L.K., and Henny, C.J. et al. (2001). Assessment of biological effects of chlorinated hydrocarbons in osprey chicks. *Environmental Toxicology and Chemistry* **20**, 866–879.

Engenheiro, E.L., Hankard, P.K., and Sousa, J.P. et al. (2005). Influence of dimethoate on acetylcholinesterase activity and locomotor function in terrestrial isopods. *Environmental Toxicology and Chemistry* **24**, 603–609.

Environment Protection Agency (EPA) (1980). *Ambient Water Quality for Mercury.* US EPA, Criteria and Standards division (EPA-600/479–049).

Environmental Health Criteria 9 (1979). *DDT and its Derivatives.* Geneva: WHO.

Environmental Health Criteria 18 (1981). *Arsenic.* Geneva: WHO.

Environmental Health Criteria 38 (1981). *Heptachlor.* Geneva: WHO.

Environmental Health Criteria 63 (1986). *Organophosphorous Insecticides: A General Introduction.* Geneva: WHO.

Environmental Health Criteria 64 *Carbamate Pesticides: A General Introduction.* WHO: Geneva, 1986.

Environmental Health Criteria 82 (1989). *Cypermethrin.* Geneva: WHO.

Environmental Health Criteria 83 (1989). *DDT and the Derivatives—Environmental Aspects.* Geneva: WHO.

Environmental Health Criteria 85 (1989). *Lead—Environmental Aspects.* Geneva: WHO.

Environmental Health Criteria 86 (1989). *Mercury—Environmental Aspects.* Geneva: WHO.

Environmental Health Criteria 88 (1989). *PCDDs and PCDFs.* Geneva: WHO.

Environmental Health Criteria 91 (1989). *Aldrin and Dieldrin.* Geneva: WHO.

Environmental Health Criteria 94 (1990). *Permethrin.* Geneva: WHO.

Environmental Health Criteria 95 (1990). *Fenvalerate.* Geneva: WHO.

Environmental Health Criteria 97 (1990). *Deltamethrin.* Geneva: WHO.

Environmental Health Criteria 101 (1990) *Methylmercury.* Geneva: WHO.

Environmental Health Criteria 116 (1990). *Tributyltin compounds.* Geneva: WHO.

Environmental Health Criteria 121 (1991). *Aldicarb.* Geneva: WHO.

Environmental Health Criteria 123 (1992). *Alpha and Beta hexachlorocyclohexanes.* Geneva: WHO.

Environmental Health Criteria 124 (1992). *Lindane.* Geneva: WHO.

Environmental Health Criteria 130 (1992). *Endrin.* Geneva: WHO.

Environmental Health Criteria 140 (1993). *Polychlorinated biphenyls and terphenyls.* Geneva: WHO.

Environmental Health Criteria 142 (1992). *Alpha-cypermethrin.* Geneva: WHO.

Environmental Health Criteria 152 (1994). *Polybrominated biphe*nyls. Geneva: WHO.

Environmental Health Criteria 153 (1994). *Carbaryl.* Geneva: WHO.

Environmental Health Criteria 197 (1997). *Demeton-S-Methyl.* Geneva: WHO.

Environmental Health Criteria 198 (1998). *Diazinon.* Geneva: WHO.

Environmental Health Criteria 202 (1998). *Non-heterocyclic polycyclic aromatic hydrocarbons.* Geneva: WHO.

Erbs, M., Hoerger, C.C., Hartmann, N. and Bucheli, T.D. (2007). Quantification of six phytoestrogens at the nanogram per liter level in aqueous environmental samples using C-13(3)-labeled internal standards. *Journal of Agricultural and Food Chemistry* **55**, 8339–8345.

Eriksson, P. and Talts, U. (2000). Neonatal exposure to neurotoxic pesticides increases adult susceptibility: a review of current findings. *Neurotoxicology* **21**, 37–47.

Ernst, W. (1977). Determination of the bioconcentration potential of marine organisms—a steady state approach. I. Bioconcentration data for 7 chlorinated pesticides in mussels and their relation to solubility data. *Chemosphere* **13**, 731–740.

Ernst, W.R., Pearce, P.A., and Pollock, T.L. (Eds.) (1989). *Environmental Effects of Fenitrothion Use in Forestry.* Environment Canada, Atlantic Region Report.

Eroschenko, V.P. (1981). Estrogenic activity of the insecticide chlordecone in the reproductive-tract of birds and mammals. *Journal of Toxicology and Environmental Health* **8**, 731–742.

Eto, M. (1974). *Organophosphorous Insecticides: Organic and Biological Chemistry.* Cleveland, OH: CRC Press.

European Commission (2003). *Proposal for a Regulation of the European Parliament Concerning the Registration, Evaluation, Authorisation and Restriction of Chemicals (REACH).* Brussels E.C.

Evers, D.C., Kaplan, J.D., Meyer, M.W. et al. (1998). Geographic trend in mercury measured in common loon feathers and blood. *Environmental Toxicology and Chemistry* **17**, 173–183.

Evers, D.C., Savoy, L.J., and DeSorbo, C.R. et al. (2008). Adverse effects from environmental mercury loads on breeding common loons. *Ecotoxicology* **17**, 69–81.

Evers, D.C., Taylor, K.M, and Major, A. et al. (2003). Common loon eggs as indicators of methylmercury availability in North America. *Ecotoxicology* **12**, 69–82.

Farabollini, F., Porrini, S., and Dessi-Fulgheri, F. (1999). Perinatal exposure to the estrogenic pollutant Bisphenol A affects behavior in male and female rats. *Pharmacology, Biochemistry, and Behavior* **64**, 687–694.

Fatoki, O.S. and Ogunfowokan, A.O. (1993). Determination of Phthalate ester plasticizers in the Aquatic environment of Southwestern Nigeria. *Environment International* **19**, 619–623.

Fatoki, O.S. and Vernon, F. (1990). Phthalate-esters in rivers of the Greater Manchester area, UK. *Science of the Total Environment* **95**, 227–232.

Fensterheim, R.J. (1993). Documenting temporal trends of polychlorinated-biphenyls in the environment. *Regulatory Toxicology and Pharmacology* **18**, 181–201.

Fent, K. and Bucheli, T.D. (1994). Inhibitors of hepatic microsomal monooxygenase system by organotins in vitro in freshwater fish. *Aquatic Toxicology* **28**, 107–126.

Fent, K., Woodin, B.R., and Stegeman, J.J. (1998). Effects of triphenyl tin and other organotins on hepatic monooxygenase system in fish. In D.R. Livingstone and J.J. Stegeman (Eds.) *Forms and Functions of Cytochrome P450,* 277–288.

Fent, K. (1996). Ecotoxicology of organotin compounds. *Critical Reviews in Toxicology* **26**, 1–117.

Ferguson, M.W.J. and Joanen, T. (1982). Temperature of egg incubation determines sex in *Alligator Mississippiensis. Nature* **296**, 850–853.

Fergusson, D. (1994). *The effects of 4-hydroxycoumarin Anticoagulant Rodenticides on Birds and the Development of Techniques for Non-destructively Monitoring Their Ecological Effects,* Ph.D. Thesis, University of Reading, UK.

Fernandez-Salguero, P.M., Hilbert, D.M., and Rudikoff, S. et al. (1996). Aryl-hydrocarbon receptor-deficient mice are resistant to 2,3,7,8-tetrachlorodibenzo-*p*-dioxin-induced toxicity. *Toxicology and Applied Pharmacology* **140**, 173–179.

Fernandez, M.F., Rivas, A., and Pulgar, R. et al. (2001). Human exposure to endocrine disrupting chemicals: The case of bisphenols. *Endocrine Disrupters* **18**, 149–169.

Fernie, K.J., Smits, J.E., and Bortolotti, G.R. et al. (2001). Reproduction success of American kestrels exposed to dietary PCBs. *Environmental Toxicology and Chemistry* **20**, 776–781.

Fest, C. and Schmidt, K.-J. (1982). *Chemistry of Organophosphorous Compounds.* Second edition, Berlin: Springer-Verlag.

Ffrench-Constant, T.A., Rochelau, J.C., and Steichen, J.C. et al. (1993). A point mutation in a *Drosophila* GABA receptor confers insecticide resistance. *Nature*, 363–449.

Filby, A.L., Neuparth, T., and Thorpe, K.L. et al. (2007a). Health impacts of estrogens in the environment, considering complex mixture effects. *Environmental Health Perspectives* **115**, 1704–1710.

Filby, A.L., Thorpe, K.L., and Tyler, C.R. (2006). Multiple molecular effect pathways of an environmental oestrogen in fish. *Journal of Molecular Endocrinology* **37**, 121–134.

Filby, A.L., Thorpe, K.L., Maack, G. et al. (2007b). Gene expression profiles revealing the mechanisms of anti-androgen and estrogen-induced feminization in fish. *Aquatic Toxicology* **81**, 219–231.

Finn, R.N. (2007). The physiology and toxicology of salmonid eggs and larvae in relation to water quality criteria. *Aquatic Toxicology* **81**, 337–354.

Finne, E.F., Cooper, G.A., and Koop, B.F. et al. (2007). Toxicogenomic responses in rainbow trout (*Oncorhynchus mykiss*) hepatocytes exposed to model chemicals and a synthetic mixture. *Aquatic Toxicology* **81**, 293–303.

Fisch, H. and Golden, R. (2003). Environmental estrogens and sperm counts. *Pure and Applied Chemistry* **75**, 2181–2193.

Flannigan, B. (1991). Mycotoxins In D'Mello J.P.F, Duffus, C.M., and Duffus, J.H. (Eds.) *Toxic Substances in Crop Plants.* London: Royal Society of Chemistry, 226–250.

Fleming, W.J. and Grue, C.E. (1981). Recovery of cholinesterase activity in 5 avian species exposed to dicrotophos, an organo-phosphorus pesticide. *Pesticide Biochemistry and Physiology* **16**, 129–135.

Folmar, L.C., Denslow, N.D., Rao, V. et al. (1996). Vitellogenin induction and reduced serum testosterone concentrations in feral male carp (*Cyprinus carpio*) captured near a major metropolitan sewage treatment plant. *Environmental Health Perspectives* **104**, 1096–1101.

Forbes, V.E. and Calow, P. (1999). Is the per capita rate of increase a good measure of population-level effects in ecotoxicology? *Environmental Toxicology and Chemistry* **18**, 1544–1556.

Forgue, S.T. et al. (1980). Direct evidence that an arene oxide is a metabolic intermediate of 2,2′,5,5′-TCB In *Abstracts of the 19th Meeting of the Society of Toxicology* (Abstract No. 383).

Fossi, M.C. and Leonzio, C. (1994). *Non-Destructive Biomarkers in Vertebrates.* Boca Raton, FL: Lewis.

Fourrnier, F., Karasow, W.H., and Kenow, K.P. et al. (2002). The oral availability and toxicokinetics of methyl mercury in common loon chicks. *Comparative Biochemistry and Physiology, Part A* **133**, 703–714.

Frohne, D. and Pfander, H.J. (2006). *Poisonous Plants* (2nd edition). U.K.: Manson Publishing.

Fry, D.M. (1995). Reproductive effects in birds exposed to pesticides and industrial-chemicals. *Environmental Health Perspectives* **103**, 165–171.

Fry, D.M. and Toone, C.K. (1981). DDT- induced feminisation of gull embryos. *Science* **2132**, 922–924.

Fuchs, P. (1967). Death of birds caused by application of seed dressings in the Netherlands. *Mededel Rijksfacultait Landbouwweetenschappen Gent* **32**, 855–859.

Gage, J.C. and Holm, S. (1976). The influence of molecular structure on the retention and excretion of PCBs by the mouse. *Toxicology and Applied Pharmacology* **36**, 555–560.

Gallant, A.L., Klaver, R.W., and Casper, G.S. et al. (2007). Global rates of habitat loss and implications for amphibian conservation. *Copeia* 967–979.

Galloway, T. and Handy, R. (2003). Immunotoxicity of organophosphorous pesticides. *Ecotoxicology* **12**, 345–363.

Garcia-Reyero, N. and Denslow, N.D. (2006). Applications of genomic technologies to the study of organochlorine pesticide-induced reproductive toxicity in fish. *Journal of Pesticide Science* **31**, 252–262.

Garcia-Reyero, N., Barber, D., and Gross, T. et al. (2006a). Modeling of gene expression pattern alteration by *p,p'*-DDE and dieldrin in largemouth bass. *Marine Environmental Research* **62**, S415–S419.

Garcia-Reyero, Barber, D.S., and Gross, T.S. et al. (2006b). Dietary exposure of large-mouth bass to OCPs changes expression of genes important for reproduction. *Aquatic Toxicology* **78**, 358–369.

Garrison, P.M., Tullis, K., and Aarts J.M.M.J.G. et al. (1996). Species specific recombinant cell lines as bioassay systems for the detection of dioxin-like chemicals. *Fundamental and Applied Toxicology* **30**, 194–203.

Gellert, R.J. (1978). Uterotrophic activity of polychlorinated biphenyls (PCB) and induction of precocious reproductive aging in neonatally treated female rats. *Environmental Research* **16**, 123–130.

Georghiou, G.P. and Saito, T. (Eds.) (1983). *Pest Resistance to Pesticides*. New York, Plenum Press.

Giam, C.S., Chan, H.S., and Neff, G.S. et al. (1978). Phthalate ester plasticizers—new class of marine pollutant. *Science* **199**, 419–421.

Gibbs, P.E. and Bryan, G.W. (1986). Reproductive failure in populations of the dog whelk caused byimposex induced by TBT from antifouling paints. *Journal of the Marine Biological Association of the United Kingdom* **66**, 767–777.

Gibbs, P.E., Pascoe, P.L., and Burt, G.R. (1988). Sex change in the female dog whelk, *Nucella lapillus*, induced by tributyltin from antifouling paints. *Journal of the Marine Biological Association of the United Kingdom* **68**, 715–731.

Gibbs, P.E. (1993). A male genital defect in the dog whelk favouring survival in a polluted area. *Journal of the Marine Biological Association of the United Kingdom* **73**, 667–668.

Gibson, R., Smith, M.D., and Spary, C.J. et al. (2005). Mixtures of estrogenic contaminants in bile of fish exposed to wastewater treatment works effluents. *Environmental Science and Technology* **39**, 2461–2471.

Giddings, J.M., Solomon, K.R., and Maund, S.J. (2001). Probabilistic risk assessment of cotton pyrethroids.II Aquatic mesocosm and field studies. *Environmental Toxicology and Chemistry* **20**, 660–668.

Giesy, J.P., Jude, D.J., and Tillitt, D.E. et al. (1997). PCDDs, PCDFs, PCBs, and 2,3,7,8-TCDD equivalents in fish from Saginaw Bay, Michigan. *Environmental Toxicology and Chemistry* **16**, 713–724.

Gilbertson, M., Fox, G.A., and Bowerman, W.W. (Eds.) (1998). *Trends in Levels and Effects of Persistent Toxic Substances in the Great Lakes*. Dordrecht: Kluwer Academic Publishers.

Gilbertson, M., Kubiak, T., and Ludwig, J. et al. (1991). Great-Lakes embryo mortality, edema, and deformities syndrome (glemeds) in colonial fish-eating birds—similarity to chick-edema disease. *Journal of Toxicology and Environmental Health* **33**, 455–520.

Gingell, R. (1976). Metabolism of ^{14}C-DDT mouse and hamster. *Xenobiotica* **6**, 15–20.

Giwercman, A., Rylander, L., and Giwercman, Y.L. (2007). Influence of endocrine disruptors on human male fertility. *Reproductive Biomedicine Online* **15**, 633–642.

Glatt, H., Engst, W., and Hagen, M. et al. (1997). The use of cell lines genetically engineered for human xenobiotic metabolising enzymes. In L.F.M. van Zutphen and M. Balls (Eds.) *Animal Alternatives, Welfare and Ethics.* Amsterdam: Elsevier 81–94.

Goodstadt, L. and Ponting, C.P. (2004). Vitamin K epoxide reductase: homology, active site and catalytic mechanism. *Trends in Biochemical Science* **29**, 289–292.

Gotelli, N.J. (1998). *A Primer of Ecology* (2nd edition). Sunderland, MA: Sinauer Associates.

Gourmelon, A. and Ahtiainen, J. (2007). Developing Test Guidelines on invertebrate development and reproduction for the assessment of chemicals, including potential endocrine active substances—The OECD perspective. *Ecotoxicology* **16**, 161–167.

Grant, A.N. (2002). Medicines for sea lice. *Pest Management Science* **58**, 521–527.

Gravel, A. and Vijayan, M.M. (2006). Salicylate disrupts interrenal steroidogenesis and brain glucocorticoid receptor expression in rainbow trout. *Toxicology Science* **93**, 41–49.

Gray, L.E. (1998). Tiered screening and testing strategy for xenoestrogens and antiandrogens. *Toxicology Letters* **103**, 677–680.

Gray, L.E., Kelce, W.R., and Monosson, E. et al. (1995). Exposure to TCDD during development permanently alters reproductive function in male Long–Evans rats and hamsters—reduced ejaculated and epididymal sperm numbers and sex accessory-gland weights in offspring with normal androgenic status. *Toxicology and Applied Pharmacology* **131**, 108–118.

Gray, L.E., Ostby, J.S., and Kelce, W.R. (1994). Developmental effects of an environmental antiandrogen—the fungicide vinclozolin alters sex differentiation of the male rat. *Toxicology and Applied Pharmacology* **129**, 46–52.

Gray, L.E., Ostby, J., and Wolf, C. et al. (1996). Effects of estrogenic, antiandrogenic and dioxin-like synthetic chemicals on mammalian sexual differentiation. *Abstracts of Papers of the American Chemical Society* **212**, 4–TOXI.

Green, R.E., Taggart, M.A., and Das, D. et al. (2006). Collapse of Asian vulture populations: Risk of mortality from residues of the veterinary drug diclofenac in carcasses of treated cattle. *Journal of Applied Ecology* **43**, 949–956.

Greig-Smith, P.W., Frampton, G., and Hardy, A.R. (1992). *Pesticides, Cereal Farming, and the Environment: the Boxworth Experiment.* London: HMSO.

Greig-Smith, P.W., Walker, C.H., and Thompson, H.M. (1992). Ecotoxicological consequences of interactions between avian esterases and organophosphorous compound. In B. Ballantyne and T.C. Marrs (Eds.) (1992). *Clinical and Experimental Toxicology of Organophosphates and Carbamates,* 295–304. Oxford, U.K.: Butterworth-Heinemann.

Grue, C.E., Hart, A.D.M., and Mineau, P. (1991). Biological consequences of depressed brain cholinesterase activity in wildlife. In Mineau, P. (Ed.) *Cholinesterase Inhibiting Insecticides—Their Impact on Wildlife and the Environment,* 151–210. Amsterdam: Elsevier.

Guez, D., Suchail, S., and Gauthier, M. et al. (2001). Contrasting effects of imidacloprid on habituation in 7- and 8-day old honeybees. *Neurobiology of Learning and Memory* **76**, 183–191.

Guillette, Jr., L.J., Pickford, D.B., and Crain, D.A. et al. (1996). Reduction in penis size and plasma testosterone concentrations in juvenile alligators living in a contaminated environment. *General and Comparative Endocrinology* **101**, 32–42.

Guillette, L.J., Jr., Gross, T.S., and Masson et al. (1994). Developmental abnormalities of the gonad and abnormal sex hormone concentrations in juvenile alligators from contaminated and control lakes in Florida. *Environmental Health Perspectives* **102**, 680–688.

Guillette, L.J., Crain, D.A., and Rooney, A.A. et al. (1995). Organization versus activation—
the role of endocrine-disrupting contaminants (EDCs) during embryonic-development
in wildlife. *Environmental Health Perspectives* **103**, 157–164.

Guillette, L.J., Crain, D.A., and Gunderson, M.P. et al. (2000). Alligators and endocrine disrupting contaminants: A current perspective. *American Zoologist* **40**, 438–452.

Guiney, P.D., Melancon, M.J., and Lech, J.J. et al. (1979). Effects of egg and sperm maturation and spawning on the elimination of a PCB in rainbow trout. *Toxicology and Applied
Pharmacology* **47**, 261–272.

Gustafsson, K., Björk, M., and Burreau, S. et al. (1999). Bioaccumulation kinetics of brominated flame retardants (polybrominated diphenyl ethers) in blue mussels (*Mytilus edulis*). *Environmental Toxicology and Chemistry* **18**, 1218–1224.

Gutleb, A.C., Appelman, J., and Bronkhorst, M. et al. (2000). Effects of oral exposure to polychlorinated biphenyls (PCBs) on the development and metamorphosis of two amphibian
species (*Xenopus laevis* and *Rana temporaria*). *Science of the Total Environment* **262**,
147–157.

Gutleb, A.C., Schriks, M., and Mossink, L. et al. (2007). A synchronized amphibian metamorphosis assay as an improved tool to detect thyroid hormone disturbance by endocrine
disruptors and apolar sediment extracts. *Chemosphere* **70**, 93–100.

Hale, R.C., La Guardia, M.J., and Harvey, E.P. et al. (2001). Polybrominated diphenyl ether
flame retardants in Virginia freshwater fishes (USA). *Environmental Science and
Technology* **35**, 4585–4591.

Halliwell, B. and Gutteridge, J.M.C. (1986). Oxygen free radicals and iron in relation to
biology and medicine—some problems and concepts. *Archives of Biochemistry and
Biophysics* **246**, 501–514.

Hamilton, G.A., Hunter, K., and Ritchie, A.S. et al. (1976). Poisoning of wild geese by carbophenothion treated winter wheat. *Pesticide Science* **7**, 175–183.

Handy, R.D., Galloway, T.S., and Depledge, M.H. (2003). A proposal for the use of biomarkers for the assessment of chronic pollution and in regulatory toxicology. *Ecotoxicology*
12, 331–343.

Hansen, D.J., Parrish, P.R., and Forester, J. (1974). Aroclor 1016-Toxicity to and Uptake by
Estuarine Animals. *Environmental Research* **7**, 363–373.

Hansen, D.J., Parrish, P.R., and Lowe, J. et al. (1971). Chronic toxicity, uptake, and retention of Aroclor-1254 in 2 estuarine fishes. *Bulletin of Environmental Contamination and
Toxicology* **6**, 113–119.

Harborne, J.B. (1993). *Introduction to Ecological Biochemistry* (4th edition). London:
Academic Press.

Harborne, J.B. and Baxter, H. (Eds.) (1993). *Phytochemical Dictionary*. London: Taylor
and Francis.

Harborne, J.R., Baxter, H., and Moss, G.P. (1996). *Dictionary of Plant Toxins* (2nd edition).
Chichester, U.K.: John Wiley.

Hardy, A.R. (1990). Estimating exposure: The identification of species at risk and routes
of exposure. In L. Somerville and C.H. Walker (Eds.) *Pesticide Effects on Terrestrial
Wildlife,* 81–98, London: Taylor & Francis.

Harries, J.E., Janbakhsh, A., and Jobling, S. et al. (1999). Estrogenic potency of effluent from
two sewage treatment works in the United Kingdom. *Environmental Toxicology and
Chemistry* **18**, 932–937.

Harries, J.E., Runnalls, T., and Hill, E. et al. (2000). Development of a reproductive performance test for endocrine disrupting chemicals using pair-breeding fathead minnows
(*Pimephales promelas*). *Environmental Science and Technology* **34**, 3003–3011.

Harries, J.E., Sheahan, D.A., and Jobling, S. et al. (1996). A survey of estrogenic activity in United Kingdom inland waters. *Environmental Toxicology and Chemistry* **15**,
1993–2002.

Harries, J.E., Sheahan, D.A., and Jobling, S. (1997). Estrogenic activity in five United Kingdom rivers detected by measurement of vitellogenesis in caged male trout. *Environmental Toxicology and Chemistry* **16**, 534–542.

Harris, C.A., Santos, E.M., and Janbakhsh, A.P. et al. (2001). Nonylphenol affects gonadotropin levels in the pituitary gland and plasma of female rainbow trout. *Environmental Science and Technology* **35**, 2909–2916.

Hart, A.D.M. (1993). Relationships between behaviour and the inhibition of acetylcholinesterase in birds exposed to organophosphorous pesticides. *Environmental Toxicology and Chemistry* **12**, 321–336.

Hassall, K.A. (1990). *The Biochemistry and Uses of Pesticides* (2nd edition). Basingstoke, U.K.: Macmillan.

Hathway, D.E. (1984). *Molecular Aspects of Toxicology*. London: Royal Society of Chemistry.

Hauser, R., Meeker, J.D., and Singh, N.P. et al. (2007). DNA damage in human sperm is related to urinary levels of phthalate monoester and oxidative metabolites. *Human Reproduction* **22**, 688–695.

Hayes, T. (2005). Comment on, Gonadal development of larval male Xenopus laevis exposed to atrazine in outdoor microcosms. *Environmental Science and Technology* **39**, 7757–7758.

Hayes, T.B., Collins, A., and Lee, M. et al. (2002). Hermaphroditic, demasculinized frogs after exposure to the herbicide atrazine at low ecologically relevant doses. *Proceedings of the National Academy of Sciences of the United States of America* **99**, 5476–5480.

Hayes, T., Haston, K., and Tsui, M. et al. (2003). Atrazine-induced hermaphroditism at 0.1 ppb in American leopard frogs (*Rana pipiens*): Laboratory and field evidence. *Environmental Health Perspectives* **111**, 568–575.

Hayes, W.J. and Laws, E.R. (1991). *Handbook of Pesticide Toxicology Vol. 2 Classes of Pesticides*. San Diego: Academic Press.

Hebert, C.E., Weseloh, D.V., and Kot, L. et al. (1994). Temporal trends and sources of PCDDs and PCDFs in the Great Lakes. Herring gull egg monitoring, 1981–1991. *Environmental Science and Technology* **25**, 1268–1277.

Hecker, M., Tyler, C.R., and Hoffmann, M. et al. (2002). Plasma biomarkers in fish provide evidence for endocrine modulation in the Elbe River, Germany. *Environmental Science and Technology* **36**, 2311–2321.

Hegdal, P.L. and Colvin, B.A. (1988). Potential hazard to Eastern Screech owls and other raptors of brodifacoum bait used for vole control in orchards. *Environmental Toxicology and Chemistry* **7**, 245–260.

Heinz, G.H., Percival, H.F., and Jennings, M.L. (1991). Contaminants in American alligator eggs from Lake Apopka, Lake Griffin, and Lake Okeechobee, Florida. *Environmental Monitoring and Assessment* **16**, 277–285.

Heinz, G.H. and Hoffman, D.J. (1998). Methylmercury chloride and selenomethione interactions on health and reproduction in mallards *Environmental Toxicology and Chemistry* **17**, 139–145.

Herbst, A.L., Ulfelder, H., and Poskanze, D.C. (1971). Adenocarcinoma of vagina—association of maternal stilbestrol therapy with tumor appearance in young Women. *New England Journal of Medicine* **284**, 878–881.

Hetherington, L.H., Livingstone, D.R., and Walker, C.H. (1996). Two and one-electron dependant reductive metabolism of nitroaromatics by *Mytilus edulis*, *Carcinus maenas* and *Asterias rubens*. *Comparative Biochemistry and Physiology* **113**, 231–239.

Higashitani, T., Tamamoto, H., and Takahashi, A. et al. (2003). Study of estrogenic effects on carp (Cyprinus carpio) exposed to sewage treatment plant effluents. *Water Science and Technology* **47**, 93–100.

Hill, B.D. and Schaalje, G.B. (1985). A two compartment model for the dissipation of delta-methrin in soil. *Journal of Agriculture and Food Chemistry* **33**, 1001–1006.

Hill, E.F. (1992). Avian toxicology of anticholinesterases. In B. Ballantyne and T.C. Marrs, (Eds.) (1992). *Clinical and Experimental Toxicology of Organophosphates and Carbamates* 272–294, Oxford, U.K.: Butterworth-Heinemann.

Hill, I.R., Matthiessen, P., and Heimbach, F. (Eds.) (1993). Guidance Document on Sediment Toxicity Tests and Bioassays for Freshwater and Marine Environments. SETAC Europe Workshop on Sediment Toxicity Assessment. Renesse, the Netherlands, November 8–10, 1993.

Hill, R.L. and Janz, D.M. (2003). Developmental estrogenic exposure in zebrafish (*Danio rerio*): I. Effects on sex ratio and breeding success. *Aquatic Toxicology* **63**, 417–429.

Hodgson, E. and Levi, P. (Eds.) (1994). *Introduction to Biochemical Toxicology* (2nd edition). Norwalk, CT: Appleton and Lange.

Hodgson, E. and Kuhr, R.J. (1990). *Safer Insecticides: Development and Use.* New York: Marcel Dekker.

Hoffman, D.J., Sileo, L., and Murray, H.C. (1984). Subchronic OP induced delayed neurotoxicity in mallards. *Toxicology and Applied Pharmacology* **75**, 128–136.

Holden, A.V. (1973). International cooperation on organochlorine and mercury residues in wildlife. *Pesticide Monitoring* **7**, 37–52.

Holm, L., Blomqvist, A., and Brandt, I. et al. (2006). Embryonic exposure to *o,p'*-DDT causes eggshell thinning and altered shell gland carbonic anhydrase expression in the domestic hen. *Environmental Toxicology and Chemistry* **25**, 2787–2793.

Holmstedt, B. (1963). Structure–activity relationships of the organophosphorous anticholinesterase agents. In G.B. Keolle (Ed.) Cholinesterases and Anticholinesterase Agents. *Handbuch der Experimentelle Phamacologie* **15**, 428–485.

Hong, W.S., Chen, S.X., and Zhang, Q.Y. et al. (2006). Sex organ extracts and artificial hormonal compounds as sex pheromones to attract broodfish and to induce spawning of Chinese black sleeper (*Bostrichthys sinensis* Lacepede). *Aquaculture Research* **37**, 529–534.

Hontela, A., Daniel, C., and Rasmussen, J.B. (1997). Structural and functional impairment of the hypothalamo- pituitary-interrenal axis in fish exposed to bleached kraft mill effluent in the St. Maurice River, Quebec. *Ecotoxicology* **6**, 1–12.

Hontela, A., Rasmussen, J.B., and Audet, C. et al. (1992). Impaired cortisol stress response in fish from environments polluted by PAHs, PCBs, and Mercury. *Archives of Environmental Contamination and Toxicology* **22**, 278–283.

Hooper, K. and McDonald, T.A. (2000). The PBDEs: An emerging environmental challenge and another reason for breast milk monitoring programs. *Environmental Health Perspectives* **108**, 387–392.

Hooper, M.J. et al. (1989). Organophosphate exposure in hawks inhabiting orchards during winter dormant spraying. *Bulletin of Environmental Contamination and Toxicology* **42**, 651–660.

Horiguchi, T., Kojima, M., and Kaya, M. et al. (2002). Tributyltin and triphenyltin induce spermatogenesis in ovary of female abalone, *Haliotis gigantea. Marine Environmental Research* **54**, 679–684.

Horiguchi, T., Nishikawa, T., and Ohta, Y. et al. (2007). Retinoid X receptor gene expression and protein content in tissues of the rock shell *Thais clavigera. Aquatic Toxicology* **84**, 379–388.

Hosokawa, M., Maki, T., and Satoh, T. (1987). Multiplicity and regulation of hepatic microsomal carboxylesterases in rats. *Molecular Pharmacology* **31**, 579–584.

Hou, Y.Y., Suzuki, Y., and Aida, K. (1999). Effects of steroid hormones on immunoglobulin M (IgM) in rainbow trout, *Oncorhynchus mykiss. Fish Physiology and Biochemistry* **20**, 155–162.

Houlahan, J.E., Findlay, C.S., and Schmidt, B.R. et al. (2000). Quantitative evidence for global amphibian population declines. *Nature* **404**, 752–755.

House, W.A., Leach, D., and Long, J.L.A. et al. (1997). Micro-organic compounds in the Humber rivers. *The Science of the Total Environment* **194/195** (special issue) 357–372.

Howald, G.R., Mineau, P., Elliott, J.E., and Cheng, K.M. (1999). Brodifacoum poisoning of avian scavengers during rat control at a seabird colony. *Ecotoxicology* **8**, 431–437.

Howell, W.M., Black, D.A., and Bortone, S.A. (1980). Abnormal expression of secondary sex characters in a population of mosquitofish, *Gambusia affinis* Holbrooki—evidence for environmentally induced masculinization. *Copeia,* 676–681.

Huckle, K.R., Warburton, P.A., Forbes, S., and Logan, C.J. (1989). Studies on the fate of flocoumafen in the Japanese quail. *Xenobiotica* **19**, 51–62.

Huggett, D.B., Brooks, B.W., and Peterson, B. et al. (2002). Toxicity of select beta adrenergic receptor blocking pharmaceuticals (beta blockers) on aquatic organisms. *Archives Environmental Contamination and Toxicology* **41**, 229–235.

Huggett, R.J., Kimerle, R.H., and Mehrle, P.M. Jr. et al. (Eds.) (1992). *Biomarkers: Biochemical, Physiological and Histological Markers of Anthropogenic Stress.* Boca Raton, FL: Lewis.

Hunt, E.G. and Bischoff, A.I. (1960). Inimical effects on wildlife of periodic DDD applications to Clear Lake California. *Fish and Game* **46**, 91–106.

Hutson, D.H. (1976). Comparative metabolism of dieldrin in the CFE rat and in two strains of mice (CF1 and LACG). *Food and Cosmetic Toxicology* **14**, 577–591.

Hutson, D.H. and Paulson, G.D. (Eds.) (1995). The mammalian metabolism of agrochemicals. *Progress in Pesticide Biochemistry and Toxicology* **8**, Chichester, U.K.: J. Wiley.

Hutson, D.H. and Roberts, T. (Eds.) (1999). *Metabolic Pathways of Agrochemicals Part 2: Insecticides and Fungicides* London: Royal Society of Chemistry.

Institute for Environmental Health (IEH) (1995). Environmental Oestrogens: Consequences to Human Health and Wildlife. Report A1 Leicester, U.K.: Institute for Environmental Health.

International Atomic Energy Agency (1972). Mercury Contamination in Man and His Environment. Technical Report Series No. 137 Wien Austria IAEA.

Iversen, S. (1991). Neurotoxicity testing. In C. Atterwill et al. (Eds.). *Alternative Methods and Their Application in Neurotoxicity Testing* 135–142.

Iwaniuk, A.N., Koperski, D.T., and Cheng, K.M. et al. (2006). The effects of environmental exposure to DDT on the brain of a songbird: Changes in structures associated with mating and song. *Behavioural Brain Research* **173**, 1–10.

Jacobs. J.M, Carmichael, N., and Cavanagh, J.B. (1977). Ultrastructural changes in the nervous system of rabbits poisoned with methyl mercury. *Toxicology and Applied Pharmacology* **39**, 249–261.

Jaeger, K., (1970). *Aldrin, Dieldrin, Endrin and Telodrin.* Amsterdam: Elsevier.

Jaensson, A., Scott, A.P., and Moore, A. et al. (2007). Effects of a pyrethroid pesticide on endocrine responses to female odours and reproductive behaviour in male parr of brown trout (*Salmo trutta L.*). *Aquatic Toxicology* **81**, 1–9.

Jakoby, W.B. (Ed.) (1980). *Enzymatic Basis of Detoxication.* New York: Academic Press.

Janssen, P.A.H., Faber, J.H., and Bosveld, A.T.C. (1998). *{Fe}male.* IBN Scientific Contributions DLO Institute for Forestry and Nature Research Wageningen Netherlands.

Jansson, B., Andersson, R., and Asplund, L. et al. (1993). Chlorinated and brominated persistent organic compounds in biological samples from the environment. *Environmental Toxicology and Chemistry* **12**, 1163–1174.

Jefferies. D.J. (1975). The role of the thyroid in the production of sublethal effects by organochlorine insecticides and PCBs In Moriarty (Ed.). *Organochlorine Insecticides: Persistent Organic Pollutants* 132–230, London: Academic Press.

Jefferies, D.J. and Parslow, J.L.F. (1976). Thyroid changes in PCB-dosed guillemots and their indication of one of the mechanisms of action of these materials *Environmental Pollution* **10**, 293–311.

Jegou, B., Soto, A., and Sundlof, S. et al. (2001). General discussion—existing guidelines for the use of meat hormones and other food additives in Europe and USA. *Apmis* **109**, S551–S556.

Jensen, K.M., Makynen, E.A., and Kahl, M.D. et al. (2006). Effects of the feedlot contaminant 17 alpha-trenholone on reproductive endocrinology of the fathead minnow. *Environmental Science and Technology* **40**, 3112–3117.

Jobling, S., Beresford, N., and Nolan, M. (2002b). Altered sexual maturation and gamete production in wild roach (*Rutilus rutilus*) living in rivers that receive treated sewage effluents. *Biology of Reproduction* 66, 272–281.

Jobling, S., Burn, R.W., and Thorpe, K.L. et al. (2008). Anti-androgens in wastewater treatment works effluents contribute to sexual disruption in wild fish living in English rivers. *Proceedings of the National Academy of Science*, submitted.

Jobling, S., Casey, D., and Rodgers-Gray, T. et al. (2003). Comparative responses of molluscs and fish to environmental estrogens and an estrogenic effluent. *Aquatic Toxicology.* **65**, 205–220.

Jobling, S., Coey, S., and Whitmore, J.G. et al. (2002a). Wild intersex roach (*Rutilus rutilus*) have reduced fertility. *Biology of Reproduction* **67**, 515–524.

Jobling, S., Nolan, M., and Tyler, C.R. (1998). Widespread sexual disruption in wild fish. *Environmental Science and Technology* **32**, 2498–2506.

Jobling, S., Reynolds, T., and White, R. et al. (1995). A variety of environmentally persistent chemicals. including some phthalate plasticizers, are weakly estrogenic. *Environmental Health Perspectives* **103**, 582–587.

Jobling, S., Williams, R., and Johnson, A. et al. (2006). Predicted exposures to steroid estrogens in U.K. rivers correlate with widespread sexual disruption in wild fish populations. *Environmental Health Perspectives* **114**, 32–39.

Johnson, I., Hetheridge, M., and Tyler, C.R. (2004). Assessment of the (Anti-) Oestrogenic and (Anti-) Androgenic Activity of Sewage Treatment Works Effluent. R&D Technical Report, Environment Agency.

Johnson, P.T.J., Lunde, K.B., and Thurman, E.M. et al. (2002). Parasite (*Ribeiroia ondatrae*) infection linked to amphibian malformations in the Western United States. *Ecological Monographs* **72**, 151–168.

Johnson, M.K. (1992). Molecular events in delayed neuropathy: Experimental aspects of neuropathy target esterase. In B. Ballantyne and T.C. Marrs, (1992). *Clinical and Experimental Toxicology of Organophosphates and Carbamates* 90–113.

Johnston, G.O, Walker, C.H., and Dawson, A. (1994). Potentiation of carbaryl toxicity to the hybrid red-legged partridge following exposure to malathion. *Pesticide Biochemistry and Physiology* **49**, 198–208.

Johnston, G.O., Collett, G., and Walker, C.H. et al. (1989). Enhancement of malathion toxicity in the hybrid red-legged partridge following exposure to prochloraz. *Pesticide Biochemistry and Physiology* **35**, 107–118.

Johnston, G.O., Walker, C.H., and Dawson, A. (1994a). Interactive effects between EBI fungicides and OP insecticides in the hybrid red-legged partridge. *Environmental Toxicology and Chemistry* **13**, 615–620.

Johnston, G.O., Walker, C.H., and Dawson, A. (1994b). Interactive effects of prochloraz and malathion in the pigeon, starling and hybrid red-legged partridge. *Environmental Toxicology and Chemistry* **13**, 115–120.

Jones, F.W. and Westmoreland, D.J. (1998). Degradation of nonylphenol ethoxylates during the composting of sludges from wool scour effluents. *Environmental Science and Technology* **32**, 2623–2627.

Jones, D.M., Bennett, D., and Elgar, K.E. (1978). Deaths of owls traced to insecticide treated timber. *Nature* 272–52.

Jooste, A.M., Du Preez, L.H., and Carr, J.A. et al. (2005a). Gonadal development of larval male Xenopus laevis exposed to atrazine in outdoor microcosms. *Environmental Science and Technology* **39**, 7759–7760.

Jooste, A.M., Du Preez, L.H., and Carr, J.A. (2005b). Gonadal development of larval male *Xenopus laevis* exposed to atrazine in outdoor microcosms. *Environmental Science and Technology* **39**, 5255–5261.

Jorgensen, S.E. (Ed.) (1990). *Modelling in Ecotoxicology.* Amsterdam: Elsevier.

Jorgenson, J.L. (2001). Aldrin and dieldrin: A review of research on their production, environmental depositions and fate, bioaccumulation, toxicology and epidemiology in the United States. *Environmental Health Perspectives* **109**, 113–139 (supplement).

Kadous, A.A, Ghiasuddin, S.M., and Matsumura, F. (1983). Differences in the picrotoxinin receptor between cyclodiene-resistant and susceptible strains of the German cockroach. *Pesticide Biochemistry and Physiology* **19**, 157.

Kaiser, T.E., Reichel, W.L., and Locke, L.H. et al. (1980). Organochlorine insecticides PCB and PBB residues and necropsy data from 29 states. *Pesticides Monitoring Journal* **13**, 145–149.

Kannan, K., Tanabe, S., and Borrell, A. et al. (1993). Isomer specific analysis and toxic evaluation of PCBs in striped dolphins affected by an epizootic in the western Mediterranean sea. *Archives Environmental Contamination and Toxicology* **25**, 227–233.

Kapustka, L.A., Williams, B.A., and Fairbrother, A. (1996). Evaluating risk predictions at population and community levels in pesticide registration. *Environmental Toxicology Chemistry* **15**, 427–431.

Karr, J.R. (1981). Assessment of biotic integrity using fish communities. *Fisheries* **6**, 21–27.

Kato, T. (1986). Sterol biosynthesis in fungi: A target for broad spectrum fungicides. In *Chemistry of Plant Protection No. 1 Sterol Biosynthesis Inhibitors and Anti-feeding Compounds.* Berlin: Springer-Verlag, 1–24.

Katsiadaki, I., Morris, S., and Squires, C. et al. (2006). Use of the three-spined stickleback (*Gasterosteus aculeatus*) as a sensitive in vivo test for detection of environmental anti-androgens. *Environmental Health Perspectives* **114**, 115–121.

Kazeto, Y., Place, A.R., and Trant, J.M. (2004). Effects of endocrine disrupting chemicals on the expression of CYP19 genes in zebrafish (*Danio rerio*) juveniles. *Aquatic Toxicology* **69**, 25–34.

Kedwards, T.J., Maund, S.J., and Chapman, P.F. (1999a and 1999b). Community level analysis of ecotoxicological field studies I Biological Monitoring and II Replicated design studies. *Environmental Toxicology and Chemistry* **18**, 149–157 and 158–166.

Keeler, R.F. and Tu, A.T. (Eds.) (1983). *Handbook of Natural Toxins Vol 1 Plant and Fungal Toxins.* New York: Marcel Dekker.

Kelce, W.R., Monosson, E., and Gamcsik, M.P. et al. (1994). Environmental hormone disruptors—evidence that Vinclozolin developmental toxicity is mediated by antiandrogenic metabolites. *Toxicology and Applied Pharmacology* **126**, 276–285.

Kelce, W.R., Stone, C.R., and Laws, S.C. et al. (1995). Persistent DDT metabolite p,p'-DDE is a potent androgen receptor antagonist. *Nature* **375**, 581–585.

Kenow, K.P., Grasman, K.A., and Hines, R.K. et al. (2007). Effects of methylmercury exposure on the immune function of common loons. *Environmental Toxicology and Chemistry* **26**, 1460–1469.

Kidd, K.A., Blanchfield, P.J., and Mills, K.H. et al. (2007). Collapse of a fish population after exposure to a synthetic estrogen. *Proceedings of the National Academy of Sciences of the United States of America* **104**, 8897–8901.

Kim, I.Y., Han, S.Y., and Kang, T.S. et al. (2005). Pyrethroid insecticides, fenvalerate and permethrin, inhibit progesterone-induced alkaline phosphatase activity in T47D human breast cancer cells. *Journal of Toxicology and Environmental Health—Part A—Current Issues* **68**, 2175–2186.

Klaasen, C.D. (1996). *Casarett and Doull's Toxicology: The Basic Science of Poisons* (5th edition) New York: McGraw-Hill.

Klasson-Wehler, E. (1989). *Synthesis of Some Radiolabelled Organochlorines and Metabolism Studies in Vivo of Two PCBs,* doctoral dissertation, University of Stockholm, Sweden.

Klasson-Wehler, E., Kuroki, H., and Athanasiadou, M. et al. (1992). Selective retention of hydroxylated PCBs in blood. In *Organohalogen Compounds Vol 10 Toxicology, Epidemiology, Risk Assessment, and Management,* Helsinki Finnish Institute of Occupational Health 121–122.

Kloeppersams, P.J., Swanson, S.M., and Marchant, T. et al. (1994). Exposure of fish to biologically treated bleached-kraft effluent 1. Biochemical, physiological and pathological assessment of rocky mountain whitefish (*Prosopium williamsoni*) and longnose sucker (*Catostomus catostomus*). *Environmental Toxicology and Chemistry* **13**, 1469–1482.

Knapen, M.H.J., Kon-Siong, G.J., and Hamulyak, K. et al. (1993). Vitamin K-induced changes in markers for osteoblast activity in urinary calcium loss. *Calcified Tissue International* **53**, 81–85.

Koeman, J.H. (Ed.) (1972). *Side Effects of Persistent Pesticides and Other Chemicals on Birds and Mammals in the Netherlands.* Report by the Working Group on Birds and Mammals of the Committee TNO for Research on Side Effects of Pesticides. TNO-Nieuws 27 527–632.

Koeman, J.H. and van Genderen, H. (1970). Tissue levels in animals and effects caused by chlorinated hydrocarbon insecticides, chlorinated biphenyls, and mercury in the marine environment along the Netherlands coast. *FAO Technical Conference on Marine Pollution.* Rome, December 1970.

Koeman, J.H. and Pennings, J.H. (1970). An orientational survey on the side effects and environmental distribution of dieldrin in a tse-tse control in S.W. Kenya. *Bulletin of Environmental Contamination and Toxicology* **5**, 164–170.

Koeman, J.H., Oskamp, A.A.G., and Veen, J. et al. (1967). Insecticides as a factor in the mortality of the sandwich tern. A preliminary communication. *Mededelingen. Rijksfaculteit Landbouwwetenschapen Gent* **32**, 841–854.

Kohling, R., Melani, R., and Koch, U. et al. (2005). Detection of electrophysiological indicators of neurotoxicity in human and rat brain slices by a three dimensional microelectrode array. *ATLA* **33**, 579–589.

Koistinen, J., Koivusaari, J., and Nuuja, I. et al. (1997). 2,3,7,8-TCDD equivalent in extracts of Baltic white-tailed sea eagles. *Environmental Toxicology and Chemistry* **16**, 1533–1544.

Könemann, H. (1981). QSAR relationships in fish toxicity studies. Part I: Relationship of 50 industrial pollutants. *Toxicology* **19**, 209–221.

Korte, F. and Arent, H. (1965). Metabolism of Insecticides IX Isolation and identification of dieldrin metabolites from urine of rabbits after oral administration of C-14 dieldrin. *Life Sciences* **4**, 2017–2026.

Krishnan, A.V., Stathis, P., and Permuth, S.F. et al. (1993). Bisphenol-a—an estrogenic substance is released from polycarbonate flasks during autoclaving. *Endocrinology* **132**, 2279–2286.

Kuhr, R.J. and Dorough, H.W. (1977). *Carbamate Insecticides.* Boca Raton, FL: CRC Press.

Kurelec, B. (1991). Personal communication.

Kuriyama, S.N., Wanner, A., and Fidalgo-Neto, A.A. (2007). Developmental exposure to low-dose PBDE-99: Tissue distribution and thyroid hormone levels. *Toxicology* **242**, 80–90.

Labenia, J.S., Baldwin, D.H., and French, B.L. et al. (2007). Behavioural impairment and increased predation mortality in cutthroat trout exposed to carbaryl. *Marine Ecology Progress* **329**, 1–11.

Landis, W.G., Moore, D.R.J., and Norton, S.B. (1998). Ecological risk assessment; looking in, looking out. In P. Douben (P) (Ed.) *Pollution Risk Assessment and Management*, Chichester, U.K., John Wiley 273–310.

Lange, I.G., Daxenberger, A., and Schiffer, B. et al. (2002). Sex hormones originating from different livestock production systems: fate and potential disrupting activity in the environment. *Anal Chimica Acta* **473**, 27–37.

Lange, A., Katsu, Y., and Ichikawa, R. et al. (2008a). Altered sexual development in roach exposed to environmental concentrations of the pharmaceutical EE2 and associated expression dynamics of aromatases and estrogens. *Toxicological Sciences* (accepted).

Lange, A., Katsu, Y., and Ichikawa, R. et al. (2008b). Sexual programming and oestrogen sensitisation in wild fish exposed to EE2. *BMC Genomics* (submitted).

Langston, W.J., Burt, G.R., and Mingjiang, Z. (1987). Tin and organotin in water, sediments, and benthic organisms of Poole Harbour. *Marine Pollution Bulletin* **18**, 634–639.

Lans, M.C., Klasson-Wehler, E., and Willemsen, M. et al. (1993). Structure-dependant competitive interaction of hydroxy-PCB, PCDD, and PCDF with transthyretin. *Chemico-Biological Interactions* **88**, 7–21.

Larsson, D.G.J., Adolfsson-Erici, M., and Parkkonen, J. et al. (1999). Ethinyloestradiol—an undesired fish contraceptive? *Aquatic Toxicology* **45**, 91–97.

Lawrence, L.J and Casida, J.E (1984). Interactions of lindane, toxaphene, and cyclodienes with brain-specific t-butyl phosphorothionate receptor. *Life Science* **35**, 171.

Leahey, J.P. (Ed.) (1985). *The Pyrethroid Insecticides,* London, Taylor and Francis.

Lebedev, A.T, Poliakova, O.V., and Karakhanova, N.K. et al. (1998). The contamination of birds with pollutants in the Lake Baikal region. *Science of the Total Environment* **212**, 153–162.

Lee, E., Ahn, M.Y., and Kim, H.J. et al. (2007). Effect of di(n-butyl) phthalate on testicular oxidative damage and antioxidant enzymes in hyperthyroid rats. *Environmental Toxicology* **22**, 245–255.

Lee, K.-S., Walker, C.H., and McCaffery, A.R. et al. (1989). Metabolism of trans cypermethrin in *H. armigera* and *H. virescens. Pesticide Biochemistry and Physiology* **34**, 49–57.

Legler, J., Broekhof, J.L.M., and Brouwer, A. et al. (2000). A novel in vivo bioassay for (xeno-)estrogens using transgenic zebrafish. *Environmental Science and Technology* **34**, 4439–4444.

Lehmann, K.P., Phillips, S., and Sar, M. et al. (2004). Dose-dependent alterations in gene expression and testosterone synthesis in the fetal testes of male rats exposed to di (n-butyl) phthalate. *Toxicological Sciences* **81**, 60–68.

Levin, M., De Guise, S., and Ross, P. (2004). Association between lymphocyter proliferation and PCBs in free-ranging harbour seal (*Phoca vitulina*) pups from British Columbia, Canada. *Environmental Toxicology and Chemistry* **24**, 1247–1252.

Lewis, D.F.V. (1996). *Cytochromes P450; Structure Function and Mechanism.* Taylor and Francis: London.

Lewis, D.F.V and Lake, B.G. (1996). Molecular modelling of CYP1A subfamily members based on an alignment with CYP 102. *Xenobiotica* **26**, 723–753.

Liney, K.E., Hagger, J.A., and Tyler, C.R. et al. (2006). Health effects in fish of long-term exposure to effluents from wastewater treatment works. *Environmental Health Perspectives* **114**, 81–89.

Liney, K.E., Jobling, S., and Shears, J.A. et al. (2005). Assessing the sensitivity of different life stages for sexual disruption in roach (*Rutilus rutilus*) exposed to effluents from wastewater treatment works. *Environmental Health Perspectives* **113**, 1299–1307.

Lipnick, R.L. (1991). Outliers: Their origin and use in the classification of molecular mechanisms of toxicity. *Science of the Total Environment* **109/110**, 131–153.

Livingstone, D.R. (1985). Responses of the detoxification/toxication enzyme system of molluscs to organic pollutants and xenobiotics. *Marine Pollution Bulletin* **16**, 158–164.

Livingstone, D.R. (1991). Organic xenobiotic metabolism in marine invertebrates. In R. Gilles (Ed.) *Advances in Comparative and Environmental Physiology* **7**, 46–185.

Livingstone, D.R., Moore, M.N., and Widdows, J. (1988). Ecotoxicology: Biological effects measurements on molluscs and their use in impact assessment. In W. Salomans, B.L. Bayne, E.K. Duursma, and U. Forstner (Eds.) *Pollution of the North Sea: An Assessment.* Berlin: Springer-Verlag 624–637.

Livingstone, D.R. and Stegeman, J.J. (Eds.) (1998). Forms and Functions of Cytochrome P450. *Comparative Biochemistry and Physiology* **121C** (special issue), 1–412.

Loeffler, I.K., Stocum, D.L., and Fallon, J.F. et al. (2001). Leaping lopsided: a review of the current hypotheses regarding etiologies of limb malformations in frogs. *Anatomical Record* **265**, 228–245.

Lotti, M. (1992). Central neurotoxicity and behavioural effects of anticholinesterases. In B. Ballantyne and T.C. Marrs (Eds.) (1992). *Clinical and Experimental Toxicology* of *Organophosphates and Carbamates* 75–83.

Ludwig, J.P., Auman, H.J., and Kurita, H. et al. (1993). Caspian tern reproduction in the Saginaw Bay ecosystem following a 100 year flood event. *Journal of Great Lakes Research* **19**, 96–108

Ludwig, J.P., Kurita-Matsuba, H.J., and Auman, M.E. et al. (1996). Deformities, PCDDs, and TCDD-equivalents in double-crested cormorants and Caspian terns of the Upper Great Lakes 1986–1991; testing a cause-effect hypothesis. *Journal of Great Lakes Research* **22**, 172–197.

Lundholm, E. (1987). Thinning of eggshells in birds by DDE: mode of action on the eggshell gland. *Comparative Biochemistry and Physiology* **88C**, 1–22.

Lundholm, E. (1997). DDE induced eggshell thinning in birds: effects of *p,p'*-DDE on the calcium and prostaglandin metabolism of the eggshell gland. *Comparative Biochemistry and Physiology* **118C**, 113–128.

Luross, J.M., Alaee, M., and Sergeant, D.B. et al. (2002). Spatial distribution of polybrominated diphenyl ethers and polybrominated biphenyls in lake trout from the Laurentian Great Lakes. *Chemosphere* **46**, 665–672.

Lutsky, B.N., Budak, M., and Koziol, P. (1975). The effects of a nonsteroid antiandrogen, flutamide, on sebaceous gland activity. *Journal Investigative Dermatology* **64**, 412–417.

Lutz, R.J., Dedrick, R.L., and Matthews, H.B. et al. (1977). A preliminary pharmacokinetic model for several chlorinated biphenyls in the rat. *Drug Metabolism and Disposition* **5**, 386–396.

Ma, M., Li, J., and Wang, Z.J. (2005). Assessing the detoxication efficiencies of wastewater treatment processes using a battery of bioassays/biomarkers. *Archives of Environmental Contamination and Toxicology* **49**, 480–487.

Ma, R., Cohen, M.B., and Berenbaum, M.R. et al. (1994). Black swallowtail alleles encode cytochrome P450s that selectively metabolise linear furanocoumarins. *Archives of Biochemistry and Biophysics* **310**, 332–340.

Machin, A.F. et al. (1975). Metabolic aspects of the toxicology of diazinon1: Hepatic metabolism in the sheep, cow, guinea-pig, rat, turkey, chicken, and duck. *Pesticide Science* **6**, 461–473.

Mackay, D. (1994). Fate models. In P. Calow (Ed.) *Handbook of Ecotoxicology* 812–831.

Mackay, D. (1991). *Multimedia Environmental Models: The Fugacity Approach.* Chelsea MI: Lewis.

Mackay, D., Shiu, W.Y., and Sutherland, R.P. (1979). Determination of air-water Henry's law constants for hydrophobic pollutants. *Environmental Science and Technology* **13**, 333–337.

Mackness, M.I., Walker, C.H, Rowlands, D.G. et al. (1982). Esterase activity in homogenates of 3 strains of rust red flour beetle. *Comparative Biochemistry and Physiology* **74C**, 65–68.

Mackness, M.I., Thompson, H.M., and Walker, C.H. (1987). Distinction between "A" esterases and arylesterases and implications for esterase classification. *Biochemical Journal* **245**, 293–296.

Maggs, J.L., Grabowski, P.S., and Park, B.K. (1983). Drug protein conjugates. 2. An investigation of the irreversible binding and metabolism of 17-alpha-ethinyl estradiol in vivo. *Biochemical Pharmacology* **32**, 301–308.

Majdic, G., Sharpe, R.M., and Oshaughnessy, P.J. et al. (1996). Expression of cytochrome P450 17 alpha-hydroxylase/C17–20 lyase in the fetal rat testis is reduced by maternal exposure to exogenous estrogens. *Endocrinology* **137**, 1063–1070.

Marcillo, Y., Ronis, M.J.J., and Porte, C. (1998). Effects of tributyl tin on the phase 1 testosterone metabolism and steroid titres of the clam. *Aquatic Toxicology* **42**, 1–13.

Markman, S., Leitner, S., and Catchpole, C. et al. (2008). *Pollutants Increase Song Complexity and the Volume of the Brain Area HVC in a Songbird.* PLoS ONE 3, e1674.

Markussen, M.D.K., Heiberg, A.C., and Fredholm, M. et al. (2008). Differential expression of P450 genes between bromadiolone-resistant and anticoagulant-susceptible Norway rats: A possible role for pharmacokinetics. *Pest Management Science* **64**, 239–248.

Marlowe, J.L. and Puga, A. (2005). Aryl hydrocarbon receptor, cell cycle regulation, toxicity, and tumorigenesis. *Journal of Cellular Biochemistry* **96**, 1174–1184.

Maron, D.M. and Ames, B.N. (1983). Revised methods for the Salmonella mutagenicity tes *Mutation Research* **113**, 173–215.

Marrs, T.C., Maynard, R.L., and Sidell, F.R. (2007). *Chemical Warfare Agents—Toxicology and Treatment* (2nd Edition). Chichester, U.K.: John Wiley and Sons.

Marsh, G., Bergman, Å., and Bladh, L.G. et al. (1998). Synthesis of p-hydroxybromodiphenyl ethers and binding to the thyroid receptor. *Organohalogen Compounds* **37**, 305–308.

Matta, M.B., Cairncross, C., and Kocan, R.M. (1997). Effect of a polychlorinated biphenyl metabolite on early life stage survival of two species of trout. *Bulletin of Environmental Contamination and Toxicology* **59**, 146–151.

Matter, J.M., McMurry, C.S., and Anthony, A.B. et al. (1998). Development and implementation of endocrine biomarkers of exposure and effects in American alligators (*Alligator mississippiensis*). *Chemosphere* **37**, 1905–1914.

Matthiessen, P. and Gibbs, P.E. (1998). Critical appraisal of the evidence for TBT-mediated endocrine disruption in molluscs. *Environmental Toxicology and Chemistry* **17**, 37–43.

Matthiessen, P., Sheahan, D., and Harrison, R. et al. (1995). Use of a Gammarus pulex bioassay to measure the effects of transient carbofuran runoff from farmland. *Ecotoxicology and Environmental Safety* **30**, 111–119.

Maynard, R.L. and Beswick, F.W. (1992). Organophosphorous compounds as chemical warfare agents. In B. Ballantyne and T.C. Marrs (Eds.) *Clinical and Experimental Toxicology of Organophosphates and Carbamates* 373–385.

McCaffery, A.R. (1998). Resistance to insecticides in heliothine Lepidoptera: a global view. *Philosophical Transactions of the Royal Society of London B* **353**, 1735–1750.

McCaffery, A.R., Gladwell, R.T., and El-Nayir, H. et al. (1991). Mechanisms of resistance to pyrethroids in laboratory and field strains of *Heliothis virescens*. *Southwestern Entomologist Supplement* **15**, 143–158.

McCarthy, A.R., Thomson, B.M., Shaw, I.C., and Abell, A.D. (2006). Estrogenicity of pyrethroid insecticide metabolites. *Journal of Environmental Monitoring* **8**, 197–202.

McCarty, J. and Secord, A.L. (1999). Reproductive ecology of tree swallows with high levels of PCB contamination. *Environmental Toxicology and Chemistry* **18**, 1433–1439.

McCormick, S.D., O'Dea, M.F., and Moeckel, A.M. et al. (2005). Endocrine disruption of parr-smolt transformation and seawater tolerance of Atlantic salmon by 4-nonylphenol and 17(beta)-estradiol. *General and Comparative Endocrinology* **142**, 280–288.

McDonald, R.A., Harris, S., and Turnbull, G. et al. (1998). Anticoagulant rodenticides in stoats and weasels in England. *Environmental Pollution* **103**, 17–23.

McDonald, T.A. (2002). A perspective on the potential health risks of PBDEs. *Chemosphere* **46**, 745–755.

McKusick, V.A. (1997). Genomics: structural and functional studies of genomes. *Genomics* **45** (2), 244–249.

McLachlan, J.A. and Arnold, S.F. (1996). Environmental estrogens. *American Scientist* **84**, 452–461.

McNicholl, P.G. and Mackay, W.C. (1975). Effect of Ddt and Ms-222 on learning a simple conditioned-response in rainbow-trout (*Salmo gairdneri*). *Journal of the Fisheries Research Board of Canada* **32**, 661–665.

Meehin, A.P. (1986). *Rats and Mice: Their Biology and Control*, East Grinstead, U.K.: Rentokil.

Meironyté, D., Norén, K., and Bergman, Å. (1999). Analysis of polybrominated diphenyl ethers in Swedish human milk. A time-related trend study, 1972–1997. *Journal of Toxicology and Environmental Health A* **58**, 101–113.

Meled, M., Thrasyvoulou, A., and Belzunces, L.P. (1998). Seasonal variation in susceptibility of *Apis mellifera* to the synergistic action of prochloraz and deltamethrin. *Environmental Toxicology Chemistry* **17**, 2517–2520.

Mellanby, K.M. (1967). *Pesticides and Pollution*, London: Collins.

Mensink, B.P., Everaarts, J.M., and Kralt, H. et al. (1996). Tributyltin exposure in early life stages induces the development of male sexual characteristics in the common whelk, *Buccinum undatum*. *Marine Environmental Research* **42**, 151–154.

Mensink, B.P. (1997). *Tributyl Tin causes Imposex in the Common Whelk Mechanism and Occurrence* NIOZ RAPPORT 1997-6 Nederlands Institut voor Onderzoek der Zee.

Mentlein, R., Ronai, A., and Robbi, M. et al. (1987). Genetic identification of rat liver carboxylesterases isolated in different laboratories. *Biochemica Biophysica Acta.* **913**, 27–38.

Merriman, J.C., Anthony, D.H.J., and Kraft, J.A. et al. (1991). Rainy River water-quality in the vicinity of bleached Kraft mills. *Chemosphere* **23**, 1605–1615.

Merson, M.H., Byers, R.E., and Kaukeinen, D.E. (1984). Residues of the rodenticide brodifacoum in voles and raptors after orchard treatment. *Journal of Wildlife Management* **48**, 212–216.

Metcalfe, T.L., Metcalfe, C.D., Kiparissis, Y. et al. (2000). Gonadal development and endocrine responses in Japanese medaka (*Oryzias latipes*) exposed to *o,p'*-DDT in water or through maternal transfer. *Environmental Toxicology and Chemistry* **19**, 1893–1900.

Meteyer, C.U., Loeffler, I.K., and Fallon, J.F. et al. (2000). Hind limb malformations in free-living northern leopard frogs (*Rana pipiens*) from Maine, Minnesota, and Vermont suggest multiple etiologies. *Teratology* **62**, 151–171.

Metzger, D., Ali, S., Bornert, J.M., and Chambon, P. (1995). Characterization of the amino-terminal transcriptional activation function of the human estrogen-receptor in animal and yeast-cells. *Journal of Biological Chemistry* **270**, 9535–9542.

Meyer, M.W. (1998). Ecological risk of mercury in the environment: the inadequacy of the best available science (Editorial). *Environmental Toxicology and Chemistry* **17**, 138–138.

Milnes, M.R., Bryan, T.A., and Katsu, Y. et al. (2008). Increased post hatching mortality and loss of sexually dimorphic gene expression in alligators (*Alligator mississippiensis*) from a contaminated environment. *Biology of Reproduction* 1 07.064915.

Mineau, P., Fletcher, M., and Glaser, L.C. et al. (1999). Poisoning of raptors with organophosphorous and carbamate pesticides with emphasis on Canada, U.S. and U.K. *Journal of Raptor Research* **33**, 1–37.

Misra, U.K. (1992). Neurophysiological monitors of anticholinesterase exposure. In B. Ballantyne and T.C. Marrs. (Eds.) *Clinical and Experimental Toxicology of the Organophosphates and Carbamates* 446–459.

Mizutani, T., Hidaka, K., Ohe, T., and Matsumoto, M. (1977). A comparative study on accumulation and elimination of tetrachlorobiphenyl isomers in mice. *Bulletin of Environmental Contamination and Toxicology* **18**, 452–461.

Monod, G. (1997). L induction du cytochrome P450 1A1. In *Biomarquers en Ecotoxicologie: Aspects Fondamentaux.* Paris: Masson, 33–51.

Monteiro, L.R. and Furness, R.W (2001). Kinetics, dose-response, excretion and toxicity of methyl mercury in free living Cory's shearwater chicks. *Environmental Toxicology and Chemistry* **20**, 1816–1824.

Monteiro, P.R.R., Reis-Henriques, M.A., and Coimbra, J. (2000a). Plasma steroid levels in female flounder (*Platichthys flesus*) after chronic dietary exposure to single polycyclic aromatic hydrocarbons. *Marine Environmental Research* **49**, 453–467.

Monteiro, P.R.R., Reis-Henriques, M.A., and Coimbra, J. (2000b). Polycyclic aromatic hydrocarbons inhibit in vitro ovarian steroidogenesis in the flounder (*Platichthys flesus L.*). *Aquatic Toxicology* **48**, 549–559.

Moore, N.W. and Walker, C.H. (1964). Organic chlorine insecticide residues in wild birds. *Nature* **201**, 1072–1073.

Morcillo, Y., Janer, G., and O'Hara, S.C.M. et al. (2004). Interaction of tributyl tin with cytochrome P450 and UDP glucuronyl transferase systems of fish: in vitro studies. *Environmental Toxicology and Chemistry* **23**, 990–996.

Morgan, M.J., Fancey, L.L., and Kiceniuk, J.W. (1990). Response and Recovery of Brain Acetylcholinesterase Activity in Atlantic Salmon (Salmo-Salar). Exposed to Fenitrothion. *Canadian Journal of Fisheries and Aquatic Sciences* **47**, 1652–1654.

Moriarty, F.M. (Ed.) (1975). *Organochlorine Insecticides: Persistent Organic Pollutants.* London: Academic Press.

Moriarty, F.M. (1968). The toxicity and sublethal effects of *p-p'*-DDT and dieldrin to *Aglais urticae* and *Chorthippus brunneus. Annals of Applied Biology* **62**, 371–393.

Moriarty, F.M. (1999). *Ecotoxicology* (3rd edition). London: Academic Press.

Morse, D.C., Klasson-Wehler, E., and Van de Pas, M. et al. (1995). Metabolism and effects of 3,3', 4, 4', TCB in pregnant and fetal rats. *Chemical and Biological Interactions* **95**, 42–56.

Muggleton, N.G., Smith, A.J., and Scott, E.A.M. et al. (2005). A long term study of the effects of diazinon on steep, the electrocorticogram and cognitive behaviour in common marmosets. *Journal of Psychopharmacology* **19**, 455–466.

Muir, D.C.G., Wagemann, R., and Hargrave, B.T. et al. (1992). Arctic marine ecosystem contamination. *Science of the Total Environment* **122**, 75–134.

Muller, J.K., Gross, T.S., and Borgert, C.J. (2007). Topical dose delivery in the reptilian egg treatment model. *Environmental Toxicology and Chemistry* **26**, 914–919.

Munkittrick, K.R., McMaster, M.E., McCarthy, L.H. et al. (1998). An overview of recent studies on the potential of pulp-mill effluents to alter reproductive parameters in fish. *Journal of Toxicology and Environmental Health-Part B-Critical Reviews* **1**, 347–371.

Nacci, D.E., Coiro, L., and Champlin, D. et al. (1999). Adaptations of wild populations of the estuarine fish *Fundulus heteroclitus* to persistent environmental contaminants. *Marine Biology* **134**, 9–17.

Nash, J.P., Kime, D.E., and van der Ven, L.T.M. et al. (2004). Long term exposures to environmental concentrations of the pharmaceutical ethynylestradiol causes reproductive failure in fish. *Environmental Health Perspectives* **112**, 1725–1733.

Natural Environment Research Council (1971). *The Sea Bird Wreck in the Irish Sea Autumn 1969*. NERC Publications C No. 4.

Nauen, R. (2006). Insecticide mode of action. The return of the ryanodine receptor. *Pest Management Science* **62**, 690–692.

Naylor, C.G. (1995). Environmental fate and safety of nonylphenol ethoxylates. *Textile Chemist and Colorist* **27**, 29–33.

Nebeker, A.V., Puglisi, F.A., and Defoe, D.L. (1974). Effect of polychlorinated biphenyl compounds on survival and reproduction of fathead minnow and flagfish. *Transactions of the American Fisheries Society* **103**, 562–568.

Nebert, D.W. and Gonzalez, F.J. (1987). P450 genes: evolution, structure, and regulation. *Annual Review Biochemistry* **56**, 945–993.

Nelson, D.R. (1998). Metazoan P450 evolution. *Comparative Biochemistry and Physiology* **121C** (special issue), 15–22.

Nelson, D.R. and Strobel, H.W. (1987). The evolution of cytochrome P 450 proteins. *Mol Biol Evol* **4**, 572–593.

Nelson, D.R., Koyman, L., and Kamataki, T. et al. (1996). P450 super family: Update on new sequences, gene mappiong, accession numbers, and nomenclature. *Pharmacogenetics* **6**, 1–42.

Neri, R.O. and Monahan, M. (1972). Effects of a novel nonsteroidal antiandrogen on canine prostatic hyperplasia. *Investigative Urology* **10**, 123–130.

Newman, M.C and Unger, M.A. (2003). *Fundamentals of Ecotoxicology* (2nd edition). Boca Raton, FL: Lewis.

Newton, I., Wyllie, I., and Freestone, P. (1990). Rodenticides in British Barn Owls. *Environmental Pollution* **68**, 101–117.

Newton, I. and Wyllie, I. (1992). Recovery of a sparrowhawk population in relation to declining pesticide contamination. *Journal of Applied Ecology* **29**, 476–484.

Newton, I. (1986). *The Sparrowhawk*. Waterhouses U.K.: Poyser Calton.

Newton, I. and Haas, M.B. (1984). The return of the sparrowhawk. *British Birds* **77**, 47–70.

Newton, I., Meek, E., and Little, B. (1978). Breeding ecology of the Merlin in Northumberland. *British Birds* **71**, 376–398.

Nicholls, D.G. (1982). *Bioenergetics: An Introduction to Chemiosmotic Theory*. London: U.K.: Academic Press.

Nimrod, A.C. and Benson, W.H. (1996). Environmental estrogenic effects of alkylphenol ethoxylates. *Critical Reviews in Toxicology* **26**, 335–364.

Nisbet, I.C.T. (1989). Organochlorines; reproductive impairment and declines in bald eagle populations; mechanisms and dose-response relationships. In B.U. Meyburg and R.D. Chancellor (Eds.) *Raptors in the Modern World*. Proceedings of the Third World Conference on Birds of Prey and Owls, Berlin 483–489.

Nolan, M., Jobling, S., and Brighty, G. et al. (2001). A histological description of intersexuality in the roach. *Journal of Fish Biology* **58**, 160–176.

Noller, K.L., Blair, P.B., and Obrien, P.C. et al. (1988). Increased occurrence of autoimmune-disease among women exposed in utero to diethylstilbestrol. *Fertility and Sterility* **49**, 1080–1082.

Noraberg, J. (2004). Organotypic brain slice cultures: an efficient and reliable method for neurotoxicological screening and mechanistic studies. *ATLA* **32**, 329–337.

Norén, K. and Meironyté, D. (2000). Certain organochlorine and organobromine contaminants in Swedish human milk in perspective of past 20–30 years. *Chemosphere* **40**, 1111–1123.

Norstrom, R.J. (1988). Bioaccumulation of polychlorinated biphenyls in Canadian wildlife. In J.-P. Crine (Ed.) *Hazards, Decontamination and Replacement of PCBs*. New York Plenum, Press.

Norstrom, R.J., McKinnon, A.E., and de Freitas, A.S.W. (1976). A bioenergetic based model for pollutant accumulation by fish. *Journal of Fisheries Research Board of Canada* **33**, 248–267.

Norstrom, R.J., Simon, M., and Weseloh, D.V. (1986). Long term trends of PCDD and PCDF contamination in the Great Lakes. *Proceedings of Dioxin 86, the Sixth International Symposium on Chlorinated Dioxins and Related Compounds* held at Fukuoka, Japan, September 1986.

Nuwaysir, E.F., Bittner, M., and Trent, J. et al. (1999). Microarrays and toxicology: the advent of toxicogenomics. *Molecular Carcinogenesis* **24**, 153–159.

Nygard, T. and Gjershaug, J.O. (2001). The effects of low levels of pollutants on the reproduction of golden eagles in Western Norway. *Ecotoxicology* **10**, 285–290.

O'Shea, T.J. and Aguilar, A. (2001). Cetacea and Sirenia. In R.F. Shore and B.A. Rattner (2001). *Ecotoxicology of Wild Mammals* 427–496.

Oaks, J.L., Gilbert, M., and Virani, M.Z. et al. (2004). Diclofenac residues as the cause of vulture population decline in Pakistan. *Nature* **427**, 630–633.

Oberdorster, E. and McClellan-Green, P. (2002). Mechanisms of imposex induction in the mud snail, *Ilyanassa obsoleta*: TBT as a neurotoxin and aromatase inhibitor. *Marine Environmental Research* **54**, 715–718.

O'Connor, D.J. and Nielsen, S.W. (1981). Environmental survey of methylmercury levels in wild mink and otter from the North Eastern USA and experimental pathology of methylmercurialism in the otter. *Proceedings, Worldwide Furbearer Conference, Frostburg, MD, USA* August 3–11 1981 1728–1745.

O'Connor, J.C., Cook, J.C., and Marty, M.S. et al. (2002). Evaluation of Tier I screening approaches for detecting endocrine-active compounds (EACs). *Critical Reviews in Toxicology* **32**, 521–549.

Oda, J. and Muller, W. (1972). Identification of a mammalian breakdown product of dieldrin. In *Environmental Quality and Safety* **1**, 248–249.

Oehlmann, J., Kloas, W., and Kusk, O. et al. (2008). A critical analysis of the biological impacts of plasticizers on wildlife. *Philosophical Transaction of the Royal Society* (accepted).

Okey, A.B. (2007). Special contribution—An aryl hydrocarbon receptor odyssey to the shores of toxicology: The Deichmann Lecture, International Congress of Toxicology-XI. *Toxicological Sciences* **98**, 5–38.

Oppenoorth, F.J. and Welling, W. (1976). Biochemistry and Physiology of Resistance. In C.F. Wilkinson (Ed.) *Pesticide Biochemistry and Physiology*. London Heyden 507–554.

Oris, J.T. and Giesy, J.P. (1986). Photoinduced toxicity of antracene to juvenile bluegill sunfish; photoperiod effects and predictive hazard evaluation. *Environmental Toxicology and Chemistry* **5**, 761–768.

Oris, J.T. and Giesy, J.P. (1987). The photoinduced toxicity of PAH to larvae of the fathead minnow. *Chemosphere* **16**, 1395–1404.

Orn, S., Holbech, H., and Madsen, T.H. et al. (2003). Gonad development and vitellogenin production in zebrafish (*Danio rerio*) exposed to ethinylestradiol and methyltestosterone. *Aquatic Toxicology* **65**, 397–411.

Osborn, D., Every, W.J., and Bull, K.R. (1983). The toxicity of trialkyl lead compounds to birds. *Environmental Pollution (Series A)* **31**, 261–275.

Ottinger, M.A., Abdelnabi, M., and Quinn, M. et al. (2002). Reproductive consequences of EDCs in birds—What do laboratory effects mean in field species? *Neurotoxicology and Teratology* **24**, 17–28.

Otto, D. and Weber, B. (Eds.) (1992). *Insecticides; Mechanism of Action and Resistance* Andover, U.K.: Intercept.

Pandian, T.J. and Sheela, S.G. (1995). Hormonal induction of sex reversal in fish. *Aquaculture* **138**, 1–22.

Panter, G.H., Hutchinson, T.H., and Hurd, K.S. et al. (2006). Development of chronic tests for endocrine active chemicals—Part 1. An extended fish early-life stage test for oestrogenic active chemicals in the fathead minnow (*Pimephales promelas*). *Aquatic Toxicology* **77**, 279–290.

Parker, P.J.-A. and Callaghan, A. (1997). Esterase activity and allele frequency in field populations of *Simulium equinum* exposed to organophosphate pollution. *Environmental Toxicology and Chemistry* **16**, 2550–2555.

Peakall, D. (1996). Disrupted patterns of behaviour in natural populations as an index of toxicity. *Environmental Health Perspectives* **104**, Supplement 2 331–335.

Peakall, D.B. (1974). Effects of di-normal-butyl and di-2-ethylhexyl phthalate on eggs of ring doves. *Bulletin of Environmental Contamination and Toxicology* **12**, 698–702.

Peakall, D.B. (1985). Behavioural responses of birds to pesticides and other contaminants *Residue Reviews* **96**, 45–77.

Peakall, D.B. (1992). *Animal Biomarkers as Pollution Indicators*. London: Chapman and Hall.

Peakall, D.B. and Fairbrother, A. (1998). Biomarkers for monitoring and measuring effects. In P.E.T. Douben (Ed.) *Pollution Risk Assessment and Management*. Chichester, U.K.: John Wiley 351–356.

Peakall, D.B. and Shugart, L.R. (Eds.) (1993). *Biomarker Research and Application in the Assessment of Environmental Health*. Berlin: Springer.

Peakall, D.B., Lincer, J.L., and Risebrough, R.W. et al. (1973). DDE-induced eggshell thinning: structural and physiological effects in three species. *Comparative and General Pharmacology* **4**, 305–313.

Peakall, D.B. (1993). DDE-induced eggshell thinning; an environmental detective story. *Environmental Reviews* **1**, 13–20.

Pelissero, C., Flouriot, G., and Foucher, J.L. (1993). Vitellogenin Synthesis in Cultured-Hepatocytes—an in vitro Test for the Estrogenic Potency of Chemicals. *Journal of Steroid Biochemistry and Molecular Biology* **44**, 263–272.

Pelissero, C., Lemenn, F., and Kaushick, S. (1991). Estrogenic effect of dietary soya bean meal on vitellogenesis in cultured siberian sturgeon *Acipenser baeri*. *General and Comparative Endocrinology* **83**, 447–457.

Pelz, H.J., Rost, S., and Huhnerberg, M. et al. (2005). The genetic basis of resistance to anticoagulants in rodents. *Genetics* **170**, 1839–1847.

Persoone, G.G., Janssen, C., and De Coen, W. (2000). *New Microbiotests for Routine Toxicity Screening and Biomonitoring*. New York: Kluwer.

Pesonen, M., Goksoyr, A., and Andersson, T. (1992). Expression of P450 1A1 in a primary culture of rainbow trout microsomes exposed to B-naphthoflavone or 2,3,7,8-TCDD. *Archives of Biochemistry and Biophysics* **292**, 228–233.

Piferrer, F. (2001). Endocrine sex control strategies for the feminization of teleost fish. *Aquaculture* **197**, 229–281.

Pijnenburg, A., Everts, J., and de Boer, J. et al. (1995). Polybrominated biphenyl and diphenylether flame retardants: analysis, toxicity, and environmental occurrence. *Reviews in Environmental Contamination and Toxicology*. **141**, 1–26.

Pilling, E.D. (1992). Evidence for pesticide synergism in the honeybee. *Aspects of Applied Biology* **31**, 43–47.

Pilling, E.D. (1993). Synergism between EBI Fungicides and a Pyrethroid Insecticide in the Honeybee. Ph.D. Thesis Southampton University, U.K.

Pilling, E.D., Bromley-Challenor, K.A.C., and Walker, C.H. et al. (1995). Mechanism of synergism between the pyrethroid insecticide lambda cyhalothrin and the imidazole fungicide prochloraz in the honeybee. *Pesticide Biochemistry and Physiology* **51**, 1–11.

Platonow, N.S. and Reinhart, B.S. (1973). Effects of polychlorinated biphenyls (Aroclor 1254) on chicken egg-production, fertility and hatchability. *Canadian Journal of Comparative Medicine—Revue Canadienne De Medecine Comparee* **37**, 341–346.

Pocar, P., Fischer, B., and Klonisch, T. et al. (2005). Molecular interactions of the aryl hydrocarbon receptor and its biological and toxicological relevance for reproduction. *Reproduction* **129**, 379–389.

Poiger, H. and Buser, H.R. (1983). Structure elucidation of mammalian TCDD metabolites In Tucker, R.E., Young, A.L., and Gray, A.P. (Eds.) *Human and Environmental Risks of Chlorinated Dioxins and Related Compounds.* New York: Plenum Press, 483–492.

Potts, G.R. (2000). The grey partridge. In D. Pain and J. Dixon (Eds.) *Bird Conservation and Farming Policy in the European Union.* London: Academic Press.

Potts, G.R. (1986). *The Partridge.* London: Collins.

Powell, W.R., Bright, S., and Bello, S.M. (2000). Developmental and tissue-specific expression of AHR1, AHR2, and ARNT2 in dioxin-sensitive and -resistant populations of the marine fish—*Fundulus heteroclitus. Toxicological Sciences* **57**, 229–239.

Prestt, I. (1970). Organochlorine pollution of rivers and the heron. *Proceedings of 11th Technical Meeting of IUCN.* New Delhi, India 1969, 95–102.

Purchase, I.F.H. (1994). Current knowledge of mechanisms of carcinogenicity: genotoxic vs. non-genotoxic. *Human and Experimental Toxicology* **13**, 17–28.

Purdom, C.E., Hardiman, P.A., and Bye, V.V.J. et al. (1994). Estrogenic effects of effluents from sewage treatment works. *Chemistry and Ecology* **8**, 275–285.

Qin, Z.F., Zhou, J.M., and Cong, L. et al. (2005). Potential ecotoxic effects of polychlorinated biphenyls on *Xenopus laevis. Environmental Toxicology and Chemistry* **24**, 2573–2578.

Qu, J.-H., Hong, X., and Chen, J.-F. et al. (2008). Fenvalerate inhibits progesterone production through cAMP-dependent signal pathway. *Toxicology Letters* **176**, 31–39.

Quicke, D.L.J. and Usherwood, P.N.R. (1990). Spider toxins as lead structures for novel pesticides In E. Hodgson and R.J. Kuhr (Eds.) *Safer Insecticides: Development and Use,* 385–452. New York: Marcel Dekker.

Rappe, C., Nygren, M., and Linstrom, G. et al. (1987). Overview on environmental fate of chlorinated dioxins and dibenzofurans sources levels and isomeric pattern in various matrices. *Chemosphere* **16**, 1603–1618.

Ratcliffe, D.A. (1967). Decrease in eggshell weight in certain birds of prey. *Nature* **215**, 208–210.

Ratcliffe, D.A. (1970). Changes attributable to pesticides in egg breakage frequency and eggshell thickness in British Birds. *Journal of Applied Ecology* **7**, 67–107.

Ratcliffe, D.A. (1993). *The Peregrine Falcon* (2nd edition). London: T & D Poyser.

Rattner, B.A., Melancon, M.J., and Custer, T.W. et al. (1993). Biomonitoring environmental contamination with pipping black-crowned night heron embryos: induction of cytochrome P450. *Environmental Toxicology and Chemistry* **12**, 1719–1732.

Rees, H.L., Waldock, R., and Matthiessen, P. et al. (2001). Improvements in the epifauna of the Crouch Estuary (United Kingdom) following a decline in TBT concentrations. *Marine Pollution Bulletin* **42**, 137–144.

Reichert, W.L., French, B.L, and Stein, J.E. et al. (1991). 32P postlabelling analysis of the persistence of bulky hydrophobic xenobiotic-DNA adducts in the liver of English sole, a marine fish. In *Proceedings of the 82nd Meeting of the American Association for Cancer Research,* Houston, TXP, 87.

Renzoni, A., Focardi, S., and Fossi, M.C. et al. (1986). Comparison between concentrations of mercury and other contaminant's in Cory's Shearwater. *Environmental Pollution* **40A**, 17–35.

Richards, P., Johnson, M., Ray, D., and Walker, C.H. (1999). Novel protein targets for organo-phosphorous compounds, In *Esterases Reacting with Organophosphorous Compounds*. E. Reiner, V. Simeon-Rudolf, and B.P. Doctor et al. (Eds.) Special Issue of *Chemico-Biological Interactions* 119–120.

Riviere, J.-L. and Cabanne, F. (1987). Animal and plant cytochrome P-450 systems. *Biochimie* **69**, 743–752.

Robertson, H.M. (2004). Genes encoding vitamin K epoxide reductase are present in Drosophila and trypanosomatid protists. *Genetics* **168**, 1077–1080.

Robertson, L.W. and Hansen, L. (Eds.) (2001). *PCBs: Recent Advances in Environmental Toxicology and Health Effects*. Kentucky: The University Press of Kentucky.

Robinson, J., Richardson, A., and Crabtree, A.N. et al. (1967a). Organochlorine residues in marine organisms *Nature* **214**, 1307–1311.

Robinson, J., Richardson, A., and Brown, V.K.H. (1967b). Pharmacodynamics of dieldrin in pigeons. *Nature* **213**, 734–735.

Rodgers-Gray, T.P., Jobling, S., and Morris, S.E. et al. (2000). Long-term temporal changes in the estrogenic composition of treated sewage effluent and its biological effects on fish. *Environmental Science and Technology* **34**, 1521–1528.

Rodgers-Gray, T.P., Jobling, S., and Kelly, C. et al. (2001). Exposure of juvenile roach (*Rutilus rutilus*) to treated sewage effluent induces dose-dependent and persistent disruption in gonadal duct development. *Environmental Science and Technology* **35**, 462–470.

Rolland, R.M. (2000). A review of chemically-induced alterations in thyroid and vitamin A status from field studies of wildlife and fish. *Journal of Wildlife Disease* **36**, 615–635.

Ronis, M.J.J. and Mason, Z. (1996). The metabolism of testosterone by the periwinkle in vitro and in vivo effects of TBT. *Marine Environmental Research* **42**, 161–166.

Ronis, M.J.J. and Walker, C.H. (1989). The microsomal monooxygenases of birds. *Reviews in Biochemical Toxicology* **10**, 301–384.

Ronis, M.J.J., Borlakoglu, J., and Walker, C.H. et al. (1989). Expression of orthologues to rat P450 1A1 and IIB1 in sea birds from the Irish sea 1978–1988: evidence for environmental induction. *Marine Environmental Research* **28**, 123–130.

Rosenberg, D.W. and Drummond, G.S. (1983). Direct in vitro effects of TBTO on hepatic cytochrome P450. *Biochemical Pharmacology* **32**, 3823–3829.

Ross, P.S., de Swart, R.L., and Reijnders, P.J.H. et al. (1995). Contaminant related suppression of delayed type hypersensitivity and antibody responses in harbor seals from the Baltic Sea. *Environmental Health Perspectives* **103**, 162.

Routledge, E.J. and Sumpter, J.P. (1996). Estrogenic activity of surfactants and some of their degradation products assessed using a recombinant yeast screen. *Environmental Toxicology and Chemistry* **15**, 241–248.

Routledge, E.J., Sheahan, D., and Desbrow, C. et al. (1998). Identification of estrogenic chemicals in STW effluent. 2. In vivo responses in trout and roach. *Environmental Science and Technology* **32**, 1559–1565.

Roy, N.K., Stabile, J., and Seeb, J.E. et al. (1999). High frequency of K-ras mutations in pink salmon embryos exposed to Exxon Valdez oil. *Environmental Toxicology and Chemistry* **18**, 1521–1528.

Safe, S. (1990). PCBs, PCDDs, PCDFs and related compounds: Environmental and mechanistic considerations which support the development of toxic equivalency figures. *CRC Critical Reviews in Toxicology* **24**, 1–63.

Safe, S. (1984). Polychlorinated biphenyls and polybrominated biphenyls: biochemistry, toxicology and mode of action *CRC Critical Reviews in Toxicology* **13**, 319–395.

Safe, S. (2001). PCBs as aryl hydrocarbon receptor agonists. In *PCBs: Recent Advances in Experimental Toxicology and Health Effects*. L.W. Robertson and L.G. Hansen (Eds.) Lexington: University of Kentucky Press, 171–178.

Salgado, V.L. (1999). Resistant target sites and insecticide discovery. In G.T. Brooks and T.R. Roberts (Eds.) *Pesticide Chemistry and Bioscience—The Food-Environment Challenge.* Cambridge, Royal Society of Chemistry 236–246.

Sancho, E., Ferrando, M.D., and Andreu, E. (1997). Response and recovery of brain acetyl-cholinesterase activity in the European eel, *Anguilla anguilla*, exposed to fenitrothion. *Ecotoxicology and Environmental Safety* **38**, 205–209.

Sandahl, J.F, Baldwin, D.H., and Jenkins, J.J. et al. (2005). Comparative thresholds for ace-tylcholinesterase inhibition and behavioural impairment in Coho Salmon exposed to chlorpyriphos. *Environmental Toxicology and Chemi*stry **24**, 136–145.

Santos, E.M., Paull, G.C., and Van Look, K.J.W. (2007). Gonadal transcriptome responses and physiological consequences of exposure to oestrogen in breeding zebrafish (*Danio rerio*). *Aquatic Toxicology* **83**, 134–142.

Santos, M.M., Reis-Henriques, M.A., and Vieira, M.N. et al. (2006). Triphenyltin and tribu-tyltin, single and in combination, promote imposex in the gastropod *Bolinus brandaris*. *Ecotoxicology and Environmental Safety* **64**, 155–162.

Scheuhammer, A.M. and Sandheinrich, M.B. (2008). Special issue on effects of methyl mer-cury on wildlife. *Ecotoxicology* **17**, 67–141.

Scheuhammer, A.M., Basu, N., and Burgess, N.M. et al. (2008). Relationships among mer-cury, selenium, and neurochemical parameters in common loons and bald eagles. *Ecotoxicology* **17**, 93–102.

Scholz, N.L., Truelove, N.K., and French, B.L. et al. (2000). Diazinon disrupts antipreda-tor and homing behaviours. *Canadian Journal of Fisheries and Aquatic Science* **57**, 1911–1918.

Schulte-Oehlmann, U., Tillmann, M., and Markert, B. et al. (2000). Effects of endocrine disrup-tors on prosobranch snails (*Mollusca gastropoda*) in the laboratory. Part II: Triphenyltin as a xeno-androgen. *Ecotoxicology* **9**, 399–412.

Schultz, S., Baral, S., and Charman, S. (2004). Diclofenac poisoning is widespread in declin-ing vulture populations across the Indian subcontinent. *Proceedings of the Royal Society of London, Series B: Biological Sciences* **271**, Supplement, 458–460.

Schuurmann, G. and Markert, B. (1998). *Ecotoxicology.* New York: Wiley-Interscience.

Schwarzenbach, R.P., Gschwend, P.M., and Imboden, D.M. (1993). *Environmental Organic Chemistry.* New York: John Wiley.

Scott, J.G., Liu, N.A., and Wen, Z. (1998). Insect cytochromes P450: diversity, insect resis-tance and tolerance to plant toxins. In D.R. Livingstone and J.J. Stegeman (Eds.) *Forms and Function of Cytochrome P450,* 147–156.

Scott, G.R. and Sloman, K.A. (2004). Effects of environmental pollutants on complex fish behaviour: integrating behavioural and physiological indicators of toxicity. *Aquatic Toxicology* **68**, 369–392.

Sebire, M., Holtorf, K., and Sanders, M. et al. (2008). Fenitrothion acts as anti-androgen and dis-rupts the reproductive behaviour of the three-spined stickleback. *Ecotoxicology* (accepted).

Secord, A.L., McCarty, J.P., Echols, K.R., and Meadows, J.C. et al. (1999). PCBs and 2,3,7,8—TCDD equivalents in tree swallows from the upper Hudson river, New York State, USA. *Environmental Toxicology and Chemistry* **18**, 2519–2525.

Sellström, U., Jansson, B., and Kierkegaard, C. et al. (1993). Polybrominated diphenyl ethers (PBDE) in biological samples from the Swedish environment. *Chemosphere* **26**, 1703–1718.

Setchell, K.D.R., Gosselin, S.J., and Welsh, M.B. et al. (1987). Dietary estrogens—a prob-able cause of infertility and liver-disease in captive cheetahs. *Gastroenterology* **93**, 225–233.

Sharpe, R.M. and Irvine, D.S. (2004). How strong is the evidence of a link between environ-mental chemicals and adverse effects on human reproductive health? *British Medical Journal* **328**, 447–451.

Sharpe, R.M. and Skakkebaek, N.E. (2003). Male reproductive disorders and the role of endocrine disruption: Advances in understanding and identification of areas for future research. *Pure and Applied Chemistry* **75**, 2023–2038.

Sheahan, D.A., Brighty, G.C., Daniel, M., and Kirby, S.J. et al. (2002). Estrogenic activity measured in a sewage treatment works treating industrial inputs containing high concentrations of alkylphenolic compounds—a case study. *Environmental Toxicology and Chemistry* **21**, 507–514.

Sheffield, S.R., Sawickaa-Kapusta, K., and Cohen, J.B. et al. (2001). Rodentia and Lagomorpha. In R.F. Shore and B.A. Rattner (Eds.) *Ecotoxicology of Wild Mammals*. Chichester, U.K.: John Wiley 215–314.

Sheldon, L.S. and Hites, R.A. (1978). Organic-compounds in Delaware River. *Environmental Science and Technology* **12**, 1188–1194.

Shimasaki, Y., Kitano, T., and Oshima, Y. (2003). Tributyl tin causes masculinisation of fish. *Environmental Toxicology and Chemistry* **22**, 141–144.

Shore, L.S., Shemesh, M., and Cohen, R. (1988). The role of estradiol and estrone in chicken manure silage in hyperestrogenism in cattle. *Australian Veterinary Journal* **65**, 68–68.

Shore, R.F., Birks, J.D.S., and Freestone, P. (1999). Exposure of non-target vertebrates to second generation rodenticides with particular reference to the polecat. *New Zealand Journal of Ecology* **23**, 199–206.

Shore, R.F. and Rattner, B.A. (2001). *The Ecotoxicology of Wild Mammals*. Chichester, U.K.: John Wiley.

Short, P. and Colborn, T. (1999). Pesticide use in the US and policy implications: A focus on herbicides. *Toxicology and Industrial Health* **15**, 240–275.

Sibly, R.M., Newton, I., and Walker, C.H. (2000). Effects of dieldrin on population growth rates of UK sparrowhawks. *Journal of Applied Ecology* **37**, 540–546.

Sideroff, S.I. and Santoluc, J.A. (1972). Behavioral and physiological effects of cholinesterase inhibitor carbaryl (1-naphthyl methylcarbamate). *Physiology and Behavior* **9**, 459–462.

Silva, E., Rajapakse, N., and Kortenkamp, A. (2002). Something from "nothing"—Eight weak estrogenic chemicals combined at concentrations below NOECs produce significant mixture effects. *Environmental Science and Technology* **36**, 1751–1756.

Simpson, E.R., Mahendroo, M.S., and Means, G.D. et al. (1994). Aromatase cytochrome P450, the enzyme responsible for estrogen biosynthesis. *Endocrine Reviews* **15**, 342–355.

Sjut, V. (Ed.) (1997). Molecular mechanisms of resistance to agrochemicals. In *Chemistry of Plant Protection* **13**, Springer: Berlin.

Skerratt, L.F., Berger, L., and Speare, R. et al. (2007). Spread of chytridiomycosis has caused the rapid global decline and extinction of frogs. *EcoHealth* **4**, 125–134.

Sleight, S.D. (1979). Polybrominatedbiphenyl: a recent environmental pollutant. In *Animals as Monitors of Environmental Pollution*. Washington, D.C.: National. Academy of Sciences, 366–374.

Smith, B.S. (1981). Male characteristics in female *Nassarius obsoletus*—variations related to locality, season and year. *Veliger* **23**, 212–216.

Smith, M.D. and Hill, E.M. (2004). Uptake and metabolism of technical nonylphenol and its brominated analogues in the roach (*Rutilus rutilus*). *Aquatic Toxicology* **69**, 359–370.

Snape, J.R., Maund, S.J., and Pickford, D.B. et al. (2004). Ecotoxicogenomics: the challenge of integrating genomics into aquatic and terrestrial ecotoxicology. *Aquatic Toxicology* **67**, 143–154.

Sodergren, A. (Ed.) (1991). *Environmental Fate and Effects of Bleached Pulp Mill Effluents*. Report No. 4031, Swedish Environmental Protection Agency.

Sohoni, P. and Sumpter, J.P. (1998). Several environmental oestrogens are also anti-androgens. *Journal of Endocrinology* **158**, 327–339.

Sohoni, P., Tyler, C.R., and Hurd, K. et al. (2001). Reproductive effects of long-term exposure to bisphenol A in the fathead minnow (*Pimephales promelas*). *Environmental Science and Technology* **35**, 2917–2925.

Solomon, K.R., Giddings, J.M., and Maund, S.J. (2001). Probabilistic risk assessment of cotton pyrethroid.I. Distributional analysis of laboratory aquatic toxicity data. *Environmental Toxicology and Chemistry* **20**, 652–659.

Somerville, L. and Greaves, M.P. (1987). *Pesticide Effects on Soil.* London: Taylor and Francis.

Somerville, L. and Walker, C.H. (1990). *Pesticide Effects on Terrestrial Wildlife.* London: Taylor and Francis.

Soto, A.M., Justicia, H., and Wray, J.W. et al. (1991). Para-nonyl-phenol: an estrogenic xenobiotic released from modified polystyrene. *Environmental Health Perspectives* **92**, 167–173.

Soto, A.M., Sonnenschein, C., and Chung, K.L. et al. (1995). The E-screen assay as a tool to identify estrogens—an update on estrogenic environmental-pollutants. *Environmental Health Perspectives* **103**, 113–122.

Stanley, P.I. and Bunyan, P.J. (1979). Hazards to wintering geese and other wildlife from the use of dieldrin, chlorfenvinphos and carbophenothion. *Proceedings of the Royal Society of London B* **205**, 31–45.

Staples, C.A. (1998). *Analysis of Surface Water Receiving Streams for Bisphenol A: Manufacturing and Processing Sites*, Technical Report to Bisphenol A TaskGroup by the Society of the Plastics Industry, Inc., Washington, D.C.

Strmac, M. and Braunbeck, T. (1999). Effects of triphenyltin acetate on survival, hatching success, and liver ultrastructure of early life stages of zebrafish (*Danio rerio*). *Ecotoxicology and Environmental* **44**, 25–39.

Stronkhurst, J., Leonards, P., and Murk, A.J. (2002). Using the dioxin receptor Calux in vitro assay to screen marine harbour sediments for a dioxin-like mode of action. *Environmental Toxicology and Chemistry* **21**, 2552–2561.

Stumpf, M., Ternes, T.A., Haberer, K., and Baumann, W. (1996). Nachweis von natürlichen und synthetischen Östrogenen in Kläranlagen und Fliessgewässern. *Vom Wasser* (in German) **87**, 251–261.

Suett, D.L. (1986). Accelerated degradation of carbofuran in previously treated soil in the United Kingdom. *Crop. Protection* **5**, 165–169.

Suffet, I.H., Brenner, L., and Cairo, P.R. (1980). GC-MS identification of trace organics in Philadelphia drinking waters during a 2-year period. *Water Research* **14**, 853–867.

Sumida, K., Ooe, N., Saito, K., and Kaneko, H. (2003). Limited species differences in estrogen receptor alpha-medicated reporter gene transactivation by xenoestrogens. *Journal of Steroid Biochemistry and Molecular Biology* **84**, 33–40.

Sumpter, J.P. and Jobling, S. (1995). Vitellogenesis as a biomarker for estrogenic contamination of the aquatic environment. *Environmental Health Perspectives* **103**, 173–178.

Sumpter, J.P. and Johnson, A.C. (2005). Lessons from endocrine disruption and their application to other issues concerning trace organics in the aquatic environment. *Environmental Science and Technology* **39**, 4321–4332.

Sun, C. and Baird, M. (1998). The determination of alkyl phenol ethoxylates in wool-scouring effluent. *Journal of the Textile Institute* **89**, 677–685.

Sun, H., Xu, X.L., and Xu, L.C. et al. (2007). Antiandrogenic activity of pyrethroid pesticides and their metabolite in reporter gene assay. *Chemosphere* **66**, 474–479.

Sundstrom, G., Hutzinger, O., and Safe, S. (1976). The metabolism of chlorobiphenyls—a review. *Chemosphere* **5**, 267–298.

Sussman, J.L., Harel, M., and Frolow, F. et al. (1991). Atomic structure of acetylcholinesterase from *Torpedo californica*: a prototypic acetylcholine-binding protein. *Science* **253**, 872–879.

Sussman, J.L., Harel, M., and Silman, I. (1993). Three-dimensional structure of acetylcholinesterase and of its complexes with anticholinesterase drugs. *Chemico-Biological Interactions* **87**, 187–197.

Suter, G.W. (Ed.) (1993). *Ecological Risk Assessment,* Boca Raton, FL: Lewis.

Swan, S.H., Brazil, C., and Drobnis, E.Z. et al. (2003a). Geographic differences in semen quality of fertile US males. *Environmental Health Perspectives* **111**, 414–420.

Swan, S.H., Kruse, R.L., and Liu, F. et al. (2003b). Semen quality in relation to biomarkers of pesticide exposure. *Environmental Health Perspectives* **111**, 1478–1484.

Taggart, M.A., Cuthbert, R., and Das, D. (2007). Diclofenac disposition in Indian cow and goat with reference to Gyps vulture population declines. *Environmental Pollution* **147**, 60–65.

Tanabe, S. and Tatsukawa, R. (1992). Chemical modernisation and vulnerability of cetaceans: increasing threat of organochlorine contaminants. In C.H. Walker and D.R. Livingstone (Eds.) *Persistent Pollutants in Marine Ecosystems,* 161–180.

Tanford, C. (1980). *The Hydrophobic Effect* (2nd edition). New York: Wiley-Interscience.

Taylor, M.R. and Harrison, P.T.C. (1999). Ecological effects of endocrine disruption: Current evidence and research priorities. *Chemosphere* **39**, 1237–1248.

Thain, J.E. and Waldock, M.J. (1986). The impact of TBT antifouling paints on molluscan fisheries. *Water Science and Technology* **18**, 193–202.

Thijssen, H.H.W. (1995). Warfarin-based rodenticides: mode of action and mechanism of resistance. *Pesticide Science* **43**, 73–78.

Thompson, H.M. (2003). Behavioural effects of pesticides in bees—their potential for use in risk assessment. *Ecotoxicology* **12**, 317–330.

Thompson, H.M. and Maus, C. (2007). The relevance of sublethal effects in honey bee testing for pesticide risk assessment. *Pest Management Science* **63**, 1058–1061.

Thompson, H.M. and Walker, C.H. (1994). Serum "B" esterases as indicators of exposure to pesticides. In M.C. Fossi and C. Leonzio (Eds.) *Non Destructive Biomarkers,* Boca Raton, FL: Lewis, 35–60.

Thorpe, K.L., Cummings, R.I., and Hutchinson, T. et al. (2003). Relative potencies and combination effects of steroidal estrogens in fish. *Environmental Science and Technology* **37**, 1142–1149.

Thorpe, K.L., Gross-Sorokin, M., and Johnson, I. et al. (2006). An assessment of the model of concentration addition for predicting the estrogenic activity of chemical mixtures in wastewater treatment works effluents. *Environmental Health Perspectives* **114**, 90–97.

Thorpe, K.L., Hutchinson, T.H., and Hetheridge, M.J. et al. (2001). Assessing the biological potency of binary mixtures of environmental estrogens using vitellogenin induction in juvenile rainbow trout (*Oncorhynchus mykiss*). *Environmental Science and Technology* **35**, 2476–2481.

Tillett, D.E.G., Ankley, D.T., and Giesy, J.P. et al. (1992). PCB residues and egg mortality in double crested cormorants from the Great Lakes. *Environmental Toxicology and Chemistry* **11**, 1281–1288.

Tilson, H.A. (1993). Neurobehavioural methods used in neurotoxicological research. *Toxicology Letters* **68**, 231–240.

Timbrell, J.A. (1999). *Principles of Biochemical Toxicology* (3rd edition). London: Taylor and Francis.

Tinbergen, N. and Vaniersel, J.J.A. (1948). Displacement reactions in the 3-spined stickleback. *Behaviour* **1**, 56–63.

Tomlin, C.D.S. (1997). *The Pesticide Manual* (11th edition). British Crop Protection Council.

Toschik, P.C, Rattner, B.A, McGowan, P.C., and Christman, M.C. et al. (2005). Effects of contaminant exposure on reproductive success of Ospreys nesting in Delaware river and bay. *Environmental Toxicology and Chemistry* **24**, 17.

Trager, W.F. (1988). Isotope effects as mechanistic probes of cytochrome P450-catalysed reactions. In *Synthesis and Application of Isotopically Labelled Compounds: Proceedings of the Third International Symposium* T.A. Baillie and J.R. Jones (Eds.) Amsterdam Elsevier 333–340.

Tran, D.Q., Ide, C.F., and McLachlan, J.A. et al. (1996). The anti-estrogenic activity of selected polynuclear aromatic hydrocarbons in yeast expressing human estrogen receptor. *Biochemical and Biophysical Research Communications* **229**, 102–108.

Turtle, E.E., Taylor, A., and Wright, E.N. et al. (1963). The effects on birds of certain chlorinated insecticides used as seed dressings. *Journal of the Science of Food and Agriculture* **14**, 567–577.

Tyler, C.R., Beresford, N., and van der Woning, M. et al. (2000). Metabolism and environmental degradation of pyrethroid insecticides produce compounds with endocrine activities. *Environmental Toxicology and Chemistry* **19**, 801–809.

Tyler, C.R., Burn, R.W., and Thorpe, K. et al. (In preparation). Anti-androgens in wastewater treatment works effluents contribute to widespread sexual disruption in fish living in English Rivers.

Tyler, C.R., Filby, A.L., and Lange, A. et al. Fish toxicogenomics. In C. Hogstrand and P. Kille (Eds.) *Advances in Experimental Biology* (in press).

Tyler, C.R., Jobling, S., and Sumpter, J.P. (1998). Endocrine disruption in wildlife: A critical review of the evidence. *Critical Reviews in Toxicology* **28**, 319–361.

Tyler, C.R. and Routledge, E.J. (1998). Oestrogenic effects in fish in English rivers with evidence of their causation. *Pure and Applied Chemistry* **70**, 1795–1804.

Tyler, C.R., Spary, C., and Gibson, R. et al. (2005). Accounting for differences in estrogenic responses in rainbow trout (*Oncorhynchus mykiss*: Salmonidae) and roach (*Rutilus rutilus*: Cyprinidae) exposed to effluents from wastewater treatment works. *Environmental Science and Technology* **39**, 2599–2607.

U.S. EPA. *An Inventory of Sources and Environmental Releases of Dioxin-Like Compounds in the United States for the Years 1987, 1995, and 2000* (EPA/600/P-03/002f, Final Report, November 2006). U.S. Environmental Protection Agency, Washington, DC, EPA/600/P-03/002F.

Uguz, C., Togan, I., Eroglu, Y., Tabak, I., Zengin, M., and Iscan, M. (2003). Alkylphenol concentrations in two rivers of Turkey. *Environmental Toxicology and Pharmacology* **14**, 87–88.

Vaillant, C., Monod, G., and Volataire, Y. et al. (1989). Measurement and induction of cytochrome P450 and monooxygenases in a primary culture of rainbow trout hepatocytes. *Comptes Redus de L'Academie des Sciences* **308**, 83–88.

Vajda, A. Schmid, H., and Groll-Knapp, E. et al. (1974). EEC changes in evoked potentials caused by insecticides. *Electroencephalography and Clinical Neurophysiology* **37**, 442.

van Aerle, R., Nolan, M., and Jobling, S. et al. (2001). Sexual disruption in a second species of wild cyprinid fish (the Gudgeon *Gobio gobio*) in United Kingdom freshwaters. *Environmental Toxicology and Chemistry* **20**, 2841–2847.

van Aerle, R., Pounds, N., and Hutchinson, T. et al. (2002). Window of sensitivity for the estrogenic effects of ethinylestradiol in early life-stages of fathead minnow, *Pimephales promelas. Ecotoxicology* **11**, 423–434.

van den Berg, M., Sanderson, T., and Kurihara, N. et al. (2003). Role of metabolism in the endocrine-disrupting effects of chemicals in aquatic and terrestrial systems. *Pure and Applied Chemistry* **75**, 1917–1932.

Van Gelder, G.A. and Cunningham, W.L. (1975). The effects of low level dieldrin exposure on the EEG and learning ability of the squirrel monkey. *Toxicology and Applied Pharmacology* **33**, 142, Abstract No. 50.

Van Straalen, N.M. (2003). Ecotoxicology becomes stress. *Environmental Science and Technology.* September 1, 2003, 324–330.

Van Straalen, N.M. and Roelofs, D. (2006). *An Introduction to Ecological Genetics.* Oxford, U.K.: Oxford University Press.

van Uitregt, V.O., Wilson, R.S., and Franklin, C.E. (2007). Cooler temperatures increase sensitivity to ultraviolet B radiation in embryos and larvae of the frog *Limnodynastes peronii. Global Change Biology* **13**, 1114–1121.

Varanasi, U., Stein, J.E., and Reichert, W.L. et al. (1992). Chlorinated and aromatic hydrocarbons in bottom sediments, fish and marine mammals in US coastal waters: laboratory and field studies of metabolism and accumulation. In C.H. Walker and D.R. Livingstone (Eds.) *Persistent Pollutants in Marine Ecosystems,* Oxford, U.K.: Pergamon Press, 83–118.

Verhallen, E.Y., van den Berg, M., and Bosveld, A.T.C. (1997). Interactive effects of the EROD—inducing potency of PHAHs in the chicken embryo hepatocyte assay. *Environmental Toxicology and Chemistry* **16**, 277–282.

Vermeirssen, E.L.M., Burki, R., and Joris, C. et al. (2005). Characterization of the estrogenicity of Swiss midland rivers using a recombinant yeast bioassay and plasma vitellogenin concentrations in feral male brown trout. *Environmental Toxicology and Chemistry* **24**, 2226–2233.

Veronesi, B. (1992). Validation of a rodent model of OPIDN. In B. Ballantyne and T.C. Marrs, *Clinical and Experimental Toxicology of Organophosphates and Carbamates,* Oxford, U.K.: Butterworth-Heinemann, 114–125.

Vethaak, A.D., Lahr, J., and Schrap, S.M. et al. (2005). An integrated assessment of estrogenic contamination and biological effects in the aquatic environment of the Netherlands. *Chemosphere* **59**, 511–524.

Vighi, M., Garlanda, M.M., and Calamari, D. (1991). QSARs for the toxicity of OP pesticides to bees. *The Science of the Total Environment* **109/110**, 605–622.

Vine, E., Shears, J., and van Aerle, R. et al. (2005). Endocrine (sexual) disruption is not a prominent feature in the pike (*Esox lucius*), a top predator, living in English waters. *Environmental Toxicology and Chemistry* **24**, 1436–1443.

Von Keyserlingk, H.C. and Willis, R.J. (1992). The GABA-activated Cl⁻ channel in insects as a target for insecticide action—a physiological study. In D. Otto and B. Weber (Eds.) *Insecticides: Mechanism of Action and Resistance.* Andover, U.K.: Intercept.

Vos, J.G., Dybing, E., and Greim, H.A. et al. (2000). Health effects of endocrine-disrupting chemicals on wildlife, with special reference to the European situation. *Critical Reviews in Toxicology* **30**, 71–133.

Wadelius, M. and Pirmohamed, M. (2007). Pharmacogenetics of warfarin: current status and future challenges. *Pharmacogenomics Journal* **7**, 99–111.

Waid, J.S. (Ed.) (1985–87). *PCBs in the Environment,* Vols I–III. Cleveland, OH: CRC Press.

Waite, M.E., Waldock, M.J., and Thain, J.E. et al. (1991). Reductions in TBT concentrations in UK estuaries following legislation in 1986 and 1987. *Marine Environmental Research* **32**, 89–111.

Wake, D.B. (1991). Declining amphibian populations. *Science* **253**, 860–860.

Wakeling, A.E. (2000). Similarities and distinctions in the mode of action of different classes of antioestrogens. *Endocrine-Related Cancer* **7**, 17–28.

Walker, C.H. (1969). Reductive dechlorination of *p,p'*-DDT by pigeon liver microsomes. *Life Science* **8**, 111–115.

Walker, C.H. (1974). Comparative aspects of the metabolism of pesticides. *Environmental Quality and Safety* **3**, 113–153.

Walker, C.H. (1975). Variations in the intake and elimination of pollutants. In *Organochlorine Insecticides: Persistent Organic Pollutants.* London: Academic Press, 73–131.

Walker, C.H. (1978). Species differences in microsomal monooxygenase activities and their relationship to biological half lives. *Drug Metabolism Reviews* **7**(2), 295–323.

Walker, C.H. (1980). Species variations in some hepatic microsomal enzymes that metabolise xenobiotics. *Progress in Drug Metabolism* **5**, 118–164.

Walker, C.H. (1981). The correlation between in vivo and in vitro metabolism of pesticides in vertebrates. *Progress in Pesticide Biochemistry* **1**, 247–286.

Walker, C.H. (1983). Pesticides and birds—mechanisms of selective toxicity. *Agriculture, Environment, and Ecosystems* **9**, 211–226.

Walker, C.H. (1987). Kinetic models for predicting bioaccumulation of pollutants in eco-sytems. *Environmental Pollution* **44**, 227–240.

Walker, C.H. (1989). The development of an improved system of nomenclature and classi-fication of esterases. In E. Reiner, W.N. Aldridge, and F.C.G. Hoskin (Eds.) *Enzymes Hydrolysing Organophosphorous Compounds*, Chichester: Ellis Harwood. 53–64.

Walker, C.H. (1990a). Persistent pollutants in fish-eating sea birds—bioaccumulation, metab-olism and effects. *Aquatic Toxicology* **17**, 293–324.

Walker, C.H. (1990b). Kinetic models to predict bioaccumulation of pollutants. *Functional Ecology* **4**, 295–301.

Walker, C.H. (Ed.) (1991). The role of enzymes in regulating the toxicity of xenobiotics: Proceedings from Colloquium of the Pharmacological Biochemistry Group Biochemical Society, 638th Meeting, Reading University, 10–12 April 1991. *Biochemical Society Transactions* **19**, 731–767.

Walker, C.H. (1994a). Comparative toxicology. In E. Hodgson and P. Levi (Eds.) *Introduction to Biochemical Toxicology,* Norwalk, CT: Appleton and Lange, 193–218.

Walker, C.H. (1994b). Interactions between pesticides and esterases in humans. In M.I. Mackness and M. Clerc (Eds.) *Esterases, Lipases, and Phospholipases: from Structure to Clinical Significance.* NATO ASI Series. Series A, Life Sciences. New York Plenum Press 91–98.

Walker, C.H. (1998a). Avian forms of cytochrome P450. In D.R. Livingstone and J.J. Stegeman (Eds.) Forms and Functions of Cytochrome P450. *Comparative Biochemistry and Physiology* **121 C**, 65–72.

Walker, C.H. (1998b). Alternative approaches and tests in ecotoxicology. *ATLA* **26**, 649–677.

Walker, C.H. (1998c). The use of biomarkers to measure the interactive effects of chemicals. *Ecotoxicology and Environmental Safety* **40**, 65–70.

Walker, C.H. (1998d). Biomarker strategies to evaluate the environmental effects of chemi-cals. *Environmental Health Perspectives* **106** (Supplement 2), 613–620.

Walker, C.H. (2003). Neurotoxic pesticides and behavioral effects upon birds. *Ecotoxicology* **12**, 307–316.

Walker, C.H. (2004). Organochlorine insecticides and raptors in Britain. In H.F. Van Emden and M. Rothschild (Eds.) *Insect and Bird Interactions,* Andover U.K: Intercept, 133–145.

Walker, C.H. (2006). Ecotoxicity testing of chemicals with particular reference to pesticides. *Pest Management Science* **62**, 571–583.

Walker, C.H., Bentley, P., and Oesch, F. (1978). Phylogenetic distribution of epoxide hydratase in different vertebrate species, strains, and tissues using three substrates. *Biochemica et Biophysica Acta* **539**, 427–434.

Walker, C.H., Hopkin, S.P., Sibly, R.M., and Peakall, D.B. (1996, 2000, 2006). *Principles of Ecotoxicology* (1st, 2nd, and 3rd editions). Boca Raton, FL CRC/Taylor and Francis.

Walker, C.H. and Jefferies, D.J. (1978). The post mortem reductive dechlorination of *p,p'*-DDT in avian tissues. *Pesticide Biochemistry and Physiology* **9**, 203–210.

Walker, C.H. and Johnston, G.O. (1989). Interactive effects of pollutants at the kinetic level: Implications for the marine environment. *Marine Environmental Research* **28**, 521–525.

Walker, C.H. and Livingstone, D.R. (Eds.) (1992). *Persistent Pollutants in Marine Ecosystems.* A Special Publication of SETAC. Oxford: Pergamon Press.

Walker, C.H. and Newton, I. (1998). Effects of cyclodiene insecticides on the sparrowhawk in Britain—a reappraisal of the evidence. *Ecotoxicology* **7**, 185–189.

Walker, C.H. and Newton, I. (1999). Effects of cyclodiene insecticides on raptors in Britain—correction and updating of an earlier paper by Walker and Newton in *Ecotoxicology* **7**, 185–189; *Ecotoxicology* **8**, 425–430.

Walker, C.H. and Oesch, F. (1983). Enzymes in selective toxicity. In *Biological Basis of Detoxication. London:* Academic Press, 349–368.

Warner, R.F., Peterson, K.K., and Borgman, L. (1966). Behavioural pathology in fish: a quantitative study of sublethal pesticide intoxication. In Pesticides in the Environment and Their Effects on Wildlife. N.W. Moore (Ed.) *Journal of Applied Ecology* **3** (Supplement), 223–248.

Watanuki, H., Yamaguchi, T., and Sakai, M. (2002). Suppression in function of phagocytic cells in common carp *Cyprinus carpio* L. injected with estradiol, progesterone or 11-ketotestosterone. *Comparative Biochemistry and Physiology C—Toxicology and Pharmacology* **132**, 407–413.

Waters, R., Jones, N.J., and Morse, H.R. (1994). DNA adducts in aquatic organisms. In Abstracts of British Toxicology Society Meeting, *Biochemical Biomarkers in Environmental Toxicology*, Churchill College, Cambridge University, p. 38.

Webb, R.E., Randolph, W.C., and Horsfall, F. Jr. (1972). Hepatic benzo(*a*)pyrene monooxygenase activity in endrin susceptible and resistant pine mice. *Life Science* **11 Part 2**, 477–484.

Weisbart, M. and Feiner, D. (1974). Sublethal effect of DDT on osmotic and ionic regulation by goldfish *Carassius auratus. Canadian Journal of Zoology—Revue Canadienne De Zoologie* **52**, 739–744.

Welch, R.M., Levin, W., and Conney, A.H. (1969). Estrogenic action of DDT and its analogs. *Toxicology and Applied Pharmacology* **14**, 358–367.

Wellejus, A., Dalgaard, M., and Loft, S. (2002). Oxidative DNA damage in male Wistar rats exposed to di-*n*-butyl phthalate. *Journal of Toxicology and Environmental Health—Part A* **65**, 813–824.

Wells, M.R., Ludke, J.L., and Varborough, J.D. (1973). *Journal of Agriculture and Food Chemistry* **21**, 428–429.

Westerlund, L., Billsson, K., and Andersson, P.L. et al. (2000). Early life stage mortality in zebrafish (*Danio rerio*) following maternal exposure to polychlorinated biphenyls and estrogen. *Environmental Toxicology and Chemistry* **19**, 1582–1588.

White, D.H., Mitchell, C.A., and Hill, E.F. (1983). Parathion alters incubation behavior of laughing gulls. *Bulletin of Environmental Contamination and Toxicology* **31**, 93–97.

White, R., Jobling, S., and Hoare, S.A. et al. (1994). Environmentally persistent alkylphenolic compounds are estrogenic. *Endocrinology* **135**, 175–182.

Whyte, J.J., Van den Heuvel, M.R.M., and Clemons, J.H.M. et al. (1998). Mammalian and teleost cell line bioassays and chemically derived 2,3,7,8-TCDD equivalent concentrations in lake trout from Lake Superior and Lake Ontario, North America. *Environmental Toxicology and Chemistry* **17**, 2214–2226.

Wiemeyer, S.N. and Porter, R.D. (1970). DDE thins eggshells of captive American kestrels. *Nature* **227**, 737–738.

Wiemeyer, S.N., Bunck, C.M., and Stafford, C.J. (1993). Environmental contaminants in bald eagles—1980–1984—and further interpretations of relationships in productivity and shell thickness. *Archives of Environmental Contamination and Toxicology* **24**, 213–244.

Williams, R.J., Brooke, D.N., and Clare, R.W. et al. (1996). *Rosemaund Pesticide Transport Studies 1987–1993,* Report No. 129. Wallingford, U.K. Institute of Hydrology (NERC).

Williams, T.D., Diab, A.M., and George, S.G. (2007). Gene expression responses of European flounder (*Platichthys flesus*) to 17-beta estradiol. *Toxicology Letters* **168**, 236–248.

Wilson, V.S., Lambright, C., Ostby, J., and Gray, L.E. (2002). In vitro and in vivo effects of 17 beta-trenbolone: A feedlot effluent contaminant. *Toxicological Sciences* **70**, 202–211.

Wobeser, G., Nielsen, N.O., and Schliefer, B. (1976). Mercury and mink 1 Experimental methylmercury intoxication. *Canadian Journal of Comparative Medicine* **40**, 34–45.

Wolfe, M.F., Atkeson, T., and Bowerman, W. et al. (2007). Wildlife indicators. In: R. Harris, D.P. Krabbenhoft, and R. Mason et al. (Eds.) *Ecosystem Response to Mercury Contamination: Indicators of Change,* CRC Press, SETAC 123–189.

Wolfe, M.F., Schwarzbach, S., and Sulaiman, R.A. (1998). The effects of mercury on wildlife: a comprehensive review. *Environmental Toxicology and Chemistry* **17**, 146–160.

Woodburn, K.B, Hansen, S.C., and Roth, G.A. et al. (2003). The bioconcentration and metabolism of chlorpyriphos by the eastern oyster. *Environmental Toxicology and Chemistry* **22**, 276–284.

Worth, A. and Balls, M. (Eds.) (2002). *Alternative Methods for Chemicals Testing: Current Status and Future Prospects.* Ispra, Italy: European Centre for the Validation of Alternative Methods(ECVAM), 71–74.

Wright, J.F. (1995). Development and use of a system for predicting the macroinvertebrate fauna found in flowing water. *Australian Journal of Ecology* **20**, 181–197.

Wurster, D.H., Wurster, C.F. Jr., and Strickland, W.N. (1965). Bird mortality following DDT spray for Dutch elm disease. *Ecology* **46**, 488–499.

Yamasaki, K., Sawaki, M., and Ohta, R. et al. (2003). OECD validation of the Hershberger assay in Japan: Phase 2 dose response of methyltestosterone, vinclozolin, and *p,p'*-DDE. *Environmental Health Perspectives* **111**, 1912–1919.

Yokota, H., Seki, M., and Maeda, M. et al. (2001). Life-cycle toxicity of 4-nonylphenol to medaka (*Oryzias latipes*). *Environmental Toxicology and Chemistry* **20**, 2552–2560.

Yunker, M.B. and Cretney, W.J. (1996). Dioxins and furans in crab hepatopancreas: uses of principle component analysis to classify congener patterns and determine linkages to contamination sources. In M. Servos, K.R. Munkittrick, J.H. Carey, and G.J. van der Kraak (Eds.) *Environmental Fate and Effects of Pulp and Paper Mill Effluents.* Delray Beach, FL: St. Lucie Press, pp. 315–325.

Zacharewski, T. (1997). In vitro bioassays for assessing estrogenic substances. *Environmental Science and Technology* **31**, 613–623.

Zala, S.M. and Penn, D.J. (2004). Abnormal behaviours induced by chemical pollution: a review of the evidence and new challenges. *Animal Behaviour* **68**, 649–664.

Zheng, W.B., Strobeck, C., and Stacey, N. (1997). The steroid pheromone 4-pregnen-17 alpha, 20 beta-diol-3-one increases fertility and paternity in goldfish. *Journal of Experimental Biology* **200**, 2833–2840.

Zhou, T., John-Alder, H.B., and Weis, P. et al. (1999). Thyroidal status of mummichogs (*Fundulus heteroclitus*) from a polluted versus a reference habitat. *Environmental Toxicology and Chemistry* **18**, 2817–2823.

Zoeller, R.T., Tyl, R.W., and Tan, S.W. (2007). Current and potential rodent screens and tests for thyroid toxicants. *Critical Reviews in Toxicology* **37**, 55–95.

Index

A

Absorption, 55
    from gut, 25, 48, 111
    of herbicides, 23
    organochlorines, 111
    potentiation mechanisms, 63
    reabsorption, 48, 54
    from water, 24
Acaricides, 77, 105
Accipiter nisus; See Sparrow hawks
Acetamidophos, 86
Acetone, 31, 48
2-Acetylaminofluroene, 50
Acetylation, conjugation reactions, 48
Acetyl carnitine, 32
Acetylcholine antagonism, 7
Acetylcholine receptors, 6, 7, 297, 299, 304, 316;
        See also Cholinergic receptors
    mercury compound affinity for, 169
    neurotoxin mechanisms of action, 298,
        299–300
Acetylcholinesterase/acetylcholinesterase
        inhibition, 316
    B esterases, 38–39
    biomarkers, 60, 86, 315
        chemicals sharing same principal mode of
            action, 308
        complex pollution problems, 250
        measuring toxicity of mixtures, 245
    carbamate-OP synergy, 216
    carbamate toxicity, 215
    mipafox and, 19
    mode of pesticide action, 56
    neurotoxin mechanisms of action, 299–300
    organophosphorus compounds
        metabolism, 199
        resistance mechanisms, 211
        toxicity, 196, 199, 202–208
    population level effects, 311
    resistance mechanisms, 10, 95, 157, 211
    toxicodynamics, 59
Acetylsalicylic acid, 23
Acid anhydrides/acid anhydrolases, 36
Aconitase hydrase, 8
Aconitine, 293
Activation
    cytochrome P450 biotransformations, 29–30
    factors determining toxicity and persistence,
        19

organophosphorus compound metabolism, 199
organophosphorus compounds, 208
PAHs, 186–187
PCBs, 144
Active transport, 23
Acute toxicity/LD$_{50}$
    anticoagulant rodenticide poisoning, 225
    anticoagulant rodenticides, 225
    cyclodienes, 123–124
    DDT and related compound ecotoxicity, 111
    hexachlorocyclohexanes/lindane, 131
    organochlorines, 111, 113
    organometallic compounds
        mercury, 170
        tributyltin oxide, 176
    organophosphorus compounds, 193, 201, 206,
        207, 208
    PAHs, 189
    pyrethroids, 236
    selectivity ratios, 61–62
    tributyltin oxide, 176
Additive toxicity
    complex pollution problems, 253
    herbicides, 263
    measuring toxicity of mixtures, 244
Adsorption
    environmental fate, 69–70
        properties affecting, 68
        in soils and sediments, 81, 82, 83
    PAHs, 185
    pyrethroid insecticides, 235
    tributyltin oxide (TBTO), 172, 174
Aerobic conditions, and detoxification, 42
A-esterases, 37–38, 39–40
    organophosphorus compound metabolism,
        198, 199
    selectivity, 62
Aflatoxins, 8, 12, 13
Agent Blue, 178
Agent Orange, 152, 257
Ageratum houstonianum, 7
Aging, enzyme, 297
    mechanistic biomarker assays, 86
    organophosphorus compound metabolism, 39,
        197, 299–300
Agricultural ecosystems, herbicides and, 257,
        258–261
Ah receptor-mediated toxicity
    bioassays, 252

complex pollution problems
  bioassays, 252
  biomarkers, 246
endocrine disruption, 268
enzyme induction, 49
PCBs, 140, 143–144, 146, 147
TCDD, 154–155, 157, 158
toxicodynamics, 57, 59
Ah receptors
  CALUX system, 156
  endocrine disruptor modes of action, 267, 268
Air
  DDT circulation, 105
  dimethyl mercury volatility, 166
  fate of pollutants from waste disposal sites,
    82
  PAH emissions, 185
  uptake routes, 76
Aldicarb, 319
  chemical properties, 207, 212, 213
  environmental fate, 213, 215
  metabolism, 213, 214
  toxicity, 215, 216
Aldrin, 102, 116
  chemical properties, 103, 116
  cytochrome P450 biotransformations, 28, 29,
    30, 31
  dirty dozen, persistent organic pollutants
    (POPs), 77, 102
  ecological effects, 124–131
  environmental fate, 119
  fish kills, 80
  metabolism, 117, 118
  structure, 101
  toxicity, 123–124
Alectoris rufa (red-legged partridge), 64
Algae toxins, 11, 12
Alicyclic hydrocarbons, 181
Aliphatic hydrocarbons, 181
Aliphatic hydroxylation, 28
Alkyl groups
  cytochrome P450 biotransformations, 29
  mercury, 57, 163, 167; *See also* Methyl
    mercury
  organophosphate metabolism, dealkylation,
    40
Alkylphenols
  complex pollution problems, biomarkers, 246
  endocrine disruption, 274, 281–282
    lessons learned, 292
    species susceptibility, 286
Alligators, endocrine disruptor case studies, 270,
    271–273
Alpha-HCH, 131
American kestrels, 139, 143
Ames test, 252
Ametryne, 258, 259

Aminopyrene, 29, 41, 50
Aminotriazole, 258
Amphibia
  endocrine disruption, 267
    assays/screening tests, 278
    atrazine, 275–276
  excretion via skin, 52
  PAH adduct formation in, 188
  storage of xenobiotics, 51
  uptake and distribution of xenobiotics, 21
Anabolic steroids, 267
Anaerobic conditions
  microbial metabolism of PCBs, 140
  reductive metabolism, 42
Androgenic activity, endocrine disruptors, 269
Androgens, Hershberger assay, 278
Animal defenses, 293; *See also* Toxicokinetics
  excretion, 51–54
  neurotoxins, 293
  against plant toxins, 8–10
  toxins, 10–11, 12
Animal models
  ethical issues, 328
  limited applicability, species differences
    and, 72
  prediction of toxicity from, 18
Animal toxicity, herbicide mode of action and, 58
Anopheles, 14, 32
Anopheles arabiensis, 130
Anopheles gambiae, 116, 130, 238
Antagonism, PCB mixtures, 157
Anthracene, 183, 185, 189
Anthropogenic sources
  arsenic, 178
  endocrine disruptors, 267
  mercury, 165, 167–168, 169
Antiantrogens, modes of action, 267
Antibiotics, 93
Anticholinesterases, 6, 14; *See also*
    Acetylcholinesterase/
    acetylcholinesterase inhibition;
    Cholinesterases
Anticoagulant rodenticides, 6, 219–228
  chemical properties, 219–221
  complex pollution problems, 245
  cytochrome P450 enzyme specialization, 10
  ecological effects, 226–228
    poisoning incidents in field, 226–227
    population genetics, 228
  environmental fate, 222–223
  mechanistic biomarker assays, 86
  metabolism, 221–222
  mode of action, 57
  toxicity, 224–226
  toxicodynamics, 59
Antiestrogenic chemicals, modes of action, 267
Antifoulants, tributyltin, 173, 179

Antifungal agents; *See* Fungicides/antifungal agents
Apis mellifera; *See* Honeybee
Application methods, herbicides, 260–261
Aquaculture, 201, 285
Aquatic animals
    cytochrome P450 evolution, 31
    excretion mechanisms, 52
Aquatic systems, 320; *See also* Surface waters
    carbamate elution into, 217
    ecological profiling, 96
    endocrine disruptors, 280
        alkylphenols, 282
        species susceptibility, 286
    environmental fate
        biological factors affecting, 71
        movement of pollutants, 76
        properties affecting, 68
        in soils and sediments, 83
    herbicides
        effects in, 261–263
        movement into surface water, 261
    models for bioconcentration and bioaccumulation, 77
    mollusks; *See* Dog whelk (*Nucella lapillus*); Mollusks; Oysters
    organochlorines, DDT and related compounds
        Clear Lake California biota, 113
        half-life in food chain, 106
        residues in food chain, 107–108
        toxicity, 111
    organomercury interactions with other pollutants, 171
    organophosphorus compounds, 201, 207, 209
    pyrethroid insecticides, 234, 235–236
        population level effects, 237–238
    uptake and distribution in, 21
    uptake and loss mechanisms, 77, 79
Ardea cinerea (herons), 106, 129–130
Arene oxides, 136
Arginine, peptide conjugations, 47
Arochlor, 133, 139, 141, 143, 150
Aromatase activity
    endocrine disruption mechanisms, 268–269, 272
    tributyltin and, 174, 175, 176, 272
Aromatic amines, 249
Aromatic hydrocarbons, 181, 182; *See also* Polycyclic aromatic hydrocarbons
Aromatic hydroxylation, 28–29
Arsenic compounds, 178–179
Aryldialkyl phosphatases, 37
Aryl esterase, 37, 40
Aryl hydrocarbon hydroxylase (AHH), 146
Aryl hydrocarbon receptor, PCDF affinity for, 153
Aryl mercury, 57

A-SCREEN, 281
Aspergillus flavus, 11–12, 13
Aspergillus parasiticus, 13
Aspirin, 23, 32
Assays, tests, assessment methods; *See also* Biomarkers
    behavioral effects, 313–316; *See also* Behavioral effects, neurotoxicity and
    complex pollution problems , 251–253
        potentiation and synergism, 253–254
        toxicity of mixtures, 244, 251–252
    endocrine disruptors, 246, 273–274, 276–278
    future prospects, 321–324
        adoption of more ecologically relevant practices, 321–323
        biomarkers, development of, 323–324
        ethical questions, 328
        field studies, 326–327
    selectivity, 61
    soil community assessment, 96
    sublethal effects, 17
    in vitro systems; *See In vitro* tests
ATPase inhibition, DDT, 110
ATP formation
    oxidative phosphorylation uncouplers, 58, 60
    tributyltin and, 174
Atrazine, 22
    complex pollution problems, 263
    endocrine disruptors, 275–276
    movement into surface water, 261, 262
    properties, 258
    structure, 259
Atropa belladonna (deadly nightshade), 5, 7
Atropine, 5, 7, 204
Avermectins, 11–12
Azamethiphos, 201
Azinphos-methyl (gusathion), 194, 195

B

Bacillus thuringiensis (BT), 13
Badger (*Meles meles*), 129
Baits, OP, 208
Barban, 258, 259
Batrachotoxin, 12
Battlefield chemical weapons, 14
Bay-region epoxides, 28, 187–188
Bees; *See* Honeybee (*Apis mellifera*)
Behavioral effects, 17
    assessment of individuals, 84
    biomarker approach, 84
    carbamates, 217
    complex pollution problems, toxic responses with shared pathways of expression, 250
    cyclodienes
        assays, 123

and raptor/predatory bird hunting ability, 128
  sublethal doses, 123
endocrine disruptors, 284, 288–290
mechanistic biomarker assays, 86
organochlorine toxicity, 111
organomercury compounds, 169, 170, 171–172
organophosphorus compounds, 205, 209
toxicodynamics, 59
Behavioral effects, neurotoxicity and, 293–317
  adoption of more ecologically relevant assessment practices, 322
  assessment approaches, 295–296, 313–316
  causal chain at different organizational levels, 308–311
    chemicals sharing same principal mode of action, 308–310
    combination of chemicals with different modes of action, 310
  mechanisms of action, 296–302
    acetylcholine receptor, 298, 299–300
    GABA receptors, 298, 299
    neuropathy target esterase, 298, 300
    sodium channels, 296, 297, 298
  nervous system effects, 302–306
    central nervous system, 305
    peripheral nervous system, 302–304
  population level, relating effects to changes at, 311–313
  whole organism level effects, 306–308
Beneficial organisms, pesticide effects on, 19, 210
Bengalese finch (Lonchura striata), 106
Benzene, 48
Benzene hexachloride (BHC); See Hexachlorocyclohexanes
Benzo[a]pyrene, 181, 182
  endocrine disruption, modes of action, 268
  immunosuppression, 189
  metabolism, 183, 184, 187
    conjugation, 44
    cytochrome P450 biotransformations, 28, 30, 31
    epoxide hydrolases, 40–41
    glutathione conjugation, 46
  models for bioconcentration and bioaccumulation, 77
  mode of action, 56
  NOEL level, 189
  octanol-water partition coefficients, 22, 183
  properties, 183
  sediment concentration, 185
  toxicodynamics, 55
Benzofluoranthenes, 185
Bergapten, 32

B-esterases, 37, 38–40
  organophosphorus compounds, 197, 198, 199, 202
  resistance mechanisms, 94
  selectivity, 62
Beta glucuronidases, 48
Beta-HCH, 131
Beta naphthoflavone, 48
Beta oxidation, methyl mercury, 165
BHC; See Hexachlorocyclohexanes; Lindane
Bile, excretion via, 48, 52
  TCDD metabolites, 153
  in terrestrial animals, 53, 54
Binding
  and environmental fate, 69–70, 72
  protein; See Proteins/protein binding
Bioaccumulation
  aquatic animals, 52
  biomagnification, 76
  cyclodienes, 120–121
  endocrine disruptors, 280
    complex exposure problems, 283, 284
    PCBs, 280
    species susceptibility, 286
  environmental fate, 71
  hepatic microsomal monooxygenase (HMO) activities and, 34
  movement of pollutants along food chains, 75–81, 108
  neurotoxins, population level effects, 312
  organometallic compounds
    mercury, 166, 167, 168
    tributyltin oxide (TBTO), 174
  PAHs, 186
  PCBs, 141
  pyrethroid insecticides, 236
  solubility properties and, 22
  storage, 50–51
Bioaccumulation factor (BAF)
  cyclodienes, 121
  models for bioconcentration and bioaccumulation, 77, 80
  movement of pollutants along food chains, 76
  PCBs, 142
Bioassays; See Assays, tests, assessment methods
Biochemical effects
  biomarker approach, 84
  PCBs, 146
Biocides, 60
  microbial level, 93
  tributyltin, 173
Bioconcentration
  endocrine disruptors, 278, 282
  environmental fate, 71
  movement of pollutants along food chains, 75, 76
  organochlorines, 112–113

PAHs, 186
PCBs, 141
pyrethroid insecticides, 235–236
Bioconcentration factor (BCF)
  cyclodienes, 120
  exchange diffusion in aquatic animals, 52
  models for bioconcentration and
      bioaccumulation, 77
  movement of pollutants along food chains, 76
  organochlorine residues, 107
  pyrethroid insecticides, 235
  tributyltin oxide (TBTO), 174
Biodegradation
  anticoagulant rodenticides, 222
  and environmental fate, 71, 76
  herbicides, 261
  organophosphorus compounds, 196
  pyrethroid insecticides, 235
  in soils and sediments, 83
Biological half-life; See Half-lives
Biological membranes; See Membranes
Bioluminescence assays, 150, 154, 156, 247, 252
Biomagnification
  cyclodiene bioaccumulation, 121
  environmental fate, 72
  herbicides, 257
  movement of pollutants along food chains,
      75–76
  organochlorines, 107, 108
  organomercury compounds, 168
  PBBs, 149
Biomagnification factor, 108
Biomarkers, 18
  anticoagulant rodenticides, 225
  carbamates, 205–206
  community and ecosystem level effects, 96
  complex pollution problems, 245–250
    combinations of chemicals with different
        modes of action, 310
    measuring toxicity of mixtures, 244–250
    toxic responses with shared pathways of
        expression, 250–251
  concept of, 60
  cyclodienes, 123
  cytochrome P450 induction, 140
  desirable features, 87–88
  distribution in communities and ecosystems
    individual effects, 84–89
    in wider ecological context, 89–90
  endocrine disruption, 278
  future prospects, 323–324
    adoption of more ecologically relevant
        assessment practices, 321–323
    development of more sophisticated testing
        methods, 323–324
    field studies, 326–327

neurotoxins; See also Behavioral effects,
    neurotoxicity and
  behavioral effects, 294, 305, 308
  brain AChE, 205–206, 209
  organophosphorus compounds, 205–206
  PAHs, 185, 190
  PCBs, 146
  and population parameters, 90, 92, 93
  prediction of ecological risks, 97
Biomonitoring, bioassays for toxicity of mixtures,
    251–252
Bioremediation, 72
  microbial metabolism of PCBs, 140
  reductive dehalogenation, 42
Biotic indices
  community and ecosystem level effects, 96
  selectivity, 60
Biotransformation; See also Metabolism
  DDT, 103
  and environmental fate, 72
  organomercury compounds, 165
  organophosphorus compounds, 195
  PAHs, 186
Bipyridyl herbicides, 257
  biomarkers, 248
  environmental fate, 69
Birds/bird species
  anticoagulant rodenticide poisoning,
      222–223, 225–227
  carbamate toxicity, 215, 217
  cyclodienes, 125
    bioaccumulation, 121, 122
    excretion, 117
    population level effects, 124
  dioxins/TCDD, ecological effects related to
      TEQs, 158, 159, 160
  endocrine disruption
    behavioral effects, 288
    case studies, 270–271
    life stages with enhanced sensitivity, 285
    species susceptibility, 286, 287
  HCH/lindane poisoning, 131
  herbicide effects on ecosystem, 260
  mechanistic biomarker assays, 86
  metabolism
    A esterase activity, 37, 38
    excretion in terrestrial animals, 52
    monooxygenase activity comparisons,
        34, 35
    peptide conjugations, 47
  oiling, 189
  organochlorines, DDT and metabolites, 107
    DDE effects, 17
    embryo feminization, 110–111
    half-life in food chain, 106
    marine food chains, 108
    population level effects, 112, 114, 115

sites of action, 110
species with eggshell thinning, 110
terrestrial food chains, 108–109
toxicity, 111
organometallic compound toxicity
    lead, 178, 179
    mercury, 166, 167, 168, 170
organophosphorus compounds
    behavioral effects, 309–310
    for control of, 200
    ecological effects, 208
    population dynamics, 210–211
    toxicity, 200–201, 207, 208
PAHs in, 186
PCBs
    bioaccumulation by seabirds, 142
    environmental fate, 140, 141, 143
    in food chain, 141
    induction of P450s as indicator, 146
    population effects, 147–149
    toxicity, 144
population level effects, 311
predatory; See Fish-eating birds; Predatory
        birds/raptors
pyrethroid toxicity, 236
selectivity, 61, 62
storage of xenobiotics, 51
toxic effects and population changes, 97
Bisphenol A, 269, 281
Black swallowtail (Papilio polyxenes), 32
Blatella germanica (German cockroach), 95
    cyclodiene resistance, 130
    organochlorine-pyrethroid cross resistance,
        116
    pyrethroid resistance, 238
Blood
    dieldrin levels in, 122
    distribution of xenobiotics in organisms, 21
    mechanistic biomarker assays, 86
Blood-brain barrier, 24, 178
Blood coagulation inhibitors, 6; See also
        Anticoagulant rodenticides
Bombardier beetles (Brachinus spp.) defense,
        10–11
Botulinum toxin, 8
Boxworth Experiment, 210
Brachinus spp. (bombardier beetle) defense, 10–11
Brain
    blood-brain barrier and, 24, 178
    mercury toxicity, 169–170, 171
Brain (acetyl)cholinesterase
    as biomarker, 205–206, 309–310
        chemicals sharing same principal mode of
            action, 308
        measuring toxicity of mixtures, 245
    carbamate-OP synergy, 216

combinations of chemicals with different
        modes of action, 310
    organophosphorus compounds and, 208, 210
Breast milk, PBDEs in, 281
Breeding
    behavioral effects
        endocrine disruptors, 289–290
        of neurotoxins, 312
    reproductive and fertility effects; See
            Reproductive and fertility effects
    seasonal population density fluctuations, 91
Brodifacoum, 219
    chemical properties, 220, 221
    ecological effects, 227
    environmental fate, 222, 223
    metabolism, 222
    mode of action, 57
    toxicity, 225, 226
    toxicodynamics, 59
Bromodiolone, 219, 223
    mode of action, 57
    resistance mechanisms, 228
Bromoxynil, 258
Butyrylcholinesterase, 38–39

C

Cabbage root fly (Erioischia brassicae), 130
Calabar bean (Physostigma benenosum), 6
Calcium flux, toxins affecting, 6, 7, 57
Calcium homeostasis
    PAH immunotoxic mechanisms, 189
    tributyltin and, 176
Calcium pump, 57
    mechanistic biomarker assays, 87
    organochlorine sites of action, 110
    toxicodynamics, 59
Calliphora erythrocephala (locust), 304
CALUX system, 150, 154, 156, 247, 252
Carassius auratus (goldfish), 123, 217
Carbamates, 194, 212–218, 319
    behavioral effects, 307, 309, 310, 313
    B esterases, 39
    biomarkers, 86, 205–206
    chemical properties, 207, 212–213
    complex pollution problems, biomarkers, 245
    ecological effects, 217–218
    effects at whole organism level, 306
    environmental fate, 69, 213, 215
    herbicide
        mode of action, 258
        structure, 259
    metabolism, 213, 214
    mode of action, 56, 297
    and organophosphorus compound toxicity,
        210
    population level effects, 311

potentiation and synergism, 33, 64
resistance mechanisms, 94
toxicity, 215–217
Carbaryl
behavioral effects, 309
chemical properties, 207, 212, 213
metabolism, 214
potentiation and synergism, 64
toxicity, 215, 216
uptake and distribution of xenobiotics, 22
Carbofuran
adsorption and elution from clay soils, 82
chemical properties, 207, 212, 213
ecological effects, 217–218
environmental fate, 213, 215, 261
metabolism, 213, 214
toxicity, 215, 216
Carbon-carbon bond hydrolysis, 36
Carbon double bonds, cytochrome P450
biotransformations, 29
Carbonium ion formation, 187
Carbon monoxide, 30, 33
Carbon-nitrogen bond hydrolysis, 36
Carbon-phosphorus bond hydrolysis, 36
Carbon tetrachloride, 41, 42, 104
Carbophenothion, 208, 210, 311, 315, 319
Carboxyl esterases, 37, 38
induction of, 50
organophosphorus compound metabolism,
198, 199
potentiation and synergism, 64
and selectivity/resistance, 62
Carcinogens/mutagens
activation of, 50
aflatoxin, 8, 13
Ames test, 252
cytochrome P450 biotransformations, 28, 30,
41, 140, 187
epoxide hydrolase detoxification of epoxides,
41
glutathione conjugation, 46
modes of action of pollutants in animals, 56
PAH metabolic products, 183, 187, 188, 191
PCB toxicity, 145
potentiation and synergism, 64
Cardiac glycosides, 8
Carrying capacity, 91
Caspian tern (*Hydroprogne caspia*), 147, 148,
158, 159
Castor oil plant (*Ricinus communis*), 7
Catalase, 248
Catalytic triad, ACh, 203
Cattle
avermectins, 11–12
feedlots as sources of endocrine disruptors,
267
Cattle ticks, 95

Causality
complex pollution problems, 243
neurotoxicity at different organizational
levels
chemicals sharing same principal mode of
action, 308–310
combination of chemicals with different
modes of action, 310
Cell culture
bioassays, 252
endocrine disruptor assays, 277
estrogen bioassays, 246
neurotoxicity assays, 303, 315
Cell division disruption, herbicide mode of
action, 58
Central nervous system agents, 6; *See also* Brain
(acetyl)cholinesterase
blood-brain barrier and, 24
mechanisms of action, 305
organomercury, 169, 172
Cetaceans, PCBs in, 143
Cevadine, 5
Charge; *See* Polarity
Charge, and environmental fate, 68, 69–70
Chemical analysis
community and ecosystem level effects, 96
complex pollution problems, 243
measuring toxicity of mixtures, 244
potentiation and synergism considerations, 244
pyrethroid ecological studies, 238
Chemical hydrolysis
carbamates, 213
OPs, 196
Chemically activated luciferase gene expression;
See CALUX system
Chemical properties; *See specific chemicals and
classes of chemicals*
Chemical stability, and environmental fate, 69
Chemical warfare (evolutionary biology), 3–15
animal toxins, 10–11, 12
human made chemical weapons, 13–14
microbial toxins, 11–13
neurotoxins, 293
plant-animal, 4–10
animal defenses, 8–10
plant toxins, 4–8
Chickens (*Gallus domestica*)
DDT half-life in, 106
endocrine disruptor studies, 266
methyl mercury studies, 166–167
Chloramphenicol metabolism, 54
Chlordane, 77, 102, 116
Chlordecone, 77, 116
Chlorfenvinphos, 196, 319
behavioral effects, 309–310
chemical properties, 196
cytochrome P450 biotransformations, 28, 29

ecological effects, 208
ecotoxicity, 209
persistence, 201
seed dressings, 208, 209
structure, 195
Chloride channel, GABA receptor, 122, 130, 297,
        303, 304
Chlorinated benzoic acids, 258
Chlorination
    degree of; See Halogenation, degree of
    PCB substituent locations (ortho, meta, para)
        and fate in environment, 141, 142
        and metabolism, 136, 139
        and toxicity, 143
Chlorine treatment, PCDD formation during
        wood pulp processing, 152
Chlorobiphenyl, 22
Chlorofluorocarbons, 68
Chloroform, 22
Chlorophenols, 131
    environmental fate, 69
    PCDD formation, 151–152
Chlorpropham, 259
Chlorpyrifos, 82, 200
Chlorpyrifos-oxon, 37
Chlorsulfuron, 258
Chlortoluron, 262, 263
Cholinergic receptors, 6, 297, 299, 304, 316
    neurotoxin mechanisms of action, 299
    neurotoxin mode of action, 7
    organophosphorus compounds, 202–208, 209
Cholinergic synapses, nicotinoids and, 311–312
Cholinesterases, 37; See also
        Acetylcholinesterase/
        acetylcholinesterase inhibition
    acetylcholinesterase and
        butyrylcholinesterase, 38–39
    behavioral effects of inhibitors, 309
    carbamate-OP synergy, 216
    carbamate toxicity, 215
    complex pollution problems
        biomarkers, 245
        combinations of chemicals with different
            modes of action, 310
    OP resistance, 201–202
    toxicodynamics, 55
Chronic toxicity, mercury, 170
Chrysanthemic acid, 231, 232
Chrysanthemum spp., 4
Chrysene, 182, 183, 185
Cinerolone, 231, 232
Cinnabar, 164
Claviceps purpurea, 13, 294
Clay/clay soils
    drying and elution of adsorbed particles, 82
    environmental fate, 68, 70, 83
    herbicide adsorption and elution, 261

Clofibrate, 32, 48, 49, 50
Clophen, 133, 147
Clostidium botulinum, 8
Clotrimazole, 48
Clotting inhibitors; See Anticoagulant
        rodenticides
CMPP, 58
Coagulation inhibitors; See Anticoagulant
        rodenticides
Cofactor requirements, microsomal enzymes, 46
Colloids, 68, 82
Columba livia (feral pigeon), 106, 123
Combustion products
    PAHs, 182, 185
    PCDD, 152
Common tern (Sterna hirundo), 144
Community level effects, 84, 91, 96–97; See
        also Distribution in communities and
        ecosystems
Compartments, environmental, 70–71
Complex mixtures, PAHs, 183
Complex pollution problems, 243–254; See also
        Mixtures
    bioassays, 251–253
    endocrine disruptor mixtures, 283–284
    herbicides, 263
    mixture toxicity measurement, 244–245
    neurotoxins
        chemicals sharing same principal mode of
            action, 308–310
        combinations of chemical with different
            modes of action, 310
    potentiation and synergism, 253–254
    shared mechanisms of action, integrated
            biomarker approach, 245–250
    toxic responses with shared pathways of
            expression, 250–251
Coniine, 5, 7
Conium maculatum (hemlock), 5, 7
Conjugation, 42–48
    acetylation, 48
    anticoagulant rodenticide metabolism, 221
    carbamates, 214
    DDT metabolism, 105
    excretion in terrestrial animals, 53, 54
    glucuronyl transferases, 42–43, 44
    glutathione-S-transferase, 43–47
    metabolism, 24, 25
    methyl mercury, 165, 166
    organophosphorus compound metabolism, 199
    peptide, 47
    pyrethroid insecticides, 234
    sulfotransferases, 43, 44
    TCDD metabolism, 153
Contact action, early pesticides, 193
Contraceptive pharmaceuticals, endocrine
        disruptor sources, 268, 274

Coplanar PCBs, 134, 135, 155
  bioassays, 252
  biomarkers, 185
  complex pollution problems, biomarkers, 246
  cytochrome P450 biotransformations, 31, 136
  enzyme induction, 48, 49, 187
  TEQs, 147
Coronene, 182
Corticosteroids, 268
Cotransport, membrane, 24
Coumarins, 10
Coumarol, 6
Covalent bonding, 55
Crassostrea gigas (Pacific oyster), 174, 176
Crassostrea virginica (Eastern oyster), 77, 200
Cross-resistance
  cyclodienes, 130
  HCH, 131
  organochlorines and pyrethroids, 116, 303
Crude oil, 181, 182
Crustaceans
  metabolic rates in marine animals, 52
  organochlorine residues in, 108
  uptake and loss mechanisms, 79
Crysanthemum cinerariaefolium, 231
Cyclodienes, 101, 102
  chemical properties, 116–117
  ecological effects, 124–131
    population numbers, 124–130
    resistance, development of, 130–131
  environmental fate, 119–122
  metabolism, 117–119
    epoxide hydrolases, 41
    excretion in terrestrial animals, 53
  mode of action, 56
  neurotoxicity assays, 303
  and population declines among predatory
    birds, 90
  population level effects, 311
  resistance mechanisms, 10, 94, 95, 157
  toxicity, 122–124
Cyclohexane, 182
Cyfluthrin, 231
Cyhalothrin, 231, 236–237
Cypermethrin, 231
  chemical properties, 232, 233
  cytochrome P450 metabolism studies, 32
  endocrine disruption, 279
  housefly monooxygenase detoxification
    mechanism, 32
  mesocosm studies, 237
  persistence, 235
  potentiation and synergism, 64
  resistance mechanisms, 94
  selectivity, 61, 62
  toxicity, 236

Cysteine conjugation
  biotransformation, 45
  methyl mercury, 165, 166
Cysteine groups
  esterases, 39–40
  mercury compound affinity for, 169
Cytochrome P450 enzymes, 26, 27–28; See also
    Monooxygenases
  anticoagulant rodenticide metabolism, 221
  anticoagulant rodenticide resistance, 228
  CALUX system, 156
  carbamate metabolism, 213
  carbamate-OP synergy, 216–217
  carcinogenic/mutagenic products, 28, 30, 41,
    50, 140, 187
  complex pollution problems, bioassays,
    252–253
  cyclodiene metabolism, 130
  DDT metabolism, 104
  endocrine disruptor modes of action, 268
  and environmental fate, 71–72
  gene families, 30–31
  hexachlorocyclohexanes/lindane, 131
  induction of, 48–50
  inhibitors, 30, 33–34, 173
  modes of action of pollutants in animals, 57
  organophosphorus compound metabolism,
    197, 198
  PAH metabolism, 183, 184–185, 186–187
  PCB metabolism, 136, 140
    induction of as indicator of PCB presence,
      146
    in marine mammals, 143
    metabolic activation, 144
  pyrethroid insecticides, 234
  reductive metabolism, 41–42
  resistance mechanisms, 94, 95
  selectivity, 62
  TCDD and, 154, 155
  terminology, 33
  toxicodynamics, 59
  tributyltin and, 173, 174, 176
Cytochrome P450 reductase, 42

D

2,4-D, 218
  fate of pollutants in soils and sediments, 81,
    82
  mode of action, 58
  properties, 258
  structure, 259
DCNB (1,2-dichloro-4-nitrobenzene), 44, 45
DDA, 104, 105
DDD (rhothane), 101
  chemical properties, 103
  DDT environmental fate, 105

DDT metabolism, 104
ecological effects, 112–113
half-life in soil, 106
persistence, 107
population level effects, 113, 311
reductive metabolism, 41
in terrestrial food chains, 109
DDE, 17, 222, 235, 322
and breeding success, 93
complex pollution problems, biomarkers, 246
DDT dehydrochlorination, 103
DDT environmental fate, 105
ecotoxicity, 111
and eggshell thinning, 110
endocrine disruption, 278
by metabolic products, 279
modes of action, 267
environmental fate, 69, 121, 235
excretion in terrestrial animals, 52–53
fish-eating birds, PCB study complications, 147, 148
interactions with mercury pollutants, 171
marine food chains, 107, 108
mechanistic biomarker assays, 86, 87
metabolism, 47, 105
mode of action, 57
movement of pollutants along food chains, 76
persistence, 106–107, 112, 120
population level effects, 90, 97, 114–115, 126, 315
in terrestrial food chains, 109
toxicity, 111
toxicodynamics, 59
DDT, 6, 102–116
behavioral effects, 307, 313
biomarkers, 85
chemical properties, 102–103
complex pollution problems, bioassays, 252
endocrine disruption, 270–272
by metabolic products, 278, 279
modes of action, 267, 269
environmental fate, 68, 105–109
fate of pollutants in soils and sediments, 81
isomers, 102
and kdr genes in populations, 238
metabolism, 104–105
enzyme induction, 48
excretion in terrestrial animals, 53
glutathione-S-transferases, 47
reductive, 41, 42
modes of action, 56, 57, 267, 269, 297, 298, 303–304
PCB study complications, 147
persistence
behavioral effects, 313
comparison with cyclodienes, 120

dirty dozen, persistent organic pollutants (POPs), 77, 102
population level effects, 90, 112–116
population genetics, 115–116
population numbers, 112–115
resistance mechanisms, 94, 95, 157, 295
selectivity, 62
sodium channel binding, 5
storage of, 51
toxicity, 109–111
comparison with cyclodienes, 124
temperature and, 236
toxicodynamics, 59
uptake and distribution of xenobiotics, 22
DDT-ase, 94
DDT-dehydrochlorinase, 95
Deadly nightshade (Atropa belladonna), 5, 7
Dealkylation
cytochrome P450 biotransformations, 28, 29
organophosphate metabolism, 40
Dechlorination
DDT, 103
PCBs, 140
Defense systems, and resistance, 94
Defoliants
Agent Orange, 152
arsenic-containing (Agent Blue), 178
Dehalogenation
metabolic enzyme systems, 72
reductive, 41
Dehydrochlorinase, DDT, 95, 104
Dehydrochlorination
DDT, 103, 104, 111
hexachlorocyclohexanes/lindane, 131
Dehydroepiandrosterone (DHEA), 269
Delayed neuropathy
mechanistic biomarker assays, 86
metabolic mechanisms, 39
neurotoxin modes of action, 297, 300, 304
organophosphorus compounds, 193, 206, 300, 305, 306
time course of toxic processes, 87
Delayed toxicity
storage in fatty tissue and, 50–51
storage of xenobiotics in yolk and, 51
Deltamethrin, 231
chemical properties, 232, 233
persistence, 235
toxicity, 236
Demeton-S-methyl, 59, 195, 196, 207
Dendrobates, 12
Depth selection in weed control, 83
Derris ellyptica, 7
Derris powder, 6, 7
Derris root, 5
Design of new pesticides, 324–326

Desulphuration, cytochrome P450 biotransformations, 29
Detoxication
  anticoagulant rodenticide resistance mechanisms, 228
  cytochrome P450 enzymes, 8–10
  DDT, 111
  endocrine disruption, species susceptibility, 286
  epoxide hydrolases and, 41
  glutathione conjugation, 46
  metabolic, 8–9
  metabolism, 25
  organophosphorus compound metabolism, 199
  potentiation and synergism, 63
  and pyrethroid resistance, 238
  reductive metabolism, 41, 42
  resistance mechanisms, 94
Developmental neurotoxicity, PHAHs, 157
Developmental stages; *See also* Embryonic/fetal development
  endocrine disruption, 284–286
  neurotoxin exposure, 296
DFP, 14
  locomotor failure, 306
  mechanistic biomarker assays, 86
  mode of action, 297
Di-2-ethylhexylphthalate, 48
Diamondback moth (*Plutella xylostella*), 95, 116, 238
Diazinon
  behavioral effects, 308–309
  metabolism, 29, 30, 197, 198, 200
  resistance mechanisms, 95
  selectivity, 61, 62
  structure, 195
  sublethal effects, 306
  toxicity, 205, 207, 208
Diazoxon metabolism, 29, 37, 198, 200
Dibenz[*a,h*]anthracene, 183, 185
Dibenzodioxins, polychlorinated, 151–160
Dibenzofuran, 152
Dibenzofurans, polychlorinated, 151–160
Dicamba, 258
Dichlobenil, 257
1,2-Dichloro-4-nitrobenzene (DCNB), 44, 45
Dichlorobiphenyl, 22, 28
Dichlorodipenyl trichloroethane; *See* DDT
Dichlorofluoromethane, 22
Dichloromethane, 46–47
1,2-Dichloro-4-nitrobenzene (DCNB), 44, 45
Dichlorophenol, 28
Dichlorvos, 201
Diclofenac, 320–321
Dicoumarol, 5, 6, 219, 220
Dieldrin, 102, 116, 210, 235, 315
  behavioral effects, 313

  bioaccumulation, 77, 121
  chemical properties, 103, 116
  delayed toxicity, storage in fat depots and, 51
  dirty dozen, persistent organic pollutants (POPs), 77, 102
  ecological effects, 124–131
  environmental fate, 68, 69, 119, 121, 235
  Farne Island ecosystem, 107
  fate of pollutants in soils and sediments, 81
  fish kills, 80
  mechanisms of toxicity, 122
  metabolism, 117, 118, 119
    cytochrome P450 biotransformations, 28, 29
    enzyme induction, 48, 49
    epoxide hydrolases, 41
  mode of action, 56, 297
  neurotoxicity assays, 303
  population level effects, 90, 311
  resistance mechanisms, 95
  and sparrow hawk population levels, 92
  storage of, 51
  structure, 101
  sublethal effects, 123
  toxicity, 123–124
  toxicodynamics, 59
  uptake and distribution of xenobiotics, 22
Dieldrin analogues, excretion in terrestrial animals, 53
Diels-Alder reaction, 116
Diet
  monooxygenase activity comparisons, 34, 35
  and organochlorine biomagnification, 108
Diethyldithiocarbamate, 30
Diethylstilbestrol, 269, 284
Difenacoum, 219
  chemical properties, 220
  ecological effects, 227
  environmental fate, 223
Diffusion
  environmental fate, biological factors affecting, 71
  excretion in aquatic animals, 52
  facilitated, 23
  in fish, 51
  uptake and loss mechanisms, 21, 22, 23, 24, 77, 79, 80, 81
  weak acids and bases, 23
Difocol, 252
Diisopropyl fluorophosphatases, 37
Diisopropyl phosphofluoridate (DPF), 37, 300
Dimethoate, 59, 61, 193, 194
  behavioral effects, 310
  structure, 195
  toxicity, 207, 208
Dimethyl-arsinic acid (Agent Blue), 178
Dimethyl mercury, 166

Dimethyl phthalate, 282
Dinitroaniline herbicides, 58
Dinitroorthocresol (DNOC), 58, 60
Dinitrophenols, 22, 60, 69, 257
Dinitro secondary butyl phenol, 247
Dinoseb, 247, 257
Dioxin equivalents, measuring toxicity of
        mixtures, 244
Dioxins/TCDD (tetrachlorodibenzodioxin), 222
    Ah receptor-mediated toxicity, 144
    bioassays, 252
    dirty dozen, persistent organic pollutants
        (POPs), 77
    ecological effects related to TEQs, 158–160
    endocrine disruption, 268, 280
    environmental fate, 69, 153–154
    formation of, 152
    herbicide contaminants, 257
    human exposure via Agent Orange, 152
    metabolism
        enzyme induction, 48, 49
        reductive, 42
    mode of action, 57
    PCDD origins, 151
    potentiation and synergism, 62
    solubility properties, 152
    toxicodynamics, 59
    uptake and distribution of xenobiotics, 22
Diphenacoum, 57
Diphenyltin (DPT), 174
Diquat, 69
Dirty dozen, 77, 102
Disinfectants, 93
Distribution and transport, in animal, 80
Distribution in communities and ecosystems,
        75–98
    biomarkers in wider ecological context, 89–90
    ecological risk prediction, new approaches,
        97–98
    effects on individuals, biomarker approach,
        84–89
    fate in soils and sediments, 81–83
    movement of pollutants along food chains,
        75–81
    population level effects, 90–95
        population dynamics, 90–93
        population genetics, 93–95
Disulfiram, 30
Disulfoton; See Disyston
Disyston, 193
    cytochrome P450 biotransformations, 29
    formulations, 201
    metabolism, 197, 199
    structure, 195
    toxicity, 207
    toxicodynamics, 59
Diuron, 83, 258, 259, 262, 263, 267, 279

DNA adduct formation, 55, 56, 184, 187, 188, 248
DNA damage, phthalates, 282
DNA methylation, 284
DNA response element, 277
DNA transcriptomics, 88
DNOC (dinitroorthocresol), 58, 60
Dog whelk (Nucella lapillus), 319, 322, 324
    breeding success, 93
    mechanistic biomarker assays, 86, 87
    population level effects, 90, 315
    toxic effects and population changes, 97
    toxicodynamics, 59
    tributyltin and, 175, 176–177
    tributyl tin mode of action, 57
Dopaminergic receptors, mercury and, 172
Dose levels
    cyclodienes
        with dressed seed consumption, 124
        sublethal, 123
    and stress, 90
Dose-response relationship
    AChE inhibition by OPs, 204
    PAHs, 190
Double bonds, cytochrome P450
        biotransformations, 29
DPT (diphenyltin), 174
Drinking water
    cyclodiene uptake routes, 130
    herbicide contamination, 261–263
    uptake routes, 76
Drosophila melanogaster (fruit fly), 130, 211
Dumpton Syndrome, 176–177

E

Earthworms
    carbamates in, 217
    organochlorine residues in, 106, 108, 109
    organophosphorus compounds, 201
Eastern oyster (Crassostrea virginica), 77, 200
EBI; See Ergosterol biosynthesis inhibitors
Ecological effects
    anticoagulant rodenticides, 226–228
        poisoning incidents in field, 226–227
        population genetics, 228
    carbamates, 217–218
    cyclodienes, 124–131
    organomercury compounds, 170–172
    organophosphorus compounds, 208–211
        population dynamics, 209–211
        population genetics, 211
        toxic effects in field, 208–209
    organotin compounds (tributyltin), 176–177
    PAHs, 189–191
    PCBs, 146–149
        physiological and biochemical effects in
            field, 146

population effects, 146–149
population genetics, 149
PCDDs and PCDFs, 158–160
pyrethroid insecticides, 237–238
population dynamics, 237–238
population genetics, 238
Ecological profiling, 96, 237–238
Ecological risk prediction, 97–98
Ecosystem level effects, 84, 96–97; *See also*
Distribution in communities and
ecosystems
Ectoparasite control
in aquaculture, 201
hexachlorocyclohexanes/lindane, 131
organophosphorus compounds, 201
pyrethroid insecticides, 235
ECVAM, 328
Edible mussel; *See Mytilus edulis*
EE2, 274, 275, 291, 320
Effluents
endocrine disruptors in; *See* Wastewater
treatment plant effluent, endocrine
disruptors in
measuring toxicity of mixtures, 244
paper mill, 152, 280, 282
Eggs
cyclodiene excretion, 117
organochlorine residues in, 106
PAHs and in fish, 189
PCB bioaccumulation, 141
storage of xenobiotics, 51
Eggshell thinning, 315, 322, 324
DDE effects, 17
endocrine disruptor case studies, 270–271
fish-eating birds, 148
mechanistic biomarker assays, 86, 87
organochlorines and, 111
and population numbers, 113–115
sites of action, 110
toxicodynamics, 59
Eider ducks (*Somateria molissima*), 51, 124
Electron transfer
biomarkers, 247
complex pollution problems, biomarkers, 249
mitochondrial, 6
monooxygenases, 27
photosystem, 58
Elimination mechanisms, 80
Elimination rates
anticoagulant rodenticides, 222
and environmental fate, 72
PCBs, 139–140
Embryonic/fetal development
DDT and, 105, 110–111
endocrine disruption, 267, 284
PCBs and, 144–145
storage of xenobiotics in yolk and, 51

Embryos, cyclodiene loss via, 117
Emmamectin benzoate, 11
Enantiomers, pyrethroid, 232
Endocrine disruption, 6, 7
anthropogenic agents and, 14
biomarkers, 60
selectivity, 61
tributyltin and, 174–176
Endocrine disruptors, 265–292, 320
behavioral effects, 288–290
biomarkers, 250
complex pollution problems
bioassays, 252
toxic responses with shared pathways of
expression, 250–251
discovery and research on, 266
ecotoxicological significance, 290–292
lessons learned and wider significance in
ecotoxicology, 290–292
mixture effects, 283–284
modes of action, 266–270
pollutant categories, 278–283
alkylphenols, 281–282
bisphenols, 281
dioxins/TCDD, 280
endogenous and pharmaceutical
estrogens, 279
natural compounds, 283
PCBs, 279–280
pesticides, 279
phthalates, 282–283
polybrominated diphenyl ethers, 280–281
screening and testing, 276–278
species susceptibility, 286–287
terminology, 270
wildlife case studies, 270–276
atrazine and frog abnormalities, 275–276
DDT, 270–272
estrogens and feminization in fish, 273–275
health effects in humans, 276
tributyltin, 272–273
windows of life with enhanced susceptibility,
284–286
Endoperoxides, 183
Endoplasmic reticulum, 25, 26, 246–247
Endosulfan, 102, 103, 116, 119
Endotoxins, *Bacillus thuringiensis* (BT), 13
Endpoints, toxic, 144
Endrin, 102, 116
chemical properties, 103, 116
dirty dozen, persistent organic pollutants
(POPs), 77, 102
environmental fate, 119
mechanisms of toxicity, 122
metabolism, 117, 118, 119
cytochrome P450 biotransformations, 31
excretion in terrestrial animals, 53

models for bioconcentration and
    bioaccumulation, 77
mode of action, 297
neurotoxicity assays, 303
structure, 101
toxicity, 123, 124
Energy flow, community and ecosystem level
    effects, 96
Energy metabolism
    biomarkers, 247–248
    complex pollution problems, biomarkers, 249
    tributyltin and, 174, 176
Enhanced toxicity, enzyme induction and, 50
Enteric parasites, 11
Enterohepatic circulation, 48, 54
Environmental fate, 67–72
    anticoagulant rodenticides, 222–223
    carbamates, 213, 215
    cyclodienes, 119–122
    DDT, 105–109
    distribution in communities and ecosystems
        movement along food chains, 75–81
        in soils and sediments, 81–84
    herbicide movement into surface waters,
        261–263
    models of, 70–71
    organomercury compounds, 166–168
    organophosphorus compounds, 196, 200–202
    organotin compounds (tributyltin), 173–174
    PAHs, 185–187
    PCBs, 140–143
    PCDDs and PCDFs, 153–154
    properties of chemicals affecting metabolism
        and disposition, 71–72
    properties of chemicals influencing fate in
        gross environment, 68–69
    pyrethroid insecticides, 234–236
Environmental risk assessment; See Risk
    assessment
Enzyme induction; See Induction of enzymes
Epoxidation
    activation, 30
    benzo[a]pyrene, 31
    cytochrome P450 biotransformations, 28, 29
    prior to glutathione conjugation, 47
    TCDD metabolism, 153
Epoxide hydrolases
    carbamate metabolism, 214
    cyclodiene metabolism, 118
    induction of, 49–50
    metabolism, 40–41
    PAH metabolism, 184
    PCB metabolism, 136, 137
Epoxides
    carbamate metabolites, 213, 214
    cyclodiene, 116
        biotransformation, 124

environmental fate, 119
    metabolism, 118
enzyme induction, 49–50
glutathione conjugation, 46
organohalogen metabolites, 47
PCBs, 136, 140
toxicodynamics, 55, 59
Equilibrium, environmental fate models, 70
Ergosterol biosynthesis
    CYP51, 33
    mode of action, 58
    toxicodynamics, 59
Ergosterol biosynthesis inhibitors (EBIs)
    complex pollution problems, biomarkers, 249
    potentiation and synergism, 63–64, 254
    pyrethroid potentiation, 234, 236–237
Ergot alkaloids, 7–8, 13, 293–294
Erioischia brassicae (cabbage root fly), 130
EROD; See Ethoxyresorufin deethylase
ES4, 38
ES15, 38
E-SCREEN, 277
Eserine (physostigmine), 6
Esfenvalerate, 237
Esterases; See also Acetylcholinesterase/
        acetylcholinesterase inhibition;
        Cholinesterases; Neuropathy target
        esterase
    carbamate metabolism, 214
    mechanistic biomarker assays, 86
    metabolism, 36–40
    mipafox and, 19
    organophosphorus compounds, 196
        metabolism, 197, 198
        resistance, 211
    pyrethroid metabolism, 19, 232, 233, 234
    resistance mechanisms, 94, 95
    selectivity, 62
Esters, pyrethroid insecticides, 232
Estrogenic activity
    DDT-related compounds, 111
    endocrine disruptors, 269
Estrogen receptors
    alkylphenol affinity, 282
    biomarkers, 246–247
    endocrine disruptor assays, 277
Estrogens
    antiestrogen chemical modes of action, 267
    bioassays, 252
    endocrine disruptors, 273–275, 279
Ethanol, 31, 48, 49
Ether bonds, hydrolytic enzymes, 36
Ethical questions, toxicity testing, 328
Ethinylestradiol, 269
Ethoxyresorufin, 49
Ethoxyresorufin deethylase (EROD)
    anticoagulant rodenticide metabolism, 222

levels at superfund site, 148–149
    PCB tissue concentration correlation, 146
    tributyltin inhibition of, 173
Ethyl mercury, 170
Ethynyl pyrene, 30
Euler-Lotka equation, 91
European Commission, REACH proposals, 77
European flat oysters (*Ostrea edulis*), 176
Evaluative models of environmental fate, 70
Evolution
    cytochrome P450 enzymes, 9, 31
    detoxification pathways, 10
Exchange diffusion
    in aquatic animals, 52
    environmental fate, biological factors
        affecting, 71
    uptake and loss mechanisms, 79, 80
Excretion
    cyclodienes, 117
    environmental fate, biological factors
        affecting, 71
    factors determining toxicity and persistence,
        20
    glutathione conjugates, 46
    metabolites, 25
    models for bioconcentration and
        bioaccumulation, 77
    peptide conjugates, 47
    toxicokinetics, 51–54
        by aquatic animals, 52
        by terrestrial animals, 52–54
    uptake and loss mechanisms, 79
Exotoxicogenomics, 88
Experimental models
    cyclodiene toxicity, 123
    prediction of toxicity from, 18
Exposure, biomarkers, 60, 85
Exposure periods, and cyclodiene
    bioaccumulation, 120

F

Facilitated diffusion, 23
FAD, 42
Falco columbarius (merlin), 127
Falco peregrinus; See Peregrine falcon
Falco tinnunculus (kestrel), 127, 129
Farm animals
    microbial toxins, poisoning by, 13
    PBBs, 149
    pesticide effects on, 19
Farne Island ecosystem, OC residues, 107
Fate in environment; See Environmental fate
Fathead minnow (*Pimephales promelas*)
    cyclodiene toxicity, 123
    endocrine disruption, 267, 274, 284

Fat mobilization, release of stored xenobiotics
        and metabolites, 50–51, 72
Fat soluble compounds; See Lipophilicity
Feeding habits
    monooxygenase activity comparisons, 34, 35
    population level effects, 311–312
Feedlots, endocrine disruptor sources, 267, 279
Feminization
    biomarkers, 246
    DDT effects on avian embryos, 110–111
    in fish, 273–275
Fenitrothion, 208, 211, 289
Fenpropathrin, 235
Fenthion, 200, 293
Fenvalerate, 231
    chemical properties, 232, 233
    endocrine disruption, 279
    persistence, 235
    selectivity, 61
    toxicity, 236
Feral pigeon (*Columba livia*), 106
Fertility effects; See Endocrine disruptors;
        Reproductive and fertility effects
Fetal development; See Embryonic/fetal
        development
Field studies, 93
    biomarker response and population level
        effects, 315, 316
    future prospects, 326–327
    PCB physiological and biochemical effects, 146
Films, surface oil, 68
Firemaster, 149
Fire retardants, 149, 150, 268, 280
Fish-eating birds, 322
    cyclodienes
        bioaccumulation, 121
        poisoning by, 129–130
    DDT and metabolites, 106, 107, 108
    monooxygenase activity comparisons, 34, 35
    organomercury compounds and, 167, 171–172
    PCBs
        biomarkers, 146
        environmental fate, 140, 141
        metabolism, 138
        population effects, 147
Fish/fish species, 320
    behavioral effects of neurotoxins, 309
    biomarker response and population level
        effects, 316
    cyclodienes
        bioconcentration factor, 120
        poisoning by, 129–130
    detoxication by, 80
    dioxins/TCDD
        dioxin-like compound resistance, 157
        ecological effects related to TEQs, 158,
            159, 160

endocrine disruption, 266
  alkylphenols, 282
  assays, 277
  assays/screening tests, 278
  behavioral effects, 288–289
  bisphenols, 281
  case studies, 273–275
  life stages with enhanced sensitivity,
    284–286
  modes of action, 267, 268, 269
  natural EDCs, 283
  PBDEs, 281
  species susceptibility, 286, 287
  triphenyltin (TPT) and, 272
environmental fate, biological factors
    affecting, 71
herbicide toxicity, 261
lindane toxicity, 131
metabolism
  comparative rates in aquatic animals, 52
  lack of cytochrome P450 enzymes, 10
  monooxygenase activity comparisons,
    34, 35
neurotoxin behavioral effects in, 306–307
OP toxicity, 207, 209
organochlorines, DDT and related
    compounds
  sites of action, 110
  toxicity, 111
organometallic compounds
  methyl mercury, 166, 167, 169
  tributyltin, 173, 176
PAHs, 186, 188, 189, 191
PCBs, 138, 141, 146
PHAH TEFs, 156
pyrethroids, 235–236
transport of xenobiotics, 72
uptake and loss mechanisms, 21, 77, 79–80
  diffusion, 52; See also Diffusion
  excretion, 51, 52
Fish hepatocyte cell lines, 252
Fish kills, 80
Flame retardants, 149, 150, 268, 280
Flavones, 10
Flavoprotein reductases, 42
Flocoumafen, 57, 59, 219, 223
  chemical properties, 220
  environmental fate, 222
  metabolism, 221–222
  toxicity, 225, 226
Flubendiamide, 7
Fluoracetate, 8
Fluoranthene, 182, 185, 189
Fluorcitrate, 8
Fluorene, 185
Flutamide, 269
Fonophos, 208

Food
  bioaccumulation factor, 76
  cyclodiene bioaccumulation, 121
  models for bioconcentration and
    bioaccumulation, 78
  and population growth rate, 91
  uptake and loss mechanisms, 77
Food chains, 17–18
  cyclodienes in, 119, 120–121
  distribution in communities and ecosystems,
    75–81
  environmental fate
    biological factors affecting, 71
    soils and sediments, 82–83
  herbicides and, 257, 260
  organochlorines in, 106, 107, 112
    aquatic/marine, 107–108
    population numbers, effects on, 112–113
    terrestrial, 108–109
  organomercury compounds, 166, 170
  PAHs, 186
  PBBs in, 149
  PCBs in, 138, 140, 141
  pyrethroid insecticides, 236
  TCDD in, 153
Food supply, complex pollution problems, 243
Forestry
  hexachlorocyclohexanes/lindane, 131
  OP applications, 208, 210
Formulation, pesticide
  carbamates, 213, 215, 216, 217–218
  and organochlorine toxicity, 111
  organophosphorus compounds, 199, 201,
    207, 208
Fox (Vulpes vulpes), 129
FRAME, 328
Free radicals; See Oxygen radicals/reactive
    oxygen metabolites
Freshwater invertebrates
  carbamate poisoning, 217
  DDT and related compounds toxicity, 111
  OP resistance, 211
Frogs
  endocrine disruption, 275–276
  tadpole predator avoidance behavior, 312
Fruit fly (Drosophila melanogaster), 130, 211
Fugacity capacity constant, 71
Fugacity modeling, 69, 70–71
Fugu vermicularis (puffer fish), 11, 12, 293
Fungal estrogens, 246, 283
Fungicides/antifungal agents
  complex pollution problems, biomarkers, 247
  dirty dozen, persistent organic pollutants
    (POPs), 77
  endocrine disruption, 268, 279
    modes of action, 267
    triphenyltin (TPT), 272

enzyme induction, 48
ergosterol biosynthesis inhibitors (EBIs),
 63–64, 234, 236–237
mode of action, 58
organomercury, 167–168, 171
population level effects, 311
potentiation and synergism, 63–64
 carbamates, 63–64
 organophosphorus compounds, 208
 pyrethroids, 234, 236–237
resistance mechanisms, 95
triphenyltin (TPT), 173
Furafylline, 30
Furanocoumarins, 32, 49
Furans, 77
Fusarium, protein synthesis inhibitors, 13
Future prospects, 319–328
design of new pesticides, 324–326
ethical questions, 328
field studies, 326–327
testing, 321–324
 adoption of more ecologically relevant
  practices, 321–323
 biomarkers, development of, 323–324

G

GABA receptors, 316
combinations of chemicals with different
 modes of action, 310
cyclodiene toxicity, 122, 123
lindane activity, 131
mode of action, 56
neurotoxicity assays, 303
neurotoxin effects at whole organism level,
 306
neurotoxin mechanisms of action, 296, 298,
 299
organochlorine sites of action, 110
resistance mechanisms, 10, 94, 95, 130, 157
toxicodynamics, 59
GABA release, avermectins and, 11
Gallus domesticus (chicken), 106, 225
Gambusia affinis (mosquito fish), 131
Gamma-HCH, 122, 131, 297; *See also* Lindane
Gastropods
endocrine disruption, 267, 272
tributyltin and, 175
Gene copy number, resistance mechanisms, 95
Gene expression
endocrine disruptors
 assays, 277
 modes of action, 269–270
induction of enzymes; *See* Induction of
 enzymes
Gene families, cytochrome P450, 30–31

Gene frequencies
cyclodiene effects, 130–131
DDT effects, 115–116
Genetically modified crops, 260
Genetics/population genetics, 93–95
anticoagulant rodenticide effects, 228
community and ecosystem level structural
 analysis, 96
DDT effects, 115–116
distribution in communities and ecosystems,
 93–95
endocrine disruptors and, 289
organophosphorus compound effects, 211
PCB effects, 149
pyrethroid insecticide effects, 238
resistance mechanisms, 10, 95
Genomics, 88, 89
Genotoxic disease syndrome, 188
Genotoxicity, 183, 188
Geochemical cycling, 70
German cockroach (*Blatella germanica*), 95
cyclodiene resistance, 130
organochlorine-pyrethroid cross resistance, 116
pyrethroid resistance, 238
Gills
excretion via, 52
uptake via, 21
Gla proteins, 86, 225
Glomerular filtration, 54
Glucuronidases-beta, 48
Glucuronide formation
enzymes, 44
TCDD metabolism, 153
Glucuronyl transferases, 42–43, 44
cyclodiene metabolism, 117
induction of, 49, 50
Glutamate receptors, 11, 12, 169, 301–302
Glutathione conjugation
carbamates, 214
DDT metabolism, 105
metabolism, 43–47
Glutathione-S-transferases, 43–47
induction of, 49, 50
organophosphorus compound metabolism,
 199, 200
resistance mechanisms, 95
Glycine conjugation, 47
Glycoacyl compounds, hydrolytic enzymes, 36
Glyphosate, 258
Goldfish (*Carassius auratus*), 123, 217
Gonadotropins, 277, 282
Grain dressings; *See* Seed dressings
Grain-eating birds
cyclodienes, 125
 bioaccumulation, 122
 population level effects, 124
OP poisoning, 208

Granule formulation
  carbamates, 213, 215, 217–218
  OPs, 201, 207, 208
Growth rate, population, 91, 92, 97
Gusathion (azinphos-methyl), 194, 195

H

Habitat, and population growth rate, 91
Half-lives
  anticoagulant rodenticides, 222
  and behavioral effects, 313
  cyclodienes, 119, 120
  DDT and related compounds, 106
  environmental fate
    properties affecting, 68
    in soils and sediments, 81
  excretion in terrestrial animals, 52–53
  models for bioconcentration and
    bioaccumulation, 77
  organometallic compounds
    mercury, 165
    methyl mercury, 167
  organophosphorus compounds, 196
  PCBs, 139–140
  pyrethroid insecticides, 235
  resistance to detoxification and, 22
  TCDD, 153
Halide bonds, hydrolytic enzymes, 36
Halogenated phenolic nitriles, 258
Halogenation, degree of
  and cyclodiene metabolism, 118
  and environmental fate, 69, 71–72
  PCBs, 133, 134, 138, 140
  and rate of loss from soil, 82
Halogens, reductive dehalogenation, 41, 42
Halothane, 42, 104
HCB, 139, 140
HCE, 53, 118
HCH; *See* Hexachlorocyclohexanes; Lindane
Heliothis virescens (tobacco bud worm); *See*
    Tobacco bud worm
Heme protein, cytochrome P450s, 26, 27–28, 248
  biotransformation, 31
  inhibitors and, 33
  reductive metabolism, 42
  tributyltin and, 173
Hemlock (*Conium maculatum*), 7
Hemolymph, 21
Henry's law, 69
Henry's law constant, 69
HEOD; *See* Dieldrin
HEOM, 118
Hepatic microsomal monooxygenases (HMO);
    See Cytochrome P450 enzymes;
    Monooxygenases
Hepatic portal system, 25

Heptachlor, 102, 116
  chemical properties, 103, 116
  cytochrome P450 biotransformations, 29, 30
  dirty dozen, persistent organic pollutants
    (POPs), 77, 102
  ecological effects, 124–131
  environmental fate, 119
  metabolism, 118, 119
  toxicity, 123, 124
Heptachlor epoxide, 210
  mechanisms of toxicity, 122
  population level effects, 90
  resistance mechanisms, 95
Herbicides, 257–263
  agricultural ecosystem impacts, 258–261
  arsenic sources, 178
  as biocide, 60
  classification and properties, 258, 259
  complex pollution problems, biomarkers, 247
  endocrine disruption, 267, 279
  environmental fate, 69–70
  fate of pollutants in soils and sediments, 81,
    82, 83
  microbial metabolism, 218
  mode of action, 58
  PCDD contamination, 151–152
  resistance mechanisms, 95
  surface water and drinking water
    contamination, 261–263
Herons (*Ardea cinerea*), 106, 129–130
Herring gull, 141
Hershberger assay, 278
Heterocyclic nitrogen compounds, cytochrome
    P450 inhibitors, 33
Hexachlorobenzene, 77
Hexachlorobiphenyl, 77
Hexachlorocyclohexanes (HCHs), 101, 131; *See*
    also Lindane
  resistance mechanisms, 130
  toxicodynamics, 59
Hexachlorocyclopentadienes, 116
Hexane, 22, 28
HHDN; *See* Aldrin
High-density lipoproteins (HDL), 37
Histological effects, biomarkers, 84
Homeostasis, dose levels of toxicants and, 90
Honeybee (*Apis mellifera*)
  behavioral effects in, 307, 311–312
  EBI potentiation of insecticides and, 63–64,
    236
  organophosphorus compounds, 209, 211
  selectivity, 61
Hot spots, 154, 188, 251
Housefly (*Musca domestica*), 32
  cyclodiene resistance, 130
  cytochrome P450 metabolism studies, 32

organochlorines, population genetics effects, 115
pyrethroid resistance, 238
resistance mechanisms, 94, 95, 295, 303
selectivity, 61, 62
Humans, 13–14
behavioral effects of neurotoxins, 308
dieldrin intoxication, 122–123
endocrine disruptor effects, 276
neurotoxicity risk assessment, 315
PBB levels, 149
PCB toxicity, 145
TCDD exposure, 152
worker exposure to pesticides, 19
Humic substances, 81
Humidity, stressors, 89
Hydration, metabolism, 24
Hydrogen cyanide, 22
Hydrolases, 80
metabolism
epoxide hydrolase, 40–41
esterases, 36–40
induction of, 49–50
reductases, 41–42
Hydrolysis
carbamates, 213, 214
conjugation products, 48
environmental fate, 69
metabolism, 24
organophosphorus compounds, 196, 197, 200
pyrethroid insecticides, 232, 234
Hydrophilic compounds; *See* Water solubility
Hydrophobicity; *See also* Lipophilicity
anticoagulant rodenticides, 219
and environmental fate, 68, 71
and fate of pollutants in soils and sediments, 81
Hydroprogne caspia (Caspian tern), 147, 148, 158, 159
Hydroxy Cl biphenyl 4-OH-3,3'4,5'-Cl-biphenyl, 57
Hydroxylamines, 42
Hydroxylaminopyrene, 41
Hydroxylation
anticoagulant rodenticide resistance mechanisms, 228
cytochrome P450 biotransformations, 28
hexachlorocyclohexanes/lindane, 131
metabolism, 24, 25
PCBs, 136, 139
and environmental fate, 140–141
pyrethroid metabolism, 233
TCDD metabolism, 153
tributyltin, 173
4-Hydroxy-3,3'4,4'-tetrachlorobiphenyl, 86
Hylemya antiqua (onion maggot), 130
Hypericin, 5, 7
Hypericum perforatum (St. John's wort), 5, 7

I

Imidacloprid, 311
Imidazoles, 30
Immune system, endocrine disruptions and, 269
Immunosuppression
organomercury compounds, 172
PAHs and, 189
PCBs, 148
PCBs and, 145–146
Imposex, 57, 315, 324
endocrine disruptors
mode of action, 267
species susceptibility, 287
mechanistic biomarker assays, 86, 87
toxicodynamics, 59
tributyltin and, 175
Index of Biotic Integrity (IBI), 96
Indicator species
biomarkers, 84
population effects, 316–317
Indirect effects
ecotoxicology, 18
establishing causality, 243
Individual effects
assessment of, 90
biomarker approach, 84–89
and population level effects, 91
Induction of enzymes
metabolism, 48–50
mixtures, 187
PAHs, 185, 187
PCB biomarkers, 146
PCDDs and PCDFs, 187
Inhalation exposure, cyclodienes, 130
Inhibitors
cytochrome P450s, 33–34
esterases inhibited by organophosphates (B esterases), 37
potentiation and synergism, 63
pyrethroid potentiation, 234
Inorganic ions, stressors, 89
Insecticides
activation of, 50
avermectin mode of action, GABA neurons, 11
dirty dozen, persistent organic pollutants (POPs), 77
toxicodynamics, 59
Insects
cyclodiene excretion, 117
cyclodiene toxicity mechanisms, 122
DDT and related compounds toxicity, 111
metabolism, 32
cytochrome P450 metabolism studies, 32
enzyme induction, 49
epoxide hydrolases, 41
A esterase activity, 37–38

excretion, 54
    glutathione conjugate excretion, 46
    peptide conjugations, 47
organochlorine toxicity, 111
organophosphorus compounds, 201
resistance mechanisms, 94
selectivity, 61, 62
Insect toxins, 11
Insect vector control, 14
Insurance areas, 210
Integrated biomarker approach, complex
        pollution problems, 245–250
Interacting compounds
    complex pollution problems, biomarkers, 253
    organomercury compounds, 171
    organophosphorus compounds, 208
    potentiation and synergism, 62–64
Interpretation of mesocosm studies, 97
Intestinal parasites, 11
Intracellular localization, uptake and distribution
        of xenobiotics, 21
Invertebrate Community Index, 96
Invertebrates, 320
    cytochrome P450 metabolism studies, 32
    endocrine disruption
        assays/screening tests, 278
        species susceptibility, 286
    herbicide toxicity, 261
    mechanistic biomarker assays, 86, 87
    metabolic rates in marine animals, 52
    metabolism in, state of research, 32
    oil spill effects, 189
    organochlorines
        in terrestrial food chains, 108–109
        toxicity, 111
    organophosphorus compounds, 209
        resistance, 211
        toxicity, 207, 209
    pyrethroid insecticides, 236
    storage of xenobiotics, 51
    toxic effects and population changes, 97
    uptake and distribution of xenobiotics, 21
In vitro metabolism, 45–46
In vitro tests; See also Cell culture
    endocrine disruption, 277–278
    neurotoxicity, 295, 303, 305, 314–316
In vivo metabolism, test systems and, 251
In vivo tests
    endocrine disruption, 278
    ethical questions, 328
    neurotoxicity, 315
Ionic compounds
    anticoagulant rodenticides, 221
    blood-brain barrier and, 24
    environmental fate, 69–70
    and storage, 51
Ionic conditions, stressors, 89

Ion transport/fluxes
    calcium pump, 57, 59, 87, 110
    organochlorine sites of action, 110
    potassium, 302
    sodium channels; See Sodium channels
    tributyltin and, 174
Ioxynil, 258
Iron, cytochrome P450s, 33–34
    heme protein, 27
    reductase metabolism, 42
Iron porphyrins, 103, 104
Isoforms
    CYP2, 10
    enzyme induction, 48–50
    glutathione-S-transferases, 46
Isomalathion, 69, 196
Isomerization, organophosphorus compounds, 196
Isomers, DDT, 102, 103
Isoproturon, 82, 261

J

Jasmolone, 231, 232
Juvenile hormone synthesis inhibition, 6, 7

K

Kanechlor, 133
kdr resistance, 115–116, 295
    mode of action, 303
    pyrethroids, 238
Kelthane, 104, 105
Kepone, 279
Kestrel (Falco tinnunculus), 127, 129
Kestrel, American, 139, 143
Ketoconazole, 48
Kidney, excretion via urine, 52, 53, 54
Killer whales, PCBs in, 143
Kinetic model, bioaccumulation, 80
Knockdown resistance, 59, 115
Knock-on effects, 91, 250, 260
Krebs cycle, 8

L

Laboratory animal models
    ethical questions, 328
    limited applicability, species differences
        and, 72
Lagomorphs, OP poisoning, 208
Lake trout, 141
Land animals; See Terrestrial animals
Landfills
    endocrine disruptors
        phthalates, 282
        in runoff, 280
    fate of pollutants in soils and sediments, 82

LC$_{50}$
cyclodienes, 123
DDT and related compound ecotoxicity, 111
PAHs, 189
LD$_{50}$; *See* Acute toxicity/LD$_{50}$
Leaching, soil, 81
Lead, 177
blood-brain barrier and, 24
neurotoxicity, 297, 301
Leaving groups, 39, 194, 196, 197, 199, 200
Leptophos, 193
delayed neuropathy, 206, 300
mechanistic biomarker assays, 86
mode of action, 297
Lethal effects
assessment of, 90
brain AChE inhibition and, 209
population level effects, 90
Levels of biological organization; *See*
Organizational levels
Lewisite, 14
Life span, and steady state, 76
Life stages, endocrine disruption, 284–286
Ligandin, 47
Lindane, 102, 131
chemical properties, 103
endocrine disruption, 279
mode of action, 297
resistance mechanisms, 130
structure, 101
uptake and distribution of xenobiotics, 22
Lineweaver-Burke plot, 78, 79, 80
Linuron, 258, 259, 267, 279
Lipids, 32
Lipid solubility, octanol-water partition
coefficients, 21–23
Lipophilicity
cyclodienes, 117–118
and environmental fate, 120
metabolites, 119
and environmental fate, 71
factors determining toxicity and persistence,
19
organometallic compounds
lead, 177
mercury, 163
PAHs, 183
PCDDs, 153
pyrethroid insecticides, 232, 235
Lipophilic xenobiotics
excretion in terrestrial animals, 52–53
metabolism, 24, 25
CYP gene families role in detoxification,
8–9
monooxygenases, 26, 31
models for bioconcentration and
bioaccumulation, 78

organochlorines, 111
PBBs, 149
PCBs, 134
storage, 50–51
toxicodynamics, 55
Lipoproteins
cyclodiene excretion, 117
storage of xenobiotics, 51
Liver
enzyme induction and, 48–49
metabolic enzyme systems, 25
metabolism, models for bioconcentration and
bioaccumulation, 78
microsomal metabolism, in vitro studies,
45–46
Livestock dips, 131, 201
Livestock yards, endocrine disruptor sources,
267, 279
Local response, response at different levels of
biological organization, 92
Location of chlorine substituents (ortho, meta,
para); *See* Chlorination
Locomotion, neurotoxin effects, 306, 310
Locust (*Locusta migratoria*), 7, 130
Locust (*Calliphora erythrocephala*), 304
Log K$_{OW}$; *See* Octanol-water partition coefficient/
log K$_{OW}$
Lonchura striata (Bengalese finch), 106
Long-term studies, 327
Luciferase assays, 252; *See also* CALUX system
Lutra canadensis (otter), 169
Lutra lutra (otter), 129, 130
Lymph, distribution of xenobiotics in organisms,
21
Lysophospholipase, 32

M

Macrocosms, 322
Macroinvertebrates, pyrethroids and, 237
Macrophytes, aquatic, 68
Malaoxon, 198
Malaria mosquito, 14, 32
Malathion, 32, 193
and carbaryl toxicity, 216
chemical properties, 196
ecotoxicity, 209
environmental fate, 69
metabolism, 38, 197, 198, 199
potentiation and synergism, 64
structure, 195
surface water pest control, 201
toxicity, 205, 207
uptake and distribution of xenobiotics, 22
Malondialdehyde, 248
Malpighian tubules, 54

Mammals
    cyclodiene resistance, 130–131
    cyclodiene toxicity, 123
    endocrine disruption; See Endocrine
        disruptors
    marine, organochlorine residues in, 107, 108
    mechanistic biomarker assays, 86
    metabolism
        A esterase activity, 37, 38
        monooxygenase activity comparisons,
            34, 35
    organophosphate toxicity, 193
    PCB environmental fate, 140
    PCB toxicity, 144–146
        immunosuppression, 145–146
        placental transfer, 144–145
    potentiation and synergism, 64
    selectivity, 62
Mammary glands
    cyclodiene excretion, 117
    DDT excretion, 105
    PBDEs in breast milk, 281
Marine environment
    OP use in, 201
    organochlorines, DDT and metabolites,
        107–108, 111
    organometallic compounds
        methyl mercury, 166
        tin; See Tributyltin
    PAH pollution, 141, 186, 189
    PCBs
        bioaccumulation by seabirds, 141, 142
        environmental fate, 140
        in mammals, 143, 148
MCPA, 218
    fate of pollutants in soils and sediments, 81, 82
    mode of action, 58
    movement into surface water, 261
    properties, 258
    structure, 259
MCPB, 58
Mechanism of action; See Modes of action
Mechanistic biomarkers, 60, 84, 85
    combinations of chemicals with different
        modes of action, 310
    measuring toxicity of mixtures, 244–245
Mecoprop, 258, 259, 262
Meles meles (badger), 129
Membranes
    metabolic enzyme systems, 25, 26
    uptake and distribution of xenobiotics, 21,
        22, 23, 24
Mercapto compounds, DDE derivatives, 47
Mercuric ion, 164
Mercury compounds, organic, 14, 163–172
    behavioral effects, 307, 312
    blood-brain barrier and, 24

ecological effects, 170–172
    environmental fate, 166–168
    metabolism, 165
    mode of action, 297, 300–302, 305
    origins and chemical properties, 163–165
    population level effects, 311
    toxicity, 168–170
Merlin (Falco columbarius), 127
Mesocosm studies, 93
    adoption of more ecologically relevant
        assessment practices, 322–323
    biomarker response and population level
        effects, 315–316
    community and ecosystem level effects, 96–97
    PAHs, 190
    pyrethroids, 237
Messenger RNA, 49
    estrogen bioassays, 247
    transcriptomics, 88
    VTG, 277
Metabolic rate, 78
Metabolism
    adaptive enzymes, 218
    animal defenses, 8–9
    anticoagulant rodenticides, 221–222, 224
    in aquatic animals, 52
    carbamates, 213, 214
        potentiation, 216
        and toxicity, 215
    complex pollution problems, bioassays,
        252–253
    cyclodienes, 117–119
        and environmental fate, 120
        epoxide hydrolases, 41
        excretion in terrestrial animals, 53
        resistance, enhanced detoxication and, 131
        and toxicity, 124
    DDT, 104–105
    endocrine disruptors
        complex exposure problems, 284
        sources of, 278
        species susceptibility, 286
    environmental fate, 71–72
    factors determining toxicity and persistence,
        19, 20
    in fish, 80
    herbicides and microbial energy source, 82
    mechanistic biomarker assays, 86
    models for bioconcentration and
        bioaccumulation, 77–78
    movement of pollutants along food chains, 75
    organometallic compounds
        mercury, 165
        tributyltin, 173
    organophosphorus compounds, 197–200
        degradation in tissues, 209–210
        potentiation and synergism, 208

oxyradical generation, biomarkers of, 248
PAHs, 183–185, 186
PCBs, 135, 136–140, 141, 144
PCDDs and PCDFs, 153
precocene mode of action, 7
pyrethroid insecticides, 232–234, 235
resistance mechanisms, 94, 95
selectivity, 62
species differences, 72
toxicokinetics, 24–50
 conjugases, 42–48
 enzyme induction, 48–50
 epoxide hydrolase, 40–41
 esterases and other hydrolases, 36–40
 general considerations, 24–26
 monooxygenases, 26–34, 35
 reductases, 41–42
uptake and loss mechanisms, 77, 79
Metabolomics, 88
Metamorphosis, 267
Metasystox, 193
Methamidophos, 206, 300
Methiocarb, 213, 217
Methomyl, 213
Methoxychlor, 101, 278, 279
Methoxyethyl groups, mercury, 163
Methylation, organohalogen metabolites, 47
Methyl bromide, 22
Methylcarbanion, 165
Methyl cholanthrene, 184, 189
Methylene dioxyphenyls, 30, 33
Methylmercuric chloride, 164
Methyl mercury, 14
acute toxicity range, 170
environmental fate, 72
fungicides, 167–168, 171
mode of action, 297
movement of pollutants along food chains, 76
Methyl parathion, 94
Methyl sufonyl metabolites of PCBs, 140
Microbial metabolism
arsenic methylation, 178–179
carbamate degradation, 217–218
endocrine disruptor production, 281–282
environmental fate, 72
 biological factors affecting, 71
 properties affecting, 68
fate of pollutants in soils and sediments, 82
herbicides and, 257
mercury methylation, 164
organophosphorus compounds, 200
PCBs, 140
pyrethroid insecticides, 235
reductive, 42
soils, 81
Microbial origin of CYP enzymes, 9

Microbial toxins, 8, 11–13
Microcosm studies, 96, 322
Microorganisms; *See also* Microbial metabolism
 bioassays, mixtures, 251, 252
 sediments, 83
Microsomal monooxygenases P450; *See*
 Cytochrome P450 enzymes;
 Monooxygenases
Microsome preparations, 78
Microtox test system, 252
Microtus pinetorum (pine mouse), 130–131
Migrating animals, 72
Milk
 cyclodiene excretion, 117
 DDT excretion, 105
 PBDEs in breast milk, 281
Milkweed, 8
Minamata incident, 169, 171
Mink (*Mustela vison*), 169, 172
Mipafox, 193
 delayed neuropathy, 206, 300
 factors determining toxicity and persistence,
 19
 locomotor failure, 306
 mechanistic biomarker assays, 86
 mode of action, 297
Mirex, 77, 102, 116
Mite control, avermectins, 11
Mitochondria
 biomarkers for uncouplers of oxidative
 phosphorylation, 247–248
 electron transport, 6
 herbicide mode of action, 257
 mode of action, 58
 tributyltin and, 176
Mitotic disruption, herbicide mode of action, 58
Mixtures; *See also* Complex pollution problems
 carbamates, 216
 endocrine disruptors, 283–284
 enzyme induction, 187
 measuring toxicity, 244–245
 neurotoxins, biomarkers of exposure, 295
 PAHs, 189
 PHAH TEQs, 156–157
 potentiation and synergism, 62–63
Mobility of animals in environment, population
 level effects, 90
Mobility of pollutant in environment; *See*
 Distribution in communities and
 ecosystems; Environmental fate
Model systems
 adoption of more ecologically relevant
 assessment practices, 322–323
 community and ecosystem level studies,
 96–97
 environmental fate, 70–71

Modes of action, 56–58, 59
  complex pollution problems
    shared mechanisms of action, integrated
      biomarker approach, 245–250
    toxic responses with shared pathways of
      expression, 250–251
  endocrine disruptors, 266–270
  genomics and, 89
  herbicides, 257, 258, 263
  neurotoxins, 296–302
    acetylcholine receptor, 298, 299
    chemicals sharing same principal mode of
      action, 308–310
    combinations of chemical with different
      modes of action, 310
    GABA receptors, 298, 299–300
    neuropathy target esterase, 298, 300
    sodium channels, 296, 297, 298
  toxicodynamics, 55
Molecular weights, 22, 24
Mollusks, 177, 319; See also Dog whelk
  endocrine disruption, species susceptibility,
    286
  models for bioconcentration and
    bioaccumulation, 77
  organochlorine residues in terrestrial food
    chains, 108–109
  organometallic compounds
    lead, 178
    tin, 57, 77, 174, 176
  organophosphorus compound toxicity, 200
  PAH bioconcentration, 186
  species differences in metabolic capacity, 72
  uptake and loss mechanisms, 77, 79
Molting disruption, 6, 7
Monarch butterfly, 8
Mono-2-ethylyhexylphthalate, 48
Monoethyl phthalate, 282
Monooxygenases, 26–34, 35, 80; See also
      Cytochrome P450 enzymes
  animal defenses, 8–10
  anticoagulant rodenticide metabolism, 221
  anticoagulant rodenticide resistance, 228
  carbamate metabolism, 214
  carbamate potentiation, 216–217
  cyclodiene metabolism, 117, 118
    and bioaccumulation, 121
    and environmental fate, 120
  OP metabolism, 40, 196, 199, 200
  PCB metabolism, 136
  potentiation and synergism, 63, 64
  precocene mode of action, 7
  pyrethroid metabolism, 19, 232
  species differences in activities, 34, 35
  in vitro metabolism studies, 45–46
Monuron, 83

Morphological changes
  biomarker approach, 84
  liver enzyme induction and, 48–49
Mortality rates, 90, 91
Mosquitoes, 14, 32
  cyclodiene resistance, 130
  cytochrome P450 metabolism studies, 32
  organochlorines
    population genetics effects, 115
    pyrethroid cross resistance, 116
    pyrethroid resistance, 166, 238
Mosquito fish (Gambusia affinis), 131
mRNA; See Messenger RNA
Multimedia models, 70–71
Multiple drug resistance (MDR) mechanism, 52
Multiple exposures, carbamates, 216
Multiple pesticide applications, 210
Musca domestica; See Housefly
Muscarinic receptors, 203
  atropine and, 204
  mercury and, 169, 171, 172, 301–302
  neurotoxin modes of action, 299
Mustard gas, 14
Mustela vison (mink), 169, 172
Mutagens; See Carcinogens/mutagens
Mutations
  acetylcholinesterase, 211
  cyclodiene resistance, 130
  knockdown resistance, 115
  PAH adduct formation and, 188–189
  resistance mechanisms, 95
  selectivity, 62
  sites of action, 59
Mya arenaria (soft clams), 77
Mycoestrogens, 246, 283
Mycotoxins, 8, 13
Mytilus edulis (edible mussel), 326–327
  cyclodienes, bioconcentration factor, 120
  excretion of xenobiotics, 52
  models for bioconcentration and
    bioaccumulation, 77
  PAH studies, 190–191
  tributyltin oxide (TBTO) bioconcentration
    factors, 174
Myzus persicae; See Peach potato aphid

N

N-Acetylaminofluorene (N-AAF), 29, 30
NADH, 248
NADPH, 42, 46
NADPH/cytochrome P450 reductase, 26, 27,
    213, 234
Naphthalene, 182, 183, 185
Naphthoflavone, 48
Naturally occurring compounds
  animal toxins, 10–11, 12

endocrine disruptors, 246, 252, 266, 268, 283
neurotoxins, 293–294, 297, 298, 316
new pesticide design, 326
plant toxins, 4–8, 93, 293
N-dealkylation, f
Neonicotinoids, 6, 297, 306, 307, 311, 319
Nerve gases, 14, 37, 196, 202, 299
Nervous system, endocrine disruptor effects,
267, 269
Neuropathy target esterase (NPE), 37
B esterases, 39
mechanistic biomarker assays, 86
neurotoxin mechanisms of action, 298, 300
organophosphorus compounds, 206
Neurotoxins/neurotoxicity, 6
avermectins, 11
behavioral effects, 17; *See also* Behavioral
effects
biomarkers, 60, 250
carbamates, 215–217
complex pollution problems, 250–251
cyclodienes
mechanisms of toxicity, 122
sublethal effects, 123, 128
knockdown resistance, 115
lead compounds, 177–178
mechanistic biomarker assays, 86
mercury, 169, 171–172, 179
modes of action, 4, 6, 7, 56
organochlorines, DDT and metabolites, 102,
109–110, 111
organophosphorus compounds
AChE inhibition, 202–208
metabolism, 199
targets, 37
PHAHs, 157
predator use of, 11
pyrethroid insecticides, 236
resistance mechanisms, 94, 95
selectivity, 62
toxicodynamics, 59
New pesticide design, 324–326
N-Hydroxyacetylaminofluorene, 29, 30
Niche metrics, 96
Nicotiana tabacum, 7
Nicotine, 6, 7, 8, 297, 304, 316
Nicotinic insecticides, 316
Nicotinic receptors, 6, 7, 297, 298, 299, 304, 316
Nicotinic synapse, 301
Nicotinoids, 311, 319
NIH shift, 136
Nitrates, stressors, 89
Nitrification, soil community assessment, 96
Nitroaromatics
complex pollution problems, biomarkers, 249
reductive metabolism, 41, 42
Nitro-derivatives, PAH, 183

Nitropyrene, 41, 249
NMDA glutamate receptors, mercury and, 169,
171
No effects levels (NOEL), PAHs, 189
Nonplanar PCBs
CYP 2 family and, 138
fate in environment, 141
Nonyl phenols, 252, 281
Nucella lapillus; *See* Dog whelk
Nucleophilic attack, glutathione conjugation, 45
Nutrient cycling, community and ecosystem level
effects, 96

O

o-chlorophenols, 151
Octanol-water partition coefficient/log K_{OW},
21–23
anticoagulant rodenticides, 220
carbamates, 207
and environmental fate, 68
organophosphorus compounds, 195
PAHs, 183
PCBs, 134, 135
pyrethroid insecticides, 233
O-dealkylation
cytochrome P450 biotransformations, 28, 29
organophosphate metabolism, 40
Oil films, surface, 68
Oil spills, 181, 189
Olive oil, 22
omics, 87, 88–89
Oncorhynchus mykiss; *See* Rainbow trout
Onion maggot (*Hylemya antiqua*), 130
OPIDN (organophosphate-induced delayed
neuropathy), 206, 300, 305
Organic matter, sediments, 83
Organic matter, soil, 68, 81
Organic soils, 82, 83
Organism level response, neurotoxicity
biomarkers, 294
Organizational levels, 92
complex pollution problems
biomarkers, 253
toxic responses with shared pathways of
expression, 250
neurotoxic effects, 308–311
chemicals sharing same principal mode of
action, 308–310
combination of chemicals with different
modes of action, 310
neurotoxicity biomarkers, 294
response to pollutants, 92, 316
Organochlorines, 101–132
behavioral effects, 307, 313
complex pollution problems, bioassays, 252
cyclodienes, 116–131

chemical properties, 116–117
ecological effects, 124–131
environmental fate, 119–122
metabolism, 117–119
toxicity, 122–124
DDT, 102–116; See also DDT
  chemical properties, 102–103
  ecological effects, 112–116
  endocrine disruption, 270–272
  environmental fate, 105–109
  metabolism, 104–105
  toxicity, 109–111
delayed toxicity, storage in fat depots and, 51
effects at whole organism level, 306
endocrine disruption, 267, 279
environmental fate, 69
  biological factors affecting, 71
  in soils and sediments, 81
hexachlorocyclohexanes, 131
mechanistic biomarker assays, 86
metabolism
  cytochrome P450 biotransformations, 28,
    29, 30, 31
  excretion in aquatic animals, 52
  excretion in terrestrial animals, 53
  glutathione-S-transferases, 47
  reductive, 41, 42
mode of action, 56, 297; See also DDT
movement of pollutants along food chains, 76
PCB study complications, 147
persistence, 235, 313
and population declines among predatory
    birds, 90
resistance mechanisms, 295
storage of, 50, 51
toxicodynamics, 59
uptake and distribution of xenobiotics
  blood-brain barrier and, 24
  octanol-water partition coefficients and, 22
Organohalogen metabolism
  glutathione conjugation, 46
  reductive, 41, 42
Organometallic compounds, 163–179
  arsenic, 178–179
  lead, 177
    mode of action, 297
    neurotoxicity, 300
  mercury, 14, 163–172
    behavioral effects, 307, 312
    ecological effects, 170–172
    environmental fate, 72, 166–168
    metabolism, 165
    mode of action, 57, 297
    neurotoxicity, 300–302
    origins and chemical properties, 163–165
    toxicity, 168–170
    toxicodynamics, 55, 59

movement of pollutants along food chains, 76
tin, 172–177, 319
  chemical properties, 172–173
  crude oil spills as vehicle for, 181
  ecological effects, 176–177
  endocrine disruption, 267, 268, 272–273,
    315
  environmental fate, 173–174
  mechanistic biomarker assays, 86
  metabolism, 173
  mode of action, 57
  toxicity, 174–176
Organophosphate-induced delayed neuropathy
    (OPDIN), 206, 300, 305, 306
Organophosphorus compounds, 194–212, 319
  atropine as antagonist, 7, 36
  battlefield use, 14
  behavioral effects, 307, 309–310, 313
  blood-brain barrier and, 24
  carbamate potentiation, 216–217
  chemical properties, 194–196
  complex pollution problems, biomarkers,
    245, 248
  ecological effects, 208–211
    population dynamics, 209–211
    population genetics, 211
    toxic effects in field, 208–209
  environmental fate, 69, 200–202
  ester bonds, 199
  factors determining toxicity and persistence, 19
  mechanistic biomarker assays, 86
  metabolism, 197–200
    B esterases, 38, 39
    cytochrome P450 biotransformations,
      29, 31
    esterases and other hydrolases, 36–37
  mode of action, 56, 297
    central nervous system, 305
    chemicals sharing same principal mode of
      action, 308
    peripheral nervous system, 304
  population level effects, 311
  potentiation and synergism, 64, 217
  resistance mechanisms, 10, 94, 95, 157
  selectivity, 61, 62
  toxicity, 202–208
  toxicodynamics, 55, 59
  tributyltin and, 173
  uptake and distribution of, 22
Ornithine, 47
Orthotrichlorophenol, 151
Ostrea edulis (European flat oysters), 176
Otter (Lutra lutra), 129, 130
Otter (Lutra canadensis), 169
Oxidation
  cytochrome P450 biotransformations, 30,
    33–34

DDT metabolism, 105
  metabolism, 24
  methyl mercury, 165
  organophosphorus compounds, 197
Oxidative activation, 50
Oxidative attack
  carbamates, 213
  halogenation and, 71–72
  PAHs, 183
  PCBs, 136, 141
Oxidative desulfuration
  cytochrome P450 biotransformations, 29, 30
  phosphorothionate metabolism, 39
Oxidative detoxication
  cyclodienes, 120
  cytochrome P450 inhibitors and, 33
  potentiation mechanisms, 217
  and pyrethroid resistance, 238
  selectivity, 62
Oxidative metabolism
  PAHs, 186–187
  pyrethroid insecticides, 234
  pyrethroids, 231
Oxidative phosphorylation
  biocides, 60
  complex pollution problems, biomarkers, 249
  mode of action
    biomarkers, 247
    herbicides, 58, 257
  tributyltin and, 174, 176
Oxime carbamates, 213, 214
Oxons
  organophosphorus compound metabolism,
    199
  phosphorothionate metabolism, 39
  properties, 196
  structure, 194, 195
  toxicodynamics, 59
Oxygen
  and cyclodiene epoxide properties, 116
  cytochrome P450 monooxygenation, 27
  and reduction reactions, 41–42
Oxygen radicals/reactive oxygen metabolites
  complex pollution problems, biomarkers,
    248, 249
  cytochrome P450 biotransformations, 28
  herbicide formation of, 58
  PAH oxidative products, 183
  PCB toxicity, 145
Oysters, 77, 174, 176, 177, 319

P

Pacific oyster (*Crassostrea gigas*), 77, 174
PAM (pyridine aldoxime methiodide), 204, 215
Paper mills; *See* Pulp and paper industry
Papilio polyxenes (black swallowtail), 32

PAPS, 44
Paraoxon, 37
Paraquat, 257
  complex pollution problems, biomarkers, 249
  environmental fate, 69, 70
  mode of action, 58
  redox cycling, 42
Parathion, 14
  bird control, 200
  metabolism, 31, 199
  resistance mechanisms, 94
  structure, 195
  toxicity, 207
  uptake and distribution of xenobiotics, 22
Paris Green, 179
Partition coefficients, 21–23, 71
Passive diffusion, 52, 81
PBO; *See* Piperonyl butoxide
PCBs; *See* Polychlorinated biphenyls
PCDDs; *See* Polychlorinated dibenzodioxins
PCDFs; *See* Polychlorinated dibenzofurans
Peach potato aphid (*Myzus persicae*), 37, 61, 62
  OP resistance, 211
  organochlorine-pyrethroid cross resistance, 116
  pyrethroid resistance, 238
  resistance development, 194
  resistance mechanisms, 94, 95
Pentachlorophenols, 69
Peptidases, 36, 44, 45
Peptide conjugation, 47
Peptides, mercury affinity, 164
Percolation, soil, 81, 82, 83
Peregrine falcon (*Falco peregrinus*), 97, 110, 113,
    114, 116, 125, 126, 127, 132, 312
Peripheral nervous system, 302–304
  effects at whole organism level, 306
  neurotoxin modes of action, 297
  organophosphorus compounds, 209
Permethrin, 231
  chemical properties, 232, 233
  endocrine disruption, 279
  metabolism, 233, 234
  persistence, 235
  toxicity, 236
Peroxidase, 248
Peroxisome proliferation, CYP4A induction and,
    48
Persistence, 17–18, 319
  anticoagulant rodenticides, 219, 222
  and behavioral effects of neurotoxins, 312, 313
  cyclodienes, 119, 120, 124
  DDE, 112
  dirty dozen, persistent organic pollutants
    (POPs), 77
  environmental fate
    biological factors affecting, 71
    properties affecting, 68

factors affecting, 19–20
fate of pollutants in soils and sediments, 83
halogenation and, 71–72
hexachlorocyclohexanes/lindane, 131
movement of pollutants along food chains, 75
organochlorines, DDT and metabolites, 102,
    106–107, 109, 111
organophosphorus compounds, 195, 196,
    200, 201
PAHs, 186
PBBs, 149
pyrethroid insecticides, 235
in soils and sediments, 81
Persistent Organic Pollutants (POPs), 77, 102
Perylene, 185
pH
    and anticoagulant solubility, 221
    and octanol-water partition coefficient, 23
    organophosphorus compounds, 196
    pyrethroid hydrolysis, 232
    stressors, 89
Phagocytosis, 24
PHAHs (polyhalogenated aromatic
        hydrocarbons); See Polyhalogenated
        aromatic hydrocarbons
Pharmaceuticals, endocrine disruptors, 268, 274,
    320
Phenanthene, 185, 189, 268
Phenobarbital, 48, 49, 50
Phenol, 22
Phenolic nitriles, halogenated, 258
Phenoxyalkanoic acid (2,4,5-T), 82, 151, 152, 257
Phenoxyalkanoic herbicides
    environmental fate, 69
    fate of pollutants in soils and sediments, 81,
        82
    formulations, 261
    mode of action, 58
    movement into surface water, 262
    properties, 258
    TCDD contamination, 151–152
Phenyl acetate, 37
Phenylmercuric acetate, 164
Phenyl mercury, 163, 167, 168
    acute toxicity range, 170
    metabolism, 165
Phenylureas, 261–262
Phorate, 193, 201, 207
Phorbol diesters, 32
Phosdrin, 200, 208
Phosphates
    hydrolysis, 37, 40
    stressors, 89
Phosphinates, 194, 196
Phosphonates, 194
Phosphorodithionates, 196

Phosphorothionates, 30, 196
    metabolism, 39
        cytochrome P450 inhibitors, 34
        enzyme activation, 50
    potentiation and synergism, 254
    structure, 194
Phosphorylation, oxidative; See Oxidative
        phosphorylation
Photochemistry
    DDT, 103
    degradation products with increased toxicity,
        69
    and environmental fate, 69
    organophosphorus compounds, 196
    PAHs, 183, 187, 189
    pyrethroids, 231
Photosynthesis inhibition
    endocrine disruptors, 279
    herbicide mode of action, 58, 258
    herbicides, 263
    toxicodynamics, 59
Phthalates
    endocrine disruption, 282–283, 286
    enzyme induction, 48
Phthalic acid diamides, 7
Phylogenetic tree, cytochrome P450 enzymes, 9
Physiology; See also Metabolism
    biomarker approach, 84, 85
    PCBs, 146
    and species susceptibility to endocrine
        disruptors, 287
Physostigmine (eserine), 6, 212, 293, 297, 316
Phytoestrogens, 246, 252, 266, 268, 283
Phytoplankton
    atrazine herbicides and, 261
    PCB concentration in food chain, 141
Phytotoxicity, herbicides effects on aquatic
        plants, 261–263
Picrotoxinin, 297, 316
Pigeon, feral (*Columba livia*), 106, 123
Pimephales promelas (fathead minnow), 123,
        267, 274, 284
Pine mouse (*Microtus pinetorum*), 130–131
Pinocytosis, 24
Piperonyl butoxide (PBO), 33, 253
    carbaryl synergists, 213, 216
    potentiation and synergism, 64
    pyrethroids, 231, 234
    resistance mechanisms, 94, 95
Pirimicarb, 213
Pirimiphos-ethyl, 62
Pirimiphos-methyl, 193
    A esterase activity, species differences in, 37
    selectivity, 62
    toxicity, 207
Pirimiphos-methyl oxon, 37

Placental transmission
  DDT, 105
  PCBs, 144–145
Planar molecules
  biomarkers, 185
  complex pollution problems, biomarkers, 246
  cytochrome P450 biotransformations, 31,
    136, 138
  enzyme induction, 187
  toxicodynamics, 59
Plankton
  atrazine herbicides and, 261
  organochlorines in, 107, 113
  PCB concentration in food chain, 141
Plant-animal chemical warfare
  animal defenses, 8–10
  plant toxins, 4–8
Plant estrogens, 246, 252, 266, 268, 283
Plants
  aquatic
    environmental fate, properties affecting,
      68
    herbicides effects on, 261–263
  extinction of farmland species, 259
  metabolic enzyme systems, 25
  oil spill effects, 189–190
  organophosphorus compound
    biotransformation, 195
  resistance mechanisms, 95
Plant toxins, 4–8
  metabolic detoxication, 32
  neurotoxicity, 293
  resistance mechanisms, 93
Plutella xylostella (diamondback moth), 95, 116,
  238
Polarity
  carbamates, 213
  cyclodienes, 116
  and environmental fate, 68, 69–70, 76
  and fate of pollutants in soils and sediments,
    81, 82, 83
  hexachlorocyclohexanes/lindane, 131
  organometallic compounds
    mercury, 163
    tin, 172
  organophosphorus compounds, 195, 196
  oxidation and, 30
  and storage, 51
  uptake and distribution of xenobiotics, 21–23
Polybrominated biphenyls (PBBs), 134, 140, 150
  endocrine disruption, 268
  environmental fate, 71
  excretion in terrestrial animals, 53
Polybrominated dimethylethers, 268, 280–281
Polychlorinated biphenyls (PCBs), 222, 322
  biomarkers, 86, 87, 185, 246
  chemical properties, 134–135

complex pollution problems, biomarkers, 246
crude oil spills as vehicle for, 181
dirty dozen, persistent organic pollutants
  (POPs), 77
ecological effects, 146–149
  physiological and biochemical effects in
    field, 146
  population effects, 146–149
  population genetics, 149
ecological effects related to TEQs, 160
endocrine disruption, 268, 279–280
environmental fate, 68–69, 72, 140–143
  biological factors affecting, 71
  in soils and sediments, 83
interactions with other pollutants
  antagonism with mixtures, 157
  mercury, 171
  potentiation and synergism, 62, 64
metabolism, 136–140
  cytochrome P450 biotransformations,
    28, 31
  enzyme induction, 48, 49, 187
  epoxide hydrolases, 41
  excretion in aquatic animals, 52
  excretion in terrestrial animals, 52–53
  glutathione-*S*-transferases, 47
  reductive, 42
mode of action, 57
PCDD formation during combustion, 152, 153
storage, 50, 51
TCDD in commercial mixtures, 152
TEF values, 156
toxicodynamics, 59
Polychlorinated dibenzodioxins (PCDDs), 151–160
  biomarkers, 185
  complex pollution problems, 246
  ecological effects, 158–160
  environmental fate, 153–154
    biological factors affecting, 71
    movement of pollutants along food
      chains, 76
  enzyme induction, 187
  interactions with other pollutants, 171
  metabolism, 153
  origins and chemical properties, 151–153
  population effects, 147
  storage, 50
  TEF values, 156
  toxicity, 154–158
Polychlorinated dibenzofurans (PCDFs), 151–160
  biomarkers, 185, 246
  complex pollution problems, 246
  ecological effects, 158–160
  environmental fate, 153–154
    biological factors affecting, 71
    movement of pollutants along food
      chains, 76

enzyme induction, 187
interactions with other pollutants, 171
metabolism, 153
origins and chemical properties, 151–153
population effects, 147
TEF values, 156
toxicity, 154–158
Polycyclic aromatic hydrocarbons (PAHs), 13, 14,
181–191, 327
complex pollution problems
bioassays, 252
biomarkers, 246
cytochrome P450 activation of mutagens/
carcinogens, 140
ecological effects, 189–191
endocrine disruption, 268
environmental fate, 69, 185–187
fate of pollutants in soils and sediments, 83
metabolism, 183–185
cytochrome P450 biotransformations,
28, 31
enzyme induction, 48, 49
epoxide hydrolases, 41
models for bioconcentration and
bioaccumulation, 78
origins and chemical properties, 182–183
structure, 182
toxicity, 187–189
toxicodynamics, 55
Polyhalogenated aromatic hydrocarbons
(PHAHs)
endocrine disruption, 268
monitoring, 154
TEF values, 155, 156–157
Polyhalogenated compounds
environmental fate, 71–72
models for bioconcentration and
bioaccumulation, 78
potentiation and synergism, 62
rate of metabolism, 80
reductive metabolism, 42
Population density
community and ecosystem level structural
analysis, 96
seasonal fluctuations, 91
Population dynamics
distribution in communities and ecosystems,
90–93
organophosphorus compounds, 209–211
prediction of ecological risks, 97
pyrethroid insecticides, 237–238
stressor impacts, 90
Population genetics; See Genetics/population
genetics
Population level effects, 14, 84, 85, 89–90, 316, 320
adoption of more ecologically relevant
assessment practices, 321–323

anticoagulant rodenticides, 228
avermectins, 12
complex pollution problems, 243
cyclodienes, 124–130
gene frequencies, 130–131
population numbers, 124–130
DDT, 112–115
gene frequencies, 115–116
population numbers, 112–115
distribution in communities and ecosystems,
90–95
population dynamics, 90–93
population genetics, 93–95
ecotoxicology, 18
endocrine disruptors, 289–290
herbicides, 260
neurotoxicity and behavioral effects, 311–313,
315
organomercury compounds, 170
organophosphorus compounds, 209–211
PCBs, 146–149
pyrethroid insecticides, 237–238
reproductive and fertility effects, 17
response at different levels of biological
organization, 92
Pores, membrane, 23, 24
Porphyria, 144
Porphyrins, 42, 103, 104
Postmortem changes
DDT reductive dechlorination, 104
organophosphorus compounds, 209
Potassium channels/potassium transport
mercury inhibition of Na/K-ATPase, 169
neurotoxins and, 302
organochlorine sites of action, 110
Potatoes, solanine, 7
Potentiation, 320; See also Synergism
carbamates, 216
complex pollution problems, 253–254
measuring toxicity of mixtures, 244
organomercury compounds, 171
organophosphorus compounds, 208
pyrethroid insecticides, 234
toxicity, 62–64
Poultry farms, endocrine disruptor sources, 279
Precocenes, 5, 6, 7
Predator-prey relationships, neurotoxin
behavioral effects, 312, 314
Predators
anticoagulant rodenticide poisoning, 222–223
beneficial insects
DDT reduction of population, 115
OP reduction of population, 210
cyclodiene bioaccumulation, 121
immobilization of prey, 11
movement of pollutants along food chains, 76

organochlorines
   beneficial insects, DDT reduction of
      population, 115
   in food chain, 106, 108
organomercury compounds, 166, 167
organophosphorus compounds for control
   of, 200
Predatory birds/raptors
   anticoagulant rodenticide poisoning,
      222–223, 225–227
   cyclodiene poisoning, 124, 125–129
   hexachlorocyclohexanes/lindane levels, 131
   lack of cytochrome P450 enzymes, 10
   metabolism in, 121
   monooxygenase activity comparisons, 34, 35
   OC residues, 106
   organomercury compounds, 168, 170
   population decline, 90
   population level effects, 311, 315
Prediction of ecological risk, 97–98
Prediction of effects of pollutants on population
   growth rates, 92
Prediction of toxicity from experimental models,
   18
Predioxin, 152
Procaine, 32
Prochloraz, 33, 208
Progesterone, 268, 277
Propane, 182
Propioconazole, 33
Propoxur, 207, 212, 213, 216
Prostaglandins, 87, 110
Proteins/protein binding
   anticoagulant rodenticides, 222
   and elimination rates, 72
   factors determining toxicity and persistence,
      19
   mercury compounds, 164, 168–169, 179, 301
   organochlorines, DDT and metabolites, 111
   PCB metabolites, 140
   storage of xenobiotics, 51
Protein synthesis inhibitors, microbial toxins, 13
Proteomics, 88
Proton translocators, 247
Psoralen, 5, 7
Puffer fish (*Fugu vermicularis*), 11, 12, 293
Pulp and paper industry
   endocrine disruptors in effluents, 280, 282
   measuring toxicity of mixtures, 244
   PCDDs in effluents, 152
Pyrazines, 10
Pyrenes, 41, 182, 183, 185
Pyrethric acid, 231, 232
Pyrethrins, 6, 316
   animal defenses, 8
   mode of action, 297
   neurotoxicity, 293

plants producing, 4, 5
structure, 232
Pyrethroid insecticides, 4, 6, 32, 231–238, 319
   behavioral effects, 307, 313
   chemical properties, 231–232, 233
   ecological effects, 237–238
      population dynamics, 237–238
      population genetics, 238
   effects at whole organism level, 306
   endocrine disruption, 268, 279
   environmental fate, 69, 234–236
   factors determining toxicity and persistence,
      19
   fate of pollutants in soils and sediments, 83
   metabolism, 32, 33, 232–234
      B esterases, 38
      enzyme induction, 49
   mode of action, 56, 297, 298
   with OPs and carbamates, 210
   population level effects, 311
   potentiation and synergism, 33, 63–64, 217, 254
   resistance mechanisms, 10, 94, 95, 116, 157,
      295
   selectivity, 61, 62
   sodium channel binding, 5
   toxicity, 236–237
   toxicodynamics, 59
Pyrethrolone, 231, 232
Pyridine aldoxime methiodide (PAM), 204, 215
Pyridines, 30

Q

Quantitative structure activity relationship
   (QSAR) models, 249, 277, 325–326
Quinine, 30
Quinones, 10
   animal defenses, 10–11
   anticoagulant rodenticides, 219, 224
   PAH oxidative products, 183

R

Rainbow trout (*Oncorhynchus mykiss*)
   cyclodiene toxicity, 123
   endocrine disruption
      cultured hepatocyte assay, 252
      sensitivity, comparative, 287
   neurotoxin behavioral effects, 309
   organochlorines, DDT and metabolites, 120
   organophosphorus compounds, 205
   PCB concentration in food chain, 141
Rat
   cyclodiene toxicity, 123
   DDT and related compounds toxicity, 111
   DDT half-life in, 106
   PCB metabolism, 139–140

REACH proposals, EC, 77, 322, 324
Reactive oxygen metabolites; *See* Oxygen
      radicals/reactive oxygen metabolites
Receptor occupancy
    endocrine disruption
        assays, 277
        modes of action, 267
    response at different levels of biological
        organization, 92
Recovery, population, 90, 128
Recruitment, and population density, 91
Red-legged partridge (*Alectoris rufa*), 64
Red spider mites, 95, 115
Red tide, 11, 12
Reduced glutathione, 131
Reduction reactions
    DDT reductive dechlorination, 103, 104
    metabolism, 24
    reductases, 41–42
Regioselectivity, 27, 28, 33
Regulatory issues, testing complex mixtures,
    253–254
Reporter genes, estrogen bioassays, 247
Reproduction, and population density, 91
Reproductive and fertility effects, 17
    breeding behavior
        endocrine disruptors, 289–290
        neurotoxins and, 312
    DDE, 93
    endocrine disruption; *See* Endocrine
        disruptors
    fish-eating birds, 148
    mechanistic biomarker assays, 86
    organochlorines, 110–111, 113–115
    organometallic compounds
        mercury, 172
        tributyltin, 174–176
    PAHs, 189
    PCBs, 148
    population decline, 17
    and population numbers, 113–115
    TCDD, TEQs and, 158, 159, 160
    toxicodynamics, 59
Reptiles
    alligators, endocrine disruptor case studies,
        270, 271–273
    cyclodiene excretion, 117
    storage of xenobiotics, 51
Residue analysis, prediction of ecological risks,
    97
Resistance
    animal defenses, 8
    anticoagulant rodenticides, 228
    cross-resistance, 303
        cyclodienes, 130
        HCH, 131
        organochlorines and pyrethroids, 116

    cyclodienes, 130–131, 303
    DDT, 296, 303
    in fish, 80–81
    mechanisms of, 10
    metabolism
        cytochrome P450 studies, 32
        monooxygenases, 32
        P450 inducible forms and, 50
    neurotoxins
        GABA receptors, 303
        sodium channels, 296
    organophosphorus compounds, 193, 197, 199,
        201–202, 211
    pesticide design, 19
    to plant toxins, 93
    population genetics, 93–94
    potentiation and synergism, 64
    pyrethroid insecticides, 296, 303
    toxicodynamics, 59
Resistance factor (selectivity ratio), 62
Respiration, inhibitors of, 8
Respiratory surfaces, aquatic animals, 21
Retinoic acid deficiency, 268
Retinoid X receptor, 272
Retinol, 86, 87
Retinol binding sites, 86, 144–145
Reversibility
    cyclodiene toxicity, 123
    dose levels of toxicants and, 90
    OP poisoning, 209
RF (resistance factor), 94
Rhesus monkey, DDT half-life in, 106
Rhothane; *See* DDD
Ricin, 7
Risk assessment
    adoption of more ecologically relevant
        assessment practices, 321–323
    future prospects
        adoption of more ecologically relevant
            assessment practices, 321–323
        development of more sophisticated testing
            methods, 323–324
    measuring toxicity of mixtures, 244
    neurotoxins
        biomarkers, 314–317
        in vitro methods, 305
    TEQs in, 157
Risk prediction, 97–98
River Invertebrate Prediction and Classification
        System (RIVPACS), 96, 237–238
Rivers; *See* Surface waters
Rodenticides; *See also* Anticoagulant
        rodenticides
    mechanistic biomarker assays, 86
    resistance mechanisms, 95
    toxicodynamics, 59

Rodents
  DDT and related compound toxicity, 111
  metabolism, B esterases, 38
Rosemaund experiment, 82, 261
Rotenone, 5, 6, 7
Runoff
  herbicide movement into surface waters, 261
  OPs in, 201
  pyrethroid contamination, 235
Rust red flour beetle (*Tribolium castaneum*), 38,
    199
Ryania species, 7
Ryanodine, 6, 7

S

Sabadilla, 5, 6
Safroles, 10
Salicylate, 268
Sampling, assessment of individuals, 84
Sandwich tern (*Sterna sandvicensis*), 125
Sarin, 14, 196, 299
Saxitoxin, 11, 12
Scavengers, anticoagulant rodenticide poisoning,
    222–223
Schoenocaulon officinale (*Veratrum sabadilla*),
    4–5
Scorpion toxins, 11
Screening, endocrine disruptors, 276–278, 281
Seabirds
  organochlorines in marine food chain, 108
  PCBs
    bioaccumulation, 141, 142
    biomarkers, 146
Seals, 148
Seasonal population density fluctuations, 91
Secondary effects, biomarkers measuring, 60
Sediments
  arsenic methylation in, 178–179
  distribution in communities and ecosystems,
    81–83
  environmental fate, 68, 69–70, 83
  mercury methylation in, 164
  organophosphorus compound degradation
    in, 196
  PAHs in, 185, 189
  pyrethroid adsorption, 235
  reductive metabolism, 42
  tributyltin oxide (TBTO) adsorption, 172, 174
Seed dressings, 216, 319
  chlorfenvinphos, 195, 196
  cyclodiene, 121, 122, 124, 128, 130
  cyclodienes, 121, 122
  hexachlorocyclohexanes/lindane, 131
  mercury, 167, 168, 170–171
  organophosphorus compounds, 201, 208,
    209, 210

Selective reabsorption, from bile to plasma, 54
Selective toxicity/selectivity, 18, 60–62
  basis of, 18
  efficacy of pesticides, 19
  mechanisms of action and, 59
  physiological basis for, 55
Selenium, mercury interactions, 171
Sequential exposures, 216
Sequestration, and delayed toxicity, 50–51
Serial sampling of individuals, 90
Sesquiterpenes, 13
Sheep dips, 105, 131
  chlorfenvinphos, 195
  pyrethroid insecticides, 235
Simazine
  complex pollution problems, 263
  environmental fate
    movement into surface water, 261, 262
    in soils and sediments, 83
  properties, 258
  structure, 259
Sinks, organochlorine residues, 109
Sites of action
  biomarker assays, 250
  complex pollution problems
    bioassays for mixtures, 251
    biomarkers, 253
  factors determining toxicity and persistence,
    19, 20
  neurotoxins, 294
  toxicodynamics, 55, 59
Sites of metabolism, factors determining toxicity
    and persistence, 19, 20
Skin
  cyclodiene uptake routes, 130
  excretion in amphibia, 52
  uptake and distribution of xenobiotics, 21
Slugs, organochlorine residues in terrestrial food
    chains, 108–109
Smothering, oil spills, 181
Snails, 108–109
Snake toxins, 11
Sodium channels, 6, 316
  neurotoxin mechanisms of action, 4–5, 11, 56,
    296, 297, 298, 303
    DDT, 4, 5, 109–110, 303–304
    pyrethroids, 4, 5, 236, 238, 303
  resistance mechanisms, 94, 95, 297, 303
    aberrant forms of, 10
    mutations, 115–116
    target insensitivity, 157
  selectivity, 62
  toxicodynamics, 59
Sodium-potassium pump, mercury inhibition, 169
Soft clams (*Mya arenaria*), 77
Soil microorganisms; *See* Microbial metabolism

Soils
    arsenic methylation in, 178–179
    carbamate movement into surface waters, 217
    community and ecosystem level effects, 96
    cyclodiene environmental fates, 119, 120
    DDT half-life in, 106
    distribution in communities and ecosystems,
        81–83
    environmental fate, 69–70
        biological factors affecting, 71
        properties affecting, 68, 70
    herbicides, 257, 261
    organochlorine residues in, 109
    organophosphorus compounds, 201
        degradation in, 196
        persistence, 200
    pyrethroid adsorption, 235
    reductive metabolism, 42
Solanine, 5, 7
Solanum tuberosum, 5
Solar radiation, and environmental fate, 69
Solubility properties; *See also* Water solubility
    and fate of pollutants in soils and sediments,
        81, 82, 83
    octanol-water partition coefficients, 21–23
    octanol-water partition coefficients and;
        See also Octanol-water partition
        coefficient/log K_{OW}
Solvents, 31
Soman, 37, 196, 299
Somateria molissima (eider ducks), 51, 124
Sparrow hawks (*Accipiter nisus*), 92, 132, 324
    cyclodiene poisoning, 116, 125–129
    DDT and metabolites, 34, 97, 106, 109, 110,
        114
    toxic effects and population changes, 97
Specialization/specificity of detoxification
        activity, 10
Species, community and ecosystem level
        structural analysis, 96
Species differences
    cyclodiene bioaccumulation, 120–121
    cyclodiene metabolism, 118
    and environmental fate, 72
    A esterase activity, 37
    excretion in terrestrial animals, 54
    hepatic microsomal monooxygenase
        activities, 34, 35
    prediction of toxicity from experimental
        models, 18
    selective toxicity, 55
    TCDD toxicity, 158
Specificity, biomarker assays, 87
Spider toxins, 11, 12
Spruce bud worm control, OP, 208
SR values, 94

Stability, chemical
    and environmental fate, 69
    organophosphorus compounds, 196
    and population level effects, 90
    pyrethroids, 231
St. Anthony's fire, 8, 13
Steady state
    bioaccumulation and bioconcentration
        factors, 76
    in fish, 80
    models for bioconcentration and
        bioaccumulation, 77, 78
Sterna hirundo (common tern), 144
Sterna sandvicensis (sandwich tern), 125
Steroids, endogenous
    endocrine disruptors, 279
    enzyme induction, 49
    metabolism
        aromatase; *See* Aromatase activity
        epoxide hydrolases, 41
    tributyltin and, 174–176
Stilbene oxide, 49–50
St. John's wort (*Hypericum perforatum*), 5, 7
Storage
    degradation products with increased toxicity,
        69
    factors determining toxicity and persistence,
        19, 20
    glutathione conjugates, 47
    toxicokinetics, 50–51
Streptomyces avermitilis, 11
Stress ecology, 89
Stressors, pollutants as, 89
Stress proteins, 250
Structural analysis, community and ecosystem
        level effects, 96
Strychnine, 5, 6, 293
Strychnos nux-vomica, 5
Sublethal effects, 17
    adoption of more ecologically relevant
        assessment practices, 322
    biomarkers, 18, 85–87
    carbamates, 217
    complex pollution problems, 245
    cyclodienes, 123, 128
    neurotoxins; *See* Behavioral effects,
        neurotoxicity and
    organomercury compounds, 170
    organophosphorus compounds, 209
    population level effects, 90
Sublimation, 68
Substituted ureas, 58, 59
Substrate specificity, monooxygenases, 27
Suicide substrates, 33, 197, 202
Sulfanilic acids, 183
Sulfatases, 48
Sulfate conjugates, 48

Sulfate formation, 44, 153
Sulfenazole, 30
Sulfhydryl groups
  esterases, 39–40
  organometallic compound affinity, 59
    lead, 178
    mercury, 164, 168–169, 179, 297, 301
Sulfometuron, 258
Sulfones
  DDE metabolism, 47
  OP metabolism, 195, 197, 199
  organohalogen metabolites, 47
Sulfonic acids, 183
Sulfonylureas, 258, 261
Sulfotransferases, 43, 44
Sulfoxides, OP, 195, 197
Sulfur
  cytochrome P450 biotransformations, 29
  cytochrome P450 inhibitors, 34
Sulfur derivatives, PAH, 183
Sulfur-nitrogen bonds, hydrolytic enzymes, 36
Sulfur-phosphorus linkages, 196
Superfund sites
  PCBs, 148–149
  TCDDs, 160
Super kdr, 115–116, 295
  mode of action, 303
  pyrethroids, 238
Superoxide anion
  biomarkers, 248, 249
  cytochrome P450 biotransformations, 28
  PCB toxicity, 145
  redox cycling, 42
Superoxide dismutase, 248
Superwarfarins, 6, 219
  ecological effects, 227
  mechanistic biomarker assays, 86
  metabolism, 222
Surface contamination, uptake routes, 76
Surface oil films, 68
Surface waters, 320; See also Aquatic systems
  carbamate movement from soil, 217
  DDT environmental fate, 105
  endocrine disruptors in, 279, 280, 283
  fate of pollutants from waste disposal sites, 83
  herbicide contamination, 261–263
  lindane discharge into, 131
  measuring toxicity of mixtures, 244
  OP use in pest control, 201
  organophosphorus compounds in, 196, 209
  pyrethroid contamination, 234–235
Sweet clover, 5, 6
Synergism, 320; See also Potentiation
  carbamates, 213, 216
  complex pollution problems, 253–254
  cytochrome P450 inhibitors and, 33

OPs, 199, 208
  pyrethroids, 231, 234, 236
  toxicity, 62–64
Systemic pesticides
  carbamates, 194
  organophosphorus compounds, 193, 195
  toxicodynamics, 59

T

2,4,5-T, 82, 151, 152, 257
Tabun, 14, 37, 196, 299
Target insensitivity, 157
TBTO (tributyltin oxide), 172
TCB; See Tetrachorobiphenyl
TCDD; See Dioxins/TCDD
TCDF, 153
TCPO (trichloropropene-1,2-oxide), 136
TEFs; See Toxic equivalency factors
Telodrin, 116, 125
Temperature
  and DDT toxicity, 59
  degradation products with increased toxicity, 69
  Henry's law, 69
  organophosphorus compound isomerization, 196
  and pyrethroid toxicity, 236
  stressors, 89
TEQs; See Toxic equivalents (TEQs)
Terrestrial animals
  cytochrome P450 enzymes, 10, 31
  environmental fate
    biological factors affecting, 71
    cyclodienes, 120
    DDT half-life in food chain, 106
  models for bioconcentration and bioaccumulation, 77, 80
  uptake and loss mechanisms, 21, 52–54, 77, 79
Terrestrial food chains
  biomagnification, 76
  organochlorine biomagnification in, 108–109
Testicular dysgenesis syndrome (TDS), 267
Testing methods; See Assays, tests, assessment methods
Testosterone, 49, 267
Tetrabromodiphenyl (TeBDE), 281
Tetracene, 182
Tetrachlorobiphenyl, 136, 137
Tetrachlorodibenzodioxin (TCDD); See Dioxins/TCDD
Tetrachlorophenols, 69
Tetrachorobiphenyl (TCB), 51, 136, 137, 150
  fate in environment, 141
  half-life, 139
  toxicodynamics, 59
  transthyretin binding by metabolites, 144

Tetraethylpyrophosphate (TEPP), 14, 306
Tetrodotoxin, 11, 12, 293
Thiazine, 58
Thimet, 207
Thioethers
    organohalogen metabolites, 47
    organophosphorus compounds, 197
Thiones, 194
    properties, 196
    structure, 195
Thyroid gland
    endocrine disruption, 279–280
        assays, 277
        modes of action, 267–268
        PBDEs, 280
    PCB toxicity, 143, 148
Thyroxine antagonism, 57
    complex pollution problems, biomarkers, 246
    mechanistic biomarker assays, 86–87
    toxicodynamics, 59
    transthyretin binding by TCB metabolites,
        144–145
Time course
    exposure periods and bioaccumulation, 120
    population growth rates, 91
    stressor impacts, 90
    toxic processes, 87
Timing of exposure to endocrine disruptors,
        life stages with enhanced sensitivity,
        284–286
Tin compounds, organic (tributyltin), 17, 172–177,
        319
    chemical properties, 172–173
    crude oil spills as vehicle for, 181
    ecological effects, 176–177
    environmental fate, 173–174
    metabolism, 173
    toxicity, 174–176
Tobacco bud worm (*Heliothis virescens*)
    organochlorine-pyrethroid cross resistance,
        116
    potentiation and synergism, 64
    pyrethroid resistance, 238
    resistance mechanisms, 94, 95
    selectivity, 61, 62
TOCP, 86
Toxaphene, 122
    behavioral effects, 123
    dirty dozen, persistent organic pollutants
        (POPs), 77, 102
    mode of action, 297
Toxic endpoints, PCBs, 144
Toxic equivalency factors (TEFs), 144, 155, 156,
        157, 158, 160, 244
    measuring toxicity of mixtures, 244
    TCDDs, 144, 155, 156–157
    TEQ calculation caveats, 157

Toxic equivalents (TEQs)
    CALUX system, 156
    mixtures, 244, 245, 246
    PCBs, 144, 147, 148
    PCDDs and PCDFs, 155–157
    2,3,7,8-TCDD ecological effects related to,
        158–160
    TEFs (toxic equivalency factors) and, 157
Toxicity, 17–64
    anticoagulant rodenticides, 219, 224–226
    carbamates, 215–217
    cyclodienes, 122–124
    factors determining toxicity and persistence,
        19–20
    increased; *See also* Synergism
        chemical degradation products with, 69
        enzyme induction and, 50
    organomercury compounds, 168–170
    organophosphorus compounds, 202–208
    organotin compounds (tributyltin), 174–176
    PAHs, 187–189
    PCDDs and PCDFs, 154–158
    potentiation and synergism, 62–64
    pyrethroid insecticides, 236–237
    selective, 60–62
    TCDD, 153
    toxicodynamics, 54–60
    toxicokinetics, 21–54; *See also* Toxicokinetics
Toxic metabolic products
    conjugate breakdown, 42
    systemic action, 193
Toxicodynamics, 54–60
    factors determining toxicity and persistence,
        20
    potentiation and synergism, 63
Toxicogenomics, 88, 89
Toxicokinetics, 21–54
    excretion, 51–54
        by aquatic animals, 52
        by terrestrial animals, 52–54
    factors determining toxicity and persistence,
        20
    lipophilic compounds, 55
    metabolism, 24–50
        conjugases, 42–48
        enzyme induction, 48–50
        epoxide hydrolase, 40–41
        esterases and other hydrolases, 36–40
        general considerations, 24–26
        monooxygenases, 26–34, 35
        reductases, 41–42
    PCBs, 139–140
    potentiation and synergism, 63; *See also*
        Potentiation; Synergism
    storage, 50–51
    uptake and distribution, 21–24
TPT (triphenyltin), 173

Transcriptomics, 88, 284
Transdihydrodiols, 41
Transport
    in environment, 69, 71, 72; *See also*
        Environmental fate
    glutathione conjugates, 47
    organophosphorus compounds, 200
    uptake and distribution of xenobiotics
        mechanisms, 23
        terrestrial versus aquatic animals, 21
Trans stilbene oxide, 50
Transthyretin, 51
    mechanistic biomarker assays, 86–87
    PCB metabolite binding, 140, 144
    toxicodynamics, 59
Trembolone, 267
Triazines
    fate of pollutants in soils and sediments, 83
    properties, 258
    structure, 259
    toxicodynamics, 59
Triazoles, 30
Triazophos, 209
Tribolium castaneum (rust red flour beetle), 38,
    199
Tributyltin (TBT), 179, 319, 322, 324
    and breeding success, 93
    endocrine disruption, 267, 268
        case studies, 272–273
        species susceptibility, 287
    mechanistic biomarker assays, 86, 87
    mode of action, 57
    and population declines in dog whelks, 90
    toxic effects and population changes, 97
    toxicodynamics, 59
Tributyltin fluoride, 176
Tributyltin oxide (TBTO), 172
Tricarboxylic acid cycle, 8
Trichloro-2,2-bis (p-chlorophenyl) ethane; *See*
    DDT
Trichlorofluoromethane, 22
Trichlorophenols
    HCH metabolites, 131
    human exposure via industrial accidents, 152
    TCDD formation, 151, 152
Trichloropropene-1,2-oxide (TCPO), 136
Trichothecenes, 13
Trifluralin, 261
Trimethylarsine, 179
Triorthocresol phosphate, 300
Triphenyltin (TPT), 173, 272
Trophic levels
    and cyclodiene bioaccumulation, 120–121
    fate of pollutants in soils and sediments, 83
    movement of pollutants along food chains,
        75, 76
    OC residues, 106

and organochlorine residues, 107, 108
organometallic compounds
    mercury, 171
    methyl mercury, 166
    PCB chlorination, 138
    PCB environmental fate, 143
    pyrethroid insecticides, 236
Tsetse fly, 14, 32, 130
Turnover, community and ecosystem level
    effects, 96

U

UDPGA, 46
UDP glucuronyl transferases, 42–43, 44
Umbellifer leaves and stems, 5, 7
United Nations Environmental Programme
    (UNEP), 77, 102
Uptake
    cyclodienes, 117–118, 130
    movement of pollutants along food chains, 76
    toxicokinetics, 21–24
Urea herbicides, 83, 263
Ureas, substituted, 58, 59
Ureides, 258, 259
Urine, excretion in, 52, 53
Uterotrophic assay, 278

V

van der Waals forces, 51, 68, 135
Vapor pressure/volatility
    anticoagulant rodenticides, 219, 220
    carbamates, 207, 213
    cyclodienes, 116
    and environmental fate, 68–69
    fate of pollutants in soils and sediments, 81
    herbicides, 261
    organophosphorus compounds, 194–195, 196
    PAHs, 183, 185
    PCBs, 134, 135
    pyrethroid insecticides, 232, 233
Vapor state, transport in, 68–69
Vector control agents, 14
Veratridine, 4–5, 6, 293, 297, 298
Veratrum sabadilla (*Schoenocaulon officinale*),
    4–5
Vinclozolin, 267, 268, 278, 279
Vinyl chloride, 22
Vitamin A (retinol) binding protein, 144–145
Vitamin A deficiency, 57, 144–145, 268
Vitamin B12, 165
Vitamin K, 219, 220
Vitamin K antagonists, 6, 57, 219
    anticoagulant rodenticides, 224–225
    biomarkers, 245–246
    complex pollution problems, biomarkers, 246

mechanistic biomarker assays, 86
metabolism, 222
toxicodynamics, 59
Vitamin K epoxide reductase, 228
Vitellogenin
    alkylphenols and, 282
    endocrine disruption, 279
        assays, 278
        complex exposure problems, 283, 284
        in fish, 273
        by natural products, 283
    phthalates and, 282
Vitellogenin synthesis, 246
Volatility; *See* Vapor pressure/volatility
Voltage-sensitive sodium channels; *See* Sodium
        channels
Vulpes vulpes (fox), 129
Vultures, 320–321

W

Warfarin, 6
    chemical properties, 220
    mechanistic biomarker assays, 86
    metabolism, 31
    mode of action, 57
    resistance mechanisms, 228
    toxicity, 225
    toxicodynamics, 59
    and vitamin K cycle, 224
Waste disposal
    fate of pollutants in soils and sediments, 82
    measuring toxicity of mixtures, 244
    PAH pollution, 185
Wastewater treatment plant effluent, endocrine
        disruptors in, 267, 268, 273, 275, 279
    alkylphenols, 282
    lessons learned, 292
    life stages with enhanced sensitivity, 284–286
Water; *See also* Aquatic systems; Surface waters
    bioconcentration factor, 76
    concentration of pollutants, 52
    cyclodienes
        environmental fate, 120
        uptake routes, 130
    environmental fate
        biological factors affecting, 71
        properties affecting, 68

herbicide contamination, 261–263
    and population growth rate, 91
    uptake and loss mechanisms, 77, 130
    vaporization from, 69
    waste disposal sites, 82
Watercress beds, malathion application, 201, 209
Water solubility
    anticoagulant rodenticides, 219, 220, 221
    carbamates, 207, 213
    conjugate transport, 25
    cyclodienes, 116
    and environmental fate, 72, 81, 82, 83
    Henry's law, 69
    herbicides, 261
    hexachlorocyclohexanes/lindane, 131
    and metabolism, 53
    octanol-water partition coefficients, 21–23
    organometallic compounds
        lead, 177
        mercury, 164
        tin, 172
    organophosphorus compounds, 194–195
    PAHs, 183
    PCBs, 134, 135
    pyrethroid insecticides, 233
    TCDD, 152
    toxicodynamics, 55
Weak acids and bases, 23
Whales, 143
Whole organism level effects, 306–308
Wood products, lindane-treated, 131
Wood pulp; *See* Pulp and paper industry

X

Xanthotoxin, 32, 49
Xenoestrogens, 268

Y

Yeast reporter gene assays, 277
Yperite, 14

Z

Zooplankton, 107, 141

Printed and bound by CPI Group (UK) Ltd, Croydon, CR0 4YY

23/10/2024

01778244-0009